T0281226

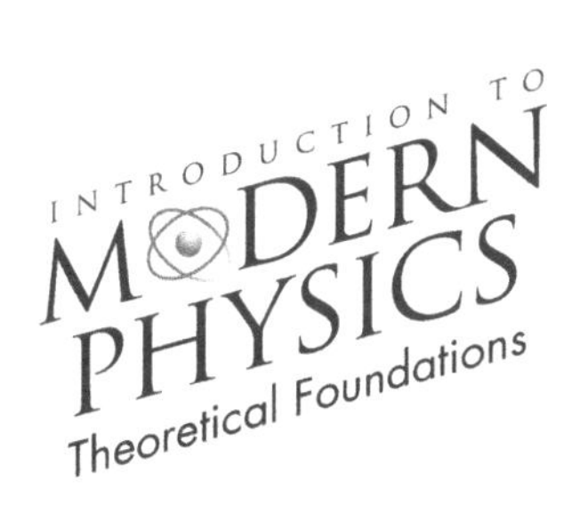
INTRODUCTION TO
MODERN
PHYSICS
Theoretical Foundations

Introduction to Modern Physics

Theoretical Foundations

John Dirk Walecka
College of William and Mary, USA

NEW JERSEY • LONDON • SINGAPORE • BEIJING • SHANGHAI • HONG KONG • TAIPEI • CHENNAI

Published by

World Scientific Publishing Co. Pte. Ltd.
5 Toh Tuck Link, Singapore 596224
USA office: 27 Warren Street, Suite 401-402, Hackensack, NJ 07601
UK office: 57 Shelton Street, Covent Garden, London WC2H 9HE

British Library Cataloguing-in-Publication Data
A catalogue record for this book is available from the British Library.

INTRODUCTION TO MODERN PHYSICS
Theoretical Foundations

ISBN-13 978-981-281-224-7
ISBN-10 981-281-224-5
ISBN-13 978-981-281-225-4 (pbk)
ISBN-10 981-281-225-3 (pbk)

Printed in Singapore.

For Ann and John

Preface

Our understanding of the physical world was revolutionized in the twentieth century—the era of "modern physics". This book is based on a sophomore, one-semester modern physics course taught twice at William and Mary, as well as on a one-quarter advanced freshman physics course taught once at Stanford. In both cases, the course was aimed at the very best students (there were many outstanding students at both places), with a goal of exposing them to the foundations and frontiers of today's physics. Of course, every effort was made to ensure that no one was left behind. Typically, students have to wade through several courses to see many of these topics, and I wanted them to have some idea of where they were going, and how things fit together, as they went along. Hopefully, they will then see more inter-relationships, and get more original insights, as they progress. I felt the courses were successful, and I know that many of the students in these courses went on to pursue careers in physics.

Physics is an *experimental science*, and it is assumed, as is usual in most schools, that such a course as this is to be accompanied by a good, thorough laboratory course in *experimental* modern physics.

In this book, the central topics of "classical physics" are first reviewed: newtonian mechanics, statistical mechanics, and electricity and magnetism. Then, after presenting several observed contradictions with classical physics, the book focuses on the following topics in modern physics: quantum mechanics; applications in atomic, nuclear, and particle physics; special relativity; relativistic quantum mechanics; general relativity; quantum fluids; and quantum fields. The aim is to cover these topics in sufficient depth that things "make sense" to students and that students achieve an elementary working knowledge of them. The coverage has expanded somewhat from the courses, and to get back down to one semester, some choices will probably have to be made.

Many problems are included, some for each chapter. While there are problems that directly amplify the material in the text, there are also a great number of them that will take dedicated readers just as far as they want to go in modern physics. The problems are not difficult, and the steps are clearly laid out. Although the book is designed so that one can, in principle, read and follow the text without doing any of the problems, the reader is strongly urged to attempt as many of them as possible in order to obtain some confidence in his or her understanding of the basics of modern physics and to hone working skills.

The book assumes the reader has had a good one-year, calculus-based freshman physics course, along with a good one-year course in calculus. While it is assumed that mathematical skills will continue to develop, several appendices are included to bring the reader up to speed on any additional mathematics required at the outset. With very few exceptions, the reader should then find the text, together with the appendices and problems, to be self-contained. The phrase "it can be shown" is anathema.

An extensive bibliography has not been attempted, although several primary references are included. Most of the key names and dates are also included in the text. To help provide historical context, there is an appendix with a chronological list of many of the theoretical physicists whose work forms the basis for the material presented here. Most of the existing modern physics texts are referenced, as are many of the author's choices for further study. Today, the existence of the world-wide web provides instant access to any required data, as well as further introduction to almost all of the topics covered here. Key websites are also referenced.

I was again delighted when World Scientific Publishing Company, which had done an exceptional job with three of my previous books, showed enthusiasm for publishing this new one. I would like to thank Dr. K. K. Phua, Executive Chairman of World Scientific Publishing Company, and my editor Ms. Lakshmi Narayanan, for their help and support on this project. I would also like to thank my colleagues Paolo Amore and Brian Serot for their reading of the manuscript.

Williamsburg, Virginia
April 14, 2008

John Dirk Walecka
Governor's Distinguished CEBAF
Professor of Physics, emeritus
College of William and Mary

Contents

Chapter 1

Introduction

At the end of the 19th century, one could take pride in the fact that the laws of physics were now understood. With newtonian mechanics, the statistical analysis of Boltzmann, and Maxwell's equations for electromagnetism, one had an excellent description of the world we see around us. At that point, a series of experiments were performed whose results simply could not be understood within this classical framework. These experiments probed a microscopic, or high-velocity, world that went far beyond our everyday perception. The resolution of these paradoxes led to the theories of quantum mechanics and special relativity, which provide the foundation of modern physics. Atomic, nuclear, particle, and condensed-matter physics are all built on this foundation. Einstein's extension of his theory of special relativity to general relativity provides the current basis for our understanding of gravitation and cosmology—physics to the farthest reaches of the universe. The theories of quantum mechanics, special relativity, and general relativity have been remarkably successful. All current experiments can be understood within this framework. The goal of this book is to introduce a reader, with an assumed knowledge of classical physics, to modern physics.[1]

It is assumed that the reader has had a good one-year, calculus-based freshman physics course, along with a good one-year course in calculus. A sufficient number of appendices are included to bring the reader up to speed on any additional mathematics required at the outset. Over 175 problems are included in the book, some for each chapter. While there are many problems that directly amplify the material in the text, there are also a great number of them that will take dedicated readers just as far as they want to go in modern physics. Although the book is designed so

[1]There is a nice irony in the parallel statement, "At the end of the 20th century, one could take pride in the fact that the laws of physics were now understood."

that one can, in principle, read and follow the text without doing any of the problems, the reader is strongly urged to attempt as many of them as possible in order to obtain some confidence in his or her understanding of the basics of modern physics. With very few exceptions, the reader should find that the text, appendices, and problems form a self-contained volume.

Chapter 2 reviews the essential elements of classical physics. We start with Newton's laws for point mechanics and then consider the continuum mechanics of the planar oscillations of a stretched string, an analysis that provides the basis for much of what is done in this book. An appendix summarizes the main features of the resulting Fourier series and Fourier integrals. With the Boltzmann distribution as starting point, the classical equipartition theorem is derived, which describes the equilibrium thermal energy in each normal-mode oscillation of a system. Two appendices present the essentials of thermodynamics and derive the Boltzmann distribution from Boltzmann's more general statistical assumptions. The essentials of electromagnetism are described and Maxwell's equations presented, with some applications to electromagnetic waves. An appendix derives two theorems of vector calculus that convert Maxwell's equations to the integral form in which they are customarily introduced.

Chapter 3 presents the key experiments whose results could not be understood in classical terms. These include the vanishing of the specific heat of a solid at low temperature, the spectrum of black-body radiation in a cavity, the photoelectric effect, Compton scattering of light from an electron, and the discrete spectral lines observed in atomic radiation. The resolution of the first two problems was obtained through Planck's thermal distribution, which reduces to the equipartition result at high temperature, but indicates a discrete excitation energy $h\nu$ of an oscillator of frequency ν at low temperature where $k_{\rm B}T/h\nu \ll 1$; here $k_{\rm B}$ is Boltzmann's constant, and h is the new Planck's constant. Einstein resolved the paradox in the photoelectric effect by assuming that the energy in a light wave actually comes in discrete packets of energy, called *photons*, each with energy $h\nu$. The Compton effect was then understood through the conservation of energy and momentum in the collision of a photon with an electron. The discrete nature of atomic radiation was explained through Bohr's revolutionary assumptions that atomic systems possess discrete stationary states in which they do not radiate, and they radiate a photon of energy $h\nu = E_i - E_f$ when making a transition between those states. His simple model of the atom involving the quantization of the angular momentum in circular orbits explained the main features of the spectrum of hydrogen.

De Broglie argued that if waves (electromagnetic radiation) exhibit particle properties (photons) then perhaps particles should manifest wave behavior. In chapter 4, the essential elements of quantum mechanics are developed starting from de Broglie's relation for matter, and culminating in the Schrödinger equation, a differential equation describing the space-time propagation of a wave associated with a particle. The absolute square of this wave function is given a probability interpretation, which together with the corresponding probability current, provides a connection with physical measurements. Many elementary one-dimensional problems are solved and interpreted, including the free particle, particle in a box, particle incident on a potential barrier, and simple harmonic oscillator. The analysis is extended to a three-dimensional box, and then angular momentum and spin are introduced. The quantum mechanics of many identical integer-spin bosons and half-integral-spin fermions is analyzed. Enough of the essential aspects of quantum mechanics is developed in the text for all of these applications. Several additional homework problems then take readers just as far as they want to go in further developing the theory. Two appendices derive the "Golden Rule" for the transition rate and solve the two-level problem in quantum mechanics.

The discussion then proceeds to further applications of the basic principles of quantum mechanics. Chapter 5 is concerned with atomic physics, where the vector model of angular momentum is developed, the Zeeman effect and spin-orbit interaction analyzed, and the Thomas-Fermi statistical model of the structure of many-electron atoms presented. The latter finds a more basic formulation in terms of the Hartree mean-field approximation, which provides valuable insight into the behavior of quantum many-body systems throughout the book. At this point, one has enough background to understand the periodic system of the elements and the chemical behavior of its various families.

Chapter 6 is concerned with nuclear physics. The concepts of baryons (neutrons and protons), mean lives, and atomic masses are presented, and β-decay described. The two-body problem is solved with a finite square-well potential, and the observed spectra of the lightest nuclei discussed. The semi-empirical mass formula, which gives the average binding energy of nuclei throughout the periodic table, is developed. It is shown how electron scattering provides a microscope for actually seeing the nucleus. As with the periodic system of the elements, the Hartree approximation provides a basis for understanding the nuclear shell model. With the introduction of a strong spin-orbit force, the shell model puts one in a position to predict properties of nuclei throughout the periodic table.

Applications to particle physics are covered in chapter 7. The relevant forces are categorized, and the concept of an antiparticle is introduced. Leptons, which have only weak and electromagnetic interactions, are discussed. The strongly interacting hadrons (mesons and baryons) are described, together with their properties of isospin, strangeness, and charm. The Yukawa interaction between two baryons, arising from the interaction with a meson field, is derived. The observed hadron multiplets are exhibited, and it is shown how hadron structure can be understood in terms of an underlying substructure of quarks. An argument is given as to why single quarks are not observed as free particles.

Feynman diagrams, together with associated Feynman rules for the scattering S-matrix, provide the language of particle physics. It is shown how one goes from the S-matrix to a transition rate and then to a cross section. Several examples are presented: quantum electrodynamics (QED) with virtual photon exchange; quantum chromodynamics (QCD) with the exchange of gluons between quarks, and gluon self-couplings; and the standard model of electroweak interactions involving the exchange of heavy weak vector bosons, both charged and neutral.

Special relativity is the topic in chapter 8, starting with a description of the Michelson-Morley experiment, which ultimately demonstrated that the speed of light is the same in any inertial frame. It was Einstein's genius to give the Lorentz transformation, which leaves the form of the wave equation for light invariant, a physical interpretation as the actual relation between space-time coordinates $(\mathbf{x}, t)$ in two different inertial frames. The consequences are profound, as time and space now become relative coordinates, changing from inertial frame to inertial frame. Both time dilation and Lorentz contraction are analyzed. Special relativity is formulated in terms of rotations in a four-dimensional Minkowski space, whose fourth component $x_4 = ict$ is imaginary. Several applications of special relativity are discussed, including an analysis of the structure of white-dwarf stars.

Chapter 9 discusses the union of the two underpinnings of modern physics, quantum mechanics and special relativity. The Dirac equation for spin-1/2 fermions is derived from general assumptions, and its implications are examined in detail. Knowledge of properties of the Dirac equation allows us to actually construct the vertices in the Feynman diagrams for QED, QCD, and the standard model of electroweak interactions.

A point particle moving without friction on a two-dimensional surface of arbitrary shape forms a paradigm for the introduction of Einstein's theory of general relativity in chapter 10. Einstein's theory is introduced through a set of three assumptions on the structure of space-time and the corresponding particle motion. The Schwarzschild solution to the Einstein field equations outside of a spherically symmetric source is introduced, and its physical implications analyzed, including time dilation and radial Lorentz contraction. The properties of the Robertson-Walker metric (with $k = 0$), corresponding to a uniform mass density throughout all space are examined as an introduction to cosmology. The leading order cosmological redshift is derived and the concept of the horizon introduced.

Quantum fluids are macroscopic systems of many identical particles whose behavior reflects the underlying quantum mechanics. In chapter 11 the properties of two such systems are analyzed. The first is superfluid ^{4}He, whose atoms are spin-zero bosons and whose properties reflect Bose condensation. The velocity field is related to the phase of the single-particle ground-state wave function, leading to the quantization of the circulation about a vortex in the superfluid. The Hartree approximation helps one to understand the interacting system. That analysis is extended through the introduction of the Gross-Pitaevskii equation, which is solved to obtain the spatial structure of a vortex. Landau's argument on the relation of the quasiparticle excitation spectrum to superfluidity is presented. The second quantum fluid is that of electrons in a superconductor, whose empirical properties are summarized. It is argued that bound pairs of electrons will yield properties similar to that of Bose systems, in particular superfluid flow. The Bethe-Goldstone equation for a pair of fermions interacting in the presence of a filled Fermi sea is derived. It is shown how an attractive interaction between pairs of particles near the Fermi surface, with opposite momenta and spins, will actually lead to such a bound state. These are known as Cooper pairs. Phonon exchange in metals provides the attractive interaction between electrons, and Cooper pairs form the underlying basis for the very successful BCS theory of superconductivity. In analogy with the quantization of circulation in a superfluid Bose system, a derivation is given of the quantization of flux in the magnetic flux tube that penetrates a type-II superconductor.

Chapter 12 introduces the notion of a quantum field, which underlies most of modern theoretical physics. To do this, we return to the starting point of the book, the analysis of the planar transverse oscillation of a string. An expansion of the displacement of the string $q(x, t)$ in normal modes reduces the energy of the string to the sum of contributions of uncoupled simple harmonic oscillators, one for each normal mode. These oscillators are readily quantized, and the quanta of the normal modes are identified as phonons. The displacement now becomes a local quantum field operator $\hat{q}(x,t)$, where the normal-mode amplitudes are the phonon creation and destruction operators. The electromagnetic field is quantized in an exactly analogous fashion. Here the quantum field operator is the Coulomb-gauge vector potential $\hat{\mathbf{A}}(\mathbf{x},t)$, and the quanta of the normal-mode excitations of the field are the transverse photons. The concept of stimulated emission into a mode where photons are already present now follows immediately. The general expression for the photon radiation rate for any quantum mechanical system is derived in the problems. The quantum field analysis is extended to that of a Dirac field, where the Fermi statistics necessitates the introduction of anticommutation relations for the creation and destruction operators. Finally, the utility of quantum fields in analyzing the behavior of many identical non-relativistic interacting particles is demonstrated.

It is worth re-emphasizing that physics is an *experimental science*, and a course such as this *must* be accompanied by a good, thorough laboratory course in experimental modern physics.

There are many good, existing books that develop the experimental and theoretical aspects of modern physics, although the theory is generally covered in less depth: for example, [Eisberg and Resnick (1985); Resnick and Halliday (1992); Ohanian (1995); Krane (1996); Serway, Moses, and Moyer (1997); Bernstein, Fishbane, and Gasiorowicz (2000); Beiser (2002); Taylor, Zafiratos, and Dobson (2003)]. It is assumed that readers will study at least one such book in parallel with this one, in order to add sufficient breadth to their knowledge of modern physics.

Progress in physics, as in all of science, does not take place in a linear fashion. For every one of the concepts and results described in this book there have been tens, if not hundreds, of wrong turns and blind alleys. On the other hand, progress does occur by building on the foundation laid by previous scientists. The goal of this book is to provide a tour of the physics foundation established in the twentieth-century. It is hoped that the reader will both enjoy and benefit from the journey. Let us begin.

Chapter 2

Classical Physics

2.1 Newton's Laws

We start with a review of Newton's three laws of classical mechanics.[1] First, define the primary inertial coordinate system as one that is at rest with respect to the fixed stars (Fig. 2.1).[2] The *momentum* of a particle is defined as the product of its mass and its velocity $\mathbf{p} \equiv m\mathbf{v}$, where we choose to denote spatial vectors with boldface type.

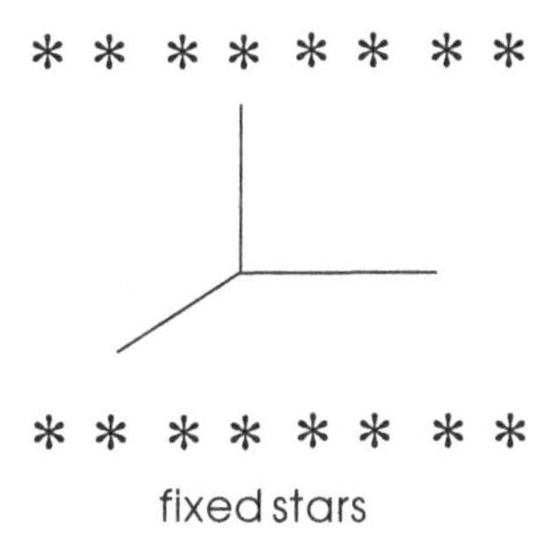

Fig. 2.1 The primary inertial coordinate system.

Newton's laws then state that in this primary inertial frame:

(1) A body at rest, or in uniform motion, remains in that state unless acted upon by a force;
(2) The time rate of change of the momentum of a particle is given by the

[1]See, for example, [Fetter and Walecka (2003)].
[2]This can be done to arbitrary accuracy.

applied force **F**

$$\frac{d\mathbf{p}}{dt} = \mathbf{F} \qquad ; \text{ Newton's 2nd law} \tag{2.1}$$

(3) For every action, there is an equal and opposite reaction.

It is important to remember that the momentum is a *spatial vector*

$$\mathbf{p} = m\mathbf{v} = m\frac{d\mathbf{x}}{dt} \qquad ; \text{ momentum} \tag{2.2}$$

and the rate of change of the momentum, which enters into Newton's second law, involves the *change* in the momentum $d\mathbf{p}$ — this can be in direction as well as length (Fig. 2.2).

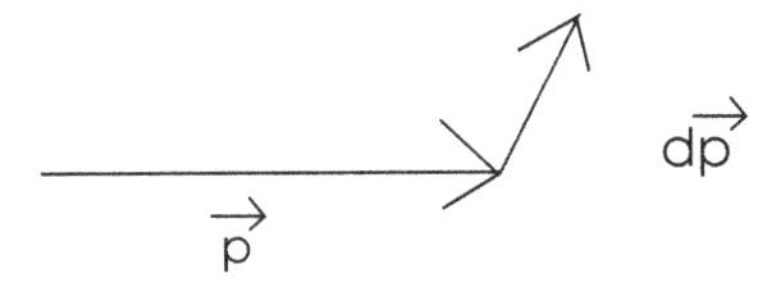

Fig. 2.2 Change in the momentum vector $d\mathbf{p}$, which enters into Newton's second law.

If the mass is constant, then Newton's second law is a statement about the particle's *acceleration*

$$\begin{aligned} \text{If} \qquad & m = \text{constant} \\ \Rightarrow \qquad & \frac{d\mathbf{p}}{dt} = m\frac{d\mathbf{v}}{dt} = m\frac{d^2\mathbf{x}}{dt^2} \qquad ; \text{ acceleration} \end{aligned} \tag{2.3}$$

It follows that any frame moving with *constant velocity* relative to an inertial frame is again an inertial frame in which Newton's laws hold (Prob. 2.14).

Newton's laws hold over a tremendous range of distances. For example (see Fig. 2.3):

- They hold down to the distance of closest approach $d \sim 10^{-13}$ cm in the scattering of an ion in the Coulomb field of the nucleus, an experiment through which Rutherford discovered the nucleus;[3]
- They hold out to, say, the Andromeda galaxy which is approximately $d \sim 1.5 \times 10^6$ light-years $\sim 10^{24}$ cm from us;[4]
- Thus they hold over a range of (at least) *37 orders of magnitude!*

[3]The details of these classic physics experiments, with which the reader should become familiar, can be found in any good existing book on modern physics, for example [Ohanian (1995)].

[4]They hold, provided the gravitational fields are not too strong.

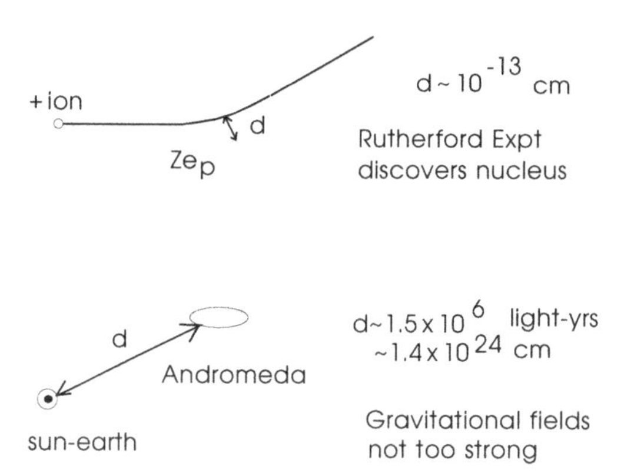

Fig. 2.3 Newton's laws hold over a tremendous range of distances.

It is the physicist's job to classify the forces. Newton's second law then tells how a particle moves under the action of these forces.

2.1.1 *Discrete Mechanics*

One-dimensional motion. Consider the motion of a particle of constant mass m under the application of a force F in one dimension (Fig. 2.4). In

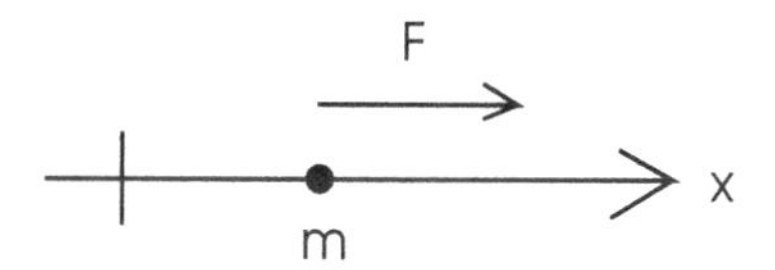

Fig. 2.4 Particle motion in one dimension.

this case, Newton's second law reads

$$m\frac{d^2x}{dt^2} = F \tag{2.4}$$

Two examples, in the z-direction, are illustrated in Fig. 2.5:

(1) A particle at the surface of the earth where the force of gravity is $F = -mg$;
(2) A particle at the end of a spring where $F = -kz$, and $z = 0$ is the equilibrium position.

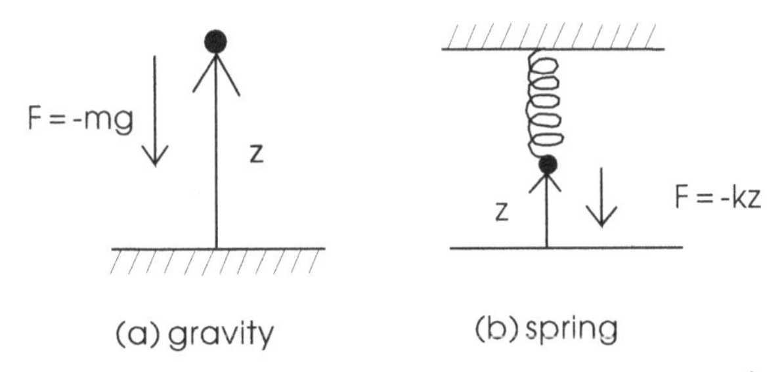

Fig. 2.5 Examples of particle motion in one dimension (here the z-direction): (a) gravity, a height z above the surface of the earth; (b) spring, where $z = 0$ is equilibrium.

Equation (2.4) is a second-order differential equation. It can be integrated numerically with a simple computer program. Divide the time interval of interest into small steps of size h (Fig. 2.6). Start from the point (x_0, t_0) with initial slope $(dx/dt)_0$. The value at the neighboring point (x_1, t_1) is then given by

$$\begin{aligned} x_1 &= x_0 + h\left(\frac{dx}{dt}\right)_0 \\ \left(\frac{dx}{dt}\right)_1 &= \left(\frac{dx}{dt}\right)_0 + h\left(\frac{d^2x}{dt^2}\right)_0 \end{aligned} \tag{2.5}$$

Equation (2.4) can now be substituted for the last term in the second line to give a closed set of finite-difference equations

$$\begin{aligned} x_1 &= x_0 + h\left(\frac{dx}{dt}\right)_0 \\ \left(\frac{dx}{dt}\right)_1 &= \left(\frac{dx}{dt}\right)_0 + h\left[\frac{F(x_0)}{m}\right] \end{aligned} \tag{2.6}$$

Now

- Step out from the initial value[5]

$$\left(x, \frac{dx}{dt}\right)_0 \xrightarrow{h} \left(x, \frac{dx}{dt}\right)_1 \xrightarrow{h} \left(x, \frac{dx}{dt}\right)_2 \xrightarrow{h} \cdots \tag{2.7}$$

- If the initial values $(x, dx/dt)_0$ are specified, then given the force $F(x)$, the solution to the finite-difference form of the second-order differential equation is determined for all subsequent times;

[5]Readers are urged to try this for themselves with some simple examples (see Prob. 2.1).

- Now take the limit as the step size $h \to 0$ for a given t. In this limit, one obtains the solution to the second-order differential equation itself.

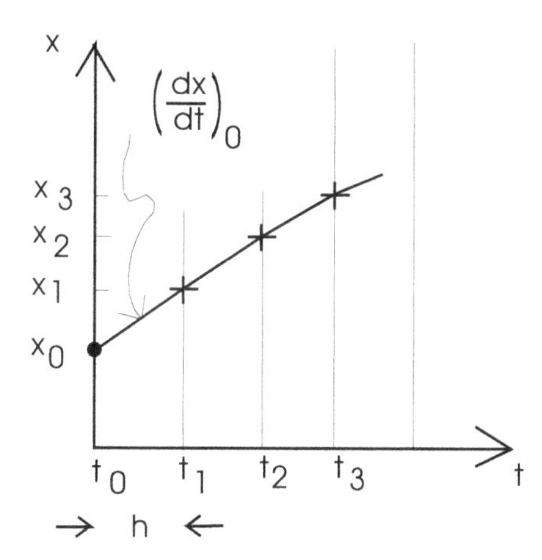

Fig. 2.6 Numerical integration of Newton's second law in one dimension with time step of size h. Here $(x, dx/dt)_0$ are the position and velocity at the initial time t_0.

Phase space. We have seen that Newton's second law gives a second-order differential equation, the solution of which requires the specification of both the value and first derivative $(x, dx/dt)_0$ at the initial time t_0. A more symmetrical formulation of classical mechanics is obtained by working in the space (p, q) of the momentum and coordinate, or *phase space.* Introduce

$$
\begin{aligned}
m\frac{dx}{dt} &\equiv p \qquad &&; \text{ momentum} \\
x &\equiv q \qquad &&; \text{ coordinate}
\end{aligned}
\tag{2.8}
$$

Determine the pair (p, q) for the particle at the time t, and plot the particle's position in phase space, as indicated in Fig. 2.7. If one specifies the initial values $(p, q)_0$ at the time t_0, then Newton's second law determines the position of the particle in phase space at all subsequent times.

We give two examples:

(1) For a free particle, the momentum is constant

$$
p = \text{constant} \qquad ; \text{ free particle} \tag{2.9}
$$

In this case, the phase space orbit is simply a straight line [Fig. 2.8(a)];

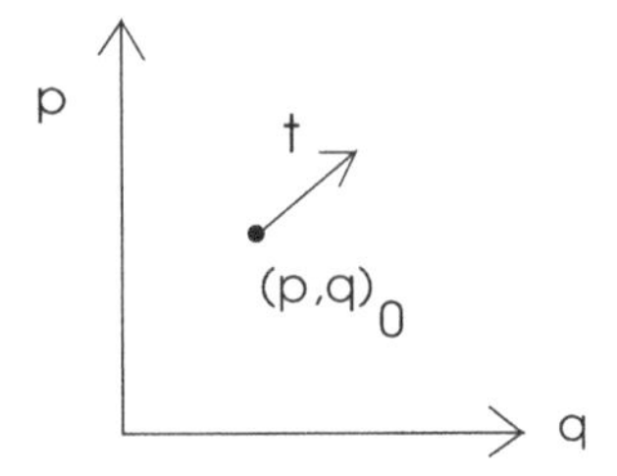

Fig. 2.7 Particle's trajectory in (p, q), or phase space.

(2) For a particle performing simple harmonic oscillations, the energy is constant

$$E = \frac{p^2}{2m} + \frac{1}{2}\kappa q^2 = E = \text{constant} \qquad ; \text{ s.h.o.} \tag{2.10}$$

In this case the phase space orbit is an ellipse [Fig. 2.8(b)].

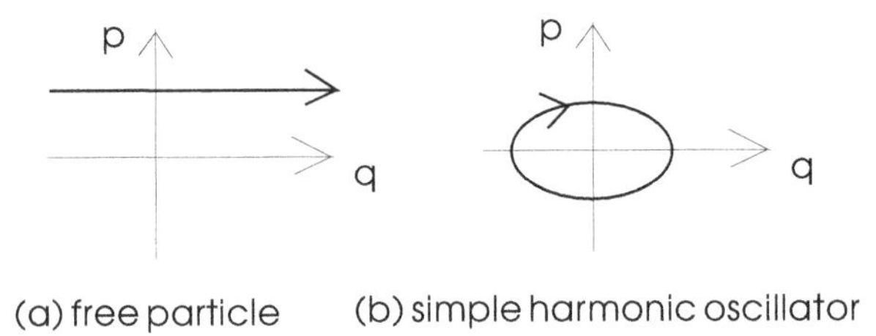

Fig. 2.8 Phase space orbits: (a) free particle; (b) simple harmonic oscillator.

Hamiltonian. The hamiltonian will play a crucial role in subsequent developments. We illustrate its role in classical mechanics. Define it as

$$H(p, q) \equiv T(p) + V(q) \qquad ; \text{ hamiltonian} \tag{2.11}$$

For the simple harmonic oscillator, for example, it follows from Eq. (2.10) that

$$H(p, q) = \frac{p^2}{2m} + \frac{1}{2}\kappa q^2 \qquad ; \text{ s.h.o.} \tag{2.12}$$

Hamilton's equations. After two centuries of classical mechanics, it is not too surprising that a more symmetric, and more elegant, formulation of Newton's laws could be found. This was done by Hamilton. Consider $H(p, q)$, and take

$$\begin{aligned} \frac{\partial H(p,q)}{\partial p} &\equiv \left[\frac{dH}{dp}\right]_{q=\text{const.}} = \frac{p}{m} = \frac{dq}{dt} \\ \frac{\partial H(p,q)}{\partial q} &\equiv \left[\frac{dH}{dq}\right]_{p=\text{const.}} = \frac{dV}{dq} = -F = -\frac{dp}{dt} \end{aligned} \tag{2.13}$$

Hence *Hamilton's equations*, which are here fully equivalent to Newton's laws, state that

$$\begin{aligned} \frac{\partial H(p,q)}{\partial p} &= \frac{dq}{dt} & &;\ \text{Hamilton's equations} \\ \frac{\partial H(p,q)}{\partial q} &= -\frac{dp}{dt} & &\Leftrightarrow \text{Newton's laws} \end{aligned} \tag{2.14}$$

The partial derivatives on the left-hand-sides (l.h.s.) of these relations, defined in Eqs. (2.13), imply that the other variable in $H(p, q)$ is to be held fixed during the differentiation. The hamiltonian, through these relations, describes how the particle moves in phase space. At the end, after the equations of motion have been computed from the hamiltonian in Eq. (2.11), one has the statement of conservation of energy

$$H = E = \text{constant} \qquad ;\ \text{energy conservation} \tag{2.15}$$

2.1.2 *Continuum Mechanics*

String. The equation of motion of a continuous string can now be obtained by applying Newton's second law to each little mass element of the string. Suppose the equilibrium stretched position of the string lies along the x-axis. Then (Fig. 2.9):

- A differentially small mass element of the string is given by $dm = \rho dx$ where

$$\rho = \frac{dm}{dx} = \text{mass density} \tag{2.16}$$

- The force transmitted along the string is given by the tension τ

$$\tau = \text{tension} \tag{2.17}$$

- Suppose the displacement of the string from equilibrium takes place in the (x, y) plane. Denote this displacement in the y-direction by $q(x, t)$;

- To simplify, assume that both ρ and τ are constant along the string.

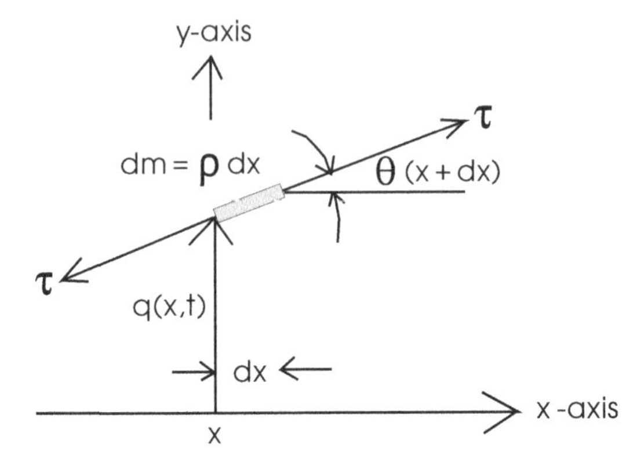

Fig. 2.9 Motion of a string with mass density ρ and tension τ. The displacement of the string in the transverse y-direction, from its equilibrium stretched position along the x-axis, is denoted by $q(x, t)$.

Now write Newton's second law for the little mass element. The mass times the acceleration at a given x is

$$\text{mass} \times \text{acceleration} = \rho dx \left[\frac{d^2 q}{dt^2}\right]_{x \text{ fixed}} \tag{2.18}$$

The force in the y-direction on the little mass element at a given instant in time is (see Fig. 2.10)

$$F_y = \tau \left[\sin\theta(x + dx) - \sin\theta(x)\right]_{t \text{ fixed}} \tag{2.19}$$

For small displacements, this expression can be *linearized* by assuming $\sin\theta \approx \tan\theta$, which is then given by the slope of the string at either end of the small element

$$\begin{aligned}
F_y &\approx \tau \left[\tan\theta(x + dx) - \tan\theta(x)\right]_{t \text{ fixed}} \\
&= \tau \left[\left.\frac{dq}{dx}\right|_{x+dx} - \left.\frac{dq}{dx}\right|_{x} \right]_{t \text{ fixed}} \\
&= \tau dx \left[\frac{d^2 q}{dx^2}\right]_{t \text{ fixed}}
\end{aligned} \tag{2.20}$$

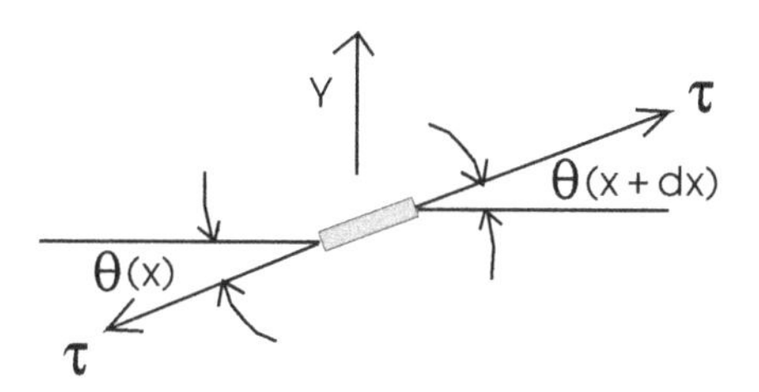

Fig. 2.10 Force in the y-direction on the little element of string in Fig. 2.9.

Newton's second law for the small mass element of the string is obtained by combining the above expressions

$$\rho\, dx \left[\frac{d^2q}{dt^2}\right]_{x \text{ fixed}} = \tau\, dx \left[\frac{d^2q}{dx^2}\right]_{t \text{ fixed}} \tag{2.21}$$

Upon cancelation of the factor dx, one obtains the one-dimensional *wave equation* for the string

$$\begin{aligned} \frac{\partial^2 q(x,t)}{\partial x^2} &= \frac{1}{c^2}\frac{\partial^2 q(x,t)}{\partial t^2} && ;\ \text{wave equation} \\ c^2 &= \frac{\tau}{\rho} && ;\ (\text{velocity})^2 \end{aligned} \tag{2.22}$$

Here the partial derivatives again imply that the other variable in $q(x,t)$ is to be held fixed during the differentiation. This is a *very important result.* The planar motion of the string is governed by the second-order partial differential wave equation for the transverse displacement of the string $q(x,t)$; and that equation has been linearized in the displacement, which is assumed to be small. The square of the wave velocity appearing in this equation is given by the ratio of the tension to the mass density of the string.

D'Alembert's Solution. Let $f(x)$ be any twice differentiable function (Fig. 2.11), and define $\bar{x} \equiv x - ct$. Then $f(\bar{x}) = f(x - ct)$ provides D'Alembert's solution to the one-dimensional wave equation

$$f(\bar{x}) = f(x - ct) \qquad ;\ \text{D'Alembert's solution} \tag{2.23}$$

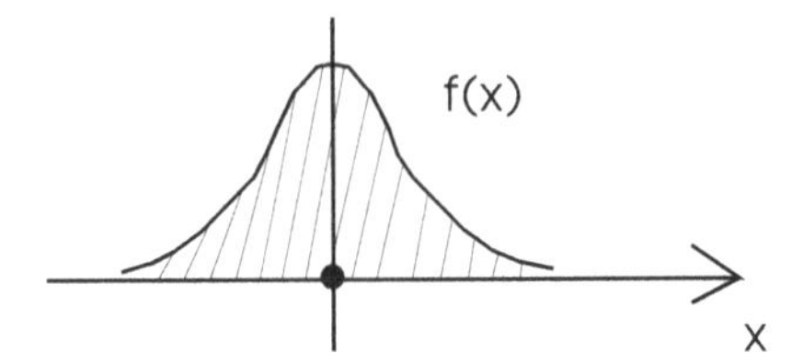

Fig. 2.11 Example of a twice differentiable function $f(x)$.

This is proven simply by using the chain rule for differentiation

$$\begin{aligned}
\frac{\partial f}{\partial x} &= \frac{df}{d\bar{x}}\frac{\partial \bar{x}}{\partial x} = f'(\bar{x}) \\
\frac{\partial^2 f}{\partial x^2} &= f''(\bar{x}) \\
\frac{\partial f}{\partial t} &= \frac{df}{d\bar{x}}\frac{\partial \bar{x}}{\partial t} = -cf'(\bar{x}) \\
\frac{\partial^2 f}{\partial t^2} &= (-c)^2 f''(\bar{x})
\end{aligned} \tag{2.24}$$

Hence $q(x,t) = f(\bar{x}) = f(x - ct)$ satisfies Eq. (2.22), and the result is established.

Let us try to *interpret* this solution. Introduce a coordinate system whose origin $\bar{O}$ moves with a velocity c along the x-axis relative to the fixed (laboratory) system with origin O, as illustrated in Fig. 2.12. The spatial coordinate in this moving frame is just $\bar{x} = x - ct$, and the function $f(\bar{x})$ retains its shape relative to this coordinate system. Hence we conclude that:

> *This pulse $f(\bar{x}) = f(x - ct)$ moves without change in shape with velocity c in the x-direction.*[6]

Readers are urged to demonstrate this for themselves by sending a small amplitude pulse of any shape down a stretched string (or spring).

Clearly, since the wave equation is linear, any linear combination of solutions again provides a solution. This is referred to as *linear superposition.* Furthermore, changing $c \to -c$ does not affect the argument in Eqs. (2.24). Thus a disturbance of the form

$$q(x,t) = f(x - ct) + g(x + ct) \qquad ; \text{ D'Alembert's solution} \quad \text{linear superposition} \tag{2.25}$$

[6]One says there is "no dispersion" of the solution.

again provides a solution to the one-dimensional wave equation for *any* twice differentiable functions (f, g). The second term has the interpretation of a pulse moving in the opposite direction.

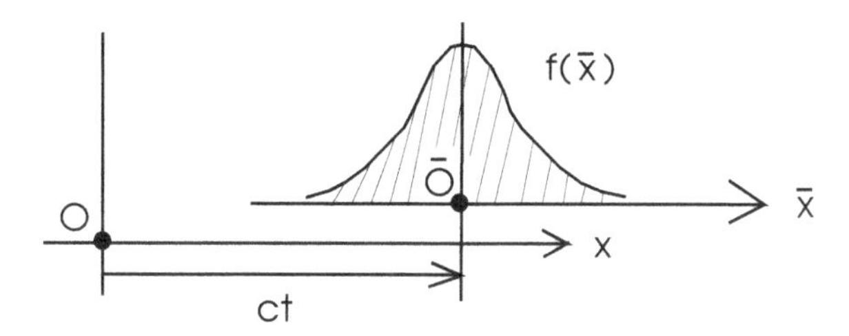

Fig. 2.12 Interpretation of D'Alembert's solution with a function $f(x)$ as illustrated in Fig. 2.11. $\bar{O}$ is the origin of a coordinate system moving with velocity c in the x-direction relative to the fixed (laboratory) frame with origin O. The function $f(\bar{x}) = f(x - ct)$ retains its shape relative to this moving coordinate system.

Wave Solution. Consider a particular form for the function $f(\bar{x})$

$$f(\bar{x}) = \cos\left[\frac{2\pi}{\lambda}(x - ct)\right] \qquad ; \text{ wave solution} \tag{2.26}$$

From the previous discussion we know that this represents a wave moving without change in shape in the x-direction with velocity c. A snapshot of this wave at a given instant in time yields the curve shown in the upper part of Fig. 2.13. The disturbance repeats itself after a spatial distance λ, which represents the *wavelength* of this disturbance.

$$\lambda \equiv \text{wavelength} \tag{2.27}$$

If one sits at a fixed position x, the disturbance oscillates in time and repeats itself after a *time period* $\varphi = \lambda/c$

$$\varphi = \frac{\lambda}{c} \qquad ; \text{ period} \tag{2.28}$$

The *frequency* ν with which the disturbance repeats itself is just the inverse of the period

$$\nu = \frac{1}{\varphi} = \frac{c}{\lambda} \qquad ; \text{ frequency}$$

$$\Rightarrow \qquad \nu\lambda = c \tag{2.29}$$

The last result $\nu\lambda = c$ is a *crucial one* for waves described by the one-dimensional wave equation.

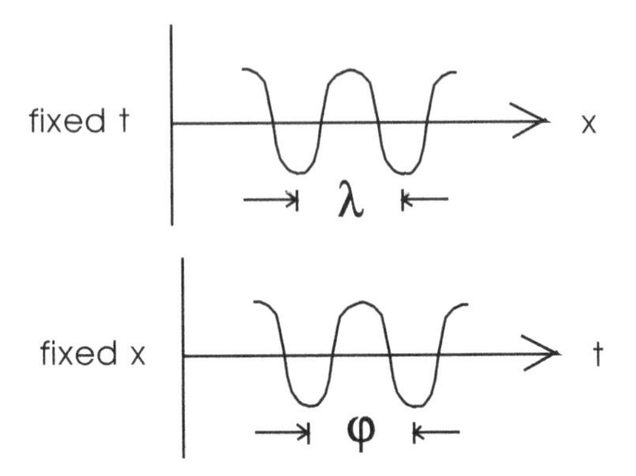

Fig. 2.13 Interpretation of the wave solution in Eq. (2.26). Snapshot at a fixed time (upper figure); time-dependent oscillation at a fixed x (lower figure).

The *wavenumber* k and *angular frequency* ω are defined as

$$k \equiv \frac{2\pi}{\lambda} \qquad ; \text{ wavenumber}$$
$$\omega \equiv 2\pi\nu = kc \qquad ; \text{ angular frequency} \tag{2.30}$$

Hence the solution in Eq. (2.26) can also be written as

$$f(\bar{x}) = \cos(kx - \omega t) \qquad ; \text{ wave solution} \tag{2.31}$$

Normal Modes. Let us look for a solution to the wave Eq. (2.22) where *everything oscillates with the same frequency.* Try

$$q(x,t) = q(x)\cos(\omega t + \delta) \qquad ; \text{ Try it} \tag{2.32}$$

Upon substitution of this expression, and cancelation of the factor $\cos(\omega t + \delta)$, the wave equation is reduced to

$$\frac{d^2 q(x)}{dx^2} = -k^2 q(x) \qquad ; \text{ Helmholtz eqn} \tag{2.33}$$

This is an ordinary second-order differential equation in the spatial coordinate x, the scalar *Helmholtz equation.*[7] It can be integrated numerically,

[7]We will meet this equation again in quantum mechanics when we look for stationary-state solutions to the Schrödinger equation.

for example, exactly as in our discussion of Newton's laws. If one specifies $(q, dq/dx)_{x_0}$, the displacement and slope at a boundary position x_0, then the solution is determined for all other values of the coordinate x. Of course, one can specify other *boundary conditions*; one needs two of them for this second-order equation.

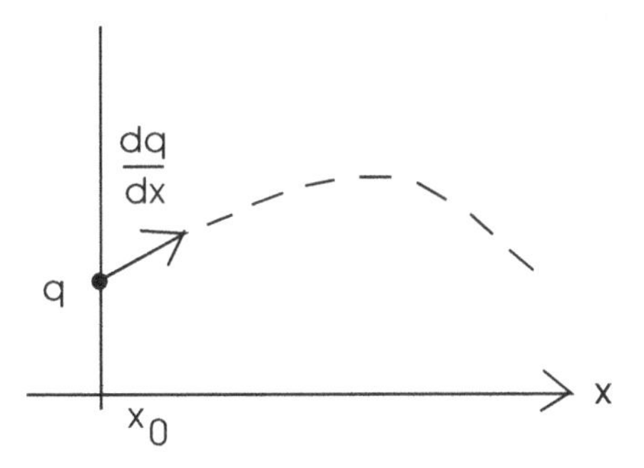

Fig. 2.14 Numerical integration of the Helmholtz equation upon specification of the boundary values $(q, dq/dx)_{x_0}$.

Fixed Endpoints. Consider a string with fixed endpoints (Fig. 2.15). In this case, if one is solving numerically as discussed above, it is necessary to start with the correct slope at the boundary where $x = 0$ so that the solution comes back to zero at the second boundary where $x = L$.

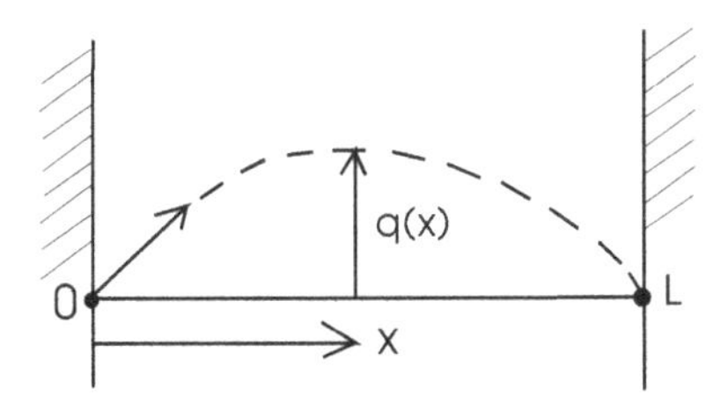

Fig. 2.15 String with fixed endpoints.

The general solution to Eq. (2.33) can be written as a linear combination of sines and cosines

$$q(x) = A \sin kx + B \cos kx \tag{2.34}$$

where (A, B) are constants. Here the solution must vanish at the endpoints, which gives the conditions

$$\begin{aligned} q(0) &= 0 & &\Rightarrow B = 0 \\ q(L) &= 0 & &\Rightarrow A \sin kL = 0 \end{aligned} \qquad (2.35)$$

The last relation in Eqs. (2.35) can evidently only be satisfied for *certain values of* k, label them k_n, with a corresponding set of frequencies $\omega_n = k_n c$

$$\begin{aligned} k_n L &= n\pi & &; n = 1, 2, 3, \cdots, \infty \\ \omega_n &= \left(\frac{\pi c}{L}\right) n & &\text{eigenvalues} \end{aligned} \qquad (2.36)$$

These are known as the *eigenvalues* for this problem. The corresponding solutions, labeled as $q_n(x)$, are known as the *eigenfunctions*.[8] We choose to normalize them with a factor $(2/L)^{1/2}$ for reasons that will become apparent. Thus

$$q_n(x) = \left(\frac{2}{L}\right)^{1/2} \sin\left(\frac{n\pi x}{L}\right) \qquad ; n = 1, 2, 3, \cdots, \infty \quad \text{eigenfunctions} \qquad (2.37)$$

With these boundary conditions, one finds a discrete set of frequencies for the normal modes; these include the fundamental frequency ω_1 and all the harmonics $2\omega_1, 3\omega_1, \cdots$. The full time-dependent, normal-mode solution is then

$$q_n(x,t) = \left(\frac{2}{L}\right)^{1/2} \sin\left(\frac{n\pi x}{L}\right) \cos\left(\omega_n t + \delta_n\right) \qquad ; \text{normal mode} \qquad (2.38)$$

where δ_n is an arbitrary phase. Each of these normal modes provides a particular solution to the wave Eq. (2.22).

Let us examine a few of the eigenfunctions. Consider the first eigenfunction with $n = 1$ as illustrated in Fig. 2.16. In this case

$$\begin{aligned} q_1(x) &= \left(\frac{2}{L}\right)^{1/2} \sin\frac{\pi x}{L} & &; n = 1 \\ \omega_1 &= \frac{\pi c}{L} & &\text{fundamental} \end{aligned} \qquad (2.39)$$

One can understand this eigenvalue by simply fitting one-half wavelength

[8]Note that $n = 0$ produces no displacement of the string, and the eigenfunctions for the negative integers are not linearly independent of those given here.

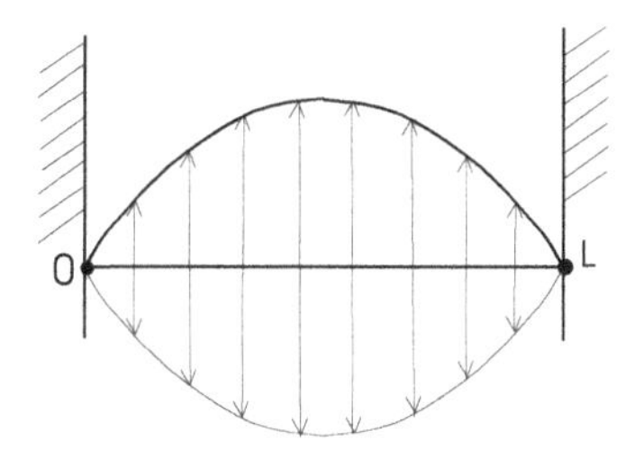

Fig. 2.16 Eigenfunction with $n = 1$.

between the fixed boundaries of the string

$$\begin{aligned} L &= \frac{\lambda}{2} & &; \text{ half wavelength} \\ \nu &= \frac{c}{\lambda} = \frac{c}{2L} \\ \omega &= 2\pi\nu = \frac{\pi c}{L} = \frac{\pi}{L}\left(\frac{\tau}{\rho}\right)^{1/2} \end{aligned} \tag{2.40}$$

Here the explicit expression for the wave velocity in Eq. (2.22) has been inserted in the last equation.

The situation for the first overtone with $n = 2$ is illustrated in Fig. 2.17.

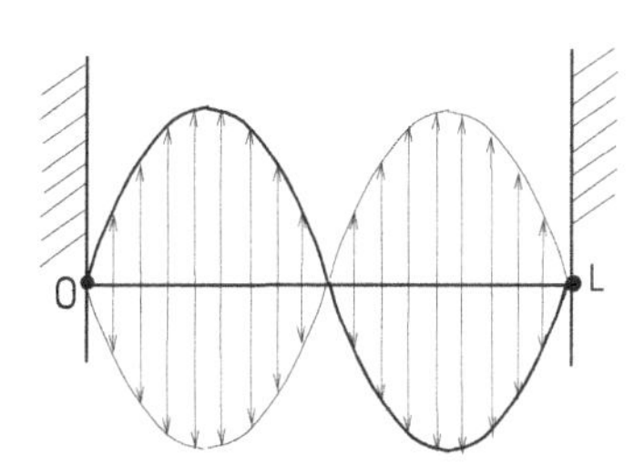

Fig. 2.17 Eigenfunction with $n = 2$.

In this case

$$\begin{aligned} q_2(x) &= \left(\frac{2}{L}\right)^{1/2} \sin\frac{2\pi x}{L} & &; \ n = 2 \\ \omega_2 &= \frac{2\pi c}{L} & &\text{first overtone} \end{aligned} \tag{2.41}$$

To understand this solution, one simply fits a *full* wavelength into the string

$$
\begin{aligned}
L &= \lambda && ;\ \text{full wavelength} \\
\nu &= \frac{c}{\lambda} = \frac{c}{L} \\
\omega &= 2\pi\nu = \frac{2\pi c}{L} && \text{etc.}
\end{aligned}
\tag{2.42}
$$

General Solution with Fixed Endpoints. Let us now construct the *full solution* to the wave Eq. (2.22) for a string with fixed endpoints. First, we again employ *linear superposition* and observe that any linear combination of the normal modes in Eq. (2.38) again yields a solution

$$
\begin{aligned}
q(x,t) &= \sum_{n=1}^{\infty} A_n q_n(x) \cos(\omega_n t + \delta_n) \\
&= \sum_{n=1}^{\infty} A_n \left(\frac{2}{L}\right)^{1/2} \sin\left(\frac{n\pi x}{L}\right) \cos(\omega_n t + \delta_n)
\end{aligned}
\tag{2.43}
$$

where (A_n, δ_n) with $n = 1, 2, \cdots, \infty$ are constants, to be determined. This expression has the following properties:

- It satisfies the wave Eq. (2.22), since each term does;
- It satisfies the boundary conditions of fixed endpoints, since each term does.

The question is, what is left? From Newton's second law for a mass point, as discussed in the previous section, we know that we must specify both the *position* and *velocity* of each little mass element in order to find a solution for the subsequent motion of the string. Thus, to complete the solution, we must match the *initial conditions*

Initial Conditions. To complete the solution in Eq. (2.43), we must specify both the *shape* of the string at the initial time, say $t = 0$, as well as the initial *velocity*. Thus both $q(x, 0)$ and $[\partial q(x,t)/\partial t]_{t=0}$ must be matched *for all* x, as illustrated in Fig. 2.18.

$$
\begin{aligned}
q(x,0) &= \sum_{n=1}^{\infty} A_n \cos\delta_n \left(\frac{2}{L}\right)^{1/2} \sin\left(\frac{n\pi x}{L}\right) \\
\left[\frac{\partial q(x,t)}{\partial t}\right]_{t=0} &= -\sum_{n=1}^{\infty} \omega_n A_n \sin\delta_n \left(\frac{2}{L}\right)^{1/2} \sin\left(\frac{n\pi x}{L}\right)
\end{aligned}
\tag{2.44}
$$

These equations indeed determine the constants (A_n, δ_n) for all n (see below). Equation (2.43) then provides the full solution to the wave equation, and hence the complete motion of the string, for all subsequent time.

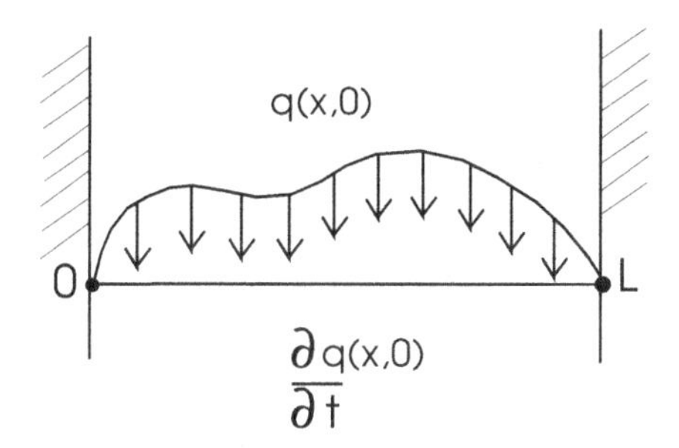

Fig. 2.18 Initial conditions for string. Both $q(x,0)$ and $[\partial q(x,t)/\partial t]_{t=0}$ must be matched.

2.1.3 *Some Mathematics*

The expressions in Eqs. (2.44) are *Fourier series* of the form

$$f(x) = \sum_{n=1}^{\infty} a_n \phi_n(x)$$
$$\phi_n(x) = \left(\frac{2}{L}\right)^{1/2} \sin\left(\frac{n\pi x}{L}\right) \tag{2.45}$$

A large class of functions can be represented in this fashion. The series are *complete* for the class of functions required for our continuous string problem.[9] To solve for the Fourier coefficients, one makes use of the *orthonormality of the eigenfunctions* (see appendix C)

$$\int_0^L \phi_m(x)\phi_n(x)dx = \frac{2}{L}\int_0^L dx\, \sin\left(\frac{m\pi x}{L}\right)\sin\left(\frac{n\pi x}{L}\right) = \delta_{mn} \tag{2.46}$$

Here δ_{mn} is the Kronecker delta defined by

$$\begin{aligned} \delta_{mn} &= 1 \qquad ; \text{ if } m = n \\ &= 0 \qquad ; \text{ if } m \neq n \end{aligned} \tag{2.47}$$

[9]See, for example, [Fetter and Walecka (2003)]. We do not here go into the niceties of operating on these series term-by-term, which depends on convergence. Suffice is to say that integration, as a smoothing function, is generally permitted.

Now multiply Eq. (2.45) by $\phi_m(x) = (2/L)^{1/2} \sin(m\pi x/L)$, and integrate on x from 0 to L using the orthonormality of the eigenfunctions

$$\int_0^L \phi_m(x) f(x)\, dx = a_m \tag{2.48}$$

This solves for the Fourier coefficients. The application of this procedure to Eqs. (2.44) yields

$$\begin{aligned} A_m \cos\delta_m &= \int_0^L q(x,0) \left(\frac{2}{L}\right)^{1/2} \sin\left(\frac{m\pi x}{L}\right) dx \\ -\omega_m A_m \sin\delta_m &= \int_0^L \left[\frac{\partial q(x,t)}{\partial t}\right]_{t=0} \left(\frac{2}{L}\right)^{1/2} \sin\left(\frac{m\pi x}{L}\right) dx \end{aligned} \tag{2.49}$$

This provides an explicit solution for the Fourier coefficients (A_m, δ_m) in terms of the initial conditions $q(x,0)$ and $[\partial q(x,t)/\partial t]_{t=0}$. Note, again, that these functions must be provided for all x. Note also that it is now clear that both $q(x,0)$ and $[\partial q(x,t)/\partial t]_{t=0}$ must be specified in order to determine the entire set of coefficients (A_m, δ_m) for all $1 \leq m \leq \infty$.

Sum Over Modes. Consider the sum over all wave numbers

$$k_n = \left(\frac{\pi}{L}\right) n \qquad ; \; n = 1, 2, 3, \cdots, \infty \tag{2.50}$$

that appears in Eq. (2.45). Suppose the string becomes very long so that $L \to \infty$. Then as n increases by an integer, the *spacing* of the wavenumbers $\Delta k = \pi/L$ becomes vanishingly small

$$\Delta k = \frac{\pi}{L} \longrightarrow 0 \qquad ; \; L \to \infty \tag{2.51}$$

Rewrite the *sum* over wavenumbers as

$$\sum_{n=1}^{\infty} f(k_n) = \frac{L}{\pi} \sum_{n=1}^{\infty} \left(\frac{\pi}{L}\right) f\left(\frac{\pi n}{L}\right) \tag{2.52}$$

Suppose that $f(k_n)$ is a smoothly varying function of its argument, as in Fig. 2.19. In this case, the *sum* over wavenumbers in Eq. (2.52) becomes an *integral* over wavenumbers

$$\begin{aligned} \sum_{n=1}^{\infty} \left(\frac{\pi}{L}\right) f(k_n) &= \sum_{n=1}^{\infty} \Delta k\, f(k_n) \\ &\longrightarrow \int_0^{\infty} dk\, f(k) \qquad ; \; L \to \infty \end{aligned} \tag{2.53}$$

and this relation is exact in the limit.

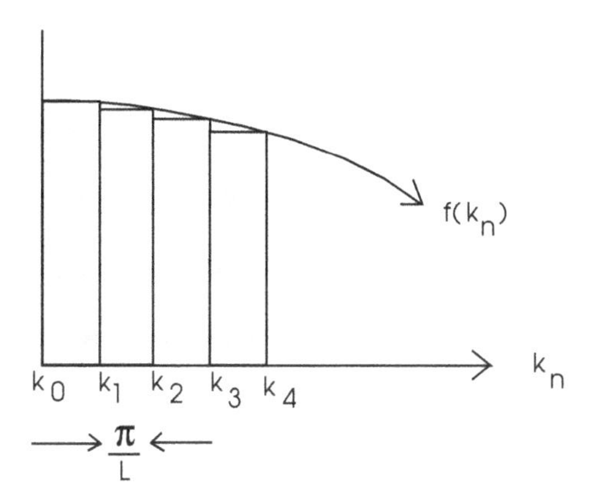

Fig. 2.19 Fourier sum in the limit $L \to \infty$, where $\Delta k = \pi/L$.

Thus we can write a general relation for the Fourier sum in Eq. (2.52)

$$\sum_{n=1}^{\infty} \longrightarrow \frac{L}{\pi}\int_0^{\infty} dk \qquad ; \; L \to \infty \tag{2.54}$$

If the function $f(k)$ is *even* about the origin in k, so that $f(-k) = f(k)$, then one can integrate over *all* wavenumbers in Eq. (2.54)

$$\sum_{n=1}^{\infty} \longrightarrow \frac{L}{2\pi}\int_{-\infty}^{\infty} dk \qquad ; \; L \to \infty \qquad f(-k) = f(k) \tag{2.55}$$

This is an *important relation* that will play a central role in what follows. Relevant properties of Fourier sums and Fourier integrals are examined in appendix C.

2.2 Statistical Physics

Consider a very large number $N \sim 10^{23}$ of particles (*e.g.* atoms) in a box of volume V in contact with a heat bath at temperature T (Fig. 2.20). One cannot hope to solve Newton's laws for such a system. Furthermore, we generally measure only a small number of bulk properties such as (P, V, T)

for the system.[10] In this case, one is forced to use *statistical methods*. Let us then review some of the principal elements of *classical statistical mechanics*.

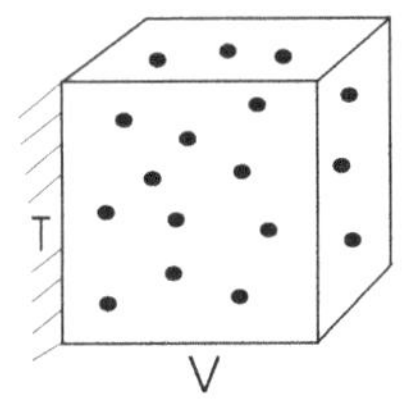

Fig. 2.20 Collection of very large number $N \sim 10^{23}$ of particles in thermal equilibrium at temperature T in a box of volume V.

2.2.1 *Classical Statistical Mechanics*

Consider a very large number N of identical, randomly prepared particles in the box with (T, V) as illustrated in Fig. 2.21. The experimental observation of a particle is equivalent to identifying one of these constituents. The complete collection of these constituents is known as an *ensemble*, and here we are focusing on the *microcanonical ensemble*.

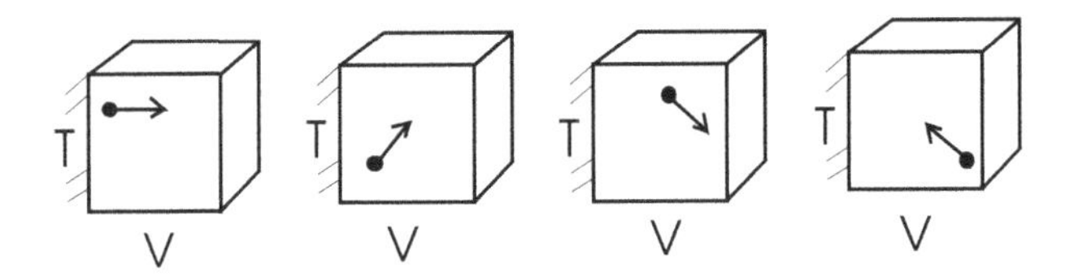

Fig. 2.21 Very large number N of identical, randomly prepared particles.

Microcanonical Ensemble. Place a point in the six-dimensional phase space $(\mathbf{p}, \mathbf{q})$, with volume element $d^3p\, d^3q = dp_1 dp_2 dp_3\, dq_1 dq_2 dq_3$, for each member of the ensemble in Fig. 2.21. This is illustrated with two dimensions in Fig. 2.22. There is a totality of N such points. Now suppose one just selects a particle *at random* from the ensemble. The differential *probability* that the particle will have a momentum between $\mathbf{p}$ and $\mathbf{p} + \mathbf{dp}$ and a coordinate position between $\mathbf{q}$ and $\mathbf{q} + \mathbf{dq}$ is just the ratio of the number of particles in the element $d^3p\, d^3q$ in phase space to the total number of

[10] Here P is the pressure.

particles

$$\frac{dN}{N} \equiv dP(\mathbf{p}, \mathbf{q}) \qquad ; \text{ differential probability of observing } (\mathbf{p}, \mathbf{q}) \tag{2.56}$$

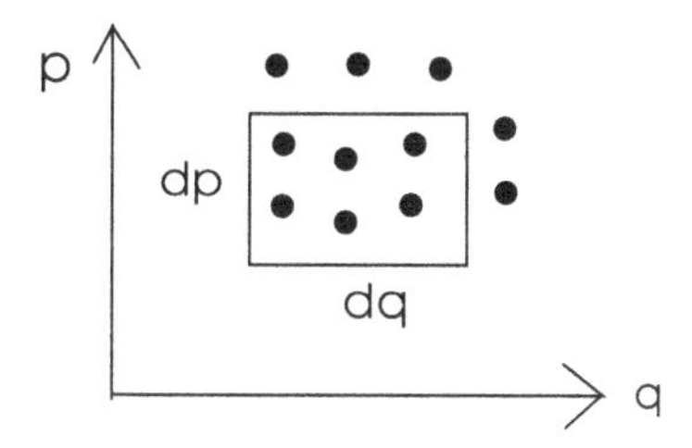

Fig. 2.22 Complete collection, the *ensemble*, of the particles shown in Fig. 2.21 as represented by points in six-dimensional phase space $(\mathbf{p}, \mathbf{q})$ with $d^3p\, d^3q = dp_1 dp_2 dp_3 dq_1 dq_2 dq_3$. Here only two dimensions are shown.

The *hamiltonian* for particle motion in a one-body potential $V(\mathbf{q})$ was introduced in Eq. (2.11)

$$H(\mathbf{p}, \mathbf{q}) = T(\mathbf{p}) + V(\mathbf{q}) \tag{2.57}$$

For the present purposes, we will start from the *assumption* that the probability of finding a particle in a given region in phase space in *thermal equilibrium* at a temperature T is given by the *Boltzmann distribution*

$$\frac{dN}{N} = dP(\mathbf{p}, \mathbf{q}) = \frac{e^{-H(\mathbf{p},\mathbf{q})/k_{\text{B}}T}\, dp_1 dp_2 dp_3\, dq_1 dq_2 dq_3}{\int \cdots \int e^{-H(\mathbf{p},\mathbf{q})/k_{\text{B}}T}\, dp_1 dp_2 dp_3\, dq_1 dq_2 dq_3} \tag{2.58}$$

Here k_{B} is Boltzmann's constant

$$k_{\text{B}} = 1.381 \times 10^{-23}\,\text{J}(^{\circ}\text{K})^{-1} \qquad ; \text{ Boltzmann's constant} \tag{2.59}$$

One can derive Eq. (2.58) from the more basic assumptions of statistical mechanics, and this is done in appendix E; however, starting from the Boltzmann distribution as an assumption is enough for us right now.

Maxwell-Boltzmann Distribution. Let us examine the implied distribution of velocities in a gas of free particles where $V(\mathbf{q}) = 0$. Here

$$\mathbf{p} = m\mathbf{v} \qquad ; \; H = \frac{\mathbf{p}^2}{2m} \tag{2.60}$$

If one is only interested in the velocity distribution, it is appropriate to integrate over all positions $\int \cdots \int dq_1 dq_2 dq_3$ in the probability distribution, and since any common factors *cancel in the ratio* in Eq (2.58), the probability distribution in the velocities then becomes

$$dP(\mathbf{v}) = \int \cdots \int dq_1 dq_2 dq_3 \, dP(\mathbf{p}, \mathbf{q})$$
$$= \frac{e^{-m\left(v_1^2+v_2^2+v_3^2\right)/2k_{\rm B}T} \, dv_1 dv_2 dv_3}{\int \cdots \int e^{-m\left(v_1^2+v_2^2+v_3^2\right)/2k_{\rm B}T} \, dv_1 dv_2 dv_3} \tag{2.61}$$

This is the celebrated *Maxwell-Boltzmann distribution*, which is well verified experimentally.

The probability distribution in Eq. (2.61) can be used to compute *mean values*. For example, the mean value of the energy $E = \mathbf{p}^2/2m$ takes the form

$$\langle E \rangle = \frac{m}{2} \frac{\int \cdots \int \left(v_1^2 + v_2^2 + v_3^2\right) e^{-m\left(v_1^2+v_2^2+v_3^2\right)/2k_{\rm B}T} \, dv_1 dv_2 dv_3}{\int \cdots \int e^{-m\left(v_1^2+v_2^2+v_3^2\right)/2k_{\rm B}T} \, dv_1 dv_2 dv_3} \tag{2.62}$$

Since the exponentials factor, and then the resulting integrals factor, this expression reduces to[11]

$$E = \frac{3m}{2} \frac{\int_{-\infty}^{\infty} v^2 e^{-mv^2/2k_{\rm B}T} \, dv}{\int_{-\infty}^{\infty} e^{-mv^2/2k_{\rm B}T} \, dv} \tag{2.63}$$

Take out the dimensions with the definition

$$u \equiv v \left(\frac{m}{2k_{\rm B}T}\right)^{1/2}$$
$$E = 3k_{\rm B}T \, \frac{\int_{-\infty}^{\infty} u^2 e^{-u^2} \, du}{\int_{-\infty}^{\infty} e^{-u^2} \, du} \tag{2.64}$$

One is left with the definite integrals

$$\int_{-\infty}^{\infty} e^{-x^2} \, dx = \sqrt{\pi} \qquad ; \int_{-\infty}^{\infty} x^2 e^{-x^2} \, dx = \frac{\sqrt{\pi}}{2} \tag{2.65}$$

Thus in a gas of free particles in thermal equilibrium at a temperature T, the mean energy per particle is given by

$$\langle E \rangle = \frac{3}{2} k_{\rm B}T \qquad ; \text{ gas of free particles mean energy/particle} \tag{2.66}$$

[11]Use $\int\int dxdy \, e^{[a(x)+b(y)]} = \int\int dxdy \; e^{a(x)} e^{b(y)} = \int dx \, e^{a(x)} \int dy \, e^{b(y)}$. With independent boundaries these integrals are factored, and common factors in the numerator and denominator cancel.

2.2.2 *Equipartition Theorem*

The classical equipartition theorem follows directly from the above analysis. It states that

> *There is a contribution to the mean energy of $k_{\rm B}T/2$ for each quadratic degree of freedom in the hamiltonian.*

The proof follows directly:

(1) The integrals factor as above;
(2) Common factors in the numerator and denominator cancel;
(3) For a generic term αq^2 in the hamiltonian, one is left with an integral of the form[12]

$$\langle \alpha q^2 \rangle = \frac{\int_{-\infty}^{\infty} dq\,(\alpha q^2)\, e^{-\alpha q^2/k_{\rm B}T}}{\int_{-\infty}^{\infty} dq\, e^{-\alpha q^2/k_{\rm B}T}} = k_{\rm B}T \frac{\int_{-\infty}^{\infty} x^2 e^{-x^2}\, dx}{\int_{-\infty}^{\infty} e^{-x^2}\, dx} = \frac{1}{2}k_{\rm B}T \tag{2.67}$$

This is the stated result.

As an example, consider the one-dimensional simple harmonic oscillator with hamiltonian

$$H(p,q) = \frac{p^2}{2m} + \frac{1}{2}\kappa x^2 \tag{2.68}$$

[12] These integrals appear so often that it is worthwhile remembering some tricks for doing them. Consider the double integral

$$I^2 \equiv \int_{-\infty}^{\infty} e^{-x^2}\, dx \int_{-\infty}^{\infty} e^{-y^2}\, dy = \int_{-\infty}^{\infty}\int_{-\infty}^{\infty} dx\, dy\, e^{-(x^2+y^2)}$$

Go to polar coordinates with $x^2 + y^2 = r^2$ and $dx\, dy \to 2\pi r\, dr$ to obtain

$$I^2 = \int_0^{\infty} 2\pi r\, dr\, e^{-r^2} = \pi \int_0^{\infty} du\, e^{-u} = \pi$$

Now consider

$$\int_{-\infty}^{\infty} e^{-\lambda x^2}\, dx = \frac{1}{\sqrt{\lambda}} \int_{-\infty}^{\infty} e^{-u^2}\, du = \sqrt{\frac{\pi}{\lambda}}$$

Differentiate with respect to λ, and then set $\lambda = 1$. This gives

$$\int_{-\infty}^{\infty} e^{-x^2}\, dx = \sqrt{\pi} \qquad ; \int_{-\infty}^{\infty} x^2 e^{-x^2}\, dx = \frac{\sqrt{\pi}}{2} \qquad ; \text{ etc.}$$

In this case, the mean energy becomes

$$\langle E\rangle = \frac{1}{2}k_{\rm B}T \times 2 = k_{\rm B}T \qquad ; \text{ mean energy/oscillator} \qquad (2.69)$$

This expression for the mean energy/oscillator is an *important result* that will play a central role in what follows.

Pressure of Ideal Gas. Let us use these results to compute the pressure exerted by an ideal gas at a temperature T. Here there is no interaction potential, and hence within the gas $V(\mathbf{q}) = 0$. The particles in the gas are moving according to the Boltzmann distribution. We will compute the pressure by computing the rate of momentum transfer to the wall as the particles elastically bounce off. The velocity of the constituents can be decomposed into its independent cartesian components $\mathbf{v} = (v_x, v_y, v_z)$, and it is the component v_x that is relevant for striking the right wall, as illustrated in Fig. 2.23; for simplicity, we start by assuming a single v_x.

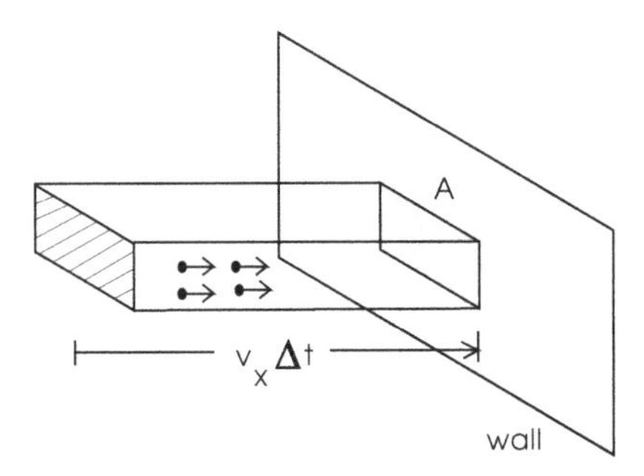

Fig. 2.23 Pressure exerted by an ideal gas. In a time interval Δt, all those particles that lie in a volume $(v_x\Delta t)A$ will hit the wall.

Let:

- $P =$ force/area be the pressure the gas exerts on the wall;
- $n = N/V$ be the particle number density;
- Δn be the number of particles striking an area of the wall A in the time Δt. In the time interval, all those particles that lie in a volume $(v_x\Delta t)A$ will hit the wall. Thus[13]

$$\Delta n = (v_x\Delta t)A\,n \times \frac{1}{2} \qquad (2.70)$$

[13]Although the velocity components (v_y, v_z) may remove particles from this volume, they bring an equal number of particles back in, and hence (v_y, v_z) play no role in this argument.

Here the last factor of 1/2 comes from the fact that only half of the particles are moving to the right and thus strike the right wall.

Since the particles are assumed to bounce off elastically, the momentum transferred to the wall is

$$\Delta\mathcal{P}/\text{particle} = 2mv_x \tag{2.71}$$

Hence the total momentum transferred to the wall by these particles is

$$\Delta\mathcal{P} = (2mv_x)(v_x\Delta t\,A)\frac{1}{2}n \tag{2.72}$$

The corresponding pressure is then computed with the aid of Newton's second law

$$P = \frac{\text{force}}{\text{area}} = \frac{1}{A}\frac{\Delta\mathcal{P}}{\Delta t} = nmv_x^2 \tag{2.73}$$

We are now in a position to take the *mean value* of this expression and use the equipartition theorem

$$\langle P\rangle = \frac{N}{V}\langle mv_x^2\rangle = \frac{N}{V}k_{\rm B}T \tag{2.74}$$

We thus derive the celebrated *ideal gas law*

$$PV = Nk_{\rm B}T \qquad ;\ \text{ideal gas law} \tag{2.75}$$

If the particle number N is expressed in units of Avagadro's number $N_{\rm A}$, that is, in *moles* $\bar{n} = N/N_{\rm A}$, then this result is written in terms of the *gas constant* R according to

$$\begin{aligned} N_{\rm A} &= 6.022\times 10^{23}\,\text{mole}^{-1} && ;\ \text{Avagadro's number} \\ N_{\rm A}\,k_{\rm B} &\equiv R = 1.987\ \text{cal/mole-}^\circ\text{K} && ;\ \text{gas constant} \\ PV &= \bar{n}RT \end{aligned} \tag{2.76}$$

2.3 Electrodynamics

First, we must deal with a choice of units. Here we use SI units where mass, distance, and time are measured in kilograms (kg), meters (m), and seconds (s). Force is then measured in Newtons (N), where $1\text{N} = 1\text{kg-m/s}^2$. Charge is measured in Coulombs (C), voltage in volts (V), and current in amperes (A), where 1A = 1C/s.[14] The relation to other commonly-employed units

[14]These are now the units of choice in most introductory physics courses.

is discussed in appendix K.

We proceed to review the basic principles of classical electrodynamics.[15]

2.3.1 *Basic Principles*

Coulomb's law states that the force between two static point charges (q_1, q_2) separated by a displacement $\mathbf{r} = \mathbf{r}_2 - \mathbf{r}_1$ [see Fig. 2.24(a)] is

$$\mathbf{F}_{21} = \frac{q_1 q_2}{4\pi\varepsilon_0}\frac{\mathbf{r}}{r^3} \qquad ; \text{ Coulomb's law}$$
$$\mathbf{r} = \mathbf{r}_2 - \mathbf{r}_1 \tag{2.77}$$

Here $\mathbf{F}_{21}$ is the force on particle 2 due to particle 1. The Coulomb force has the same $1/r^2$ spatial dependence as Newton's law of gravitation: however, the electrostatic interaction can be either attractive or repulsive according to the signs of the charges. The constant of proportionality appearing in this expression is determined to be

$$\frac{1}{4\pi\varepsilon_0} = 8.988 \times 10^9\,\frac{\text{Nm}^2}{C^2} \tag{2.78}$$

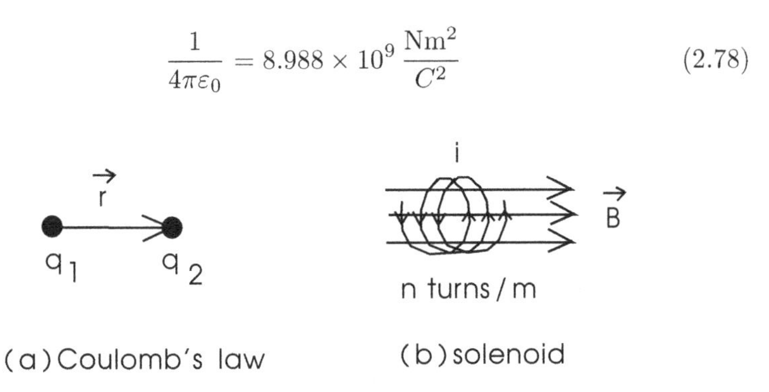

Fig. 2.24 Configuration for: (a) Coulomb's law; (b) Solenoid.

A cylindrical coil of wire with a current i running through it forms a *solenoid* [Fig 2.24(b)]. There is a magnetic field $\mathbf{B}$ in the interior running down the axis, with a direction given by the right-hand rule. If the coil consists of n turns/meter, then the magnitude of the magnetic field is

$$B = \mu_0 n i \qquad ; \text{ solenoid}$$
$$n \text{ turns/meter} \tag{2.79}$$

[15] For texts on classical electrodynamics, see [Griffiths (1998); Jackson (1998)].

The constant of proportionality in this case is

$$\frac{\mu_0}{4\pi} = 1.000 \times 10^{-7}\,\frac{\mathrm{Ns}^2}{\mathrm{C}^2} = 1.000 \times 10^{-7}\,\frac{\mathrm{Tm}}{\mathrm{A}} \tag{2.80}$$

Magnetic fields are thus measured in *Tesla* (T), where

$$1\,\mathrm{T} = 1\frac{\mathrm{kg}}{\mathrm{Cs}} = 10^4\,\mathrm{Gauss} \tag{2.81}$$

The velocity of light $c_{\mathrm{light}} \equiv c$ is given in terms of the constants (ε_0, μ_0) by[16]

$$c = \frac{1}{\sqrt{\varepsilon_0 \mu_0}} = 2.998 \times 10^8\,\frac{\mathrm{m}}{\mathrm{s}} \tag{2.82}$$

The *Lorentz force* acting on a particle with charge q moving in electric and magnetic fields $(\mathbf{E}, \mathbf{B})$ is given in these units by

$$\mathbf{F} = q\,(\mathbf{E} + \mathbf{v} \times \mathbf{B}) \qquad ; \text{ Lorentz force} \tag{2.83}$$

2.3.2 *Some Applications*

We review some simple applications of these results:

(1) *Circular motion in a uniform magnetic field.* If the charge $q > 0$ and $(\mathbf{v}, \mathbf{B})$ are oriented as in Fig. 2.25, then the force $q\mathbf{v}\times\mathbf{B}$ points along $\mathbf{r}$ in that figure, and since the magnetic force is perpendicular to the velocity $\mathbf{v}$, only the direction of $\mathbf{v}$ changes. The result is circular motion, with $\mathbf{r}$ pointing to the center of the circle. A combination of the Lorentz force and Newton's second law for circular motion with radius r then gives

$$ma = \frac{mv^2}{r} = qvB \qquad ; \text{ circular motion}$$
$$\Rightarrow \qquad r = \frac{mv}{qB} \tag{2.84}$$

At fixed field strength, the radius of the circle is proportional to the momentum.

[16]We use the symbol c for both the velocity of light and the generic velocity in the wave equation. The relevant velocity should be clear from the context.

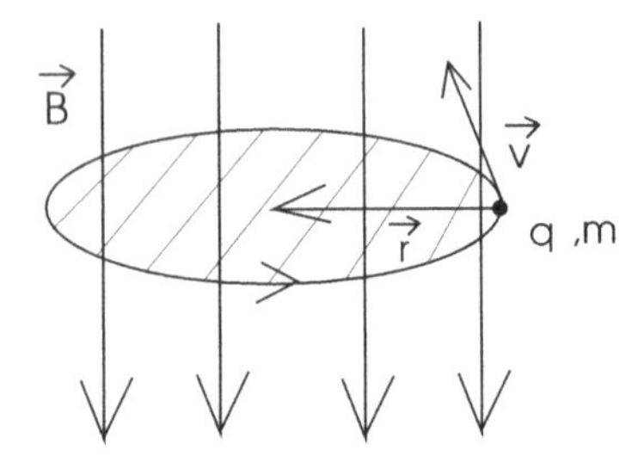

Fig. 2.25 Circular motion in a uniform magnetic field. Here $q > 0$, **B** is perpendicular to the plane of the orbit, and **v** lies in it. The vector **r** here points to the center of the circle.

(2) *Velocity selector.* Consider a particle of charge q moving with velocity **v** through crossed **E** and **B** fields as illustrated in Fig. 2.26. The Lorentz force on the particle is given by Eq. (2.83), and the electric and magnetic forces will just balance in this configuration if

$$\begin{aligned} E &= vB \\ \Rightarrow \qquad v &= \frac{E}{B} \end{aligned} \tag{2.85}$$

Only particles with this velocity will be transmitted through the device — particles with any other velocity will be deflected out of the beam, and one has a *velocity selector.*

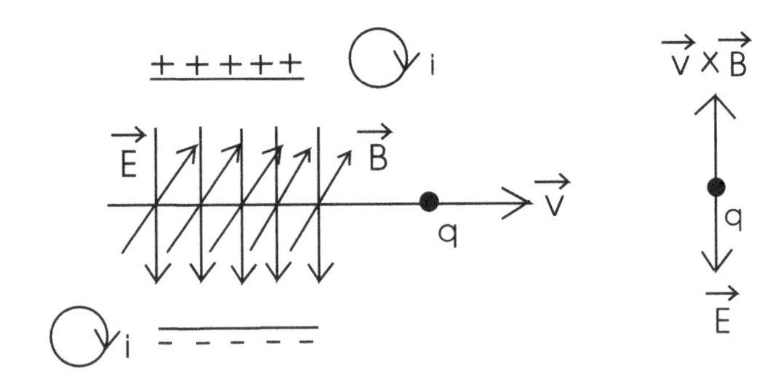

Fig. 2.26 Field configuration for velocity selector.

The velocity selector can be combined with the momentum selector in (1) to produce a *mass spectrograph.* If the particle is sent into a uniform magnetic field after exiting the velocity selector, as illustrated in Fig. (2.27), then Eq. (2.84) can be solved for the mass to give

$$m = \left(\frac{qB}{v}\right) r \tag{2.86}$$

The radius of curvature thus determines the mass.[17]

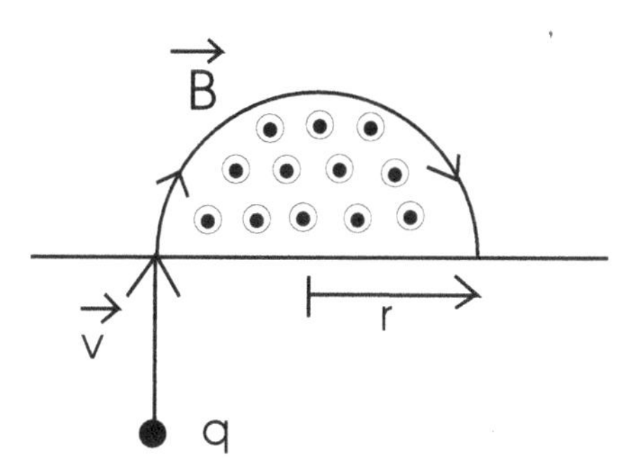

Fig. 2.27 Combination of the velocity selector and momentum selector to produce a mass spectrograph. The radius of the orbit of a particle with given velocity in a uniform magnetic field determines its momentum, and hence its mass. Here $q > 0$ and $\mathbf{B} = B\hat{\mathbf{e}}_z$ where $\hat{\mathbf{e}}_z$ is a unit vector coming out of the page.

(3) *Circular motion in a Coulomb field.* For a particle of charge $-q$ and mass m performing circular motion at a radius r about a fixed charge $+q$, a combination of Newton's second law and Coulomb's law gives

$$\frac{mv^2}{r} = \frac{q^2}{4\pi\varepsilon_0}\frac{1}{r^2} \qquad ; \text{ circular motion} \tag{2.87}$$

The angular momentum is

$$\mathbf{l} = \mathbf{r} \times \mathbf{p} = (rmv)\,\hat{\mathbf{e}}_z \qquad ; \text{ angular momentum} \tag{2.88}$$

Here $\hat{\mathbf{e}}_z$ is a unit vector perpendicular to the plane of the motion in Fig. 2.28. A combination of these two results gives

$$\frac{l^2}{m} = \left(\frac{q^2}{4\pi\varepsilon_0}\right) r \tag{2.89}$$

The radius goes as l^2. We will return to this relation when we discuss the Bohr atom.

[17] Of course, one has to know q. Fortunately, q only changes in integer units of the basic charge e.

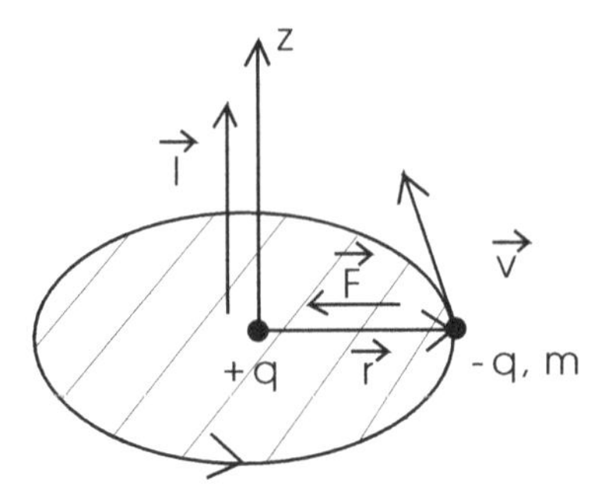

Fig. 2.28 Circular motion in a Coulomb field.

2.3.3 *Maxwell's Equations*

In addition to being created by charge and current densities $(\rho, \mathbf{j})$, electric fields are created by oscillating magnetic fields, and *vice versa.* The laws governing these fields are summarized by *Maxwell's equations.* In free space with sources $(\rho, \mathbf{j})$, in these units, in differential form they are[18]

$$
\begin{aligned}
\nabla \cdot \mathbf{E} &= \frac{\rho}{\varepsilon_0} && ;\ \text{Maxwell's equations} \\
\nabla \cdot \mathbf{B} &= 0 && \text{free space with sources } (\rho, \mathbf{j}) \\
\nabla \times \mathbf{E} &= -\frac{\partial \mathbf{B}}{\partial t} \\
\nabla \times \mathbf{B} &= \mu_0 \mathbf{j} + \frac{1}{c^2}\frac{\partial \mathbf{E}}{\partial t}
\end{aligned} \tag{2.90}
$$

The simplest expression for ∇ is to write it, and an arbitrary vector field $\mathbf{v}(\mathbf{x}, t)$, in terms of cartesian coordinates

$$
\begin{aligned}
\nabla &= \hat{\mathbf{e}}_x \frac{\partial}{\partial x} + \hat{\mathbf{e}}_y \frac{\partial}{\partial y} + \hat{\mathbf{e}}_z \frac{\partial}{\partial z} \\
\mathbf{v}(\mathbf{x}, t) &= \hat{\mathbf{e}}_x\, v_x(\mathbf{x}, t) + \hat{\mathbf{e}}_y\, v_y(\mathbf{x}, t) + \hat{\mathbf{e}}_z\, v_z(\mathbf{x}, t)
\end{aligned} \tag{2.91}
$$

[18]For future reference, we note that in these units

$$
\begin{aligned}
a_0 &= \frac{\hbar^2}{m_e(e^2/4\pi\varepsilon_0)} = 0.5292\ \text{Å} && ;\ \text{Bohr radius} \\
\alpha &= \left(\frac{e^2}{4\pi\varepsilon_0}\right)\frac{1}{\hbar c} = \frac{1}{137.0} && ;\ \text{fine-structure constant}
\end{aligned}
$$

where $(\hat{\mathbf{e}}_x, \hat{\mathbf{e}}_y, \hat{\mathbf{e}}_z)$ are a set of constant, orthonormal, cartesian unit vectors. Then the divergence and curl of $\mathbf{v}$ are given by[19]

$$\nabla \cdot \mathbf{v} = \frac{\partial v_x}{\partial x} + \frac{\partial v_y}{\partial y} + \frac{\partial v_z}{\partial z}$$

$$\nabla \times \mathbf{v} = \hat{\mathbf{e}}_x \left(\frac{\partial v_z}{\partial y} - \frac{\partial v_y}{\partial z} \right) + \hat{\mathbf{e}}_y \left(\frac{\partial v_x}{\partial z} - \frac{\partial v_z}{\partial x} \right) + \hat{\mathbf{e}}_z \left(\frac{\partial v_y}{\partial x} - \frac{\partial v_x}{\partial y} \right) \tag{2.92}$$

The *energy flux* is given by the *Poynting vector*

$$\mathbf{S} = \frac{1}{\mu_0} \mathbf{E} \times \mathbf{B} \qquad ; \text{ Poynting vector} \tag{2.93}$$

Here $\mathbf{S} \cdot d\mathbf{A}$ is the energy flowing across the infinitesimal area $d\mathbf{A}$ per unit time (compare Fig. 2.29).

2.3.4 *Electromagnetic Radiation*

Solution in free space. There are solutions to Maxwell's equations in the absence of $(\rho, \mathbf{j})$ where the electric and magnetic fields oscillate against each other and move through free space with velocity c. This is *electromagnetic radiation* that propagates with frequencies ranging from those in the longest radio waves, through the visible spectrum, to the highest energy gamma rays.

One particular solution to Maxwell's equations in free space is the following (see Fig. 2.29)

$$\begin{aligned} E_y &= A \cos k(x - ct) \qquad ; \text{ electromagnetic plane wave} \\ cB_z &= A \cos k(x - ct) \end{aligned} \tag{2.94}$$

This solution also satisfies the one-dimensional wave equation, just like the string

$$\frac{\partial^2 \phi(x,t)}{\partial x^2} = \frac{1}{c^2} \frac{\partial^2 \phi(x,t)}{\partial t^2} \qquad ; \text{ one-dimensional wave equation} \quad \phi = (E_y, cB_z) \tag{2.95}$$

We know from our previous discussion that the solution in Eqs. (2.94) represents a wave that propagates without change in shape along the x-axis with

[19]The relation to Maxwell's equations in integral form, through which they are usually presented in introductory E&M courses (Gauss' law, Ampere's law, Faraday's law, *etc.*), is discussed in appendix F and Probs. 2.9-2.11.

velocity c. The field configuration for the plane wave solution in Eqs. (2.94) is shown in Fig. (2.29), along with the direction of the corresponding Poynting vector $\mathbf{S}$.

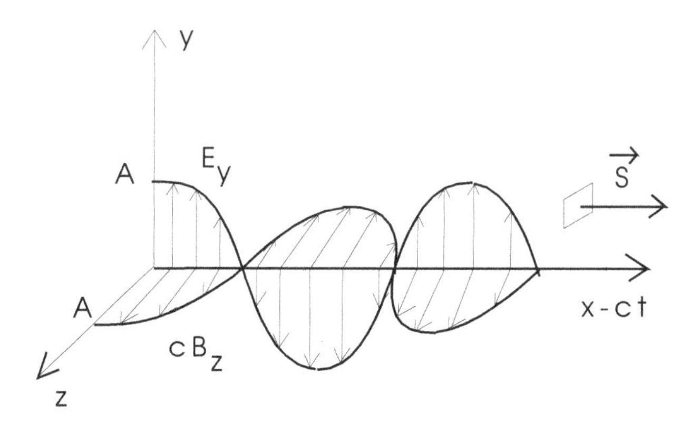

Fig. 2.29 Field configuration in Eqs. (2.94) describing an electromagnetic plane wave traveling to the right with velocity c. The direction of the corresponding Poynting vector $\mathbf{S}$ is also shown.

Solution between metal plates. Consider an electromagnetic oscillation in free space confined between parallel metal plates separated by a distance L as in Fig. (2.30). If it is assumed that the plates are perfect conductors with a conductivity $\sigma \to \infty$, then there can be no electric field $\mathbf{E}$ in the metal plates, and hence $\mathbf{E}$ must vanish on the boundary. One can verify that the following *standing wave* provides a solution to Maxwell's equations that satisfies the proper boundary conditions

$$\begin{aligned} E_y &= A \sin \frac{n\pi x}{L} \cos \frac{n\pi ct}{L} && ;\ \text{electromagnetic standing wave} \\ -cB_z &= A \cos \frac{n\pi x}{L} \sin \frac{n\pi ct}{L} && n = 1, 2, \cdots, \infty \\ \omega_n &= k_n c = \left(\frac{\pi c}{L}\right) n && \end{aligned} \tag{2.96}$$

These are *normal modes* where everything oscillates with the same frequency, for example

$$\frac{\partial^2 E_y}{\partial t^2} = -\omega_n^2 E_y \qquad ;\ \text{normal modes} \tag{2.97}$$

The corresponding eigenvalues and eigenfunctions are then those in Eqs. (2.96). The normal-mode standing waves between the plates, with

$n = 1, 2, \cdots, \infty$, form an infinite, discrete set of simple harmonic oscillators.

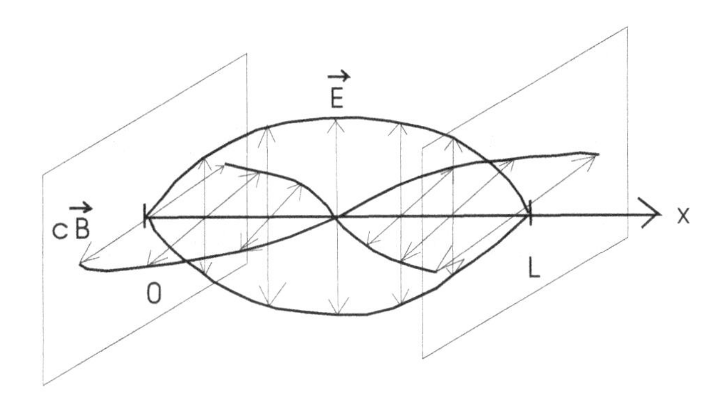

Fig. 2.30 Field configuration in Eqs. (2.96) describing an electromagnetic standing wave between two perfectly conducting parallel metal plates separated by a distance L. Here $n = 1$.

2.3.5 *Source of Radiation*

The source of electromagnetic radiation is *accelerated charge.*

- The amplitude of the radiation from a charge q is proportional to the charge and to the acceleration;
- The amplitude depends on the component of the acceleration perpendicular to the direction of observation.

Thus

$$\mathcal{A} \propto \left(q\frac{d^2\mathbf{r}}{dt^2}\right)_\perp \tag{2.98}$$

We give two examples:

(1) Consider the radiation from a dipole antenna. If the source is an oscillating charge undergoing simple harmonic motion in the z-direction with

$$\frac{d^2z}{dt^2} = -\omega^2 z \qquad ; \text{ s.h.o.}$$
$$\text{accelerated charge} \tag{2.99}$$

then the radiation intensity ($\propto \mathcal{A}^2$) has the pattern sketched in Fig. 2.31.

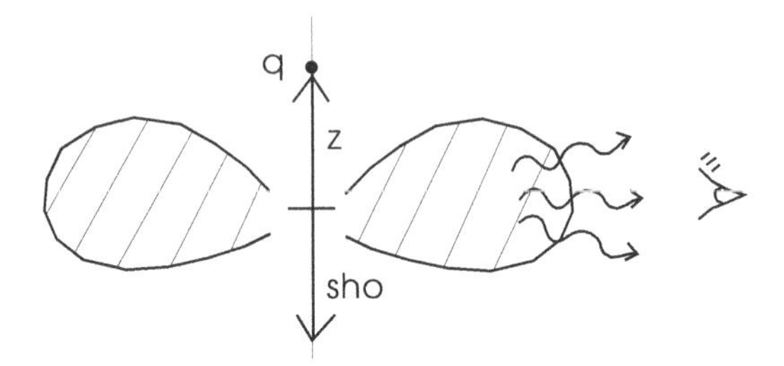

Fig. 2.31 Radiation pattern from a dipole antenna with a radiating charge undergoing simple harmonic motion in the z-direction.

(2) Consider the current in a storage ring. Here the acceleration of the charge is always pointed toward the center of the ring and the radiation pattern is that illustrated in Fig. 2.32. The radiation is in a direction tangent to the orbit, and this is the operating principle of *light sources.*[20]

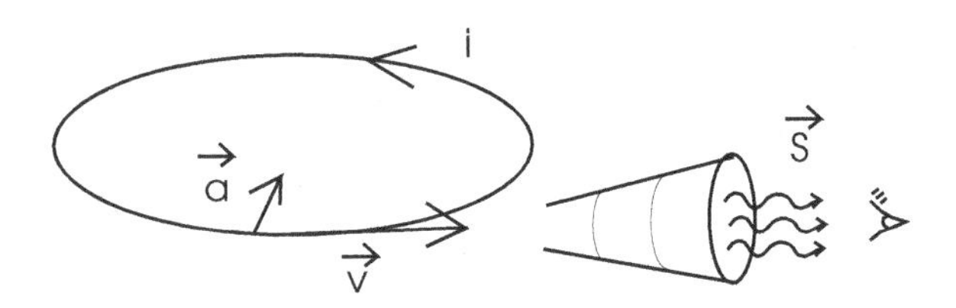

Fig. 2.32 Radiation pattern from a charge in a storage ring.

[20]See [Jackson (1998)] for the actual focused angular distribution of the radiation from a relativistic particle.

Chapter 3

Some Contradictions

At the end of the nineteenth century one had *classical mechanics*, with Newton's laws and particle mechanics, *electromagnetic theory*, with Maxwell's equations and wave phenomena, and Boltzmann's *statistical physics*. These were tremendous achievements that described well what is observed in the world around us. At that point, however, a series of phenomena and experiments were discovered that simply could not be understood in these classical terms.

3.1 Specific Heat of Solids

Consider the simplest classical mean-field model of a monatomic solid at a temperature T. Assume N independent, three-dimensional, simple harmonic oscillators (Fig. 3.1).[1]

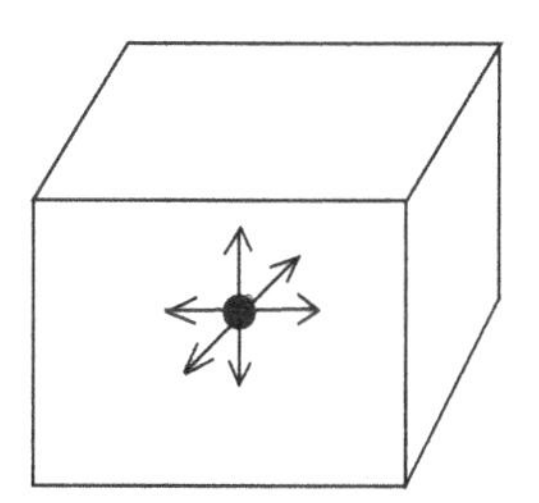

Fig. 3.1 Classical mean-field approximation for a monatomic solid as N independent, three-dimensional, simple harmonic oscillators at a temperature T.

The total internal energy of the solid follows immediately from the

[1]Or, equivalently, $3N$ one-dimensional s.h.o.

equipartition theorem

$$E = 3N\langle\varepsilon\rangle = 3Nk_{\rm B}T \qquad ; \text{ equipartition theorem} \tag{3.1}$$

The *specific heat* at constant volume is given by[2]

$$C_v = \frac{đQ}{dT} = \frac{dE}{dT} \tag{3.2}$$

where the last equality follows from the first law of thermodynamics. Thus

$$C_v = 3Nk_{\rm B} \tag{3.3}$$

If we confine the discussion to one mole of solid where $N = N_{\rm A}$, then

$$\begin{aligned} C_v &= 3N_{\rm A}k_{\rm B} = 3R \qquad ; \text{ molar specific heat} \\ &= 5.961 \text{ cal/mole-}^{\rm o}\text{K} \end{aligned} \tag{3.4}$$

This is the law of Dulong and Petit. Experimentally, this result holds well at high temperature (see Fig. 3.2),[3] but it fails completely at low temperatures where

$$C_v \rightarrow 0 \qquad ; \text{ as } T \rightarrow 0 \tag{3.5}$$

The question is, why does this happen?

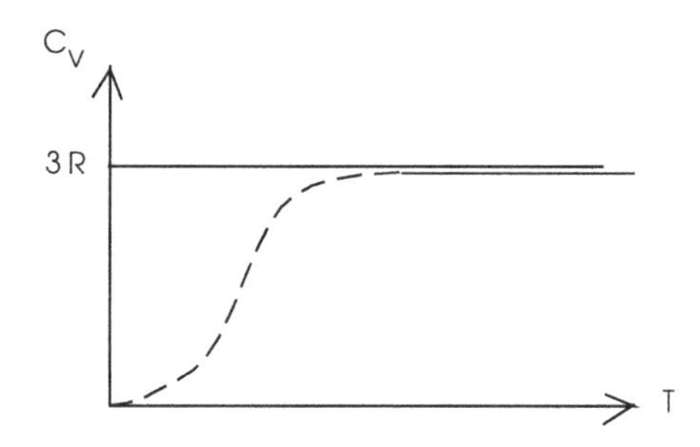

Fig. 3.2 Sketch of the observed molar specific heat C_v for a monatomic solid.

One can significantly improve this one-body mean-field model by considering the *normal mode oscillations of the whole solid.* Imagine the boundaries are clamped. Then the problem is in direct analogy to the string with

[2]We use the terms "specific heat" and "heat capacity" interchangeably.
[3]We assume, of course, that the solid remains a solid for all the temperatures considered here.

fixed endpoints considered in the previous chapter (Fig. 3.3). The allowed wave numbers in the three directions (x, y, z) are

$$k_x = \frac{\pi}{L}n_x \qquad ; k_y = \frac{\pi}{L}n_y \qquad ; k_z = \frac{\pi}{L}n_z \qquad ; n_i = 1, 2, 3, \cdots$$
$$i = x, y, z \tag{3.6}$$

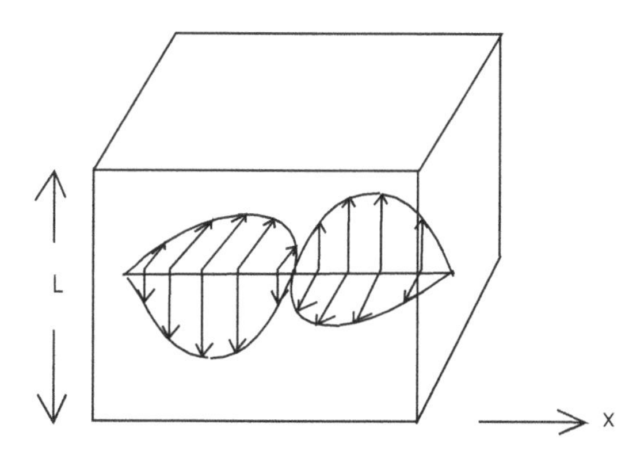

Fig. 3.3 Illustrative normal-mode oscillations of the solid with clamped boundaries in direct analogy to the standing waves of a string with fixed endpoints. The two transverse vibrations for a given k_x are illustrated.

Each normal mode is a simple harmonic oscillator, as is the case for the string where all elements of the string oscillate with the same normal-mode frequency

$$q_n(x, t) = q_n(x) \cos(\omega_n t + \delta_n) \qquad ; \text{ normal mode on string}$$
$$\frac{\partial^2 q_n(x,t)}{\partial t^2} = -\omega_n^2\, q_n(x, t) \tag{3.7}$$

The total energy of the solid is then given by the equipartition theorem as

$$E = (\text{number of normal modes}) \times k_{\text{B}}T \tag{3.8}$$

The problem is thus reduced to counting the number of normal modes that can be fit into the solid.

To count the normal modes in the limit of large L, we can make use of the arguments in Eqs. (2.50)-(2.55). Define the vector $\mathbf{k}$ as

$$\mathbf{k} \equiv (k_x, k_y, k_z) = \frac{\pi}{L}(n_x, n_y, n_z) \tag{3.9}$$

Suppose one is interested in the sum over allowed wavenumbers of a function $f(\mathbf{k})$. Then by utilizing those previous results one has

$$\sum_{n_x}\sum_{n_y}\sum_{n_z} f(\mathbf{k}) = \left(\frac{L}{\pi}\right)^3 \sum_{n_x}\Delta k_x \sum_{n_y}\Delta k_y \sum_{n_z}\Delta k_z\, f(\mathbf{k})$$
$$\to \left(\frac{L}{\pi}\right)^3 \int_0^\infty dk_x \int_0^\infty dk_y \int_0^\infty dk_z\, f(\mathbf{k}) \qquad ; L\to\infty \qquad (3.10)$$

If $f(\mathbf{k})$ is actually $f(\mathbf{k}^2)$ so that it is even in all three components, this expression may be rewritten as $(1/2)^3$ times the integral over all $\mathbf{k}$

$$\sum_{n_x}\sum_{n_y}\sum_{n_z} f(\mathbf{k}^2) \to \left(\frac{L}{2\pi}\right)^3 \int_{-\infty}^\infty dk_x \int_{-\infty}^\infty dk_y \int_{-\infty}^\infty dk_z\, f(\mathbf{k}^2)$$
$$; L\to\infty \qquad (3.11)$$

Since this is now just an integral in three-dimensional wavenumber (or k-) space, it may be rewritten in spherical coordinates as

$$\sum_{n_x}\sum_{n_y}\sum_{n_z} f(\mathbf{k}^2) \to \left(\frac{L}{2\pi}\right)^3 \int\int\int d^3k\, f(\mathbf{k}^2) \qquad ; L\to\infty$$

clamped boundaries

$$= \left(\frac{L}{2\pi}\right)^3 \int_0^\infty 4\pi k^2 f(k^2)\, dk \qquad (3.12)$$

To count the number of modes, one can now simply set $f(\mathbf{k}^2) = 1$. Since $L^3 = V$, where V is the volume of the solid, one can just read off from Eq. (3.12) that the number of normal modes $d\mathcal{N}$ with wave number of magnitude between k and $k + dk$ is given by

$$d\mathcal{N} = \frac{V}{(2\pi)^3} 4\pi k^2 dk \qquad (3.13)$$

There may always be different *types* of normal modes for a given $\mathbf{k}$. In the solid, for example, there can be transverse oscillations in the two distinct transverse planes (Fig. 3.3), as well as longitudinal oscillations of the atomic spacings along the direction of propagation. In the counting of normal modes, one must therefore also include a *degeneracy factor g*

$$g \equiv \text{degeneracy factor} \qquad (3.14)$$

For the oscillations of the solid, $g = 3$ to count the two transverse and one longitudinal modes

$$g = 3 \qquad ; \text{ oscillations of solid} \tag{3.15}$$

Thus, in summary, the number of normal modes $d\mathcal{N}$ with wave number of magnitude between k and $k + dk$ and degeneracy g in a solid of volume V with clamped boundaries is given by

$$d\mathcal{N} = g\frac{V}{(2\pi)^3}4\pi k^2 dk \qquad ; \text{ number of normal modes} \tag{3.16}$$

This is an important result, and the arguments leading to it should be studied carefully by the reader.[4]

One can go over from wave number to frequency using the result from the wave equation that

$$k = \frac{\omega}{c_s} = \frac{2\pi\nu}{c_s} \tag{3.17}$$

where c_s is the speed of propagation of the wave, in this case the speed of sound in the solid.[5] Hence

$$\frac{d\mathcal{N}}{V} = g\frac{4\pi\nu^2\,d\nu}{c_s^3} \qquad ; \text{ number of normal modes} \tag{3.18}$$

We now invoke our previous result from the *equipartition theorem* in Eq. (3.8), and thus

$$\frac{dE}{V} = k_{\rm B}T\left[g\frac{4\pi\nu^2\,d\nu}{c_s^3}\right]$$
$$\equiv u(\nu, T)\,d\nu \qquad ; \text{ energy density} \tag{3.19}$$

The last relation defines the energy density $u(\nu, T)$.

In a string, ultimately formed from discrete atoms, it makes no sense to talk about a disturbance whose wavelength is shorter than the interparticle spacing. Thus there is a *minimum wavelength* $\lambda_{\rm min}$, or equivalently, a *maximum normal-mode frequency* $\nu_{\rm max} = c_s/\lambda_{\rm min}$, for the propagation of waves on a string. The same holds for a solid of discrete atoms. What is this frequency cut-off $\nu_{\rm max}$? One can argue that the total number of

[4]For an alternate derivation of this important result, see Prob. 3.11.

[5]In fact, the velocities of propagation of the transverse and longitudinal waves are different — we ignore that complication here; however, see Prob. 3.4.

normal modes must reproduce the total number of degrees of freedom $3N$, and hence[6]

$$\int_0^{\nu_{\max}} \frac{d\mathcal{N}}{V} = \frac{3N}{V} \tag{3.20}$$

From Eqs. (3.18)–(3.20), one then concludes

$$E = 3Nk_{\rm B}T \tag{3.21}$$

This just *reproduces the result of Dulong and Petit at all* T, and the more sophisticated treatment of the oscillations of the solid has produced no improvement of this result — one is faced with the same problem as before.

3.2 Black-Body Radiation

Consider a cavity formed from some material that is heated to a temperature T. If heated hot enough, the walls will actually glow in the visible spectrum. Now put a small hole in a wall of the closed cavity, and observe the radiation coming out (Fig. 3.4). Make a measurement of the *energy*

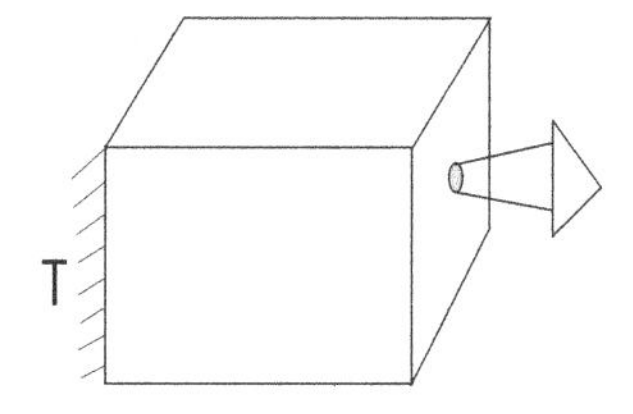

Fig. 3.4 Observation of the energy density $u(\nu, T)$ in a cavity heated to a temperature T through the energy flux $c\,u(\nu, T)/4$ streaming out of it.

flux of the radiation at a given frequency.[7] The energy flux will be equal to the energy density in the cavity $u(\nu, T)$ times the velocity of propagation of the radiation, namely c, the speed of light, times a geometric factor of

[6]It is a theorem from mechanics that the total number of degrees of freedom is unchanged when one goes over to normal modes, which are the uncoupled simple harmonic oscillations of the entire system [Fetter and Walecka (2003)]; here, one might just use the argument that this treatment should reproduce Dulong and Petit as $T \to \infty$.

[7]Historically, the study of black-body radiation actually came before that of the specific heat of solids. In the author's opinion, a presentation in this order makes it much easier to understand the results.

1/4 taking into account that portion of the light moving towards the hole (see appendix G). Thus

$$\text{energy flux} = \frac{1}{4} c\, u(\nu, T) \tag{3.22}$$

It is found experimentally that the observed energy density is a universal function independent of the shape of the cavity, nature of the wall material, *etc.* Let us then consider the simplest case of a cubical box with perfectly conducting metallic walls. There will be standing waves in the cavity as illustrated in Fig. 2.30, each of which is a normal mode where everything oscillates with the same frequency. The electric field must vanish on the boundaries, and again, we have a direct analogy to the standing waves on a string with fixed endpoints. The use of the equipartition theorem and the counting of modes then proceeds as in previous section. The only difference is that the degeneracy factor g is now 2, corresponding to the two possible planes of oscillation of the transverse electric field in Fig. 2.30.

$$g = 2 \qquad ; \text{ two transverse modes} \tag{3.23}$$

Thus the number of modes per unit volume is again given by

$$\frac{d\mathcal{N}}{V} = g\frac{4\pi\nu^2\, d\nu}{c^3} \qquad ; \text{ number of normal modes} \tag{3.24}$$

and the energy density in the cavity is given by the classical equipartition theorem as

$$\begin{aligned} \frac{dE}{V} &= k_{\rm B}T\left[g\frac{4\pi\nu^2\, d\nu}{c^3}\right] \\ &\equiv u(\nu, T)\, d\nu \qquad ; \text{ energy density} \end{aligned} \tag{3.25}$$

This is the Rayleigh-Jeans law. It is sketched in Fig. 3.5. There are several comments:

- In contrast to the situation with the specific heat of solids where there is a natural cut-off to the normal-mode frequencies arising from the fact that one cannot talk about disturbances with wavelength smaller than the interatomic spacing, there is *no natural cutoff to the normal-mode frequencies of the standing electromagnetic normal modes in a cavity.* Thus the classical Rayleigh-Jeans law says that the energy density should continue to grow as ν^2 without bound. This is known as the *ultraviolet catastrophe*;

- While the Rayleigh-Jeans law reproduces experiment at low ν, the experimental spectrum is observed to turn over and go to zero as ν gets large

$$u(\nu, T) \to 0 \qquad ; \text{ as } \nu \to \infty \tag{3.26}$$

- Why does this happen? There is no understanding at all within the framework of classical physics.

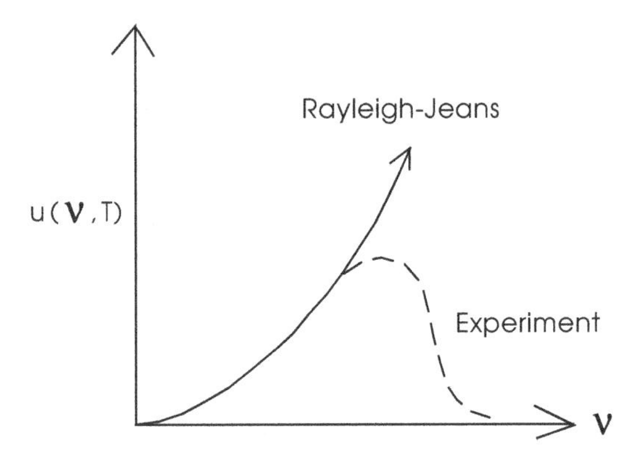

Fig. 3.5 Sketch of black-body energy spectrum $u(\nu, T)$: Rayleigh-Jeans law and experimental observation.

3.2.1 *Planck's Hypothesis*

In order to obtain an empirical fit to the black-body spectrum, Planck replaced the energy/oscillator of $k_{\rm B}T$ in the classical equipartition theorem by the following expression [Planck (1901)]

$$\langle \varepsilon \rangle = \frac{h\nu}{e^{h\nu/k_{\rm B}T} - 1} \qquad ; \text{ Planck's energy/oscillator} \tag{3.27}$$

This expression reproduces the equipartition theorem result of $k_{\rm B}T$ at low frequencies where $h\nu \ll k_{\rm B}T$, and it exhibits an exponential decrease at high frequencies where $h\nu \gg k_{\rm B}T$

$$\begin{aligned} \frac{h\nu}{e^{h\nu/k_{\rm B}T} - 1} &\to k_{\rm B}T && ; \ h\nu \ll k_{\rm B}T \\ &\to h\nu\, e^{-h\nu/k_{\rm B}T} && ; \ h\nu \gg k_{\rm B}T \end{aligned} \tag{3.28}$$

Here h is a constant, *Planck's constant,* with dimensions of (energy)× (time), and from a fit to the data it has a value

$$h = 6.626 \times 10^{-34}\,\text{Js} \qquad ; \text{ Planck's constant} \qquad (3.29)$$

When combined with the expression for the number of normal modes in Eq. (3.24) and (3.23), this gives an empirical black-body energy density of

$$u(\nu, T) = \frac{8\pi\nu^2}{c^3}\frac{h\nu}{e^{h\nu/k_{\rm B}T} - 1} \qquad ; \text{ Planck's expression} \qquad (3.30)$$

A dimensionless form of this distribution can be extracted by writing

$$u(\nu, T) = h\left(\frac{k_{\rm B}T}{hc}\right)^3 u(x)$$
$$u(x) \equiv \frac{8\pi x^3}{e^x - 1} \qquad ; \; x \equiv \frac{h\nu}{k_{\rm B}T} \qquad (3.31)$$

This black-body energy density is plotted in Fig. 3.6.

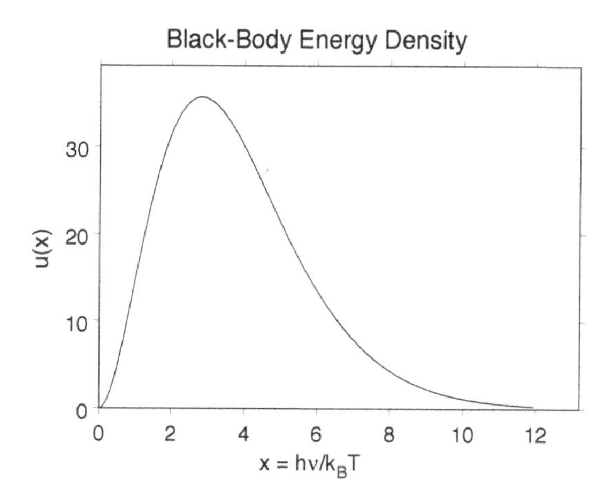

Fig. 3.6 Black-body energy density $u(\nu, T) = h(k_{\rm B}T/hc)^3 u(x)$ where $x \equiv h\nu/k_{\rm B}T$.

In order to understand his result, Planck came up with the following "derivation".

> *Assume the energy of each oscillator of frequency ν can only occur in quantum steps of $h\nu$.*

He again used the Boltzmann weighting factor (appendix E) to compute the mean energy of an oscillator at a temperature T

$$\begin{aligned}
\langle\varepsilon\rangle &= \frac{\sum_{n=0}^{\infty} nh\nu\, e^{-nh\nu/k_{\rm B}T}}{\sum_{n=0}^{\infty} e^{-nh\nu/k_{\rm B}T}} \qquad ; \text{ s.h.o.} \\
&= -\frac{\partial}{\partial(1/k_{\rm B}T)} \ln\left(\sum_{n=0}^{\infty} e^{-nh\nu/k_{\rm B}T}\right) \\
&= \frac{\partial}{\partial(1/k_{\rm B}T)} \ln\left(1 - e^{-h\nu/k_{\rm B}T}\right) \\
&= \frac{1}{1 - e^{-h\nu/k_{\rm B}T}}\, h\nu\, e^{-h\nu/k_{\rm B}T} \qquad (3.32)
\end{aligned}$$

Here the geometric series

$$\begin{aligned}
\sum_{n=0}^{\infty} e^{-nh\nu/k_{\rm B}T} &= 1 + \left(e^{-h\nu/k_{\rm B}T}\right) + \left(e^{-h\nu/k_{\rm B}T}\right)^2 + \cdots \\
&= \frac{1}{1 - e^{-h\nu/k_{\rm B}T}} \qquad (3.33)
\end{aligned}$$

has been summed in arriving at the third line in Eqs. (3.32). Thus, through this assumption and calculation, one indeed arrives at the *Planck distribution*

$$\langle\varepsilon\rangle = \frac{h\nu}{e^{h\nu/k_{\rm B}T} - 1} \qquad ; \text{ Planck distribution} \qquad (3.34)$$

The situation is illustrated in Fig. 3.7. For $k_{\rm B}T \gg h\nu$, the oscillator energy distribution appears continuous, and one reproduces the result of the classical equipartition theorem; for $k_{\rm B}T \ll h\nu$, there is not enough thermal energy to excite the first oscillator level, and the thermal excitation energy of the oscillator falls exponentially to zero.

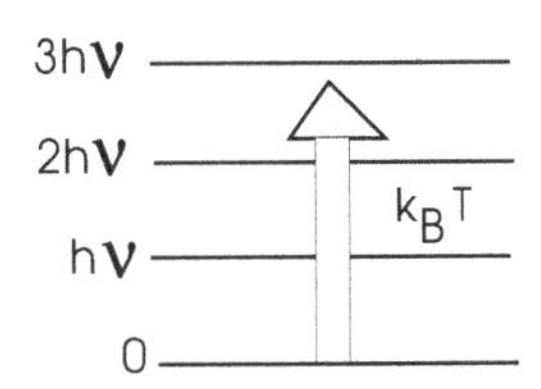

Fig. 3.7 Thermal energy $k_{\rm B}T$ relative to oscillator spacing $h\nu$ in Planck's derivation.

3.3 Specific Heat — Revisited

3.3.1 *Einstein Model*

A little later, Einstein (1907) applied Planck's ideas to the specific heat of a solid. He assumed, for simplicity, that there is a single frequency ν_0 for each of the oscillators in Fig. 3.1. With the Planck distribution, the internal energy of the solid in Eq. (3.1) is then modified to

$$E = 3N\left[\frac{h\nu_0}{e^{h\nu_0/k_{\rm B}T}-1}\right] \qquad ;\ \text{Einstein model} \qquad (3.35)$$

The new molar specific heat is calculated immediately, just as before,

$$C_v = \frac{dE}{dT} = 3N_{\rm A}k_{\rm B}\left[\frac{x^2e^x}{(e^x-1)^2}\right] \qquad ;\ x \equiv \frac{h\nu_0}{k_{\rm B}T} \qquad (3.36)$$

This reproduces Dulong and Petit at high temperature, and it now solves the problem of the specific heat going to zero at low temperatures

$$C_v \rightarrow 3N_{\rm A}k_{\rm B}\left(\frac{h\nu_0}{k_{\rm B}T}\right)^2 e^{-h\nu_0/k_{\rm B}T} \qquad ;\ T\rightarrow 0 \qquad (3.37)$$

But this model does its job too well! The specific heat now vanishes *exponentially* at low T, while empirically it is observed to go to zero only as T^3.

3.3.2 *Debye Model*

Debye (1912) extended Einstein's ideas by using a more realistic oscillation spectrum for the solid. Our previous treatment of the normal-mode oscillations now pays dividends for our understanding of this work. A combination of the number of normal modes in the frequency interval $d\nu$ in Eq. (3.18) with the Planck distribution for the thermal energy in each mode gives an internal energy of the solid as

$$\begin{aligned}
\frac{E}{V} &= \int_0^{\nu_{\rm max}} g\frac{4\pi\nu^2\,d\nu}{c_s^3}\left[\frac{h\nu}{e^{h\nu/k_{\rm B}T}-1}\right] \qquad ;\ \text{Debye model} \\
\frac{3N}{V} &= \int_0^{\nu_{\rm max}} g\frac{4\pi\nu^2\,d\nu}{c_s^3} \qquad (3.38)
\end{aligned}$$

Here Eq. (3.20) has again been used in the second line, whose role now is to relate $(\nu_{\rm max}/c_s)^3$ to the number density N/V.

The *ratio* of the two expressions in Eqs. (3.38) now gives

$$\frac{E}{3N} = \frac{h\nu_{\rm max}}{1/3}\int_0^1 x^3\,dx \frac{1}{e^{h\nu_{\rm max}x/k_{\rm B}T}-1} \qquad ; \; x \equiv \frac{\nu}{\nu_{\rm max}} \tag{3.39}$$

Introduce the *Debye temperature*

$$\theta_{\rm D} = \frac{h\nu_{\rm max}}{k_{\rm B}} \qquad ; \text{ Debye temperature} \tag{3.40}$$

With a change of variables to $u = \theta_{\rm D}\, x/T$, Equation (3.39) can then be rewritten as

$$E = 9Nk_{\rm B}\theta_{\rm D}\left(\frac{T}{\theta_{\rm D}}\right)^4 \int_0^{\theta_{\rm D}/T} u^3\,du \frac{1}{e^u - 1} \qquad ; \text{ Debye theory} \tag{3.41}$$

This provides a remarkably successful one-parameter description of the specific heat of monatomic solids, as well as the celebrated Debye T^3 law at low temperatures (Fig. 3.8).[8]

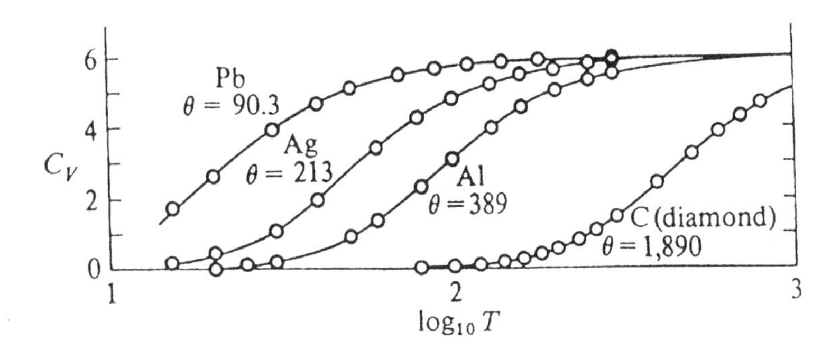

Fig. 3.8 Observed molar heat capacity in cal/mole-$^{\rm o}$K of some monatomic solids and fits with various values of the Debye temperature θ_D in $^{\rm o}$K. From [Fetter and Walecka (2003a)].

3.4 Photoelectric Effect

When light of frequency ν is incident on appropriate materials, it is possible to free electrons from the material and produce an electric current. This

[8]See Prob. 3.2 for the corresponding molar specific heat.

is known as the *photoelectric effect.* Suppose one has a set-up as shown in Fig. 3.9(a) where a retarding voltage is applied which just nulls the current. The maximum kinetic energy (K.E.) of the electrons can be determined from this retarding voltage, and the experimental results from such an experiment are sketched in Fig. 3.9(b). There are no electrons at all until

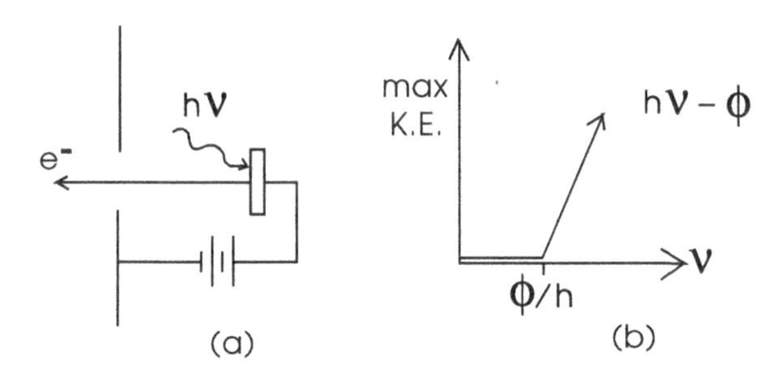

Fig. 3.9 Photoelectric effect: sketches of (a) experimental setup; (b) experimental results.

the frequency reaches a certain threshold, and then the maximum K.E. of the emitted electrons rises linearly with the frequency ν

$$\text{max K.E.} = h\nu - \phi \tag{3.42}$$

Here ϕ is a constant (the "work function") which depends on the material, and the slope in ν is the same Planck's constant!

Suppose one makes a simple classical model of this process. Imagine an electron with charge and mass (e, m) on a spring subject to the electric field in a plane electromagnetic wave as illustrated in Fig. 3.10. At a given x, the time dependence of the field can be taken as $\mathcal{E}_0 \cos{(\omega t)}$ (see Fig. 2.13). Newton's second law then gives for the motion of the particle

$$m\frac{d^2y}{dt^2} = -\kappa y + e\mathcal{E}_0 \cos{(\omega t)} \tag{3.43}$$

Let us look for the driven solution to Eq. (3.43) with $y = y_0 \cos{(\omega t)}$. Substitution, and cancelation of the factor $\cos{(\omega t)}$, gives

$$\begin{aligned} y &= y_0 \cos{(\omega t)} \\ y_0 &= \frac{e\mathcal{E}_0}{m}\frac{1}{\omega_0^2 - \omega^2} \qquad ; \; \omega_0^2 \equiv \frac{\kappa}{m} \end{aligned} \tag{3.44}$$

Now compute the *time-average energy* of the electron[9]

$$\langle E\rangle = \overline{\frac{1}{2}m\dot{y}^2 + \frac{1}{2}\kappa y^2} = \frac{(e\mathcal{E}_0)^2}{4m}\left[\frac{\omega^2+\omega_0^2}{(\omega^2-\omega_0^2)^2}\right] \tag{3.45}$$

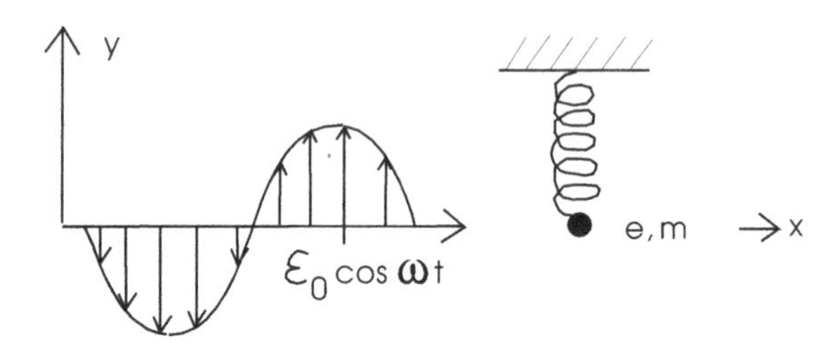

Fig. 3.10 Classical model of charge on a spring driven by an electromagnetic wave. At a given x, the time dependence of the field can be taken as $\mathcal{E}_0 \cos(\omega t)$ (see Fig. 2.13).

We make two observations:

- This result in Eq. (3.45) is proportional to $\mathcal{E}_0^2$, and hence the maximum kinetic energy of the subsequently freed electron should be proportional to the *intensity* of the electromagnetic wave;
- This result does not rise with frequency at high ν, rather it *falls off* as $1/\nu^2$.

Both of these results that the maximum K.E. of the electrons should go as $\mathcal{E}_0^2/\nu^2$ are *completely wrong*!

3.4.1 *Einstein's Analysis*

In order to explain this paradox, Einstein (1904) took Planck seriously. Einstein assumed the following:

- An electromagnetic wave of frequency ν consists of discrete packets of energy called *photons*, each with energy $h\nu$;

$$E = h\nu \qquad ; \text{ photon} \tag{3.46}$$

- A photon can transfer its energy to an electron in the material;
- The *intensity* of the wave then determines the *number* of photons, and therefore the number of electrons.

---

[9]Use $\overline{\sin^2(\omega t)} = \overline{\cos^2(\omega t)} = 1/2$.

Thus, once the photons have enough energy to overcome the work function ϕ of the material, electrons will be emitted with an energy $h\nu - \phi$, and the intensity of the electromagnetic wave determines not the maximum K.E. of the electrons, but the magnitude of the current. The slope in Fig. 3.9 should then indeed be Planck's constant. Thus, based on the phenomenological observations of Planck, Einstein was able to provide a complete understanding of the photoelectric effect.

3.5 Compton Scattering

The particle nature of the photon was confirmed by Compton (1924) who studied the light scattered from electrons in atoms. He found a *frequency shift* in the scattered light that was a function of both the incident frequency and the scattering angle $f(\nu, \theta)$. While again mysterious from a classical point of view,[10] that shift could be understood through the conservation of energy and momentum in a collision between two particles. In order to follow the argument, we must first determine the *momentum* of the photon.

There is a *radiation pressure* P exerted by light arising from the momentum flux of the light into a surface (Fig. 3.11)

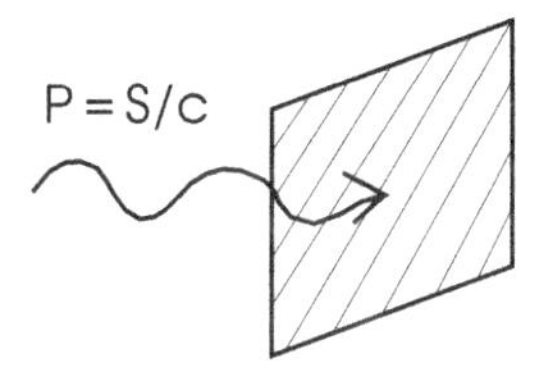

Fig. 3.11 Radiation pressure $P = S/c$ for light where S is the energy flux given by the Poynting vector.

$$\text{radiation pressure} = \text{momentum flux} = P \tag{3.47}$$

The corresponding energy flux is given by the Poynting vector

$$\text{energy flux} = \text{Poynting vector} = S \tag{3.48}$$

In Maxwell's classical theory of light, these two quantities are related by

$$P = \frac{1}{c}S \qquad ; \text{ radiation pressure} \tag{3.49}$$

---

[10]See Prob. 3.6.

Since the energy carried by a photon is $h\nu$, the corresponding momentum it carries must be $h\nu/c$

$$p = \frac{h\nu}{c} \qquad ; \text{ photon momentum} \tag{3.50}$$

Let us now analyze the photon-electron collision through the conservation of momentum and energy (Fig. 3.12)[11]

$$\begin{aligned} \mathbf{p}_e &= \mathbf{p}_0 - \mathbf{p}_1 && ; \text{ momentum conservation} \\ h\nu_0 &= h\nu_1 + \frac{\mathbf{p}_e^2}{2m} && ; \text{ energy conservation} \end{aligned} \tag{3.51}$$

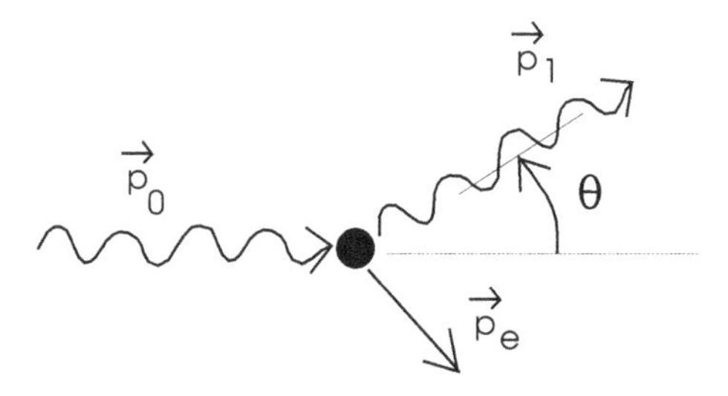

Fig. 3.12 Compton scattering. Here $(\mathbf{p}_0, \mathbf{p}_1)$ are the initial and final photon momenta, and $\mathbf{p}_e$ is the recoil momentum of the electron, which is originally at rest. The magnitude of the photon momentum is related to its energy by $p = h\nu/c$. The scattering angle is θ.

Substitution of the first relation in the second leads to

$$\begin{aligned} h(\nu_0 - \nu_1) &= \frac{1}{2m}(\mathbf{p}_0 - \mathbf{p}_1)^2 = \frac{1}{2m}\left(\frac{h}{c}\right)^2 (\nu_0^2 + \nu_1^2 - 2\nu_0\nu_1 \cos\theta) \\ \nu_0 - \nu_1 &= \frac{h}{mc^2}\nu_0\nu_1 \left[(1 - \cos\theta) + \frac{(\nu_0 - \nu_1)^2}{2\nu_0\nu_1}\right] \end{aligned} \tag{3.52}$$

Since the energy shift is small, the last term can be neglected for all θ of interest

$$\frac{(\nu_0 - \nu_1)^2}{2\nu_0\nu_1} \ll 1 \tag{3.53}$$

[11]Here we use non-relativistic kinematics for the electron. Later, this calculation will be re-done using proper relativistic kinematics; it turns out that the Compton formula in Eq. (3.54) then holds without approximation (see Prob. 8.13).

The frequency of the light is related to its wavelength by $\nu_0 = c/\lambda_0$ and $\nu_1 = c/\lambda_1$. Hence one arrives at the lovely, simple Compton formula for the shift in wavelength

$$\lambda_1 - \lambda_0 = \frac{h}{mc}(1 - \cos\theta) \qquad ; \text{ Compton formula} \qquad (3.54)$$

This is in complete agreement with the experimental results. The quantity h/mc is known as the *Compton wavelength*, and for an electron has the value

$$\frac{h}{m_e c} = 2.427 \times 10^{-12}\,\text{m} \qquad ; \text{ Compton wavelength} \quad (3.55)$$

The results on Compton scattering confirmed the particle nature of the photon introduced by Einstein in his explanation of the photoelectric effect. Today, individual photons are routinely detected in nuclear and particle physics experiments.

It is useful to summarize the conclusions reached so far:

(1) Light manifests both wave and particle properties;
(2) Light consists of a collection of photons, each with energy $h\nu$ and momentum $h\nu/c$;
(3) The intensity of the light determines the number of photons.

3.6 Atomic Spectra

It is the discrete spectra from radiating atoms that provides the *most obvious paradox* in classical physics, and the empirical Balmer formula (1885) for hydrogen, for example, existed well before the turn of the century.[12] We briefly provide some background.[13]

It was Thomson (1897) who discovered the electron through the behavior of cathode rays (today's TV). Rutherford (1910) then discovered the nucleus through his analysis of the scattering of α-particles. Large deflections were seen, indicating that the atomic mass (characterized by the atomic weight) and atomic charge (characterized by the atomic number Z) were concentrated at the center of the atom (Fig. 3.13). This led to the development of his model of the atom.

---

[12] See Prob. 3.8.

[13] Again, the student should become conversant with these classic experiments by studying a good existing book on modern physics, for example [Ohanian (1995)].

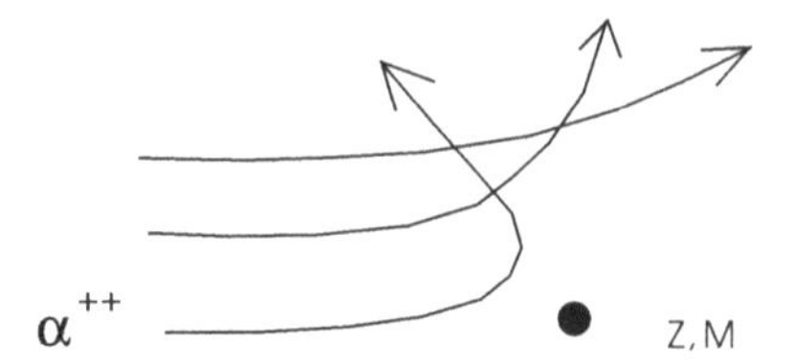

Fig. 3.13 Observed large-angle scattering of α-particles interpreted by Rutherford as coming from the point-like nucleus of the atom. Here the nuclear charge is Ze_p where Z is the atomic number, and the nuclear mass is $M \approx Am_p$ where A is the nucleon number.

3.6.1 *Rutherford Atom*

Rutherford (1911) developed a model of the atom where electrons with negative charge $e = -e_p$ orbit around the nucleus with positive charge Ze_p at the center of the atom, as illustrated in Fig. 3.14.[14] For circular orbits with radius r and velocity $v = r\omega$, one has from a combination of Newton's second law and Coulomb's law

$$\frac{mv^2}{r} = mr\omega^2 = \left(\frac{Ze^2}{4\pi\varepsilon_0}\right)\frac{1}{r^2} \tag{3.56}$$

This equation can be solved for the angular velocity

$$\omega^2 = \frac{1}{m}\left(\frac{Ze^2}{4\pi\varepsilon_0}\right)\frac{1}{r^3} \tag{3.57}$$

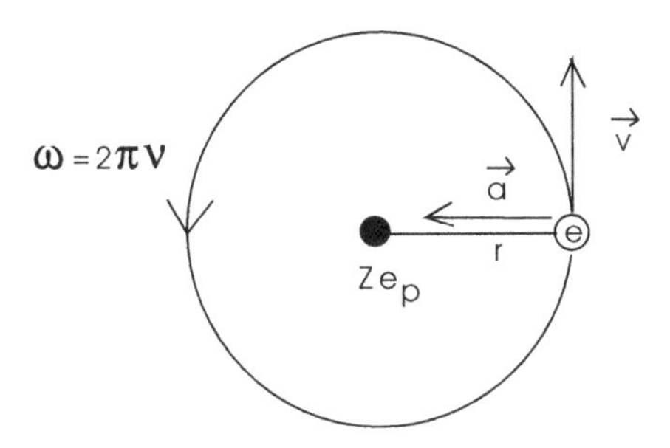

Fig. 3.14 Circular orbits of radius r and angular velocity $\omega = 2\pi\nu$ in the Rutherford atom. The small nucleus with charge Ze_p lies at the center of the atom and contains almost all of its mass.

[14]Here e_p, a positive quantity, is the charge on the proton. Note for hydrogen, $Z = 1$.

But we now observe the following:

- Classically, this Rutherford atom is *unstable*;
- A charge moving in a circle is *accelerated*;
- The electron should *continuously radiate energy and spiral into the nucleus*;
- It is evident from Eq. (3.57) that there should be a *continuous range of emitted frequencies* as the electron spirals in to smaller r.

This situation simply cannot be understood in classical terms.

3.6.2 *Bohr Atom*

In order to understand the atom, Bohr (1913) made a series of three assumptions:

(1) Atoms can have *discrete stationary states*, in which they do not radiate;
(2) Atoms can make *transitions* between states with energies (E_i, E_f) and radiate a photon of energy

$$h\nu = E_i - E_f \tag{3.58}$$

(3) The angular momentum l in the atom is *quantized*[15]

$$l = \frac{nh}{2\pi} \equiv n\hbar \qquad ; \; n = 1, 2, \cdots, \infty \tag{3.59}$$

Let us investigate some consequences of the Bohr assumptions. Recall from Fig. 2.28 that the angular momentum vector $\mathbf{l}$ is perpendicular to the plane of the motion, and in a circular orbit it has a magnitude

$$l = mrv \tag{3.60}$$

Hence upon multiplying Eq. (3.56) by r^3, introducing this relation on the l.h.s., and then solving for r, one has

$$r = \frac{l^2}{m(Ze^2/4\pi\varepsilon_0)} = \frac{n^2}{Z}\frac{\hbar^2}{m(e^2/4\pi\varepsilon_0)} \tag{3.61}$$

[15]Note the introduction here of the commonly used quantity $\hbar \equiv h/2\pi$. The reduced Compton wavelength of the electron is correspondingly frequently employed

$$\hbar = 1.055 \times 10^{-34}\ \text{Js}$$
$$\hbar/m_e c = 3.862 \times 10^{-13}\ \text{m}$$

Here the last equality now follows from the Bohr quantization condition. Thus

$$\begin{aligned} r &= \frac{n^2}{Z}a_0 && ;\ n = 1, 2, \cdots, \infty \\ a_0 &\equiv \frac{\hbar^2}{m_e(e^2/4\pi\varepsilon_0)} && ;\ \text{Bohr radius} \\ &= 0.5292 \times 10^{-10}\,\text{m} = 0.5292\,\text{Å} && \end{aligned} \tag{3.62}$$

The second relation defines the *Bohr radius*, which sets the scale for the size of atoms. The radius of the circular Bohr orbits then goes as $\propto n^2/Z$.

The *energy* in a Bohr orbit is given by

$$\begin{aligned} E &= \frac{1}{2}mv^2 - \frac{Ze^2}{4\pi\varepsilon_0}\frac{1}{r} \\ &= \left(\frac{1}{2} - 1\right)\frac{Ze^2}{4\pi\varepsilon_0}\frac{1}{r} \end{aligned} \tag{3.63}$$

where Eq. (3.56) has been used in the second line. Upon the insertion of the first of Eqs. (3.62) this becomes

$$E_n = -\frac{1}{2}Z^2\frac{e^2}{4\pi\varepsilon_0}\frac{1}{a_0}\frac{1}{n^2} \tag{3.64}$$

In summary, the energy in the circular orbits of the Bohr atom is given by[16]

$$\begin{aligned} E_n &= -\frac{1}{2}Z^2\alpha^2 m_e c^2 \frac{1}{n^2} && ;\ \text{Bohr atom} \\ & && \quad n = 1, 2, \cdots, \infty \\ \alpha &\equiv \frac{1}{\hbar c}\frac{e^2}{4\pi\varepsilon_0} = \frac{1}{137.0} && ;\ \text{fine-structure constant} \\ m_e c^2 &= 0.5110\,\text{MeV} && ;\ \text{electron rest mass} \end{aligned} \tag{3.65}$$

A combination of these relations gives

$$E_n = -\frac{13.61\,\text{eV}}{n^2}Z^2 \qquad ;\ \text{Bohr atom} \tag{3.66}$$

The Bohr spectrum and hypotheses satisfactorily explain the primary structure of hydrogen with $Z = 1$ as illustrated in Fig. 3.15.

---

[16]Note $1\,\text{eV} = 1.602 \times 10^{-19}\,\text{J}$.

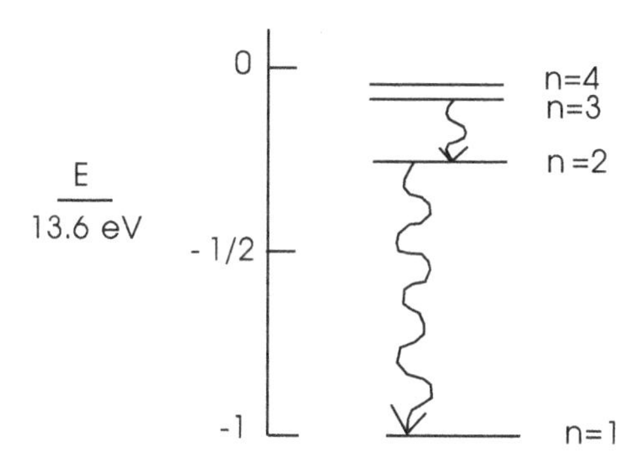

Fig. 3.15 Bohr spectrum for hydrogen with principal radiative transitions satisfying $h\nu = E_n - E_{n-1}$ (note that there are many other radiative transitions in hydrogen).

Chapter 4

Quantum Mechanics

4.1 Matter Waves

DeBroglie (1924) asked himself the following question:

> *If waves (E&M) have particle properties (photons), why don't particles have wave properties?*

We have seen that for photons

$$p = \frac{h\nu}{c} = \frac{h}{\lambda} \qquad ; \text{ photons} \tag{4.1}$$

DeBroglie conjectured that this relation might also hold true for *particles*, which would allow an assignment of a wavelength (the "DeBroglie wavelength") to a particle of given momentum

$$p = \frac{h}{\lambda} \qquad ; \text{ particles–DeBroglie} \tag{4.2}$$

This has some immediate consequences:

(1) Suppose one requires that an integral number of DeBroglie wavelengths fit around a circular orbit of radius r in the Bohr atom (Fig. 4.1). Then

$$\begin{aligned}
2\pi r &= n\lambda && ; \; n = 1, 2, 3, \cdots \infty \\
l &= mvr = p\frac{n\lambda}{2\pi} = \frac{nh}{2\pi} && \\
l &= n\hbar && ; \text{ Bohr model}
\end{aligned} \tag{4.3}$$

This gives precisely Bohr's condition on the quantization of angular momentum!

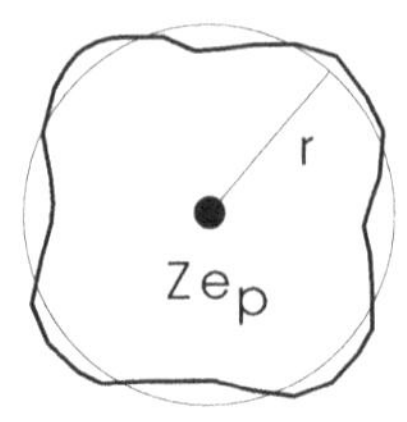

Fig. 4.1 Put integral number of DeBroglie wavelengths around a circular orbit of radius r in the Bohr atom.

(2) Let us put in some numbers. First, use $E = p^2/2m$ to write

$$\lambda = \frac{h}{p} = \frac{h}{\sqrt{2mE}} = \frac{h}{mc}\left(\frac{mc^2}{2E}\right)^{1/2} \tag{4.4}$$

Then suppose one has an electron of energy $E = 100\,\text{eV}$. With the aid of our previous constants one finds

$$\begin{aligned} \frac{h}{m_e c} &= 2.43 \times 10^{-12}\,\text{m} \\ m_e c^2 &= .511 \times 10^6\,\text{eV} \\ E = 100\,\text{eV} \quad \Rightarrow \qquad \lambda &= 1.23 \times 10^{-8}\,\text{cm} \end{aligned} \tag{4.5}$$

This gives $\lambda = 1.23\,\text{Å}$, which is of the order of the interatomic spacing in crystals. Consequently, one should see *Bragg reflection* when sending an electron of this energy through a crystal.

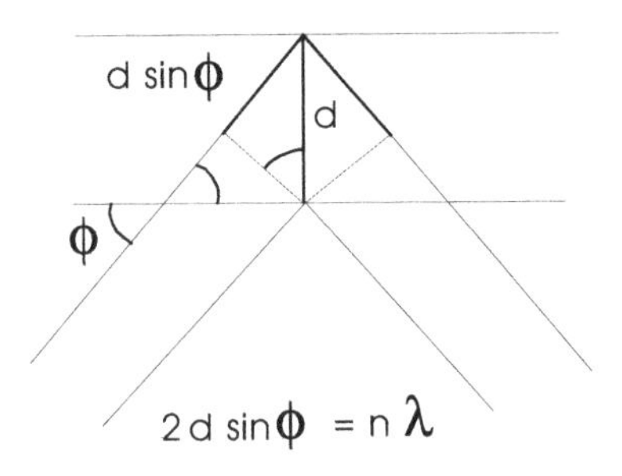

Fig. 4.2 Condition for Bragg reflection from a crystal lattice.

Recall the condition for Bragg reflection. If d is the inter-plane spacing, then the difference in pathlength between crystal planes for an incident plane wave at a grazing angle ϕ is given by (Fig. 4.2)

$$\Delta = 2d \sin\phi \qquad ; \text{ difference in pathlength} \tag{4.6}$$

The condition for *constructive interference* is then

$$2d \sin\phi = n\lambda \qquad ; \ n = \text{integer} \quad \text{Bragg reflection} \tag{4.7}$$

4.2 Davisson-Germer Experiment

Davisson and Germer (1927) carried out one of the key experiments in the development of quantum mechanics. They took electrons from an oven, let them impinge on a crystal, and looked for Bragg diffraction maxima (Fig. 4.3).

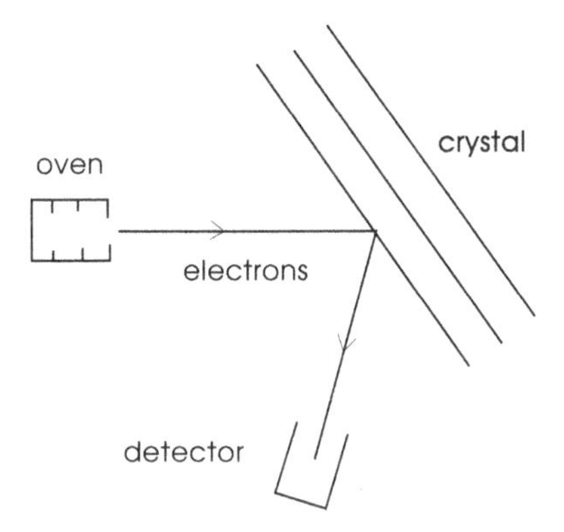

Fig. 4.3 Sketch of Davisson-Germer experiment.

They observed a diffraction pattern, and quantum mechanics was on!

4.3 Schrödinger Equation

Schrödinger set out with the following goals:

- To express DeBroglie's observation and Bohr's hypotheses in a more unified, consistent mathematical form;
- To understand Bohr's stationary states, and corresponding discrete spectra, through analogy to the standing waves on a string [compare

Figs. (2.16, 2.17)].

To understand this work, we return to the material in section 2.1.[1]

4.3.1 *One-Dimensional Wave Equation*

Consider the one-dimensional wave equation. We can represent a wave traveling without change in shape with velocity c in the positive x-direction as

$$\phi(x,t) = \exp\{ik(x-ct)\} \qquad ; \; k = \frac{2\pi}{\lambda} \tag{4.8}$$

This expression evidently satisfies the wave equation

$$\frac{\partial^2\phi(x,t)}{\partial x^2} = \frac{1}{c^2}\frac{\partial^2\phi(x,t)}{\partial t^2} \tag{4.9}$$

For the actual disturbance, we can just use the real part ("Re") of this expression

$$\mathrm{Re}\,\phi(x,t) = \cos k(x-ct) \tag{4.10}$$

Since the classical wave equation is real, $\mathrm{Re}\,\phi(x,t)$ also satisfies that equation.

The wave equation is linear, and one can use the *linear superposition* of solutions to generate another solution[2]

$$\phi(x,t) = \exp\{ik_1(x-ct)\} + \exp\{ik_2(x-ct)\} \qquad ; \; \text{linear superposition} \tag{4.11}$$

One can generalize this result to an expression of the form

$$\phi(x,t) = \int_{-\infty}^{\infty} dk\, a(k) \exp\{ik(x-ct)\} \qquad ; \; \text{again a solution} \tag{4.12}$$

where $a(k)$ is a set of coefficients to be determined from the initial pulse shape at, say, $t = 0$

$$\phi(x,0) = \int_{-\infty}^{\infty} dk\, a(k) \exp\{ikx\} \qquad ; \; \text{initial pulse} \tag{4.13}$$

We make two comments:

---

[1] It is essential for what follows that the reader become fully conversant with the material in section 2.1 and appendix A.

[2] Compare Prob. 2.3.

- We indeed know from our previous discussion that this pulse moves to the right with velocity c without change in shape (Fig. 4.4);

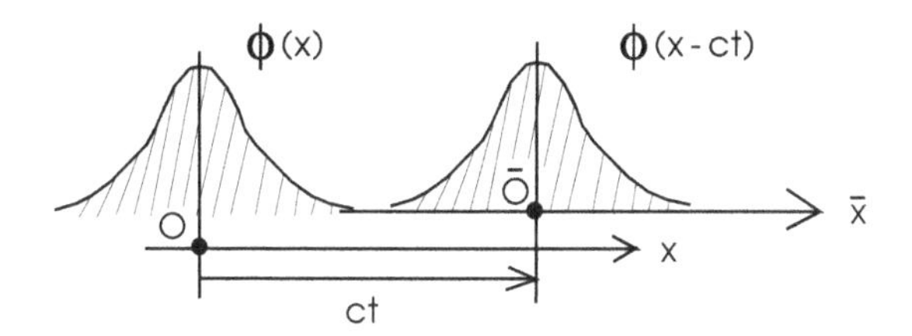

Fig. 4.4 Propagation of solution to the one-dimensional wave equation of a pulse $\phi(x-ct)$ whose initial shape at $t=0$ is $\phi(x)$.

- This provides us with enough flexibility to match a pulse of arbitrary shape at the initial time. We will see later how to find the coefficients $a(k)$ for an arbitrary $\phi(x,0)$.[3]

Let us rewrite the plane wave in Eq. (4.8) as

$$\exp\{ik(x-ct)\} = \exp\{i(kx-\omega_k t)\} \tag{4.14}$$

For a *photon*

$$\begin{aligned} E &= h\nu & ;\ E &= \hbar\omega_k & ;\ \text{photon} \\ p &= \frac{h\nu}{c} = \frac{h}{\lambda} & ;\ p &= \hbar k & \end{aligned} \tag{4.15}$$

Therefore, for a photon,

$$\exp\{i(kx-\omega_k t)\} = \exp\left\{2\pi i\left(\frac{px}{h}-\frac{Et}{h}\right)\right\} = \exp\left\{\frac{i}{\hbar}(px-Et)\right\} \tag{4.16}$$

Schrödinger tried to work in analogy for a free, massive particle. Here the relation between the energy and momentum is

$$E = \frac{p^2}{2m} \qquad ;\ \text{massive particle} \tag{4.17}$$

He wrote an expression for the wave packet of a free, massive particle as

$$\Psi(x,t) = \int_{-\infty}^{\infty} dk\, a(k)\exp\left\{\frac{i}{\hbar}(px-Et)\right\} \tag{4.18}$$

---

[3]Compare Prob. 4.2.

where

$$
\begin{aligned}
p &= \hbar k && \text{; de Broglie} \\
E &= \hbar\omega_k = \frac{(\hbar k)^2}{2m} && \text{; Schrödinger}
\end{aligned}
\tag{4.19}
$$

A combination of these relations gives

$$
\begin{aligned}
\Psi(x,t) &= \int_{-\infty}^{\infty} dk\, a(k) \exp\{i\,(kx - \omega_k t)\} \qquad \text{; Schrödinger wave packet} \\
\omega_k &= \frac{\hbar k^2}{2m}
\end{aligned}
\tag{4.20}
$$

Note that the exponent in Eq. (4.20) is no longer of the form $k(x-ct)$ with constant c.

4.3.2 *Phase Velocity*

The phase velocity of a wave is the velocity of a surface of *constant phase* (Fig. 4.5). Thus for each k in Eq. (4.20)

$$
\begin{aligned}
kx - \omega_k t &= \text{constant} \\
v_{\text{phase}} = \frac{dx}{dt} &= \frac{\omega_k}{k} = \frac{\hbar k}{2m} = \frac{1}{2}\frac{p}{m} = \frac{1}{2} v_{\text{particle}}
\end{aligned}
\tag{4.21}
$$

Here v_{particle} is the *particle velocity*. Thus the phase velocity of the wave is one-half the particle velocity, and one might well be tempted to simply quit at this point!

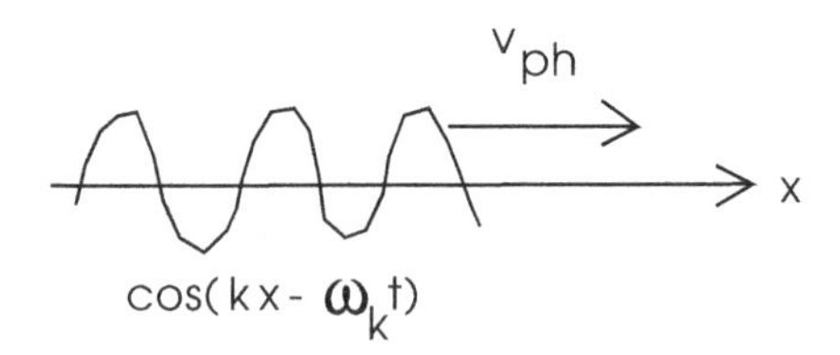

Fig. 4.5 Phase velocity of a wave: $kx - \omega_k t =$ constant, for a given k, implies $v_{\text{phase}} = dx/dt = \omega_k/k$. The illustration is for $k > 0$.

The velocity with which a general wave packet actually *propagates*, however, is the *group velocity*, and we must calculate this quantity.

4.3.3 *Group Velocity*

Suppose one has a *carrier wave* with wave number $k_0 > 0$ and corresponding angular frequency ω_0

$$\exp\{i(k_0 x - \omega_0 t)\} = \text{carrier wave} \tag{4.22}$$

This wave extends in this form for all x. One must superpose waves in an interval about k_0 to produce a wave packet that is *localized in* x. Suppose one has an amplitude $a(k - k_0)$ that is peaked at k_0 and includes a small interval about k_0 as illustrated in Fig. 4.6.

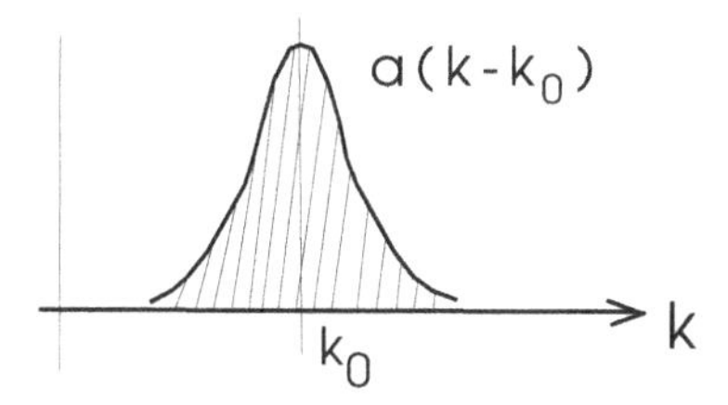

Fig. 4.6 Amplitude for a representative wave packet localized in k-space about k_0.

In this case

$$\Psi(x,t) = \int_{-\infty}^{\infty} dk\, a(k - k_0) \exp\{i(kx - \omega_k t)\} \tag{4.23}$$

Now make a Taylor series expansion of the angular frequency about k_0

$$\begin{aligned} k &\equiv k_0 + (k - k_0) \\ \omega_k &= \omega_{k_0} + (k - k_0)\left(\frac{\partial \omega_k}{\partial k}\right)_0 + \cdots \end{aligned} \tag{4.24}$$

With the introduction of the variable $u = k - k_0$, and the retention of just the indicated terms in the second of Eqs. (4.24), the wave packet in Eq. (4.20) takes the form

$$\begin{aligned} \Psi(x,t) &= \exp\{i(k_0 x - \omega_{k_0} t)\} \int_{-\infty}^{\infty} du\, a(u) \exp\{iu(x - v_{\rm gp} t)\} \\ &= \exp\{i(k_0 x - \omega_{k_0} t)\}\, \mathcal{A}(x - v_{\rm gp} t) \end{aligned} \tag{4.25}$$

Here the *group velocity* is defined by

$$v_{\rm gp} \equiv \left(\frac{\partial \omega_k}{\partial k}\right)_0 \qquad ; \text{ group velocity} \tag{4.26}$$

There is an *amplitude modulation* of the carrier wave given by

$$\mathcal{A}\left(x - v_{\rm gp}t\right) = \int_{-\infty}^{\infty} du\, a(u) \exp\left\{iu\left(x - v_{\rm gp}t\right)\right\} \quad \text{; amplitude modulation} \tag{4.27}$$

The form of $\mathcal{A}$ depends on the localizing function $a\,(k - k_0)$, and this amplitude modulation moves with the *group velocity* (see Fig. 4.7).

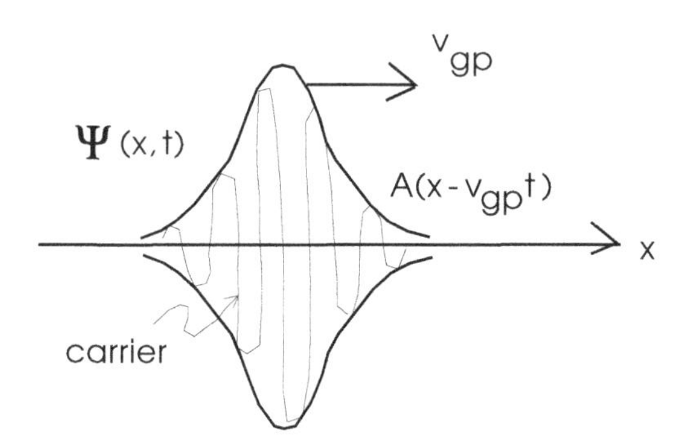

Fig. 4.7 Group velocity of a localized wave packet: the amplitude modulation $\mathcal{A}(x - v_{\rm gp}t)$ of $\Psi(x,t)$ [Eq. (4.25)] moves with the group velocity $v_{\rm gp} = (\partial\omega_k/\partial k)_0$.

Thus with Schrödinger's choice of ω_k in Eq. (4.20), one has

$$v_{\rm gp} = \left(\frac{\partial\omega_k}{\partial k}\right)_0 = \frac{\hbar k_0}{m} = \frac{p_0}{m} = v_{\rm particle} \tag{4.28}$$

The *group velocity of the wave packet is identical to the particle's velocity*, and it appears that one is, indeed, on the right track.[4]

4.3.4 *Interpretation*

How does one *interpret* this new, complex wave function $\Psi(x,t)$ in Eq. (4.20)? It certainly should be large where the particle is, and small where it is not! Let us then

Take $|\Psi(x,t)|^2$ *as a measure of where the particle is.*

[4]Note that for a solution to the classical wave equation with $\omega_k = ck$, the phase velocity and group velocity are identical.

With the wave packet in Eq. (4.25), one has (see Fig. 4.8)

$$|\Psi(x,t)|^2 = |\mathcal{A}(x - v_{\rm gp}t)|^2 \tag{4.29}$$

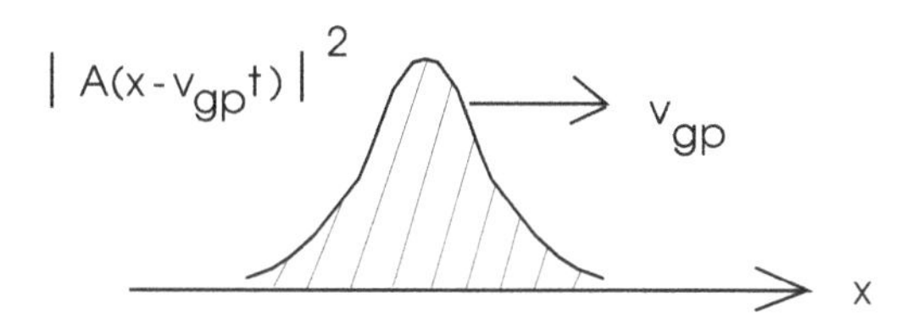

Fig. 4.8 $|\Psi(x,t)|^2$ for the wave packet in Eq. (4.25), as given in Eq. (4.29).

Notice the following:

- This envelope moves with the particle velocity

$$v_{\rm gp} = v_{\rm particle} \tag{4.30}$$

- *The carrier wave has disappeared from this relation!*

4.3.5 *Differential Equation*

Let us ask what *differential equation* the wave packet in Eqs. (4.20) satisfies, as this will provide us with a firmer mathematical foundation. Just take the following derivatives

$$\begin{aligned}
i\hbar\frac{\partial}{\partial t}\Psi(x,t) &= \int_{-\infty}^{\infty} dk\, a(k)\,(\hbar\omega_k)\exp\{i(kx - \omega_k t)\} \\
-\frac{\hbar^2}{2m}\frac{\partial^2}{\partial x^2}\Psi(x,t) &= \int_{-\infty}^{\infty} dk\, a(k)\left(\frac{\hbar^2k^2}{2m}\right)\exp\{i(kx - \omega_k t)\}
\end{aligned} \tag{4.31}$$

It follows from the second of Eqs. (4.20) that

$$i\hbar\frac{\partial}{\partial t}\Psi(x,t) = -\frac{\hbar^2}{2m}\frac{\partial^2}{\partial x^2}\Psi(x,t) \qquad ;\ \text{Schrödinger equation} \tag{4.32}$$

This is the *Schrödinger equation*, which provides the foundation of quantum mechanics [Schrödinger (1926)]. We make some observations on this result:

- In contrast to the classical wave equation, this equation is inherently complex, as an i appears on the l.h.s.;

- This quantum wave equation is first-order in the time derivative, as opposed to the classical string problem where Newton's second law gives a differential wave equation that is second-order in the time;
- It follows from this last remark that, in contrast to the classical string where both the initial displacement and velocity must be specified for all x to determine the subsequent motion, it is enough here to specify the initial wave function $\Psi(x, 0)$ for all x to determine it for all subsequent times.

From Eq. (4.20), the initial wave function is given by

$$\Psi(x, 0) = \int_{-\infty}^{\infty} dk\, a(k)\, e^{ikx} \tag{4.33}$$

We clearly need to investigate the properties of expressions of this form.

4.3.6 *More Mathematics*

Let us return to the string problem, and consider a string wrapped around a cylinder and free to perform transverse motion along a direction parallel to the axis of the cylinder (Fig. 4.9).

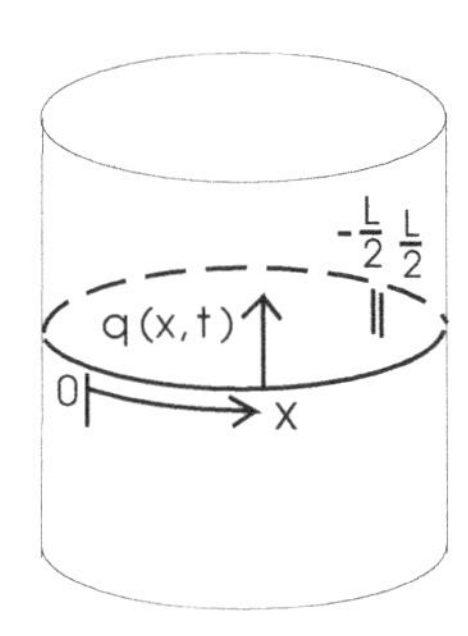

Fig. 4.9 String of length L stretched around a cylinder free to oscillate without friction in a direction parallel to the axis of the cylinder. Illustration of periodic boundary conditions.

The solutions to the wave equation in this case, and their properties, are derived in Prob. 2.4

$$\phi_n = \frac{1}{\sqrt{L}} e^{ik_n x} \qquad ; \text{ eigenfunctions}$$
$$k_n = \frac{2\pi}{L} n \qquad ; \text{ periodic boundary conditions}$$
$$n = 0, \pm 1, \pm 2, \cdots, \pm\infty \tag{4.34}$$

The eigenfunctions are orthonormal on the interval $-L/2 \leq x \leq L/2$

$$\int_{-L/2}^{L/2} \phi_m^\star(x)\phi_n(x)\, dx = \delta_{mn} \tag{4.35}$$

Consider a general linear superposition of the normal modes of the string in this case

$$q(x,t) = \sum_{n=-\infty}^{n=\infty} A(k_n) \frac{1}{\sqrt{L}} e^{ik_n x} \cos(\omega_n t + \delta_n) \tag{4.36}$$

To match the displacement of the string at the initial time $t = 0$ one has

$$q(x,0) \equiv q(x) = \sum_{n=-\infty}^{n=\infty} C(k_n) \frac{1}{\sqrt{L}} e^{ik_n x}$$
$$C(k_n) \equiv A(k_n) \cos\delta_n \tag{4.37}$$

The first relation in Eqs. (4.37) can be inverted with the use of the orthonormality of the eigenfunctions

$$C(k_n) = \frac{1}{\sqrt{L}} \int_{-L/2}^{L/2} e^{-ik_n x} q(x)\, dx \tag{4.38}$$

In the limit $L \to \infty$, the sum in Eq. (4.37) can be converted to an integral following the arguments in Eqs. (2.51)–(2.54)

$$\Delta k = \frac{2\pi}{L}$$
$$q(x) = \frac{L}{2\pi} \sum_{n=-\infty}^{n=\infty} \Delta k\, C(k_n) \frac{1}{\sqrt{L}} e^{ik_n x}$$
$$C(k_n) = \frac{1}{\sqrt{L}} \int_{-L/2}^{L/2} e^{-ik_n x} q(x)\, dx \tag{4.39}$$

Now define

$$a(k) \equiv \left(\frac{L}{2\pi}\right)^{1/2} C(k_n) \tag{4.40}$$

Then in the limit $L \to \infty$, the relations in Eqs. (4.39) become

$$\begin{aligned} q(x) &= \frac{1}{\sqrt{2\pi}} \int_{-\infty}^{\infty} dk\, a(k) e^{ikx} && ;\ L \to \infty \\ a(k) &= \frac{1}{\sqrt{2\pi}} \int_{-\infty}^{\infty} dx\, q(x) e^{-ikx} && \text{Fourier transform pair} \end{aligned} \tag{4.41}$$

The initial displacement $q(x)$ and the amplitude of the wave packet $a(k)$ form a *Fourier transform pair*. Appropriate properties of Fourier transforms are discussed in appendix C.

4.3.7 *Continuity Equation*

To summarize where we are. Consider a free, non-relativistic particle of mass m. We assume:

(1) There is a wave function $\Psi(x, t)$ for the system;
(2) This wave function satisfies the Schrödinger equation

$$i\hbar \frac{\partial}{\partial t} \Psi(x, t) = -\frac{\hbar^2}{2m} \frac{\partial^2}{\partial x^2} \Psi(x, t) \qquad ;\ \text{free particle} \tag{4.42}$$

(3) The interpretation of the wave function is that $|\Psi(x, t)|^2\, dx$ gives the *probability* that the particle lies between x and $x + dx$.

We have to ask ourselves if this interpretation makes sense.[5] If $|\Psi(x, t)|^2$ is to serve as a *probability density*, it must satisfy certain conditions. It certainly is positive definite. Consider, then, the *time development* of $|\Psi(x, t)|^2$

$$i\hbar \frac{\partial}{\partial t}(\Psi^\star \Psi) = \Psi^\star \left(i\hbar \frac{\partial}{\partial t} \Psi\right) + \left(i\hbar \frac{\partial}{\partial t} \Psi^\star\right) \Psi \tag{4.43}$$

The complex conjugate of Eq. (4.42) gives

$$-i\hbar \frac{\partial}{\partial t} \Psi^\star(x, t) = -\frac{\hbar^2}{2m} \frac{\partial^2}{\partial x^2} \Psi^\star(x, t) \tag{4.44}$$

[5]The interpretation is due to [Born (1926)].

Substitution of this and Eq. (4.42) into (4.43) gives

$$
\begin{aligned}
i\hbar\frac{\partial}{\partial t}(\Psi^\star\Psi) &= -\frac{\hbar^2}{2m}\left[\Psi^\star\frac{\partial^2}{\partial x^2}\Psi - \left(\frac{\partial^2}{\partial x^2}\Psi^\star\right)\Psi\right] \\
&= -\frac{\hbar^2}{2m}\frac{\partial}{\partial x}\left[\Psi^\star\frac{\partial}{\partial x}\Psi - \left(\frac{\partial}{\partial x}\Psi^\star\right)\Psi\right] \qquad (4.45)
\end{aligned}
$$

We have thus derived a (one-dimensional) *continuity equation* for the probability density

$$
\begin{aligned}
\rho &= |\Psi|^2 && ;\ \text{probability density} \\
S_x &= \frac{\hbar}{2im}\left[\Psi^\star\frac{\partial}{\partial x}\Psi - \left(\frac{\partial}{\partial x}\Psi^\star\right)\Psi\right] && ;\ \text{probability current} \\
\frac{\partial \rho}{\partial t} + \frac{\partial S_x}{\partial x} &= 0 && ;\ \text{continuity eqn} \qquad (4.46)
\end{aligned}
$$

This analysis allows us to identify the *probability current* S_x, to which we shall return shortly.

Assume that the wave function $\Psi(x,t)$ represents a *localized disturbance* for all t under consideration, as illustrated in Fig. 4.10.

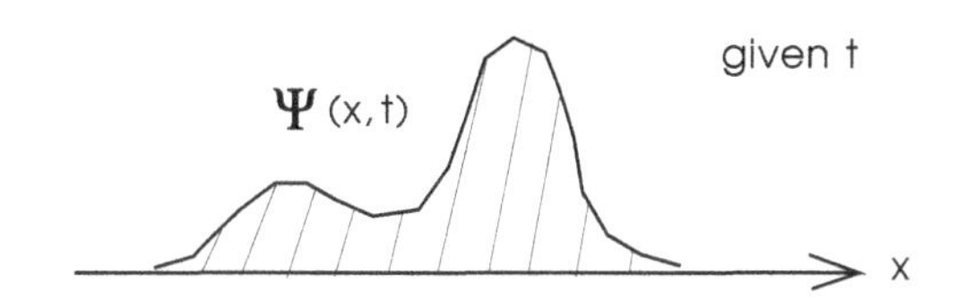

Fig. 4.10 Sketch of a wave function $\Psi(x,t)$ that represents a *localized* disturbance.

Now integrate the continuity equation over all x

$$
\begin{aligned}
\int_{-\infty}^{\infty}\frac{\partial \rho}{\partial t}dx &= \frac{d}{dt}\int_{-\infty}^{\infty}\rho(x,t)\,dx \\
&= -\int_{-\infty}^{\infty}\frac{\partial S_x}{\partial x}\,dx = -S_x(x,t)\Big|_{x=-\infty}^{x=\infty} \\
&= 0 \qquad ;\ \text{localized disturbance} \qquad (4.47)
\end{aligned}
$$

It thus follows from the continuity equation that

$$
\int_{-\infty}^{\infty}\rho(x,t)\,dx = \text{constant} \qquad ;\ \text{constant of the motion} \quad (4.48)
$$

This is essential to our interpretation of $\rho = |\Psi|^2$ as a probability density, since we know that one free electron remains one free electron for all time.

The Schrödinger equation is a linear, homogeneous equation, and it says nothing about the *normalization* of Ψ. Consistent with a probability interpretation, we choose to normalize Ψ so that

$$\int_{-\infty}^{\infty} |\Psi(x,t)|^2\, dx = 1 \qquad ; \text{ normalization} \tag{4.49}$$

Now suppose one instead just integrates the continuity equation over a small region from $x_0 - \varepsilon$ to $x_0 + \varepsilon$. In this case one finds

$$\frac{d}{dt}\left[2\varepsilon\rho(x_0,t)\right] = S_x\Big|_{x_0-\varepsilon} - S_x\Big|_{x_0+\varepsilon} \tag{4.50}$$

The l.h.s. is the time rate of change of the probability that the particle lies in the region $x_0 - \varepsilon \leq x \leq x_0 + \varepsilon$. The first term on the r.h.s. then has an interpretation as the probability flowing *into* this region per unit time, and the second term as the probability flowing *out* (see Fig. 4.11). The probability current will play a key role in our subsequent interpretation of the theory.

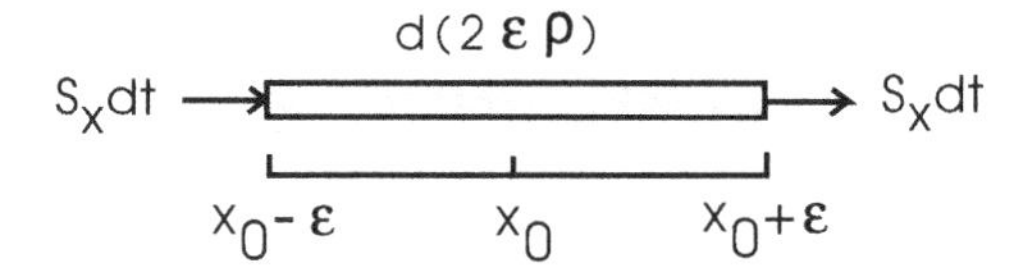

Fig. 4.11 Identification of the probability flux from the continuity equation.

4.3.8 *General Solution for Free Particle*

Consider an arbitrary wave packet, as illustrated in Fig. 4.10. It can be represented as a linear superposition of plane waves

$$\Psi(x,t) = \frac{1}{\sqrt{2\pi}} \int_{-\infty}^{\infty} dk\, a(k) \exp\{i\,(kx - \omega_k t)\} \qquad ; \text{ free particle} \tag{4.51}$$

where the angular frequency and wavenumber are related by the *dispersion relation* in the second of Eqs. (4.20)

$$\omega_k = \frac{\hbar k^2}{2m} \qquad ; \text{ dispersion relation} \tag{4.52}$$

This solution has the following properties:

- It satisfies the Schrödinger equation, since each plane wave does;
- With the appropriate choice of $a(k)$, it satisfies the *boundary conditions* that the particle be localized with $\Psi \to 0$ as $x \to \pm\infty$ (Fig. 4.10);
- It remains to satisfy the *initial condition*, and as noted above, since the Schrödinger equation is linear in $\partial/\partial t$, this means that it must reproduce $\Psi(x,0)$ for all x.

$$\Psi(x,0) = \frac{1}{\sqrt{2\pi}} \int_{-\infty}^{\infty} dk\, a(k)\, e^{ikx} \qquad \text{; initial condition} \qquad (4.53)$$

Since this is just a Fourier transform, it can be inverted using the relations in Eqs. (4.41)

$$a(k) = \frac{1}{\sqrt{2\pi}} \int_{-\infty}^{\infty} dx\, \Psi(x,0)\, e^{-ikx} \qquad \text{; Fourier transform} \quad (4.54)$$

This relation allows one to determine the coefficients $a(k)$ for an arbitrary initial wave function $\Psi(x,0)$.

4.3.9 *Interpretation (Continued)*

Recall from Eqs. (4.19) that the momentum of the particle is given by $p = \hbar k$. In analogy with our assumption concerning the probability density in coordinate space, we will assume that

> *The quantity $|a(k)|^2\, dk$ is the probability of finding the particle with momentum between $\hbar k$ and $\hbar k + d(\hbar k)$.*

For example, suppose one has a coordinate space wave function $\Psi(x,t)$ that at the initial time $t = 0$ has the form shown in Fig. (4.12). Take

$$\begin{aligned} \Psi(x,0) &= \frac{1}{\sqrt{2d}} \qquad &&; -d \le x \le d \\ &= 0 &&; \text{otherwise} \end{aligned} \qquad (4.55)$$

This wave function is clearly normalized[6]

$$\int_{-\infty}^{\infty} dx\, |\Psi(x,0)|^2 = 1 \qquad (4.56)$$

[6]And Eq. (4.48) implies that this normalization is preserved in time as the wave function develops according to the Schrödinder equation.

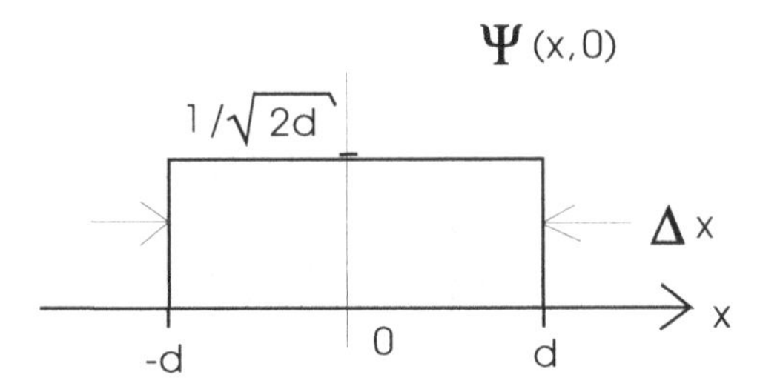

Fig. 4.12 Example of a coordinate space wave function $\Psi(x,t)$, which at the initial time $t = 0$ has this form.

The Fourier transform of this function is readily calculated

$$a(k) = \frac{1}{\sqrt{2\pi}} \frac{1}{\sqrt{2d}} \underbrace{\int_{-d}^{d} dx\, e^{-ikx}}_{\frac{1}{-ik}\left(e^{-ikd} - e^{ikd}\right)}$$

$$a(k) = \sqrt{\frac{d}{\pi}} \frac{1}{kd} \frac{1}{2i} \left(e^{ikd} - e^{-ikd}\right) \tag{4.57}$$

Hence

$$a(k) = \sqrt{\frac{d}{\pi}} \frac{\sin kd}{kd} \tag{4.58}$$

This result is sketched in Fig. 4.13.[7]

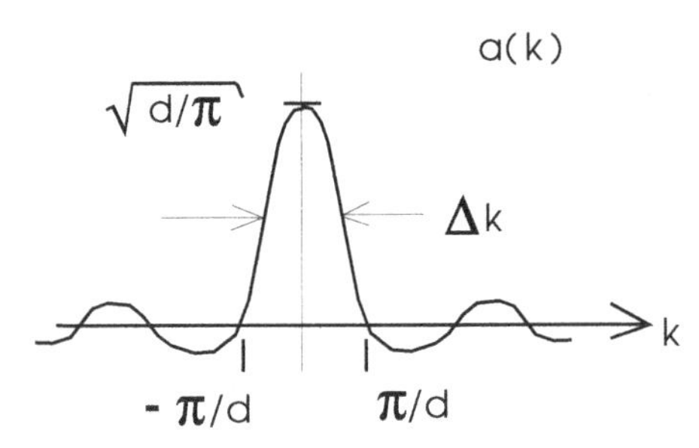

Fig. 4.13 Fourier transform $a(k)$ of the wave function in Eq. (4.55) and Fig. 4.12.

We now observe that the width of the wave packet in coordinate space in Fig. 4.12 is $\Delta x \sim 2d$, while the width of the Fourier transform in momentum

[7]Note that $\sin x/x \to 1$ as $x \to 0$.

space in Fig. 4.13 is $\Delta k \sim \pi/d$. Hence their *product* satisfies

$$\begin{aligned} & \Delta x \cdot \Delta k \sim 2\pi \\ \Rightarrow \quad & \Delta x \cdot \Delta p \sim 2\pi\hbar \sim h \qquad ; \text{ Heisenberg uncertainty principle} \end{aligned} \tag{4.59}$$

This is simply a property of Fourier transforms. In order to make the quantity d small and have a well-localized wave packet in coordinate space, one needs a broad range of k, and *vice versa.* These widths *cannot simultaneously be made arbitrarily small.* This is known as the *Heisenberg uncertainty principle* [Heisenberg (1927)], and it implies a profound break with classical physics where the starting point in phase space is a precise specification of both a particle's position *and* its momentum.[8]

In contrast to the classical string problem where a small-amplitude pulse propagates without change in shape, because of the momentum spread, the quantum wave packet for a free, localized particle *spreads* with time. This is illustrated in Probs. 4.11 and 4.12.

4.3.10 *Include Forces*

So far we have just considered a free particle. We would now like to include forces. In order to do this, we first give a "heuristic" derivation of the previous results. Write the classical statement of the conservation of energy (here, as before, $q \equiv x$)

$$\begin{aligned} H(p,q) &= E = \text{constant} \\ H(p,q) - E &= 0 \end{aligned} \tag{4.60}$$

Let this expression *operate on a wave function*

$$[H(p,q) - E]\,\Psi(q,t) = 0 \tag{4.61}$$

Now make the following replacements

$$\begin{aligned} p &\rightarrow \frac{\hbar}{i}\frac{\partial}{\partial q} \\ E &\rightarrow i\hbar\frac{\partial}{\partial t} \end{aligned} \tag{4.62}$$

[8] If one defines Δx as the root-mean square deviation from the expectation value of x, and similarly for k, then it is a rigorous property of Fourier transforms that $\Delta x \cdot \Delta k \geq 1/2$. Thus, the precise statement of the Heisenberg uncertainty principle is (see Prob. 4.10)

$$\Delta x \cdot \Delta p \geq \hbar/2 \qquad ; \text{ Heisenberg uncertainty principle}$$

For a free particle, this gives

$$H = T = \frac{p^2}{2m} \rightarrow -\frac{\hbar^2}{2m}\frac{\partial^2}{\partial q^2} \tag{4.63}$$

This argument leads to the Schrödinger Eq. (4.32), and *suggests* that a force can be included by simply using the appropriate expression for $H(p, q)$

$$H = T(p) + V(q) = \frac{p^2}{2m} + V(q) \tag{4.64}$$

This leads to the following extension of the Schrödinger equation in the presence of a force

$$i\hbar\frac{\partial \Psi}{\partial t} = \left[-\frac{\hbar^2}{2m}\frac{\partial^2}{\partial q^2} + V(q)\right]\Psi(q,t) \qquad ; \text{ Schrödinger eqn with } V(q) \tag{4.65}$$

The *justification* of this last result lies in showing that a well-localized wave packet obeys Newton's Laws in the appropriate limit. This is the *correspondence principle* of quantum mechanics — one must reproduce classical physics in the classical domain. We satisfied the correspondence principle for a free particle with $V = 0$ by showing that the group velocity of the wave packet is equal to the particle velocity $v_{\text{gp}} = v_{\text{particle}}$ [see Eq. (4.30) and Fig. (4.8)]. We will be content here to simply extend the assumption in Eq. (4.42) and leave the demonstration of the correspondence principle in this case as a problem (Prob. 4.9). The proof that the continuity Eq. (4.46) holds without modification in the presence of a real $V(q)$ is also left as a problem (Prob. 4.1).

4.3.11 *Boundary Conditions*

Since we are dealing with a new quantity $\Psi(x, t)$ in a new realm of physics, we need some guiding principles to deduce the appropriate boundary conditions to go along with the differential equation governing its behavior. We start from the following observations:

(1) *Physics*, as contained in the probability density and probability current (ρ, S_x), *must be continuous and single-valued*;[9]

[9] Note that these physical quantities are *bilinear* in the wave function (they go as $\sim \Psi^\star\Psi$), so the deduction of the boundary conditions to be applied to the linear Schrödinger equation is a non-trivial matter. Indeed, a rush to apply intuitive, classical boundary conditions can lead one to overlook interesting and important new physics.

(2) Acceptable solutions to the Schrödinger equation must satisfy the principle of *superposition*, that is, the sum of two acceptable solutions must again be an acceptable solution.

We can then derive the following boundary conditions for the Schrödinger equation:

I. If V is finite, the implications are straightforward, and these principles immediately lead to the conclusion that

$$\Psi \text{ and } \frac{\partial \Psi}{\partial x} \text{ must be continuous if } V \text{ is finite} \tag{4.66}$$

II. At a *wall*, that is, at an infinite potential barrier (Fig. 4.14),

$$\begin{aligned} \Psi &= 0 && ; \text{ at wall } (\infty \text{ barrier}) \\ \frac{\partial \Psi}{\partial x} &= \text{anything} \end{aligned} \tag{4.67}$$

We will return shortly to the proof of this second result.

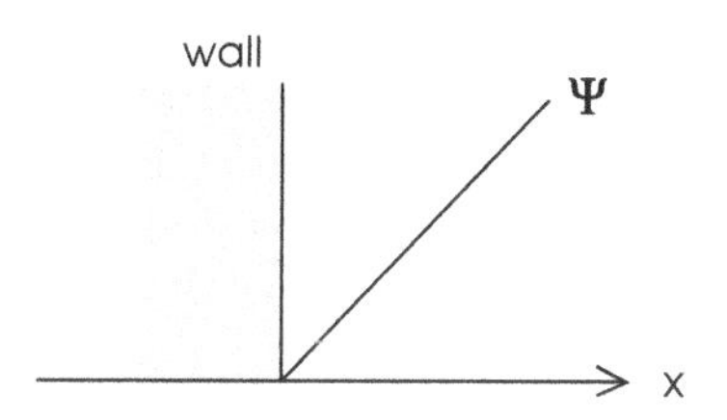

Fig. 4.14 Boundary condition at a wall.

4.3.12 *Stationary States*

We are now in a position to look for *normal-mode* solutions to the Schrödinger equation, just as we did for the classical string. In a normal mode, everything oscillates with the *same frequency*. Thus we will look for solutions to the complex Eq. (4.65) of the form

$$\Psi(x,t) = \psi(x) \exp\left\{-\frac{i}{\hbar}Et\right\} \qquad ; \text{ normal mode} \tag{4.68}$$

Substitution into Eq. (4.65), and cancelation of the exponential, leads to

$$\left[-\frac{\hbar^2}{2m}\frac{d^2}{dx^2} + V(x)\right]\psi(x) = E\,\psi(x)$$

; stationary-state Schrödinger eqn (4.69)

It is evident from Eq. (4.61) that E is the energy of the state; correspondingly, the *eigenvalue* E of $H(p,x)$ will be real, as it should be (see Prob. 4.6).

Now note that in the normal mode the probability density and probability current are *independent of time*

$$\rho = |\Psi(x,t)|^2 = |\psi(x)|^2 \qquad ;\ \text{normal mode}$$
$$S_x = \frac{\hbar}{2im}\left\{\psi^\star(x)\frac{\partial\psi(x)}{\partial x} - \left[\frac{\partial\psi(x)}{\partial x}\right]^\star\psi(x)\right\} \qquad \text{time-independent!} \qquad (4.70)$$

This analysis has led to *stationary states* of a quantum system that are precisely in accord with *the first of Bohr's hypotheses*! Clearly, the Schrödinger approach has merit. Note that the equation for the normal-mode amplitudes, the stationary-state Schrödinger Eq. (4.69), is similarly independent of time (as was the case with the string).

We now have enough background to turn to several applications of quantum mechanics. Further formal developments, although not essential to the applications we shall discuss, are examined in Probs. (4.3)–(4.12). The reader is urged to work as many of these as he or she can, in order to establish a good grounding in quantum mechanics.[10]

4.4 Solution to Some One-Dimensional Problems

Now that one has the differential Eq. (4.69), the boundary conditions in Eqs. (4.66)–(4.67), and the interpretation through the continuity Eq. (4.46), it is possible to start solving some interesting problems.

4.4.1 *Particle in a One-Dimensional Box*

Consider a particle of mass m in a one-dimensional box. Here the potential vanishes inside the box and the boundary conditions are that the wave function must vanish at both walls. In this case, the stationary-state Schrödinger Eq. (4.69) takes the form of the scalar Helmholtz equation,

[10]For good textbooks on quantum mechanics, see [Griffiths (1998); Schiff (1968); Landau and Lifshitz (1981); Merzbacher (1997)].

with appropriate boundary conditions

$$\frac{d^2\psi(x)}{dx^2} = -k^2\,\psi(x) \qquad ;\ k^2 \equiv \frac{2mE}{\hbar^2}$$
$$\psi(0) = \psi(L) = 0 \qquad ;\ \text{B.C.} \tag{4.71}$$

The situation is illustrated in Fig. (4.15). The problem is now exactly analogous to the string with fixed endpoints! We can simply make use of our previous analysis.

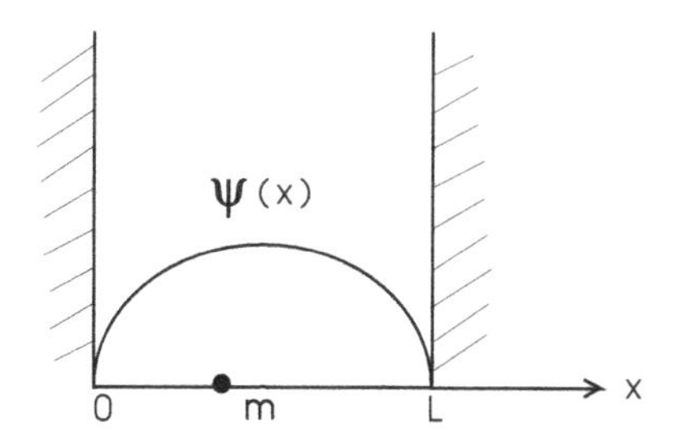

Fig. 4.15 Particle in a one-dimensional box.

The normalized eigenfunctions and the eigenvalues are given by

$$\psi_n(x) = \left(\frac{2}{L}\right)^{1/2} \sin\frac{n\pi x}{L} \qquad ;\ \text{eigenfunctions}$$
$$n = 1, 2, 3, \cdots, \infty$$
$$k_n = \frac{\pi n}{L} \qquad ;\ \text{eigenvalues} \tag{4.72}$$

The energies of the states follow from the definition of k^2 in Eqs. (4.71)

$$E_n = \frac{\hbar^2\pi^2}{2mL^2}\,n^2 \qquad ;\ n = 1, 2, 3, \cdots, \infty \tag{4.73}$$

The eigenfunctions are orthonormal and satisfy

$$\int_0^L dx\,\psi_m^\star(x)\psi_n(x) = \delta_{mn} \tag{4.74}$$

The first few eigenfunctions $\psi_n(x)$, and well as the implied probability density $|\psi_n(x)|^2$, are sketched in Fig. 4.16.

We make several comments:

- These are *stationary states* that have *definite energy*;

- The probability density and probability current (ρ, S_x) are *independent of time* in these states;
- In fact, the probability current vanishes if ψ is real;
- $\rho = |\psi|^2$ is the *particle density* in these modes. This is a quantity that can be observed experimentally, for example, through electron scattering from it (if the target particle is charged);
- Note the presence of *nodes* in the excited states where the wave function, and hence the probability density, vanishes. The particle will never be found at these positions in these states!
- This simple problem illustrates the *essentials of bound states in quantum mechanics.*

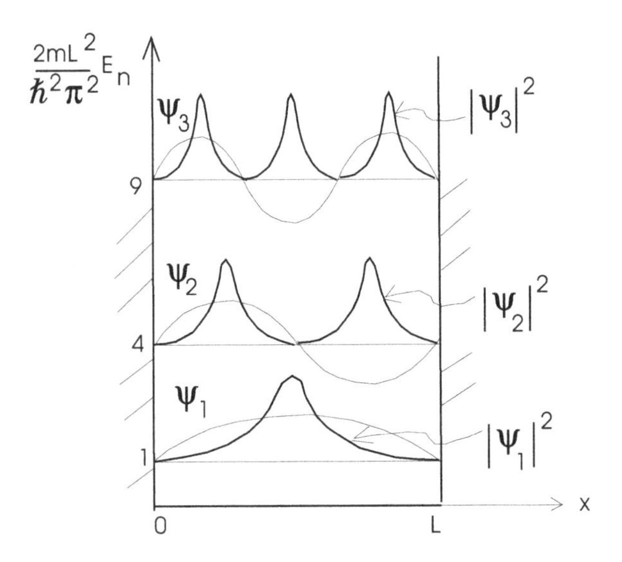

Fig. 4.16 Sketch of the first few eigenfunctions $\psi_n(x)$, and the implied probability densities $|\psi_n(x)|^2$, for a particle in a one-dimensional box.

4.4.2 *Potential Barrier in One-Dimension*

Suppose one prepares a particle with a definite energy and lets it impinge on a potential barrier in one-dimension. To solve this problem, one again looks for normal modes, or *stationary states*, which are the states of definite energy

$$\Psi(x,t) = \psi(x)\,\exp\left\{-\frac{i}{\hbar}Et\right\} \qquad ;\ \text{normal modes} \qquad (4.75)$$

This leads to the time-independent Schrödinger equation

$$\left[-\frac{\hbar^2}{2m}\frac{d^2}{dx^2} + V(x)\right]\psi(x) = E\,\psi(x) \qquad ;\ \text{Schrödinger equation}\ (4.76)$$

Here E is the energy.

4.4.2.1 *Scattering State*

We now introduce the notion of a *scattering state.* This is a state that is prepared with a definite momentum $p_{\text{inc}} = \hbar k$ and energy $E_{\text{inc}} \equiv E = \hbar^2 k^2/2m$. Here E is something we specify at the beginning, for example, by taking particles from an accelerator; it is *not* an eigenvalue determined as in the previous problem. The *incident wave function*, representing the initially prepared state of definite momentum, then takes the form[11]

$$\psi_{\text{inc}}(x) = Ne^{ikx} \qquad ;\ \text{incident wave function} \qquad (4.77)$$

If there were no potential barrier, this would be the appropriate solution to the Schrödinger equation for all x.

We shall eventually get at the physics in the scattering problem by taking the *ratio of probability fluxes.* Recall from our discussion of the probability current that

> *$S_x(x)$ is the rate at which the particle probability moves past the point x in the x-direction.*

If one is only interested in the *ratio* of fluxes, which is, after all, something that is experimentally observable, then the overall normalization of the wave function is irrelevant. Consequently, we shall simply set the normalization constant $N = 1$.

Let us, then, compute the probability flux corresponding to the incident

[11]The reader should convince himself or herself that this is indeed an eigenstate of momentum.

wave in Eq. (4.77)

$$S_x^{\rm inc} = \frac{\hbar}{2im}\left[\psi^\star \frac{d\psi}{dx} - \left(\frac{d\psi}{dx}\right)^\star \psi\right]_{\rm inc}$$
$$= \frac{\hbar}{2im}\left[e^{-ikx}\left(ike^{ikx}\right) - \left(ike^{ikx}\right)^\star e^{ikx}\right] = \frac{\hbar k}{m} \tag{4.78}$$

Hence, the incident probability flux can be written as

$$S_x^{\rm inc} = |\psi_{\rm inc}|^2 \frac{\hbar k}{m} = \rho\, v_{\rm particle} \qquad ;\ \text{incident flux} \tag{4.79}$$

where $|\psi_{\rm inc}|^2 \equiv \rho$. This has a nice, familiar interpretation and analogy with classical fluid mechanics.

Now include an additional potential barrier of the form

$$\begin{aligned} V &= 0 \qquad ;\ x < 0 \qquad ;\ \text{region I} \\ &= V_0 \qquad ;\ x > 0 \qquad ;\ \text{region II} \end{aligned} \tag{4.80}$$

Assume to start with that the incident energy is greater than the barrier height $E_{\rm inc} > V_0$ as illustrated in Fig. 4.17.

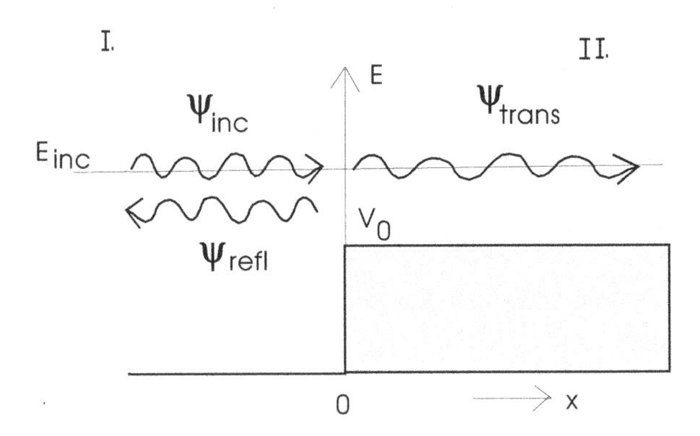

Fig. 4.17 Transmission and reflection of a prepared state from a potential barrier with $E_{\rm inc} \equiv E > V_0$. In addition to the incident wave $\psi_{\rm inc} = e^{ikx}$ in region I, there will be a reflected wave re^{-ikx}. There will also be a transmitted wave $t\,e^{i\kappa x}$ with $\hbar^2\kappa^2/2m = E - V_0$ in region II.

There will now be an additional *reflected wave* re^{-ikx} in region I, where r is an amplitude, yet to be determined. The probability flux of this reflected wave is calculated exactly as above (note the sign)

$$\begin{aligned} \psi_{\text{refl}} &= re^{-ikx} && \text{; reflected wave in I} \\ S_x^{\text{refl}} &= -|r|^2 \rho\, v_{\text{particle}} && \text{; reflected flux} \end{aligned} \tag{4.81}$$

By linear superposition, the *total wave function in region I is the sum of the incident and reflected waves*[12]

$$\psi_I = \psi_{\text{inc}} + \psi_{\text{refl}} \qquad \text{; total wave in I} \tag{4.82}$$

In addition to the reflected wave in region I, there will be a *transmitted wave* in region II of the form $\psi_{\text{trans}} = t\, e^{i\kappa x}$, where $\hbar^2\kappa^2/2m \equiv E - V_0$. The transmitted flux is calculated exactly as above

$$\begin{aligned} \psi_{\text{trans}} &= t\, e^{i\kappa x} && \text{; transmitted wave in II} \\ S_x^{\text{trans}} &= \frac{\hbar\kappa}{m}|t|^2 \rho && \text{; transmitted flux} \\ \frac{\hbar^2\kappa^2}{2m} &\equiv E - V_0 \end{aligned} \tag{4.83}$$

4.4.2.2 *Reflection and Transmission Coefficients*

We are now in a position to define *transmission and reflection coefficients* as the magnitude of the ratios of the transmitted and reflected fluxes to the incident flux

$$\begin{aligned} \mathcal{T} &\equiv \left| \frac{S_x^{\text{trans}}}{S_x^{\text{inc}}} \right| = \frac{\kappa}{k}|t|^2 && \text{; transmission coefficient} \\ \mathcal{R} &\equiv \left| \frac{S_x^{\text{refl}}}{S_x^{\text{inc}}} \right| = |r|^2 && \text{; reflection coefficient} \end{aligned} \tag{4.84}$$

Some comments:

- The overall wave function normalization clearly cancels in these ratios, as advertised;
- These coefficients give the *ratio* of probability fluxes for a single particle;

[12]This is a very important point, and the reader is asked to ponder it carefully. The incident and reflected fluxes can be separated by simply pointing a detector in the right direction; however, it is the *total* wave function that satisfies the Schrödinger equation, and which must satisfy the boundary conditions.

- If one repeats the experiment over and over again with $N_{\rm inc}$ particles, then the number of reflected and transmitted particles will be given by

$$
\begin{aligned}
\frac{N_{\rm refl}}{N_{\rm inc}} &= \mathcal{R} = |r|^2 \\
\frac{N_{\rm trans}}{N_{\rm inc}} &= \mathcal{T} = \frac{\kappa}{k}|t|^2
\end{aligned}
\tag{4.85}
$$

- These are just the quantities one *measures in a scattering experiment*!

In order to determine the amplitudes (r, t), it is necessary to satisfy the boundary conditions between regions I and II. The wave function in the two regions is given by

$$
\begin{aligned}
\psi_I &= \psi_{\rm inc} + \psi_{\rm refl} = e^{ikx} + r\,e^{-ikx} & &; \ x \leq 0 \\
\psi_{II} &= \psi_{\rm trans} = t\,e^{i\kappa x} & &; \ x \geq 0
\end{aligned}
\tag{4.86}
$$

Since the potential is finite, the boundary conditions from Eq. (4.66) are that the *wave function and its derivative must be continuous at the interface.* Thus

$$
\begin{aligned}
\psi_I(0) &= \psi_{II}(0) & &\Rightarrow & 1 + r &= t \\
\psi_I'(0) &= \psi_{II}'(0) & &\Rightarrow & ik(1 - r) &= i\kappa t
\end{aligned}
\tag{4.87}
$$

Division of the second equations by ik, and addition to the first, leads to

$$
2 = \left(1 + \frac{\kappa}{k}\right) t \tag{4.88}
$$

Hence

$$
t = \frac{2k}{k + \kappa} \qquad ; \ r = \frac{k - \kappa}{k + \kappa} \tag{4.89}
$$

where the second equation follows immediately from the first of Eqs. (4.87).

The reflection and transmission coefficients are then given by Eqs. (4.84)

$$
\mathcal{R} = \frac{(k - \kappa)^2}{(k + \kappa)^2} \qquad ; \ \mathcal{T} = \frac{4\kappa k}{(k + \kappa)^2} \tag{4.90}
$$

Evidently

$$
\mathcal{R} + \mathcal{T} = 1 \qquad ; \ \text{conservation of probability} \tag{4.91}
$$

This is simply the statement of conservation of probability — if a particle goes in, it must come out somewhere.

So far it has been assumed that $E_{\rm inc} = E > V_0$. What happens if the incident energy is *less* than the barrier height so that $E_{\rm inc} = E < V_0$? In this case $\kappa^2 < 0$, and there are two roots for κ in the last of Eqs. (4.83), $\pm i|\kappa|$. On physical grounds, one should retain only the root $+i|\kappa|$, so that there is a *decreasing* wave under the potential (see Fig. 4.18). This is equivalent to making the following replacements in the above analysis

$$\kappa \rightarrow +i\kappa \equiv +i\left[\frac{2m(V_0 - E)}{\hbar^2}\right]^{1/2} \qquad ; \; E = E_{\rm inc} < V_0$$
$$\psi_{\rm trans} \rightarrow t\, e^{-\kappa x} \tag{4.92}$$

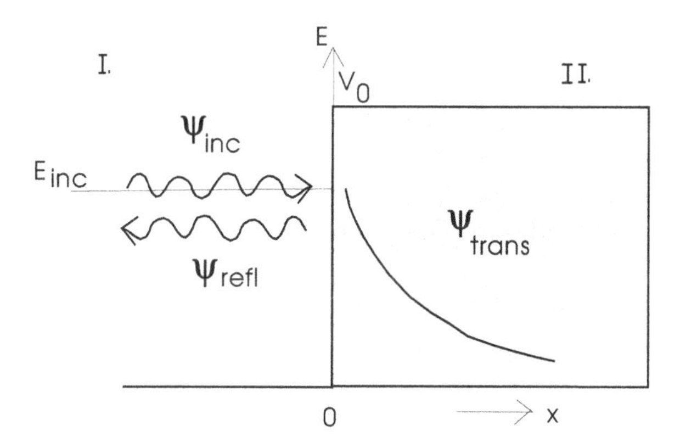

Fig. 4.18 Transmission and reflection of a prepared state from a potential barrier with $E_{\rm inc} = E < V_0$. In addition to the incident wave $\psi_{\rm inc} = e^{ikx}$ in region I, there will be a reflected wave re^{-ikx}. There will also be a transmitted wave $t\,e^{-\kappa x}$ with $\kappa = \sqrt{2m(V_0 - E)/\hbar^2}$ in region II.

Since the transmitted wave function is now real, the transmitted probability flux vanishes, and hence in this case

$$S_x^{\rm trans} = 0 \qquad\qquad \Rightarrow \quad \mathcal{T} = 0 \tag{4.93}$$

The amplitude of the reflected wave is now $r = (k - i\kappa)/(k + i\kappa)$, which has magnitude $|r| = 1$. It may therefore be written as a pure phase

$$r = \frac{k - i\kappa}{k + i\kappa} \equiv e^{i\phi} \qquad \Rightarrow \quad \mathcal{R} = 1 \tag{4.94}$$

Thus, in this case, one has pure reflection of the incident wave at the potential barrier, with a phase shift depending on the incident energy.

But now observe that the solution to this problem has yielded a non-zero $\psi_{\rm trans}$ inside the barrier! After a distance $x = d$ into the barrier, one has

$$\left|\psi_{\rm trans}(d)\right|^2 = \left|\frac{2k}{k+i\kappa}\right|^2 \exp\left\{-2d\left[\frac{2m(V_0-E)}{\hbar^2}\right]^{1/2}\right\} \quad ; \; x = d \qquad (4.95)$$

We make two observations concerning this result:

(1) One now finds the particle in a region where it is absolutely *forbidden to be classically*, since classically, the particle's energy is given by

$$E = \frac{1}{2}m\dot{x}^2 + V(x)$$
$$\Rightarrow \qquad [E - V(x)] = \frac{1}{2}m\dot{x}^2 \geq 0 \qquad (4.96)$$

The kinetic energy is non-negative, which leads to the last inequality.

(2) If the barrier is of finite width, the solution to the problem, obtained by solving the Schrödinger equation in three different regions and matching boundary conditions (see Prob. 4.16), implies that there will be a non-zero transmitted flux. Hence the particle can *penetrate through a barrier* in quantum mechanics, even though it is forbidden to even get into the barrier in classical mechanics.[13]

There are many application of barrier penetration observed today in physics, for example[14]

- Electrons in a wire made by joining different materials;
- Electrons in a channel with a voltage step;
- α-particle emission from nuclei through the Coulomb barrier; etc.

4.4.3 *Boundary Condition at a Wall*

We are now in a position to derive the boundary condition at a wall, which was stated previously without proof in Eq. (4.67). Let $V_0 \to \infty$ for any

[13]Note that barrier penetration arises here from the rather innocuous-looking boundary conditions on the wave function imposed in Eqs. (4.87). One might have been tempted to say that since classical physics prevents barrier penetration, one should really impose the boundary condition that $\psi = 0$ at an interface where $V_0 > E_{\rm inc}$. This, however, would force one to deal with a slope that is discontinuous at the boundary (compare Fig. 4.14). Again, simple classical considerations on the boundary conditions can cause one to miss important physics.

[14]See any good existing book on modern physics for a discussion of applications of barrier penetration, for example [Ohanian (1995)].

fixed distance d inside the barrier in Eq. (4.95). It is evident that

$$\psi_{II}(d) \rightarrow 0 \qquad ; \; V_0 \rightarrow \infty \qquad d > 0 \tag{4.97}$$

Thus the boundary condition at the wall is

$$\psi_{II}(0) = \psi_I(0) = 0 \qquad ; \text{ at wall} \tag{4.98}$$

This gives Eq. (4.67), as illustrated in Fig. 4.14. Note that in this same limit

$$r \rightarrow -1 \qquad ; \text{ same limit} \tag{4.99}$$

Thus there is total reflection of the incident wave at the wall, with a phase shift of π.

4.4.4 *Simple Harmonic Oscillator*

Consider the one-dimensional simple harmonic oscillator with a potential

$$V(x) = \frac{1}{2}\kappa x^2 \qquad ; \; \omega \equiv \left(\frac{\kappa}{m}\right)^{1/2} \qquad ; \text{ s.h.o.} \tag{4.100}$$

Again, first look for normal-mode, stationary-state solutions to the Schrödinger equation, which reduces that equation to its time-independent form

$$\Psi(x,t) = \psi(x)\exp\left\{-\frac{i}{\hbar}Et\right\}$$

$$\left\{\frac{d^2}{dx^2} + \frac{2m}{\hbar^2}[E - V(x)]\right\}\psi(x) = 0 \tag{4.101}$$

This differential equation, together with the appropriate boundary conditions, can be solved analytically. The actual solution is a technical matter, which we will leave for a later course in quantum mechanics; however, the analytic results can be found in any good text on the subject, for example [Schiff (1968)]. We can understand, qualitatively, the nature of the solutions by comparing with our analysis of a particle in a box. One wants those solutions that die off as $x \rightarrow \pm\infty$, and both ψ and ψ' must be continuous as one moves through the potential. There will be some barrier penetration, and the wave functions will leak through the potential barrier to some extent at either end. Thus the wave functions and probability densities can be sketched as in Fig. 4.19.

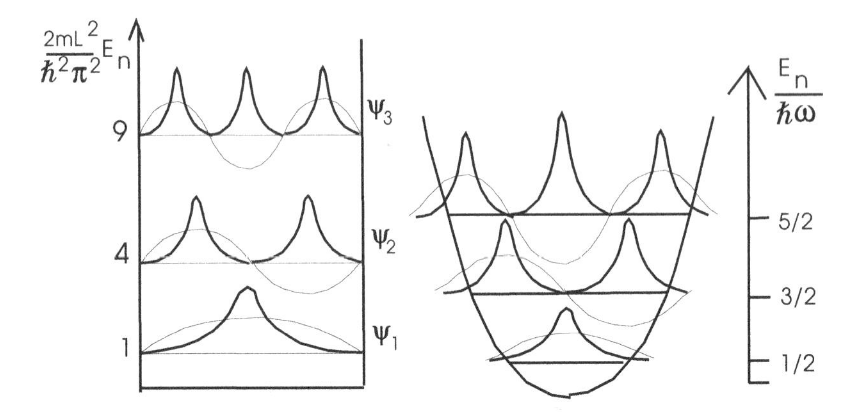

Fig. 4.19 Sketch of spectrum, wave functions, and probability densities for the simple harmonic oscillator as inferred from particle in a box.

There are only certain values of E for which the wave function will indeed die out in both directions, and the eigenvalue spectrum for the simple harmonic oscillator is

$$E_n = \hbar\omega\left(n + \frac{1}{2}\right) \qquad ; \; n = 0, 1, 2, \cdots, \infty \qquad ; \text{ s.h.o.} \quad (4.102)$$

This is exactly the spectrum of the s.h.o. assumed by Planck in his derivation in Eqs. (3.32)! In fact, one can obtain the harmonic oscillator spectrum without ever solving a differential equation. It follows directly using only general properties of the operators involved. The reader is strongly urged to work through this derivation in Probs. 4.17 and 4.18 — the results play an important role in subsequent developments.

4.5 Three Dimensions

So far, we have worked with the string and Schrödinger equation in one spatial dimension. It is essential now to extend this analysis to a higher number of dimensions, and in order to do this, we return to classical continuum mechanics.

4.5.1 *Classical Continuum Mechanics*

Membrane. Consider a two-dimensional membrane, and recall the definition of *surface tension*, $\sigma = \text{force/length}$, as the stretching force per unit length acting in the transverse direction in a stretched surface (Fig. 4.20).

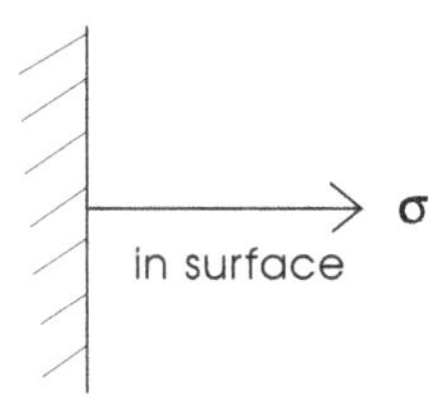

Fig. 4.20 Definition of surface tension, $\sigma = \text{force/length}$, as the stretching force per unit length acting in the transverse direction in a stretched surface.

Let μ be the mass density of the membrane. Then[15]

$$\begin{aligned} \sigma &= \text{force/length} \qquad &&; \text{ surface tension} \\ \mu &= \text{mass/area} \qquad &&; \text{ mass density} \end{aligned} \tag{4.103}$$

If the membrane is displaced by a distance $q(x, y, t)$ in the z-direction from its equilibrium, flat configuration in the (x, y)-plane (Fig. 4.21), then the mass times the acceleration in the z-direction of a small element of the membrane $dxdy$ is given by

$$\text{mass} \times \text{acceleration} = (\mu dxdy)\frac{\partial^2 q}{\partial t^2} \tag{4.104}$$

This little displaced element experiences a surface tension force F_x on both sides where F_x is perpendicular to the y-axis and lies in the surface, and a second force F_y on both sides where F_y is perpendicular to the x-axis and also lies in the surface (Fig. 4.20). The treatment of these components is just like that of the one-dimensional string in Eqs. (2.19)–(2.20), and these forces each give rise to an *additive* contribution to the net force F_z in the z-direction

$$\begin{aligned} F_z &= \sigma dy \left(\left.\frac{\partial q}{\partial x}\right|_{x+dx} - \left.\frac{\partial q}{\partial x}\right|_{x} \right) + \sigma dx \left(\left.\frac{\partial q}{\partial y}\right|_{y+dy} - \left.\frac{\partial q}{\partial y}\right|_{y} \right) \\ &= \sigma \, dxdy \left(\frac{\partial^2 q}{\partial x^2} + \frac{\partial^2 q}{\partial y^2} \right) \end{aligned} \tag{4.105}$$

[15] For simplicity, we assume that both of these quantities are constant.

The expressions in Eqs. (4.104) and (4.105) are to be equated according to Newton's second law, and with a cancelation of the factor of $dxdy$, one arrives at the *two-dimensional (2-D) wave equation*

$$\left(\frac{\partial^2}{\partial x^2}+\frac{\partial^2}{\partial y^2}\right)q(x,y,t)=\frac{1}{c^2}\frac{\partial^2 q(x,y,t)}{\partial t^2} \qquad ;\ \text{2-D wave equation}$$
$$c^2\equiv\frac{\sigma}{\mu} \qquad ;\ (\text{velocity})^2 \qquad (4.106)$$

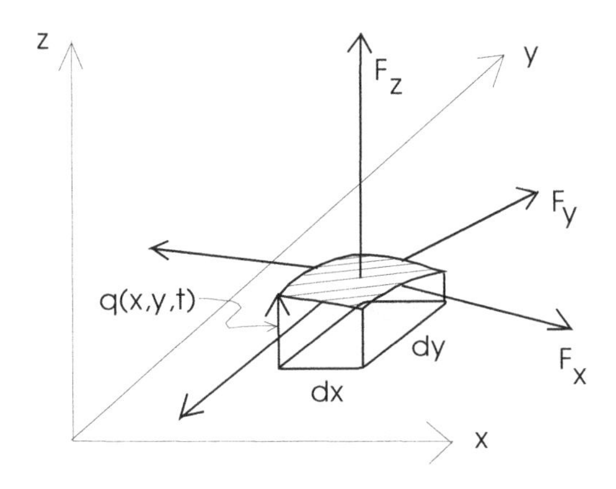

Fig. 4.21 Displacement $q(x,y,t)$ in the z-direction of a two-dimensional classical membrane which has a stretched, flat, equilibrium configuration in the (x,y)-plane. The restoring force comes from the surface tension acting in the surface of the displaced membrane. F_x acts in a direction perpendicular to the y-axis, and F_y in a direction perpendicular to the x-axis.

The generalization of the wave equation to any number of dimensions is now immediate

$$\nabla^2\, q(\mathbf{x},t)=\frac{1}{c^2}\frac{\partial q(\mathbf{x},t)}{\partial t^2} \qquad ;\ \text{wave equation} \qquad (4.107)$$

Here ∇^2 is the *laplacian*

$$\nabla^2=\frac{\partial^2}{\partial x^2} \qquad ;\ \text{1-D} \qquad ;\ \text{laplacian}$$
$$=\frac{\partial^2}{\partial x^2}+\frac{\partial^2}{\partial y^2} \qquad ;\ \text{2-D}$$
$$=\frac{\partial^2}{\partial x^2}+\frac{\partial^2}{\partial y^2}+\frac{\partial^2}{\partial z^2} \qquad ;\ \text{3-D} \qquad (4.108)$$

4.5.2 *Schrödinger Equation*

In direct analogy, the Schrödinger equation is now immediately extended to three dimensions as

$$i\hbar\frac{\partial\Psi(\mathbf{x},t)}{\partial t} = \left[-\frac{\hbar^2}{2m}\nabla^2 + V(x)\right]\Psi(\mathbf{x},t) \quad ; \text{ 3-D Schrödinger eqn} \tag{4.109}$$

If one seeks normal-mode, stationary-state solutions to this equation, it is again converted to its time-independent form

$$\Psi(\mathbf{x},t) = \psi(\mathbf{x})\exp\left\{-\frac{i}{\hbar}Et\right\} \qquad ; \text{ normal modes}$$

$$\left[\nabla^2 + \frac{2m}{\hbar^2}(E-V)\right]\psi(\mathbf{x}) = 0 \tag{4.110}$$

The generalization of the continuity equation, which provides the basis for our interpretation of the theory, is also directly obtained for real $V(x)$ as[16]

$$\begin{aligned} \rho &= |\Psi|^2 && ; \text{ probability density} \\ \mathbf{s} &= \frac{\hbar}{2im}\left[\Psi^\star\boldsymbol{\nabla}\Psi - (\boldsymbol{\nabla}\Psi^\star)\Psi\right] && ; \text{ probability current} \\ \frac{\partial\rho}{\partial t} &+ \boldsymbol{\nabla}\cdot\mathbf{s} = 0 && ; \text{ continuity eqn} \end{aligned} \tag{4.111}$$

Here $\boldsymbol{\nabla}$ is the gradient operator defined as

$$\boldsymbol{\nabla} = \hat{\mathbf{e}}_x\frac{\partial}{\partial x} + \hat{\mathbf{e}}_y\frac{\partial}{\partial y} + \hat{\mathbf{e}}_z\frac{\partial}{\partial z} \tag{4.112}$$

where $(\hat{\mathbf{e}}_x, \hat{\mathbf{e}}_y, \hat{\mathbf{e}}_z)$ are a set of orthonormal, cartesian unit vectors.

We observe that Eq. (4.109) has the form

$$i\hbar\frac{\partial\Psi(\mathbf{x},t)}{\partial t} = H(\mathbf{p},\mathbf{x})\Psi(\mathbf{x},t) \qquad ; \text{ Schrödinger eqn} \tag{4.113}$$

where the hamiltonian and momentum are given by

$$\begin{aligned} H(\mathbf{p},\mathbf{x}) &= \frac{\mathbf{p}^2}{2m} + V(x) && ; \text{ hamiltonian} \\ \mathbf{p} &= \frac{\hbar}{i}\boldsymbol{\nabla} && ; \text{ momentum} \end{aligned} \tag{4.114}$$

---

[16]See Prob. 4.20 (note that $\nabla^2 = \boldsymbol{\nabla}\cdot\boldsymbol{\nabla}$). Strictly speaking, $\mathbf{s}$ is now the current *density*, however, we shall always refer to $\mathbf{s}$ as the probability current (compare appendix K).

4.5.3 *Particle in a Three-Dimensional Box*

Consider the problem of a particle of mass m moving in the 3-D box in Fig. 4.22.

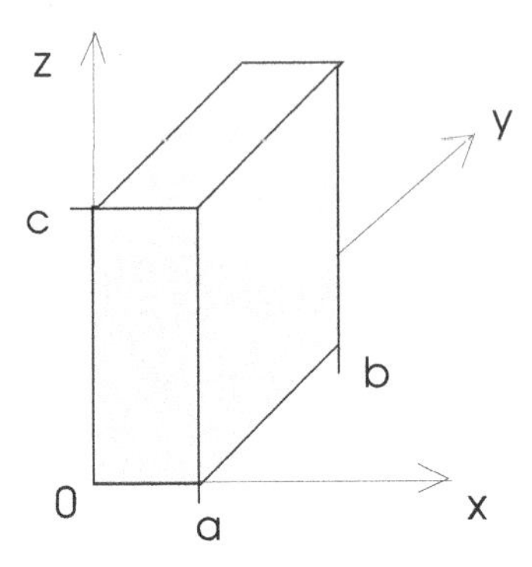

Fig. 4.22 Particle in a 3-D box with sides of length (a, b, c) in the (x, y, z) directions.

The wave function must vanish at each of the walls, and the normal-mode amplitudes are an immediate generalization of those in a 1-D box

$$\psi_{n_1 n_2 n_3}(\mathbf{x}) = \left(\frac{8}{abc}\right)^{1/2} \sin\frac{n_1 \pi x}{a} \sin\frac{n_2 \pi y}{b} \sin\frac{n_3 \pi z}{c}$$
$$; \mathbf{n} = (n_1, n_2, n_3) \qquad ; n_i = 1, 2, 3, \cdots, \infty \qquad (4.115)$$

These wave functions:

- Satisfy the boundary conditions, as they vanish at each of the walls;
- Satisfy the Schrödinger equation — just substitute into Eq. (4.110);
- The substitution into the Schrödinger equation allows one to identify the eigenvalues as

$$\frac{2m}{\hbar^2} E_{n_1 n_2 n_3} = \left(\frac{n_1^2}{a^2} + \frac{n_2^2}{b^2} + \frac{n_3^2}{c^2}\right) \pi^2 \qquad ; \text{eigenvalues} \qquad (4.116)$$

The wave functions, and corresponding probability densities, are sketched for two cases in Figs. 4.23 and 4.24. Note the *nodal plane* in the second case.

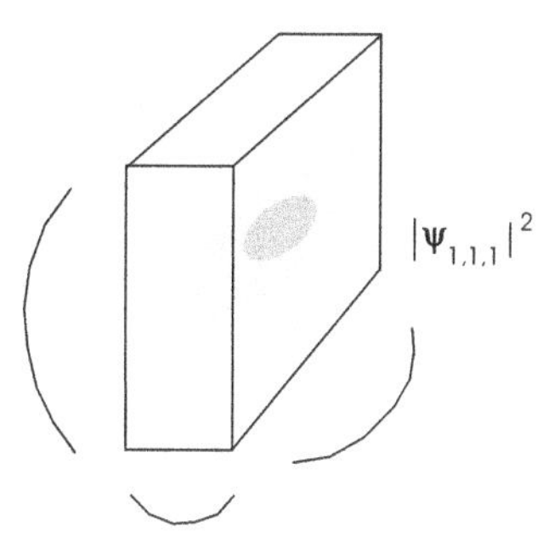

Fig. 4.23 Sketch of the wave function and probability density in the (1, 1, 1) ground state for a particle in the box in Fig. 4.22.

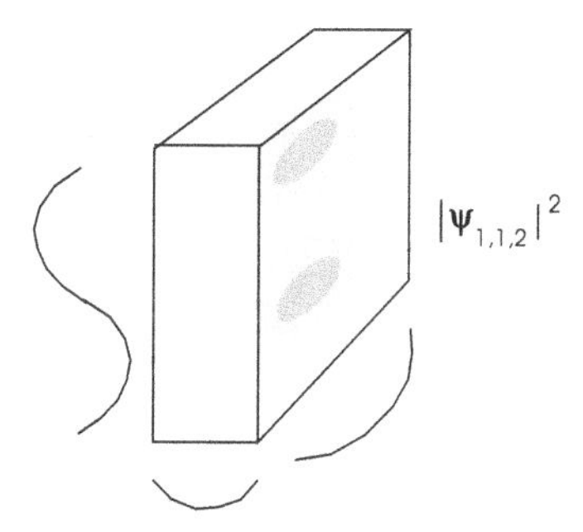

Fig. 4.24 Sketch of the wave function and probability density in the (1, 1, 2) excited state for a particle in the box in Fig. 4.22. Note the nodal plane at $z = c/2$ in this case.

4.5.4 *Free Particle — Periodic Boundary Conditions*

Consider a free particle of mass m moving in a cubical box of side L in three dimensions, where periodic boundary conditions are applied in all three (x, y, z) directions. The eigenfunctions are again an immediate extension of the one-dimensional case

$$\psi(\mathbf{x}) = \frac{1}{\sqrt{L^3}}\, e^{ik_x x} e^{ik_y y} e^{ik_z z} \tag{4.117}$$

To satisfy the periodic boundary conditions, one must have $k_i = 2\pi n_i/L$ with $n_i = 0, \pm 1, \pm 2, \cdots, \pm\infty$ in each direction. These solutions can then be written as

$$\psi_{\mathbf{k}}(\mathbf{x}) = \frac{1}{\sqrt{L^3}} e^{i\mathbf{k}\cdot\mathbf{x}} \qquad ; \mathbf{x} = (x, y, z)$$

$$\mathbf{k} = (k_x, k_y, k_z)$$

$$\mathbf{k} = \frac{2\pi}{L}(n_x, n_y, n_z) \qquad ; n_i = 0, \pm 1, \pm 2, \cdots, \pm\infty \qquad (4.118)$$

The eigenvalues are obtained by substituting this expression into the time-independent Schrödinger Eq. (4.110) for a free particle with $V = 0$

$$E_k = \frac{\hbar^2 \mathbf{k}^2}{2m} \qquad ; \text{eigenvalues} \qquad (4.119)$$

Evidently, the particle's momentum is given by

$$\mathbf{p} = \hbar\mathbf{k} \qquad ; \text{momentum} \qquad (4.120)$$

in this case.

4.6 Comments on the Structure of Quantum Mechanics

We summarize the results obtained so far concerning the structure of quantum mechanics.[17]

Energy. The search for normal modes [Eq. (4.68)] of the time-dependent Schrödinger Eq. (4.65) leads to the *time-independent Schrödinger equation*

$$H(p, x)\psi(x) = E\psi(x) \qquad ; \text{time-independent S-eqn} \qquad (4.121)$$

We make several observations:

- The momentum in the hamiltonian $H(p, x)$ is to be replaced by the differential operator

$$p \rightarrow \frac{\hbar}{i}\frac{\partial}{\partial x} \qquad ; \text{differential operator} \qquad (4.122)$$

- The time-independent Schrödinger Eq. (4.121) then becomes a differential equation;
- Appropriate *boundary conditions* [Eqs. (4.66)–(4.67)] are to be applied to this differential equation;
- The resulting *eigenvalues* E are the *observed energies*;
- The corresponding $\psi(x)$ are the *eigenfunctions*;

[17]Many of these topics are examined in more detail in Probs. 4.3–4.12.

- In these stationary states, the probability density and probability current are *independent of time* [Eq. (4.70)];
- The general solution to the time-dependent Schrödinger equation is then obtained as a *linear superposition* of normal-modes.

Momentum. Plane waves form the *eigenstates of momentum*

$$\begin{aligned} \psi_k(x) &= \frac{1}{\sqrt{L}} e^{ikx} \\ p\,\psi_k(x) &= \hbar k\,\psi_k(x) \end{aligned} \tag{4.123}$$

Here

- p is the differential operator in Eq. (4.122);
- The eigenvalues $\hbar k$ are the *observed momenta*;
- In three dimensions, these relations take the form

$$\begin{aligned} \psi_{\mathbf{k}}(\mathbf{x}) &= \frac{1}{\sqrt{L^3}} e^{i\mathbf{k}\cdot\mathbf{x}} \\ \mathbf{p}\,\psi_{\mathbf{k}}(\mathbf{x}) &= \hbar\mathbf{k}\,\psi_{\mathbf{k}}(\mathbf{x}) \end{aligned} \tag{4.124}$$

- In this expression

$$\mathbf{p} = \frac{\hbar}{i}\boldsymbol{\nabla} \tag{4.125}$$

  where $\boldsymbol{\nabla}$ is the gradient operator in Eq. (4.112);
- Periodic boundary conditions lead to the allowed wavenumbers $\mathbf{k}$ given in the last of Eqs. (4.118).

4.7 Angular Momentum

The angular momentum in quantum mechanics is defined as[18]

$$\begin{aligned} \boldsymbol{\mathcal{L}} &= \mathbf{r}\times\mathbf{p} = \frac{\hbar}{i}\mathbf{r}\times\boldsymbol{\nabla} \\ &\equiv \hbar\,\mathbf{L} \end{aligned} \tag{4.126}$$

---

[18] The reference [Edmonds (1974)] provides an invaluable guide to the theory of angular momentum in quantum mechanics; see also appendix B in [Fetter and Walecka (2003a)].

The quantity **L** in the second line defines the angular momentum in units of $\hbar$. Thus

$$\mathbf{L} = \frac{1}{i}\mathbf{r} \times \boldsymbol{\nabla} \tag{4.127}$$

Two-Dimensional Motion. Consider motion in the (x, y) plane. Introduce polar coordinates (Fig. 4.25)

$$x = \rho\cos\phi \qquad ; \; y = \rho\sin\phi \tag{4.128}$$

The z-component of angular momentum then takes the form

$$L_z = \frac{1}{i}\left(x\frac{\partial}{\partial y} - y\frac{\partial}{\partial x}\right) \tag{4.129}$$

The chain rule of differentiation gives

$$\begin{aligned}\frac{\partial}{\partial\phi}f(x,y) &= \frac{\partial f}{\partial x}\frac{\partial x}{\partial\phi} + \frac{\partial f}{\partial y}\frac{\partial y}{\partial\phi} \\ &= -\rho\sin\phi\,\frac{\partial f}{\partial x} + \rho\cos\phi\,\frac{\partial f}{\partial y} \\ &= x\frac{\partial f}{\partial y} - y\frac{\partial f}{\partial x}\end{aligned} \tag{4.130}$$

Hence, the z-component of the angular momentum is given in polar coordinates by

$$L_z\,\psi(\rho,\phi) = \frac{1}{i}\frac{\partial}{\partial\phi}\psi(\rho,\phi) \tag{4.131}$$

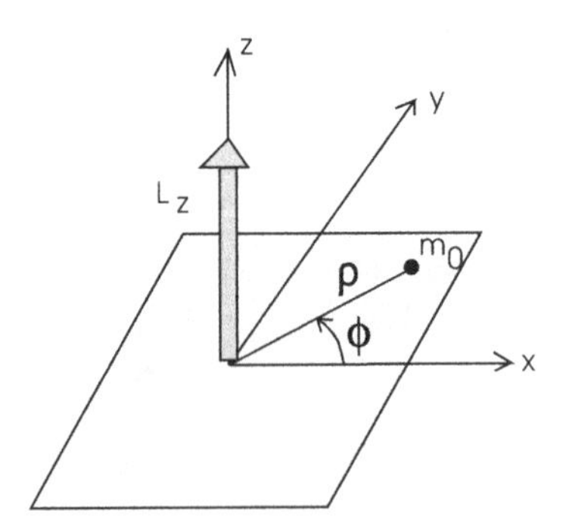

Fig. 4.25 Angular momentum associated with motion in the two-dimensional (x, y) plane. The particle's rest mass is now denoted by m_0.

Suppose one has a central-force problem in two dimensions with a potential $V(\rho)$. It is then appropriate to look for a solution to the time-independent Schrödinger equation in *separated form*

$$\psi(\rho, \phi) = R(\rho)\Phi(\phi) \qquad \text{; separated form with } V(\rho) \qquad (4.132)$$

To determine the effect of L_z on this wave function, one must examine

$$L_z\,\Phi(\phi) = \frac{1}{i}\frac{\partial}{\partial\phi}\Phi(\phi) \qquad \text{; angular wave function} \qquad (4.133)$$

The solutions to Eq. (4.133), the *eigenfunctions* in ϕ, are given by

$$\Phi(\phi) = \frac{1}{\sqrt{2\pi}}\,e^{im\phi} \qquad \text{; eigenfunctions} \qquad (4.134)$$

If one applies *periodic boundary conditions* to these eigenfunctions (see Fig. 4.25), then[19]

$$\Phi(\phi + 2\pi) = \Phi(\phi) \qquad \text{; periodic boundary conditions}$$
$$\Rightarrow \qquad m = 0, \pm 1, \pm 2, \cdots \qquad (4.135)$$

It follows that

$$L_z\,\Phi(\phi) = m\,\Phi(\phi)$$
$$\mathcal{L}_z = \hbar m \qquad \text{; eigenvalues} \qquad (4.136)$$

We observe that this result:

- Gives the quantization of angular momentum;
- Was one of Bohr's hypotheses;
- Provides a proper version of DeBroglie's argument in Eqs. (4.3).

Three-Dimensional Motion. For the central-force problem in three dimensions with a potential $V(r)$, one goes to spherical coordinates (Fig. 4.26). The wave function can again be written in separated form

$$\psi(r, \theta, \phi) = R(r)Y_{lm}(\theta, \phi) \qquad \text{; separated solution} \qquad (4.137)$$

Spherical Harmonics. While the radial wave functions $R(r)$ depend on the form of the potential $V(r)$, the well-behaved angular wave functions, the *spherical harmonics* $Y_{lm}(\theta, \phi)$, are common to any central-force problem. Their derivation is a technical matter, which students will see in more advanced courses; however, their properties can be found in any book on

[19]See Prob. 4.27 for a more thorough discussion of the boundary conditions here.

classical mechanics (see [Fetter and Walecka (2003)]), or quantum mechanics (see [Schiff (1968); Edmonds (1974)]), and the student should become familiar with them.

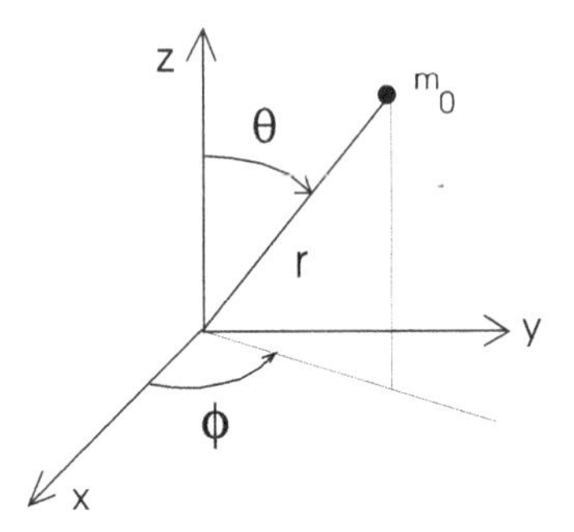

Fig. 4.26 Spherical coordinates for the central force problem with $V(r)$ in three dimensions. The particle's rest mass is now denoted by m_0.

The quantum numbers (l, m) are integers, and the spherical harmonics satisfy the following equations (see Fig. 4.27)

$$\begin{aligned} \mathbf{L}^2\, Y_{lm}(\theta,\phi) &= l(l+1)\, Y_{lm}(\theta,\phi) && ;\ l = 0, 1, 2, \cdots, \infty \\ L_z\, Y_{lm}(\theta,\phi) &= m\, Y_{lm}(\theta,\phi) && -l \leq m \leq l \end{aligned} \tag{4.138}$$

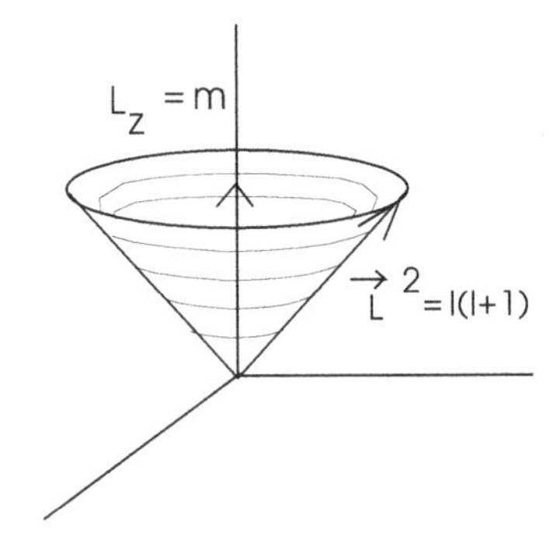

Fig. 4.27 Illustration of quantum numbers appearing in the spherical harmonics $Y_{lm}(\theta,\phi)$. There is an uncertainty relation between the components of the angular momentum (L_x, L_y, L_z), and only one of them, here L_z, can be precisely specified along with $\mathbf{L}^2$ (see Prob. 4.25). One can picture the angular momentum, of magnitude $|\mathbf{L}| = \sqrt{l(l+1)}$, as precessing about the z-axis.

The first few spherical harmonics are given by

$$
\begin{aligned}
Y_{00} &= \frac{1}{\sqrt{4\pi}} & ; \; Y_{20} &= \sqrt{\frac{5}{16\pi}}\,(3\cos^2\theta - 1) \\
Y_{10} &= \sqrt{\frac{3}{4\pi}}\cos\theta & ; \; Y_{2,\pm 1} &= \mp\sqrt{\frac{15}{8\pi}}\sin\theta\cos\theta\, e^{\pm i\phi} \\
Y_{1,\pm 1} &= \mp\sqrt{\frac{3}{8\pi}}\sin\theta\, e^{\pm i\phi} & ; \; Y_{2,\pm 2} &= \sqrt{\frac{15}{32\pi}}\sin^2\theta\, e^{\pm 2i\phi}
\end{aligned}
\tag{4.139}
$$

The spherical harmonics are orthonormal

$$
\int_0^{2\pi} d\phi \int_0^{\pi} \sin\theta\, d\theta\, Y^{\star}_{lm}(\theta,\phi) Y_{l'm'}(\theta,\phi) = \delta_{ll'}\delta_{mm'} \tag{4.140}
$$

They are also complete for well-behaved functions on the unit sphere.

Radial Equation. Assume a central potential $V(r)$. Define

$$
R(r) \equiv \frac{u(r)}{r} \qquad ; \text{ radial wave function} \tag{4.141}
$$

The radial equation for $u(r)$ that follows from the Schrödinger Eq. (4.110) and the wave function in Eq. (4.137) is then given by[20]

$$
\frac{d^2u(r)}{dr^2} + \left\{\frac{2m_0}{\hbar^2}[E - V(r)] - \frac{l(l+1)}{r^2}\right\} u(r) = 0 \;\; ; \text{ radial equation} \tag{4.142}
$$

where m_0 now denotes the particle's rest mass. The term in $l(l+1)/r^2$ provides the *angular momentum barrier* familiar from classical mechanics.

S-Waves $(l = 0)$. For s-waves with $l = 0$, the angular momentum barrier disappears, and the radial equation reduces to

$$
\frac{d^2u(r)}{dr^2} + \left\{\frac{2m_0}{\hbar^2}[E - V(r)]\right\} u(r) = 0 \qquad ; \text{ s-waves } (l = 0) \tag{4.143}
$$

Since a behavior $R(r) \propto \text{constant}/r$ is too singular as $r \to 0$,[21] one has the following boundary condition on $u(r)$

$$
u(0) = 0 \qquad ; \text{ B.C. at } r = 0 \tag{4.144}
$$

Equation (4.143) is just a one-dimensional problem again, for the function $u(r)$ in the radial coordinate r, where $0 \leq r \leq \infty$, and Eq. (4.144) presents a "wall" boundary condition on $u(r)$ at $r = 0$.

---

[20]The expression for the laplacian in spherical coordinates, and the corresponding derivation of the radial part of the Schrödinger Eq. (4.110), are discussed in Prob. 4.26.

[21]See Prob. 4.27.

4.8 Point Coulomb Potential

Consider the solutions to the radial Eq. (4.142) in the case of an attractive point Coulomb potential

$$V(r) = -\frac{Ze^2}{4\pi\varepsilon_0}\frac{1}{r} \qquad ; \text{ point Coulomb potential} \qquad (4.145)$$

This forms the basis for the *hydrogen atom*, and is the starting point for all of *atomic physics*. We make several observations concerning the solutions in this case:

- We here examine the *bound states* with $E < 0$

$$E < 0 \qquad ; \text{ bound states} \qquad (4.146)$$

- The behavior of an acceptable solution as $r \to 0$, where the angular momentum barrier dominates, as well as that for $r \to \infty$, where both the potential and angular momentum barrier are negligible, can be determined directly from the differential Eq. (4.142)

$$\begin{aligned} u(r) &\propto r^{l+1} && ; r \to 0 \\ &\propto \exp\left\{-\left(\frac{2m_0|E|}{\hbar^2}\right)^{1/2} r\right\} && ; r \to \infty \end{aligned} \qquad (4.147)$$

- One can now simply integrate the differential equation numerically, starting at the origin and integrating outward. It will only be possible to match on to this exponentially decreasing radial function for certain *eigenvalues* of the energy $E_{\bar{n}l}$;
- In fact, this problem can be solved analytically. The analytic solutions present a nice exercise in special functions, and they can be found in any good book on quantum mechanics, for example [Schiff (1968)];[22]
- We recall that in newtonian mechanics the bound orbits in a $1/r$ potential are *ellipses* with the origin at a focus (Fig. 4.28), and there is a *dynamical degeneracy* in the classical problem in that the energy of the bound state depends only on the semi-major axis of the ellipse and not on the angular momentum (see [Fetter and Walecka (2003)])

$$E = -\frac{Ze^2}{4\pi\varepsilon_0}\frac{1}{2a} \qquad ; \text{ classical result} \quad a \text{ is semi-major axis} \qquad (4.148)$$

[22]The analytic results for a few of the low-lying states are examined in Prob. 4.28.

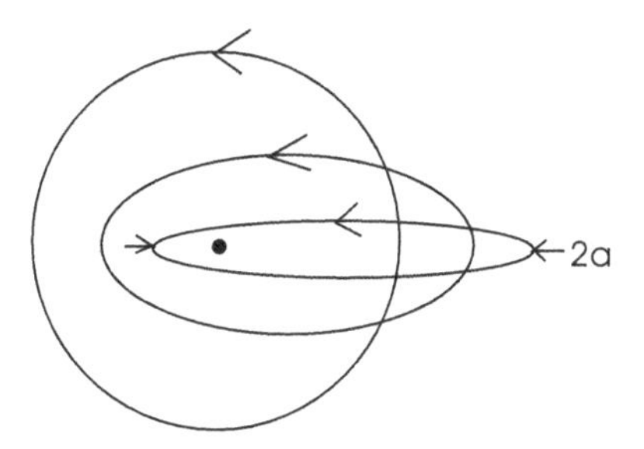

Fig. 4.28 Bound elliptical orbits in newtonian mechanics in the potential of Eq. (4.145). The origin lies at a focus of the ellipse. There is a dynamical degeneracy, as all orbits with the same semi-major axis have the same energy.

This degeneracy shows up in quantum mechanics in that one finds the same eigenvalue for several different values of the angular momentum quantum number l;[23]

- The eigenvalue spectrum in the point Coulomb potential is the following

$$E_{\bar{n}l} = -\frac{1}{2}\frac{Z^2\alpha^2}{\bar{n}^2}m_0c^2 \qquad ; \ \bar{n} = 1, 2, 3, \cdots, \infty$$
$$l = 0, 1, 2, \cdots, \bar{n}-1 \tag{4.149}$$

- The states with the maximum value of l for a given $\bar{n}$ represent the *circular orbits*

$$l = \bar{n} - 1 \qquad ; \text{ circular orbits} \tag{4.150}$$

- We note that Eq. (4.149) is exactly the Bohr formula of Eq. (3.65)!
- The *eigenfunctions* corresponding to the eigenvalues in Eqs. (4.149) will be labeled by

$$\psi_{\bar{n}lm}(r, \theta, \phi) = R_{\bar{n}l}(r)Y_{lm}(\theta, \phi) \tag{4.151}$$

The radial equation is independent of m and only depends on l; $\bar{n}$ is the additional radial quantum number;[24]

- The spectrum, and a few of the radial wave functions, are sketched in Fig. 4.29.

---

[23]The degeneracy in m with $-l \leq m \leq l$ is present in any three-dimensional central-force problem in quantum mechanics; the eigenvalues do not depend on the *orientation* of the angular momentum in space (see Fig. 4.27).

[24]The spectroscopic notation for states with $l = 0, 1, 2, 3, 4, 5, 6 \cdots$ is $s, p, d, f, g, h, i, \cdots$ *etc.*

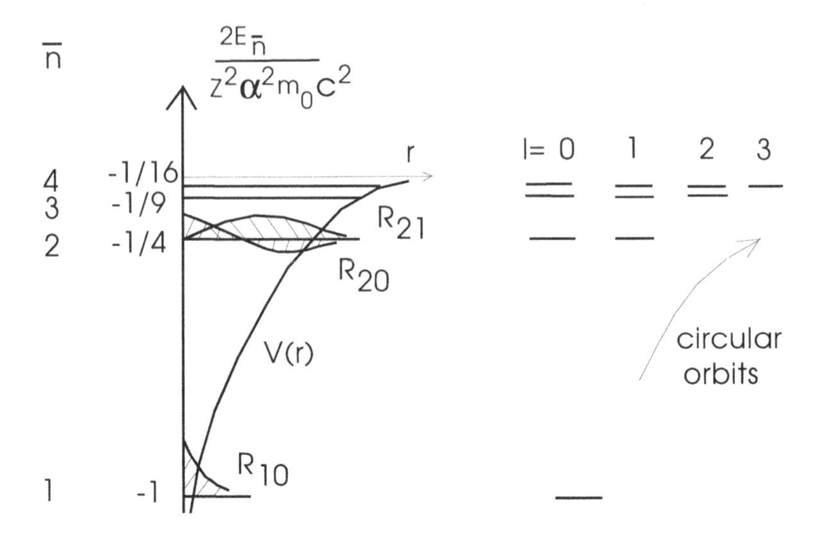

Fig. 4.29 Sketch of the eigenvalue spectrum, dynamical degeneracy, and first few radial wave functions for the attractive point Coulomb potential. Recall that each of the l-states has an additional degeneracy in m with $-l \leq m \leq l$.

- The eigenfunctions satisfy the following orthonormality relation

$$\int_0^\infty r^2 dr \int_0^\pi \sin\theta \, d\theta \int_0^{2\pi} d\phi \;\psi^\star_{\bar{n}lm}(r,\theta,\phi)\psi_{\bar{n}'l'm'}(r,\theta,\phi) = \delta_{\bar{n}\bar{n}'}\delta_{ll'}\delta_{mm'} \tag{4.152}$$

- It is evident from the first of Eqs. (4.147) that only the s-states with $l = 0$ will yield a finite value of $R = u/r$ at the origin. The analytic expression for the square of the corresponding (normalized) wave function at the origin is [Schiff (1968)]

$$|\psi_{\bar{n}00}(0)|^2 = \frac{1}{\pi}\left(\frac{Z\alpha}{\bar{n}}\right)^3\left(\frac{m_0c}{\hbar}\right)^3 \tag{4.153}$$

This result has many important applications.

4.9 Spin

The concept of spin in quantum mechanics is due to Pauli (1924) and Goudsmit and Uhlenbeck (1925). It was observed that many particles have an *intrinsic angular momentum* $\mathbf{s}$, exactly analogous to the orbital angular momentum discussed previously. The quantities $(\mathbf{s}^2, s_z)$ for a particle take the values

$$\begin{aligned}\mathbf{s}^2 &= s(s+1) && ;\ \text{spin-s} \\ s_z &= -s, -s+1, \cdots, +s && \end{aligned} \tag{4.154}$$

where s is either an integer or a half-integer. There is an accompanying *intrinsic magnetic moment* that, for a particle with mass and charge (m_0, e), is given in units of the appropriate *magneton* by

$$\begin{aligned}\boldsymbol{\mu} &= g_s \frac{e}{|e|}\mu_M \mathbf{s} && ;\ \text{intrinsic magnetic moment} \\ \mu_M &= \frac{|e|\hbar}{2m_0} && ;\ \text{magneton}\end{aligned} \tag{4.155}$$

Here g_s is known as the *gyromagnetic ratio.* The two most important magnetons are[25]

$$\begin{aligned}\frac{|e|\hbar}{2m_e} &= 9.274 \times 10^{-24}\,\text{J/Tesla} && ;\ \text{Bohr magneton} \\ \frac{|e|\hbar}{2m_p} &= 5.051 \times 10^{-27}\,\text{J/Tesla} && ;\ \text{nucleon magneton}\end{aligned} \tag{4.156}$$

The interaction energy of the intrinsic moment with a magnetic field $\mathbf{B}$ that determines the z-direction is given by (see Fig. 4.30)

$$\begin{aligned}H' &= -\boldsymbol{\mu}\cdot\mathbf{B} \\ &= -g_s\frac{e}{|e|}\mu_M B s_z\end{aligned} \tag{4.157}$$

The magnetic moment can be measured, for example, by determining the frequency of a radiative transition between two states with different s_z in a known magnetic field. For a spin-1/2 particle, this frequency is given by (Fig. 4.31)

$$\Delta E = h\nu = g_s \mu_M B \tag{4.158}$$

[25] It is also useful to have these magnetons in cgs units

$$\begin{aligned}|e|\hbar/2m_e c &= 9.274 \times 10^{-21}\,\text{erg/Gauss} \\ |e|\hbar/2m_p c &= 5.051 \times 10^{-24}\,\text{erg/Gauss}\end{aligned}$$

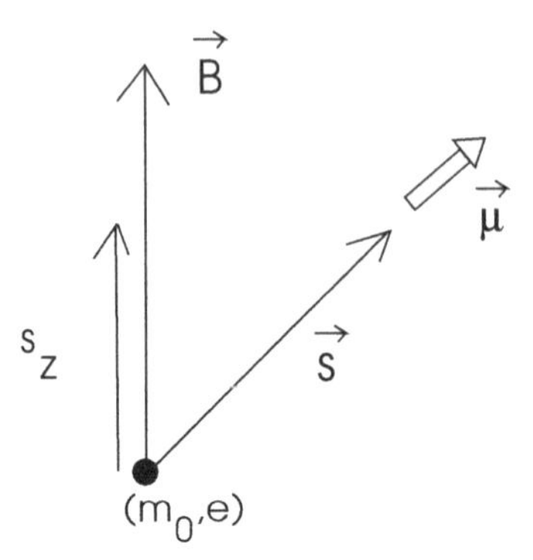

Fig. 4.30 Interaction of a particle with mass and charge (m_0, e), spin $\mathbf{s}$, and intrinsic magnetic moment $\boldsymbol{\mu}$, with a magnetic field $\mathbf{B}$ that determines the z-direction.

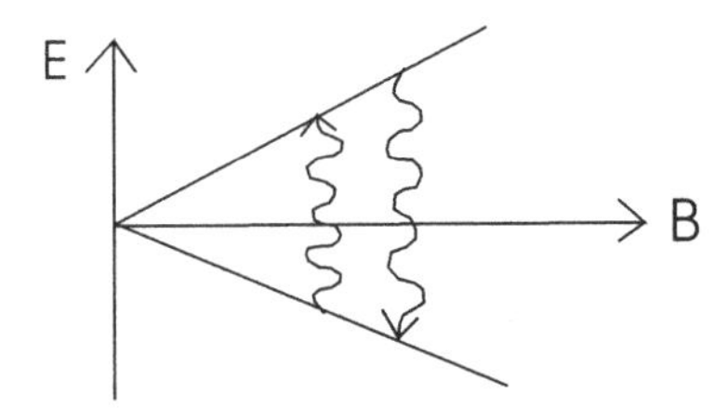

Fig. 4.31 Radiative transitions between the states with $s_z = \pm 1/2$ for a spin-1/2 particle in a magnetic field B. Here E is the energy of the state.

Some measured values of the g-factor are given in Table 4.1.

Table 4.1 Some measured g-factors as defined in Eq. (4.155). Note the second line is sg_s. For the neutron $e/|e| \equiv 1$ and $\mu_{\rm M}$ is the nucleon magneton.

Particle	Electron	Proton	Neutron
s	$1/2$	$1/2$	$1/2$
sg_s	$1.001, 159, 652, 187, \cdots$	$2.792, 847, \cdots$	$-1.913, 042, \cdots$

One of the great achievements in modern physics is the calculation of the g-factor for the electron, to amazing accuracy, within the framework of *quantum electrodynamics* (QED) — see later.

4.10 Identical Particles

We next consider a collection of *identical particles* in quantum mechanics. Most of the systems of physical interest in every-day life are quantum many-body systems (*e.g.* atoms, nuclei, molecules, solids, *etc.*).

4.10.1 *Connection Between Spin and Statistics*

There is an important relation, which can be derived in relativistic quantum field theory,[26] between the *spin* of particles and the *statistics* that they obey.

- Particles with half-integral spin (*e.g.* electrons, muons, neutrons, protons, *etc.*) obey the *Pauli exclusion principle* (1925):

  It is not possible to put more than one particle in any state.

  Such particles are known as *fermions*.
- Particles with integer spin (*e.g.* photons, pions, ^{4}He atoms, *etc.*) can have *any number of particles in any state.* Such particles are called *bosons*.

4.10.2 *Non-interacting, Spin-1/2 Fermions ("Fermi Gas")*

Consider a uniform system of many identical, non-interacting, spin-1/2 fermions occupying a cubical box of side L (Fig. 4.32).

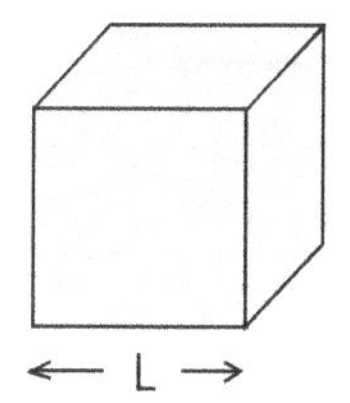

Fig. 4.32 A uniform system of spin-1/2 fermions occupying a large cubical box with side L.

Imagine that this box is simply one of many identical boxes stacked together throughout space (Fig. 4.33). Now observe that if the box is large enough, the quantities of physical interest in any one box should be the

[26] See [Bjorken and Drell (1965)].

same as in any other. This serves as a justification for the application of *periodic boundary conditions* to the system illustrated in Fig. 4.32.

The single-particle wave functions in this case are just those of Eqs. (4.118)

$$\psi_{\mathbf{k}\lambda}(\mathbf{x}) = \frac{1}{\sqrt{L^3}} e^{i\mathbf{k}\cdot\mathbf{x}}\,\eta_\lambda \qquad ; \; \mathbf{x} = (x, y, z)$$

$$\mathbf{k} = (k_x, k_y, k_z)$$

$$\mathbf{k} = \frac{2\pi}{L}(n_x, n_y, n_z) \qquad ; \; n_i = 0, \pm 1, \pm 2, \cdots, \pm\infty \qquad (4.159)$$

Here we choose to keep track of the spin of the particle, which can take two values $s_z = \pm 1/2$, with the simple wave functions

$$\eta_\uparrow = \begin{pmatrix} 1 \\ 0 \end{pmatrix} \qquad ; \; \eta_\downarrow = \begin{pmatrix} 0 \\ 1 \end{pmatrix} \qquad ; \text{ spin-1/2 wave functions} \qquad (4.160)$$

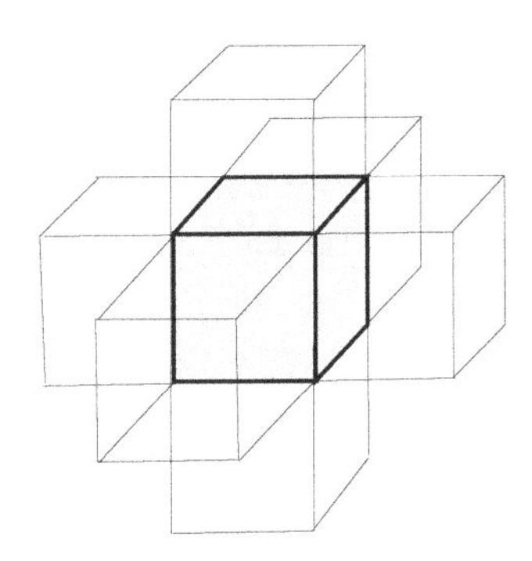

Fig. 4.33 A large number of boxes, identical to the one in Fig. 4.32, stacked together throughout space (only a few are illustrated here)— basis for the application of periodic boundary conditions in Fig. 4.32.

With periodic boundary conditions, the integers n_i can have either sign, and the spacing between the wave numbers is given by

$$\Delta k_i = \frac{2\pi}{L} \qquad ; \text{ wavenumber spacing} \qquad (4.161)$$

The momentum and energy of a particle in one of the states in Eq. (4.159) are given by

$$\mathbf{p} = \hbar \mathbf{k} \qquad ; \; E_k = \frac{\hbar^2 \mathbf{k}^2}{2m} \tag{4.162}$$

To get the ground state of this many-fermion system, where the Pauli exclusion principle applies, one fills the single-particle levels up to a maximum wavenumber, the *Fermi wavenumber* $k_{\rm F}$ (Fig. 4.34). The spin degeneracy of this spin-1/2 *Fermi gas* is $g = 2$.

$$g = 2 \qquad ; \text{ spin degeneracy} \tag{4.163}$$

The total number of particles is obtained by simply counting the number of filled levels

$$N = \sum_{\mathbf{k}\lambda}^{k_{\rm F}} 1 = g \sum_{\mathbf{k}}^{k_{\rm F}} 1 \tag{4.164}$$

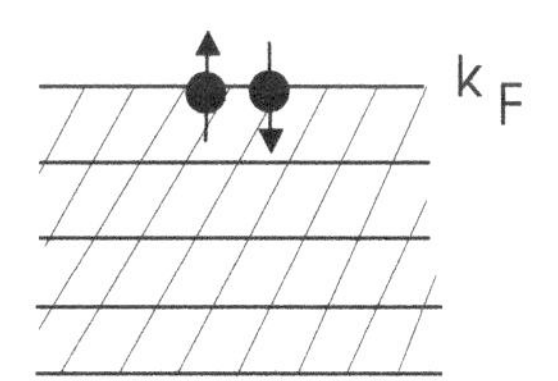

Fig. 4.34 Ground state of the non-interacting, spin-1/2 Fermi gas. The single-particle levels are filled to a maximum wavenumber $k_{\rm F}$. The spin degeneracy here is $g = 2$.

In the limit of a very large box where $L \to \infty$, the sum over wavenumbers can be replaced by an integral. We have now been through this argument many times. With periodic boundary conditions and the level spacing in Eq. (4.161), one has for the sum over a small region in k-space (see Fig. 4.35), here denoted with a prime,

$$\sum_{\mathbf{k}}{}' = \sum_{k_x}{}'\sum_{k_y}{}'\sum_{k_z}{}' = \frac{dk_x dk_y dk_z}{(2\pi/L)^3} = \frac{V}{(2\pi)^3} d^3k \quad ; \text{ small region} \tag{4.165}$$

Hence we arrive at the important result

$$\sum_{\mathbf{k}} \to \frac{V}{(2\pi)^3} \int d^3k \qquad ; \text{ p.b.c.,} \quad L \to \infty \tag{4.166}$$

Particle Number. In counting the number of states, one simply integrates the relation in Eq. (4.166) out to a radius $k_{\rm F}$, and the particle number in Eq. (4.164) is therefore obtained as

$$N = \frac{gV}{(2\pi)^3}\int_0^{k_{\rm F}} d^3k = \frac{gV}{(2\pi)^3}\frac{4\pi}{3}k_{\rm F}^3 \tag{4.167}$$

Thus the *particle density* $n = N/V$ in the Fermi gas is given by

$$n = \frac{N}{V} = \frac{gk_{\rm F}^3}{6\pi^2} \qquad ; \text{ particle density} \tag{4.168}$$

This equation relates the *particle density to the Fermi wavenumber* $k_{\rm F}$.

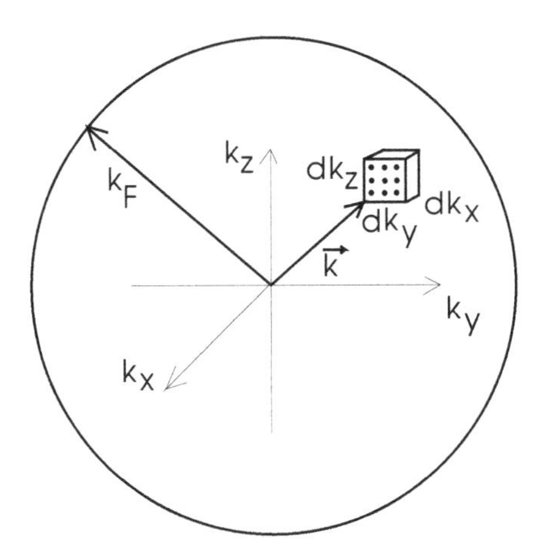

Fig. 4.35 Conversion of the sum over wavenumbers to an integral over wavenumbers in the limit $L \to \infty$ with p.b.c. The level spacing is $\Delta k_i = 2\pi/L$, and the number of levels in the interval dk_i is then $dk_i/(2\pi/L)$.

Energy. The energy of the Fermi gas is obtained by summing the energy of the single-particle states over the filled levels

$$\begin{aligned} E &= \frac{gV}{(2\pi)^3}\int_0^{k_{\rm F}} \frac{\hbar^2k^2}{2m}\, d^3k \\ &= \frac{gV}{(2\pi)^3}\frac{\hbar^2}{2m}\int_0^{k_{\rm F}} 4\pi k^4\, dk = \frac{gV}{(2\pi)^3}\frac{\hbar^2}{2m}\left(\frac{4\pi k_{\rm F}^5}{5}\right) \end{aligned} \tag{4.169}$$

Hence the *energy/particle* in the Fermi gas is given by

$$\frac{E}{N} = \frac{3}{5}\varepsilon_{\rm F} \qquad ; \; \varepsilon_{\rm F} \equiv \frac{\hbar^2 k_{\rm F}^2}{2m} \quad \text{Fermi energy} \tag{4.170}$$

The quantity $\varepsilon_{\rm F}$ is known as the *Fermi energy*.[27]

Pressure. A Fermi gas exerts a *pressure* because as the size of the box with side L decreases, each wavelength inside the box is also decreased, and hence the energy of each particle is increased. The pressure work done *by* a gas on expansion against a piston of area A when it is moved a distance dl is given by (see Fig. D.2)

$$đW = PA\,dl = PdV \tag{4.171}$$

The *first law of thermodynamics* with pressure-volume work states that (see appendix D)

$$dE = đQ - PdV \qquad ; \text{ first law of thermodynamics} \tag{4.172}$$

At $T = 0$, the system will be in its ground state. The *second law of thermodynamics* implies that the heat flow $đQ = TdS$ vanishes at $T = 0$. Thus for a system of given N

$$\begin{aligned} dE &= -PdV \\ \text{or;} \qquad P &= -\left(\frac{\partial E}{\partial V}\right)_N \qquad ; \; T = 0 \end{aligned} \tag{4.173}$$

It is merely necessary to express the energy in terms of the density, and the pressure exerted by the Fermi gas can then be determined by differentiation. The first step is readily accomplished by using Eqs. (4.168) and (4.170) to write

$$\begin{aligned} k_{\rm F} &= \left(\frac{6\pi^2}{g}n\right)^{1/3} \\ \frac{E}{N} &= \frac{3}{5}\frac{\hbar^2}{2m}\left(\frac{6\pi^2}{g}\frac{N}{V}\right)^{2/3} \end{aligned} \tag{4.174}$$

[27] Equations (4.168) and (4.170) are *local* relations involving only the particle density n at a given point, and one would not expect them to depend on the boundary conditions far away on the surface of the big box. In fact, the same relations between $(n, k_{\rm F}, E/N)$ are obtained with rigid walls (see Prob. 4.29).

Differentiation with respect to V, at fixed N, then gives

$$P = N\frac{2}{5}\frac{\hbar^2}{2m}\left(\frac{6\pi^2}{g}N\right)^{2/3}\frac{1}{V^{5/3}} \tag{4.175}$$

Thus

$$P = \frac{2}{5}\frac{\hbar^2}{2m}\left(\frac{6\pi^2}{g}\right)^{2/3} n^{5/3} \qquad ; \text{ Fermi gas } (T=0) \tag{4.176}$$

This equation relates the *pressure exerted by a Fermi gas to its density.*

4.10.3 *Non-Interacting Bosons ("Bose Gas")*

With a collection of bosons there is no restriction on occupation number, and the ground state is obtained by simply placing all the particles in the state with $\mathbf{k} = 0$ (Fig. 4.36). Hence at $T = 0$, both the energy and pressure of the Bose gas vanish[28]

$$P = E = 0 \qquad ; \text{ Bose gas } (T=0) \tag{4.177}$$

$\vec{k} = 0$

Fig. 4.36 Ground state of a Bose gas.

4.10.4 *Quantum Statistics ($T \neq 0$)*

The single-particle distribution functions for the non-interacting Fermi and Bose systems at *finite temperature* take the form (see Fig. 4.37)

$$n_k = \left[\exp\left(\frac{\varepsilon_k - \mu}{k_B T}\right) + 1\right]^{-1} \qquad ; \text{ fermions}$$

$$n_k = \left[\exp\left(\frac{\varepsilon_k - \mu}{k_B T}\right) - 1\right]^{-1} \qquad ; \text{ bosons} \tag{4.178}$$

We make several comments on Eqs. (4.178):

- Here ε_k is the single-particle energy, and μ is the chemical potential (see appendix D);

[28]In a large box with rigid walls, the Bose gas will actually exert a small pressure in its ground state, but this pressure vanishes as $L \to \infty$ (see Prob. 4.30).

- The chemical potential is to be chosen so that when summed over all single-particle states, n_k yields the correct total number of particles N;

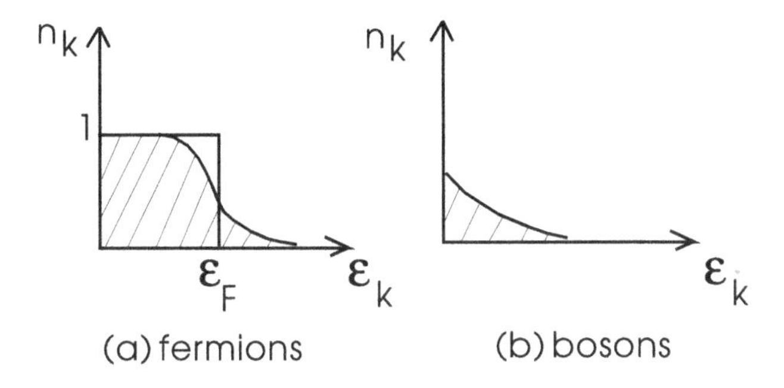

Fig. 4.37 Single-particle distributions n_k for non-interacting particles at finite temperature $T \neq 0$: (a) fermions; (b) bosons. Here ε_k is the single-particle energy.

- It is the thermal energy k_BT that causes particles to be distributed over the single-particle levels in this fashion (compare Fig. 3.7);
- At high temperature, both of the expressions in Eqs. (4.178) reduce to the Boltzmann distribution;[29]
- Note that it is only the sign in the denominator that changes in Eqs. (4.178), but this makes all the difference in the world!
- At $T = 0$, the Fermi distribution yields a step-function at the Fermi energy $\mu = \varepsilon_F$, as previously employed;
- There is a phase transition at $T = T_0$ in the non-interacting Bose system of fixed N (see Prob. 4.31). At that temperature, a finite fraction of the bosons begin to occupy the single-particle state with $\mathbf{k} = 0$ ("Bose condensation").
- Photons are massless bosons, and their number is not conserved. With $\varepsilon_k = \hbar\omega_k = h\nu$ and vanishing chemical potential, the second of Eqs. (4.178) yields the *Planck distribution* in Eq. (3.34);[30]
- These results are derived and discussed in, for example, chap. 2 of

[29] In the classical limit where $T \rightarrow \infty$, the chemical potential of an ideal gas behaves as $\mu/k_BT \rightarrow -\infty$ (see Probs. E.3 and E.4). Therefore, both of the expressions in Eqs. (4.178) reproduce the Boltzmann distribution in this limit.

[30] The total number of photons in a black-body is not conserved. It follows from thermodynamics that the chemical potential of the photons must then vanish at all T, otherwise an appropriate change in their number at fixed (T, V) could lower the free energy of the system [see Eq. (D.11)]. Correspondingly, there is no T_0 for a collection of photons.

[Fetter and Walecka (2003a)]. For the present purposes, we will take the distributions in Eqs. (4.178) as our *definition of quantum statistics.*

4.10.5 *Wave Functions*

It is a principle of non-relativistic quantum mechanics, from which the Pauli exclusion principle follows, that the many-particle wave function for a collection of identical *fermions* must be *antisymmetric* under the interchange of any two particles; in contrast, the many-particle wave function for a collection of identical *bosons* must be *symmetric* under the interchange of any two particles. These many-particle wave functions are constructed and examined in appendix H. The fermion wave functions are Slater determinants, and some theorems concerning matrix elements calculated with Slater determinants, as well as several important applications, can be found in that appendix.

Chapter 5

Atomic Physics

We proceed to discuss the application of the previous developments to various areas of modern physics, and we start with *atomic physics*.[1] The first topic is the vector model for the addition of angular momenta.[2]

5.1 Vector Model for Addition of Angular Momenta

We use lower case letters to stand for *any* angular momentum, and Latin letters signify units of $\hbar$. Write the sum of two angular momenta as (Fig. 5.1)

$$\mathbf{j} = \mathbf{l} + \mathbf{s} \qquad ; \; \mathbf{l}^2 = l(l+1) \qquad \mathbf{s}^2 = s(s+1) \tag{5.1}$$

The allowed quantum numbers for $\mathbf{j}$, again an angular momentum, are

$$\begin{aligned} \mathbf{j}^2 &= j(j+1) \\ -j \leq m_j &\leq j \qquad ; \text{ integer steps} \end{aligned} \tag{5.2}$$

Now square the vector relation in Eq. (5.1)

$$\begin{aligned} &\mathbf{j}^2 = \mathbf{l}^2 + \mathbf{s}^2 + 2\mathbf{l}\cdot\mathbf{s} \\ \Rightarrow \qquad &\mathbf{l}\cdot\mathbf{s} = \frac{1}{2}\left(\mathbf{j}^2 - \mathbf{l}^2 - \mathbf{s}^2\right) = \frac{1}{2}\left[j(j+1) - l(l+1) - s(s+1)\right] \end{aligned} \tag{5.3}$$

[1] See [Born (1989); Kuhn (1969); Sobel'man (1972)].

[2] The theory of angular momentum in quantum mechanics is very lovely and powerful [Edmonds (1974)]. The vector model reproduces the general results in its region of applicability, and it will suffice for our purposes.

In a similar fashion one obtains

$$\mathbf{j} - \mathbf{l} = \mathbf{s}$$
$$\Rightarrow \qquad \mathbf{j} \cdot \mathbf{l} = \frac{1}{2}\left(\mathbf{j}^2 + \mathbf{l}^2 - \mathbf{s}^2\right) = \frac{1}{2}\left[j(j+1) + l(l+1) - s(s+1)\right] \tag{5.4}$$

and

$$\mathbf{j} - \mathbf{s} = \mathbf{l}$$
$$\Rightarrow \qquad \mathbf{j} \cdot \mathbf{s} = \frac{1}{2}\left(\mathbf{j}^2 + \mathbf{s}^2 - \mathbf{l}^2\right) = \frac{1}{2}\left[j(j+1) + s(s+1) - l(l+1)\right] \tag{5.5}$$

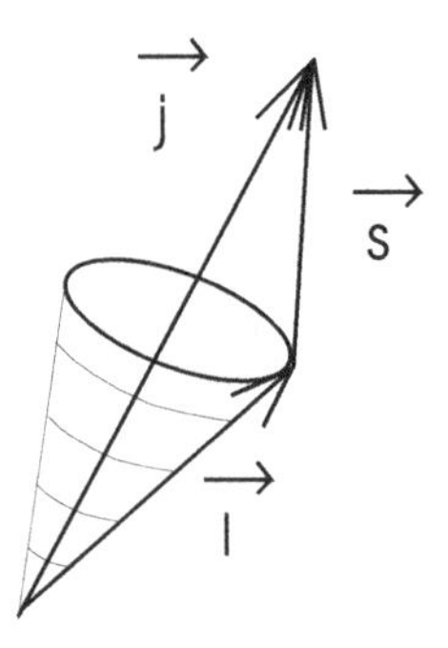

Fig. 5.1 Addition of two angular momenta **l** and **s** to give the new angular momentum $\mathbf{j} = \mathbf{l} + \mathbf{s}$. In the vector model, the angular momenta $(\mathbf{l}, \mathbf{s})$ can be viewed as precessing about **j**. The transverse components average out and their only *effective* component is the component along **j**.

What are the allowed values of the angular momentum quantum number j? It must again be positive, and the allowed values of j must differ by an integer. It is evident from Fig. (5.1) that the maximum value of j will be $l + s$, while the minimum value will be $|l - s|$. Thus the answer is[3]

$$l + s \geq j \geq |l - s| \qquad ; \text{ integer steps} \tag{5.6}$$

For an electron with $s = 1/2$ this reduces to

$$j = l \pm 1/2 \qquad ; \text{ electron}$$
$$j = 1/2 \qquad ; \text{ if } l = 0 \tag{5.7}$$

[3]At this stage, the reader should be content to simply accept and memorize this relation; however, one *can* check that the total number of states for a given (l, s) is preserved with this prescription, that is $\sum_{j=|l-s|}^{l+s}(2j+1) = (2l+1)(2s+1)$.

One can picture the two angular momenta $(\mathbf{l}, \mathbf{s})$ in Fig. 5.1 as precessing about the resultant $\mathbf{j}$. Their transverse components average to zero, and

> *In this vector model, it is only the component of the angular momentum* $\mathbf{l}$ *or* $\mathbf{s}$ *along* $\mathbf{j}$ *that remains effective.*

Thus in the vector model

$$\mathbf{l}_{\text{eff}} = \left(\frac{\mathbf{l}\cdot\mathbf{j}}{\mathbf{j}^2}\right)\mathbf{j} \qquad ; \ \mathbf{s}_{\text{eff}} = \left(\frac{\mathbf{s}\cdot\mathbf{j}}{\mathbf{j}^2}\right)\mathbf{j} \qquad ; \ \text{vector model} \tag{5.8}$$

As an application, suppose one has a magnetic moment that has both an orbital and spin part

$$\boldsymbol{\mu} = \mu_l \mathbf{l} + \mu_s \mathbf{s} \qquad ; \ \text{magnetic moment} \tag{5.9}$$

Then the *effective magnetic moment* is

$$\boldsymbol{\mu}_{\text{eff}} = \frac{1}{\mathbf{j}^2}\left[\mu_l(\mathbf{l}\cdot\mathbf{j}) + \mu_s(\mathbf{s}\cdot\mathbf{j})\right]\mathbf{j} \quad ; \ \text{effective magnetic moment} \tag{5.10}$$

It follows from Eqs. (5.4) and (5.5) that

$$\begin{aligned}\boldsymbol{\mu}_{\text{eff}} = \frac{1}{2j(j+1)} &\{\mu_l\left[j(j+1) + l(l+1) - s(s+1)\right] + \\ &\mu_s\left[j(j+1) + s(s+1) - l(l+1)\right]\}\mathbf{j}\end{aligned} \tag{5.11}$$

In *summary*, we will make use of the following relations from the general theory of angular momentum:

(1) The total angular momentum has the following quantum numbers

$$\begin{aligned}\mathbf{j}^2 &= j(j+1) \\ -j \leq m_j &\leq j \qquad ; \ \text{integer steps}\end{aligned} \tag{5.12}$$

(2) As the sum of two angular momenta, the total angular momentum takes the following values

$$\begin{aligned}&\mathbf{j} = \mathbf{l} + \mathbf{s} \qquad ; \ \mathbf{l}^2 = l(l+1) \\ &\qquad\qquad\qquad\ \ \mathbf{s}^2 = s(s+1) \\ \Rightarrow \quad &|l - s| \leq j \leq l + s \qquad ; \ \text{integer steps}\end{aligned} \tag{5.13}$$

(3) The *vector model* states that the effective angular momenta are the components along $\mathbf{j}$

$$\mathbf{l}_{\text{eff}} = \left(\frac{\mathbf{l}\cdot\mathbf{j}}{\mathbf{j}^2}\right)\mathbf{j} \qquad ; \ \mathbf{s}_{\text{eff}} = \left(\frac{\mathbf{s}\cdot\mathbf{j}}{\mathbf{j}^2}\right)\mathbf{j} \qquad ; \ \text{vector model} \tag{5.14}$$

and the vector model gives *correct results* in its region of applicability.

5.1.1 *Larmor's Theorem*

There is a classical result known as *Larmor's theorem* that relates the magnetic dipole moment of a system to its orbital angular momentum. Suppose one has a distribution of mass $m(\mathbf{r}) = m\rho(\mathbf{r})$ with a velocity field $\mathbf{v}(\mathbf{r})$, and a distribution of charge $\rho_C(\mathbf{r}) = e\rho(\mathbf{r})$ with the identical $\rho(\mathbf{r})$ (Fig. 5.2).

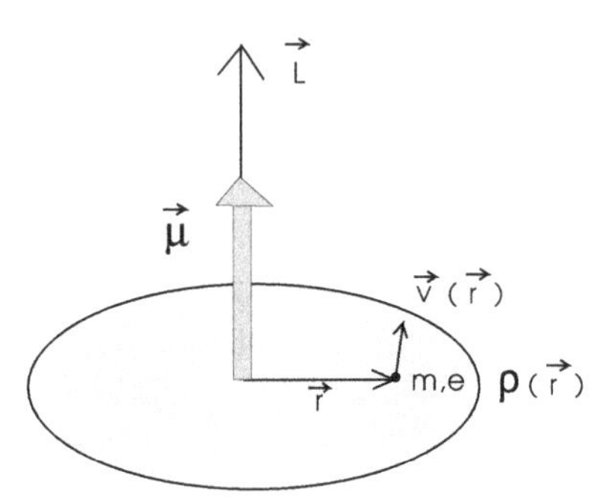

Fig. 5.2 Configuration assumed in the proof of Larmor's theorem. There is a mass distribution $m(\mathbf{r}) = m\rho(\mathbf{r})$ with a velocity field $\mathbf{v}(\mathbf{r})$, and a distribution of charge $\rho_C(\mathbf{r}) = e\rho(\mathbf{r})$ with an identical $\rho(\mathbf{r})$.

The total angular momentum of the system is given by

$$\begin{aligned} \boldsymbol{\mathcal{L}} &= \int d^3x\ (\mathbf{r} \times \mathbf{p}) \\ &= \int d^3x\ [\mathbf{r} \times m\rho(\mathbf{r})\mathbf{v}(\mathbf{r})] \qquad ;\ \text{angular momentum} \qquad (5.15) \end{aligned}$$

The magnetic dipole moment of this system is

$$\begin{aligned} \boldsymbol{\mu} &= \int d^3x\, \frac{1}{2}\, (\mathbf{r} \times \mathbf{j}) \\ &= \frac{1}{2} \int d^3x\ [\mathbf{r} \times e\rho(\mathbf{r})\mathbf{v}(\mathbf{r})] \qquad ;\ \text{magnetic moment} \qquad (5.16) \end{aligned}$$

These quantities are evidently *proportional to each other*, and one immediately deduces Larmor's theorem for this classical system of flowing mass and charge

$$\boldsymbol{\mu} = \frac{e}{2m}\boldsymbol{\mathcal{L}} = \frac{e\hbar}{2m}\mathbf{L} \qquad ;\ \text{Larmor's theorem} \qquad (5.17)$$

One can combine this result with that of Eqs. (4.155) and (4.156) to write the orbital and spin parts of the magnetic moment of the electron as

$$\boldsymbol{\mu}_l = \frac{e\hbar}{2m_e}\mathbf{l} = \mu_{\rm B}\frac{e}{|e|}\mathbf{l} \qquad ; \text{ electron}$$
$$\boldsymbol{\mu}_s = g_s\frac{e}{|e|}\mu_{\rm B}\,\mathbf{s} \tag{5.18}$$

Here $\mu_{\rm B}$ is the Bohr magneton

$$\mu_{\rm B} = \frac{|e|\hbar}{2m_e} \qquad ; \text{ Bohr magneton} \tag{5.19}$$

and g_s is the gyromagnetic ratio (see Table 4.1).

5.1.2 *Effective Magnetic Moment*

The expressions in Eqs. (5.18) can be combined with the vector model results in Eqs. (5.9)–(5.11) to yield the following expression for the effective magnetic moment of an electron

$$\boldsymbol{\mu}_{\rm eff} = \frac{e}{|e|}g_{\text{Landé}}\,\mu_{\rm B}\,\mathbf{j} \qquad ; \text{ electron}$$
$$g_{\text{Landé}} = \frac{1}{2j(j+1)}\{[j(j+1)+l(l+1)-s(s+1)]+ g_s\,[j(j+1)+s(s+1)-l(l+1)]\} \tag{5.20}$$

The quantity $g_{\text{Landé}}$ is known as the *Landé g-factor.*[4]

The *experimental magnetic moment* of a system is defined by lining the system up as well as possible along the z-axis, in which case $j_z = m_j = j$ (Fig. 5.3).

$$j_z = j \qquad ; \text{ experimental moment} \tag{5.21}$$

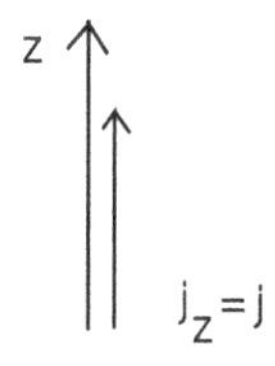

Fig. 5.3 The *experimental magnetic moment* of a system is defined by lining the system up as well as possible along the z-axis, in which case $j_z = m_j = j$.

[4]See Prob. 5.4.

Hence the static magnetic moment of the system is given by

$$\mu = \frac{e}{|e|} g_{\text{Landé}}\, \mu_{\text{B}}\, j \qquad ; \text{ static magnetic moment} \qquad (5.22)$$

5.2 Zeeman Effect

Suppose one has an atom with a valence electron that has an effective magnetic moment $\boldsymbol{\mu}_{\text{eff}}$. If this atom is placed in a static magnetic field $\mathbf{B}$ that determines the z-direction, then there is an additional interaction energy (see Fig. 5.4)

$$\begin{aligned} H' &= -\boldsymbol{\mu}_{\text{eff}} \cdot \mathbf{B} \\ &= -\frac{e}{|e|} g_{\text{Landé}}\, \mu_{\text{B}} B m_j \qquad ; -j \leq m_j \leq j \text{ in integer steps} \end{aligned} \qquad (5.23)$$

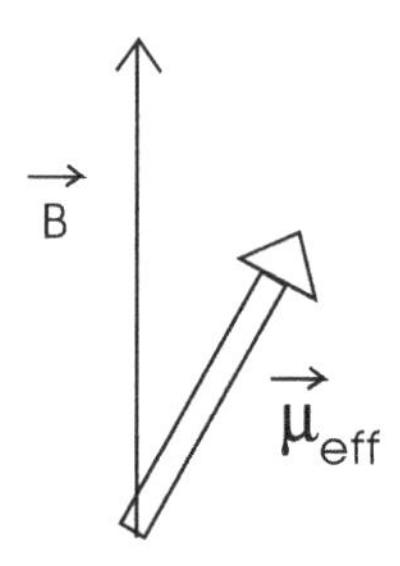

Fig. 5.4 Interaction of an effective magnetic moment $\boldsymbol{\mu}_{\text{eff}}$ with a magnetic field $\mathbf{B}$ as given in Eq. (5.23).

We note the following:

- The quantum number m_j runs from $-j$ to j in integer steps. There are $2j + 1$ values of m_j, and thus the original level with quantum numbers (j, m_j) will split into $(2j + 1)$ components in the magnetic field (Fig. 5.5);
- Since the effective magnetic moment in Eq. (5.23) only depends on the quantum numbers $(lsjm_j)$, there is *no dependence of this Zeeman splitting on the radial wave function of the valence electron*;[5]
- The slope of the splitting as a function of B allows one to determine $g_{\text{Landé}}$.

[5]The effective magnetic moment is an *integral* property of the atom.

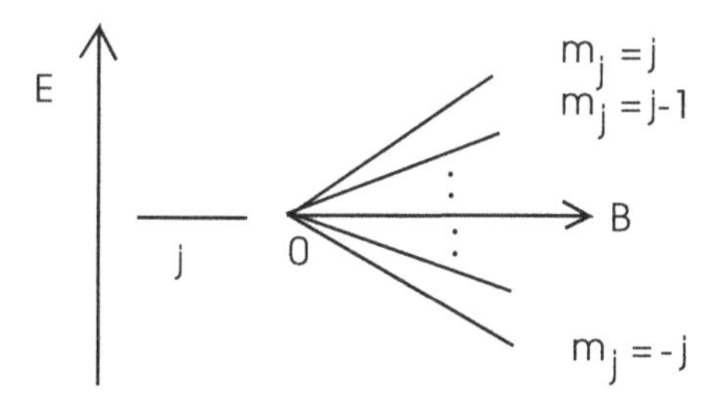

Fig. 5.5 Zeeman splitting of a level with quantum numbers $(lsjm_j)$ and the effective magnetic moment $\boldsymbol{\mu}_{\text{eff}}$ of Eq. (5.23) in a static magnetic field $\mathbf{B}$. The level splits into $(2j+1)$ components with $-j \leq m_j \leq j$ in integer steps.

5.3 Spin-Orbit Interaction

Consider an atom with a valence electron moving with a velocity $\mathbf{v}$ in the laboratory frame. The charged nucleus appears to the electron to be moving with a velocity $-\mathbf{v}$ (Fig. 5.6). Thus, at its position, the electron sees a current $\propto -Ze_p\mathbf{v}$ at a location $-\mathbf{r}$ and a corresponding magnetic field $\mathbf{B}_{\text{eff}} \propto Ze_p\mathbf{r} \times \mathbf{v}$. This gives an effective field $\mathbf{B}_{\text{eff}} \propto Ze_p\mathbf{l}$. The intrinsic magnetic moment of the electron will interact with this effective magnetic field according to

$$H' = -\boldsymbol{\mu}_{\text{spin}} \cdot \mathbf{B}_{\text{eff}} \equiv +V_{\text{so}}(r)\,\mathbf{s}\cdot\mathbf{l} \qquad ;\ \text{spin-orbit interaction} \quad (5.24)$$

We now note the following:

- This is a *relativistic effect*,[6] and one needs a relativistic theory of the electron, as provided by the Dirac equation (see later), to get the proper $V_{\text{so}}(r)$;
- Once one has the interaction in Eq. (5.24), the quantity $\mathbf{l}\cdot\mathbf{s}$ can be evaluated using the result in Eq. (5.3);
- One can use the probability distribution of the electron to compute the mean value of $V_{\text{so}}(r)$ (see Prob. 4.5). The probability density is obtained as the absolute square of the electron's wave function

$$\psi_{\bar{n}lm}(r,\theta,\phi) = R_{\bar{n}l}(r)Y_{lm}(\theta,\phi) \qquad (5.25)$$

[6]See Prob. 5.1.

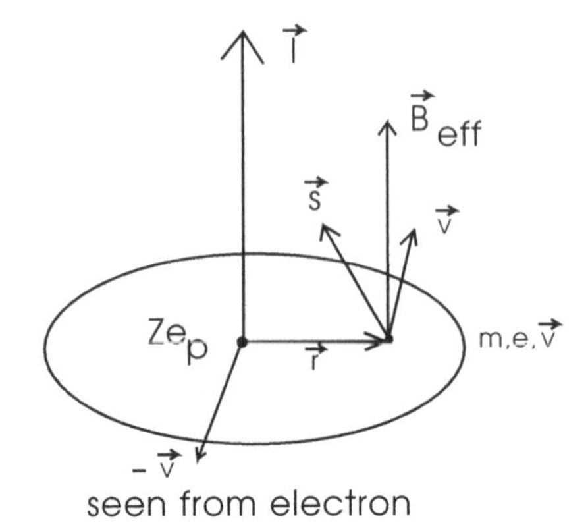

Fig. 5.6 Effective magnetic field $\mathbf{B}_{\text{eff}}$ as seen at the location of a valence electron due to the current of the moving nucleus.

- A combination of these results leads to[7]

$$\langle H' \rangle = \langle V_{\text{so}}(r) \rangle \frac{1}{2} [j(j+1) - l(l+1) - s(s+1)]$$

$$\langle V_{\text{so}}(r) \rangle = \int |R_{\bar{n}l}(r)|^2 V_{\text{so}}(r) r^2 \, dr \tag{5.26}$$

- For an electron in the central Coulomb field of an atom, one has $j = l \pm 1/2$, and the energy shift from the originally degenerate l states becomes

$$\Delta\varepsilon_{j=l+1/2} = \langle V_{\text{so}}(r) \rangle \frac{1}{2} l \qquad ; \; j = l + \frac{1}{2}, \quad s = \frac{1}{2}$$

$$\Delta\varepsilon_{j=l-1/2} = \langle V_{\text{so}}(r) \rangle \left[-\frac{1}{2}(l+1) \right] \quad ; \; j = l - \frac{1}{2}, \quad s = \frac{1}{2} \tag{5.27}$$

- The *splitting* of these two levels is thus given by

$$\varepsilon_{j=l+1/2} - \varepsilon_{j=l-1/2} = \frac{1}{2}(2l+1) \langle V_{\text{so}}(r) \rangle \tag{5.28}$$

  This gives rise to the *fine structure* in atomic physics;
- The states remain degenerate in m_j;
- The fine structure of the $2p$-states in hydrogen is sketched in Fig. 5.7.

---

[7] We note that the volume element in spherical coordinates is $d^3r = r^2 \sin\theta \, dr d\theta d\phi \equiv r^2 d\Omega$, and $\int d\Omega \, Y_{lm}^\star Y_{l'm'} = \delta_{ll'}\delta_{mm'}$. The angular integral in the expectation value of $V_{\text{so}}(r)$ is thus diagonal in m and independent of its value. Here $\langle V_{\text{so}}(r) \rangle$ is positive.

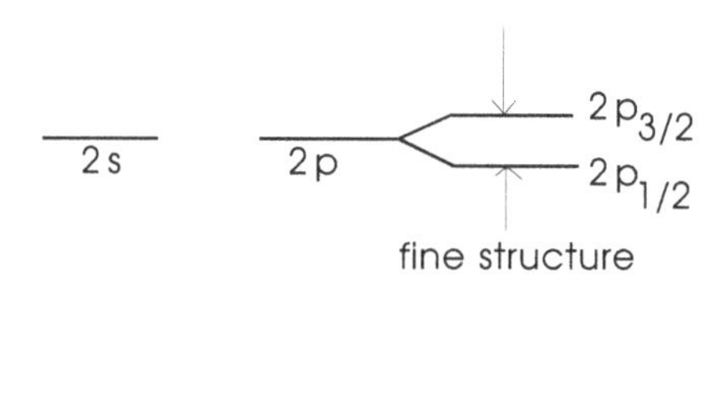

Fig. 5.7 Fine structure of the $2p$-state in hydrogen arising from the spin-orbit interaction (not to scale — see Fig. 5.14). The notation for the states is $(\bar{n}l)_j$.

5.4 Thomas-Fermi Theory

Most atoms are many-electron systems, with dynamics (at least in the non-relativistic limit) governed by the many-electron Schrödinger equation, and one must develop methods for dealing with such systems. One of the simplest, most informative, and most useful approaches is that due to Thomas and Fermi. Here one employs the previous results for the Fermi gas and makes a *local density approximation.* In this approximation, the system is viewed locally as a Fermi gas with properties derived from the local electron density $n(r)$.[8]

5.4.1 *Thomas-Fermi Equation*

The *pressure* exerted by the Fermi gas then follows immediately from Eq. (4.176)[9]

$$P = \frac{2}{5}\frac{\hbar^2}{2m}(3\pi^2)^{2/3}n^{5/3} \tag{5.29}$$

The electron density will vary from point to point in the atom, increasing as one goes in toward the nucleus, and there will be a hydrostatic force arising from the pressure gradient.

Hydrostatic Equilibrium. Consider a small volume element d^3x and a pressure $P(x)$ that varies in the x-direction. The net force in the x-direction

[8] We assume a spherically symmetric atom.

[9] Here $m = m_e$, and the degeneracy $g = 2$.

on this volume element is given by (Fig. 5.8)

$$F_x = P(x)dydz - P(x+dx)dydz = -\frac{dP}{dx}dxdydz \tag{5.30}$$

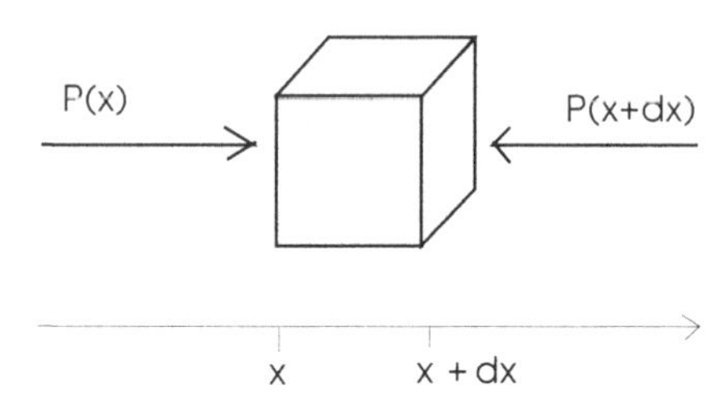

Fig. 5.8 Force on the volume element d^3x arising from a gradient to the pressure $P(x)$ in the x-direction. The area of the transverse face is $dydz$.

Hence, quite generally, if there is a pressure gradient in a fluid, there will be a pressure force on each little volume element of

$$\mathbf{F} = -\boldsymbol{\nabla} P\, d^3x \qquad ;\ \text{pressure force} \tag{5.31}$$

In the atom, there will be an electric field $\boldsymbol{\mathcal{E}}(r)$ at the position r arising from the nuclear charge, shielded by the interior electrons. Each little volume element in the electron cloud has a charge $en\, d^3x$. This charge is attracted toward the nucleus with a force

$$\mathbf{F} = en\boldsymbol{\mathcal{E}}\, d^3x \qquad ;\ \text{electric force} \tag{5.32}$$

In hydrostatic equilibrium, the pressure force and electric force must balance (see Fig. 5.9). Thus

$$en\boldsymbol{\mathcal{E}} - \boldsymbol{\nabla} P = 0 \qquad ;\ \text{hydrostatic equilibrium} \tag{5.33}$$

In our problem, everything is in the radial direction so this expression reduces to

$$en\mathcal{E}_r - \frac{\partial P}{\partial r} = 0 \qquad ;\ \text{hydrostatic equilibrium} \tag{5.34}$$

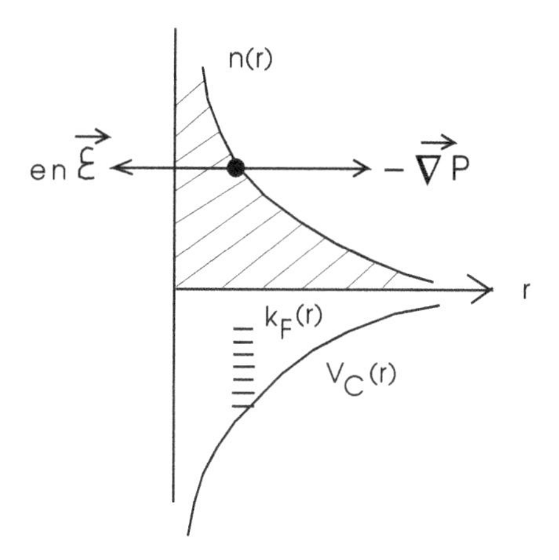

Fig. 5.9 Local hydrostatic equilibrium in the Thomas-Fermi theory of the atom. Here $n(r)$ is the local electron density, $k_F(r)$ is the local Fermi wavenumber, and $(\boldsymbol{\mathcal{E}}, P)$ are the local electric field and pressure. $V_C(r) = -Ze^2/4\pi\varepsilon_0 r$ is the attractive nuclear Coulomb interaction, which is then shielded by the electrons.

Electric Field. The electric field is derivable from the electrostatic potential

$$\boldsymbol{\mathcal{E}} = -\boldsymbol{\nabla}\Phi$$
$$\mathcal{E}_r = -\frac{\partial \Phi}{\partial r} \tag{5.35}$$

Equations (5.34) and (5.29) can then be combined to obtain a local relation between the electrostatic potential and the electron density

$$|e|n\frac{\partial \Phi}{\partial r} = \frac{\partial P}{\partial r} = \frac{2}{3}\frac{\hbar^2}{2m}(3\pi^2)^{2/3}n^{2/3}\frac{\partial n}{\partial r}$$
$$\text{or;} \qquad \frac{\partial \Phi}{\partial r} = \frac{\hbar^2}{2m|e|}(3\pi^2)^{2/3}\frac{\partial}{\partial r}n^{2/3} \tag{5.36}$$

This differential relation is immediately integrated to give

$$\Phi(r) = \frac{\hbar^2}{2m|e|}(3\pi^2)^{2/3}n(r)^{2/3} \tag{5.37}$$

Here the boundary condition of a *neutral atom* has been employed

$$n(r), \Phi(r) \to 0 \qquad ; \; r \to \infty$$
$$\text{neutral atom} \tag{5.38}$$

We now have one relation between the electrostatic potential $\Phi(r)$ and the electron number density $n(r)$ coming from the condition of hydrostatic equilibrium. It remains to examine the implications of electrostatics.

It is a basic result from Gauss' theorem that there is no electric field *inside* a spherically symmetric charge distribution. In addition, *outside* of a spherically symmetric charge distribution, the total charge inside acts as a point charge at the origin. Thus, at a radial distance r inside the atom, one has a radial electric field (see Fig. 5.10)

$$\begin{aligned}\mathcal{E}_r &= \frac{Ze_p}{4\pi\varepsilon_0 r^2} + \frac{e}{4\pi\varepsilon_0 r^2}\int_0^r n(s)4\pi s^2\,ds \\ &= -\frac{\partial\Phi}{\partial r}\end{aligned} \tag{5.39}$$

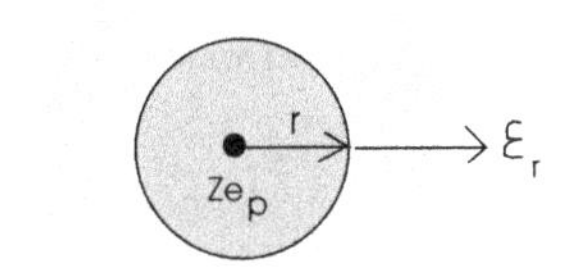

Fig. 5.10 Electrostatics with a spherically symmetric charge distribution.

The relation in Eq. (5.39) can be rearranged, and then differentiated with respect to r to give

$$\begin{aligned}\frac{\partial}{\partial r}\left(r^2\frac{\partial\Phi}{\partial r}\right) &= -\frac{1}{4\pi\varepsilon_0}4\pi r^2 en(r) \\ \text{or;}\qquad \frac{1}{r^2}\frac{\partial}{\partial r}\left(r^2\frac{\partial\Phi}{\partial r}\right) &= -\frac{en(r)}{\varepsilon_0}\end{aligned} \tag{5.40}$$

This result can, in turn, be written as[10]

$$\frac{1}{r}\frac{\partial^2}{\partial r^2}(r\Phi) = \frac{|e|n(r)}{\varepsilon_0} \tag{5.41}$$

[10] Note that

$$\frac{1}{r^2}\frac{\partial}{\partial r}\left(r^2\frac{\partial\Phi}{\partial r}\right) = \frac{2}{r}\frac{\partial\Phi}{\partial r} + \frac{\partial^2\Phi}{\partial r^2} = \frac{1}{r}\frac{\partial^2}{\partial r^2}(r\Phi)$$

Electrostatics thus provides a second relation between $\Phi(r)$ and $n(r)$. The result for $n(r)$ in Eq. (5.37) can be substituted into Eq. (5.41) to give a differential equation for the shielded Coulomb potential itself

$$\begin{aligned}\frac{1}{r}\frac{\partial^2}{\partial r^2}(r\Phi) &= \frac{|e|}{\varepsilon_0}\left[\frac{2m|e|}{\hbar^2}\frac{1}{(3\pi^2)^{2/3}}\right]^{3/2}\Phi^{3/2} \\ &\equiv \kappa\Phi^{3/2}\end{aligned} \tag{5.42}$$

Some algebra then leads to

$$\begin{aligned}\kappa &= \frac{|e|}{3\pi^2\varepsilon_0}\left[\frac{2m|e|}{\hbar^2}\right]^{3/2} \\ &= \frac{8\sqrt{2}}{3\pi}\frac{1}{4\pi\varepsilon_0}\frac{1}{|e|^{1/2}}\left[\frac{me^2}{\hbar^2}\right]^{3/2} \\ &= \frac{8\sqrt{2}}{3\pi}\frac{1}{4\pi\varepsilon_0}\frac{1}{|e|^{1/2}}\left[\frac{4\pi\varepsilon_0}{a_0}\right]^{3/2} \\ &= \frac{8\sqrt{2}}{3\pi}\frac{1}{a_0^2}\left[\frac{1}{|e|/4\pi\varepsilon_0 a_0}\right]^{1/2}\end{aligned} \tag{5.43}$$

Here the Bohr radius a_0 has been introduced in the third line

$$a_0 = \frac{\hbar^2}{m_e e^2/4\pi\varepsilon_0} \qquad ; \text{ Bohr radius} \tag{5.44}$$

The final result for κ in Eqs. (5.43) now has the correct dimensions of $[\Phi^{-1/2}L^{-2}]$.

We know that as $r \to 0$, the electrostatic potential must simply be given by the Coulomb potential arising from the point nuclear charge. Hence Thomas-Fermi theory takes the form of a second order, non-linear differential equation for the electrostatic potential, together with two boundary conditions

$$\begin{aligned}\frac{1}{r}\frac{\partial^2}{\partial r^2}(r\Phi) &= \kappa\Phi^{3/2} && ; \text{ electrostatic potential} \\ r\Phi &\to 0 && ; r \to \infty \qquad ; \text{ neutral atom} \\ \Phi &\to \frac{Ze_p}{4\pi\varepsilon_0 r} && ; r \to 0 \qquad ; \text{ nucleus at origin}\end{aligned} \tag{5.45}$$

Dimensionless Form. We are now in a position to take the dimensions out of the problem. Define

$$\phi \equiv \frac{\Phi}{|e|/4\pi\varepsilon_0 a_0} \qquad ; \rho \equiv \frac{r}{a_0} \tag{5.46}$$

The differential Eq. (5.45) can then be rewritten as

$$\frac{1}{\rho}\frac{\partial^2}{\partial\rho^2}(\rho\phi) = \frac{8\sqrt{2}}{3\pi}\phi^{3/2} \equiv \left(\frac{\phi}{b}\right)^{3/2} \tag{5.47}$$

Here b is the dimensionless constant

$$b = \frac{1}{2}\left(\frac{3\pi}{4}\right)^{2/3} \tag{5.48}$$

The second boundary condition in Eqs. (5.45) becomes

$$\rho\phi \to Z \qquad ; \ \rho \to 0 \tag{5.49}$$

Now take the constants out of the problem by defining a new set of dimensionless variables

$$\phi \equiv \frac{Z}{\rho}\chi \qquad ; \ \rho \equiv \frac{b}{Z^{1/3}}x \tag{5.50}$$

Substitution into Eq. (5.47) then gives

$$\frac{Z}{b^3}\frac{1}{x}\frac{d^2}{dx^2}(Z\chi) = \left[\frac{Z}{b\rho}\chi\right]^{3/2} = \left[\frac{Z^{4/3}}{b^2}\frac{\chi(x)}{x}\right]^{3/2} \tag{5.51}$$

Hence, Thomas-Fermi theory finally takes the form of the following dimensionless, second-order, non-linear differential equation and boundary conditions

$$\begin{aligned} \sqrt{x}\frac{d^2}{dx^2}\chi(x) &= [\chi(x)]^{3/2} && ; \text{ Thomas-Fermi equation} \\ \chi(0) &= 1 && ; \text{ nucleus at origin} \\ \chi(\infty) &= 0 && ; \text{ neutral atom} \end{aligned} \tag{5.52}$$

Here $\chi(x)$ is now a *universal function* determined once-and-for-all for any atom with any Z. It is known as the *Thomas-Fermi screening function.* A combination of the above results expresses the Coulomb potential in the atom as

$$\begin{aligned} \Phi &= \frac{Ze_p}{4\pi\varepsilon_0 r}\chi(x) && ; \text{ screened Coulomb potential} \\ x &= \frac{Z^{1/3}r}{ba_0} && \chi(x) \text{ is universal function of } x \end{aligned} \tag{5.53}$$

The electron density is determined in terms of Φ through Eq. (5.37), which can be rewritten as (Prob. 5.2)

$$n(r) = \frac{Z}{4\pi}\left(\frac{Z^{1/3}}{ba_0}\right)^3\left[\frac{\chi(x)}{x}\right]^{3/2} \quad ; \text{ electron density}$$

$$4\pi n(r)r^2\,dr = Z\,[\chi(x)]^{3/2}\,x^{1/2}\,dx \quad ; \text{ electrons in shell} \tag{5.54}$$

The last expression is the number of electrons in the shell between r and $r + dr$.

5.4.2 *Binding Energy of Atom*

Consider the total binding energy of an atom in this Thomas-Fermi theory. There are three contributions:

Kinetic Energy. First compute the kinetic energy of the electrons. Since the energy per unit volume in a Fermi gas is $E/V = 3\varepsilon_{\rm F}n/5$ [see Eq. (4.170)], this contribution is

$$\begin{aligned} T &= \int\left(\frac{3}{5}\varepsilon_{\rm F}\right)4\pi n(r)r^2\,dr & (5.55)\\ &= \frac{3}{5}\frac{\hbar^2}{2m}\int\left[3\pi^2 n\right]^{2/3}4\pi n(r)r^2\,dr \\ \frac{T}{e^2/8\pi\varepsilon_0 a_0} &= \frac{3}{5}(3\pi^2)^{2/3}\left(\frac{Z}{4\pi}\right)^{2/3}\frac{Z^{2/3}}{b^2}\int\frac{\chi(x)}{x}Z\,[\chi(x)]^{3/2}\sqrt{x}\,dx \end{aligned}$$

Thus the kinetic energy takes the form

$$\frac{T}{e^2/8\pi\varepsilon_0 a_0} = Z^{7/3}\frac{6}{5b}\int_0^\infty [\chi(x)]^{5/2}\frac{dx}{\sqrt{x}} \tag{5.56}$$

Interaction With Nucleus. The electron cloud has an attractive Coulomb interaction with the nuclear charge

$$\begin{aligned} V_{\rm eN} &= -\int\frac{Ze^2}{4\pi\varepsilon_0 r}4\pi n(r)r^2\,dr \\ \frac{V_{\rm eN}}{e^2/8\pi\varepsilon_0 a_0} &= -2Z\frac{Z^{1/3}}{b}\int\frac{1}{x}Z\,[\chi(x)]^{3/2}\sqrt{x}\,dx & (5.57) \end{aligned}$$

Thus this contribution to the total energy of the atom takes the form

$$\frac{V_{\rm eN}}{e^2/8\pi\varepsilon_0 a_0} = -Z^{7/3}\frac{2}{b}\int_0^\infty [\chi(x)]^{3/2}\frac{dx}{\sqrt{x}} \tag{5.58}$$

Interaction Between Electrons. The total Coulomb energy of the electrons if located at positions $(\mathbf{x}_1, \mathbf{x}_2, \cdots, \mathbf{x}_N)$ is

$$V_{\text{ee}} = \frac{e^2}{4\pi\varepsilon_0}\frac{1}{2}\sum_{i\neq j=1}^{N}\frac{1}{|\mathbf{x}_i - \mathbf{x}_j|} \tag{5.59}$$

The factor of $1/2$ occurs because each interaction is to be counted only once. On the other hand, the Coulomb potential at the point $\mathbf{x}$ coming from these electrons is

$$\Phi_{\text{ee}}(\mathbf{x}) = \frac{e}{4\pi\varepsilon_0}\sum_{j=1}^{N}\frac{1}{|\mathbf{x} - \mathbf{x}_j|} \tag{5.60}$$

In the atom, the quantity $\Phi_{\text{ee}}(r)$ can be obtained from the *total* Coulomb potential $\Phi(r)$ by simply subtracting out that part coming from the nuclear charge

$$\Phi_{\text{ee}}(r) = \Phi(r) - \frac{Ze_p}{4\pi\varepsilon_0 r} \tag{5.61}$$

It follows that in the atom

$$\begin{aligned} V_{\text{ee}} &= \frac{1}{2}\int e\Phi_{\text{ee}}(r)4\pi n(r)r^2\,dr \\ &= -\frac{|e|}{2}\int \Phi(r)4\pi n(r)r^2\,dr - \frac{1}{2}V_{\text{eN}} \end{aligned} \tag{5.62}$$

Hence

$$V_{\text{ee}} + \frac{1}{2}V_{\text{eN}} = -\frac{e^2}{8\pi\varepsilon_0 a_0}\frac{Z^{4/3}}{b}\int \frac{\chi(x)}{x} Z\left[\chi(x)\right]^{3/2}\sqrt{x}\,dx \tag{5.63}$$

Thus the contribution of the Coulomb repulsion between the electrons to the total energy of the atom takes the form

$$\frac{V_{\text{ee}} + V_{\text{eN}}/2}{e^2/8\pi\varepsilon_0 a_0} = -Z^{7/3}\frac{1}{b}\int_0^\infty \left[\chi(x)\right]^{5/2}\frac{dx}{\sqrt{x}} \tag{5.64}$$

Total Energy The total energy is obtained by summing the three contributions

$$\begin{aligned} E &= T + V_{\text{eN}} + V_{\text{ee}} \\ &= \frac{e^2}{8\pi\varepsilon_0 a_0}\frac{Z^{7/3}}{b}\left[\left(\frac{6}{5} - 1\right)\int_0^\infty \left[\chi(x)\right]^{5/2}\frac{dx}{\sqrt{x}} - \int_0^\infty \left[\chi(x)\right]^{3/2}\frac{dx}{\sqrt{x}}\right] \end{aligned} \tag{5.65}$$

The final result is

$$E = -Z^{7/3}\frac{\mathcal{R}}{b}\left\{\int_0^\infty [\chi(x)]^{3/2}\,\frac{dx}{\sqrt{x}} - \frac{1}{5}\int_0^\infty [\chi(x)]^{5/2}\,\frac{dx}{\sqrt{x}}\right\}$$
$$; \text{ total energy of atom} \qquad (5.66)$$

Here we have introduced the *Rydberg* [see Eqs. (3.65) and (3.66)]

$$\mathcal{R} \equiv \frac{e^2}{4\pi\varepsilon_0}\frac{1}{2a_0} = 13.61\,\text{eV} \qquad ; \text{ Rydberg} \qquad (5.67)$$

The result in Eq. (5.66) scales as $Z^{7/3}$, and one can write

$$E = -C_{\text{TF}}Z^{7/3} \qquad (5.68)$$

Now C_{TF} is a universal constant for all atoms calculated from the solution to the Thomas-Fermi Eq. (5.52).

We observe that in this approach to the structure of the many-electron atom, the total energy is written as a *functional* of the electron density[11]

$$E = \int F(n)\,d^3r \qquad ; \text{ density functional}$$
$$F(n) = c_0 n^{5/3}[1 + c_1 V_{\text{C}}(r)n^{-2/3} + \cdots] \quad ; \text{ Thomas-Fermi} \qquad (5.69)$$

One can consider the first equation to be an exact one, and develop from it a density-functional theory for a wide variety of many-electron systems [Kohn (1999)]. Thomas-Fermi gives the first two terms in the high-density expansion in the second line. There will be additional terms in this expansion. For example, one may have a (dimensionless) term $(1/n)(\hbar/mc)(\partial n/\partial r)$ which would become important either at low density or where there is a rapid spatial variation of $n(r)$. Clearly, considering the electron system to be a local, degenerate Fermi gas, as one does in T-F theory, will be in error at vanishingly small electron density.

5.4.3 *Numerical Results*

The Thomas-Fermi Eq. (5.52) can be integrated numerically by starting at the origin where $\chi(0) = 1$ and choosing a slope $\chi'(0)$. One then just steps out in x, and readjusts $\chi'(0)$ until a solution is found that decreases

[11]A *functional* is a function of a function.

to zero for large x. Numerical results obtained using Mathcad11 and the Runge-Kutta algorithm are shown in Fig. 5.11.[12]

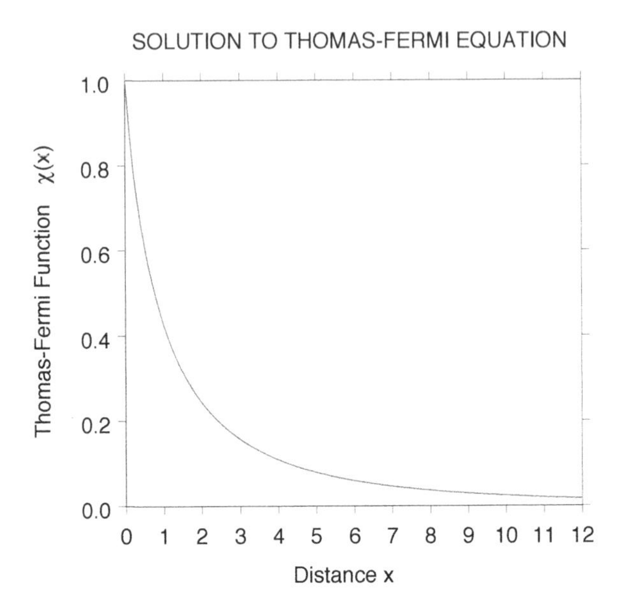

Fig. 5.11 Numerical solution to the Thomas-Fermi Eq. (5.52) plotted out to $x = 12$.

The result for $C_{\rm TF}$ in Eqs. (5.68) and (5.66) is then obtained from numerical integration of the solution $\chi(x)$. We find

$$C_{\rm TF} = 20.9\,{\rm eV} \tag{5.70}$$

A comparison of the expression in Eq. (5.68), and this value of $C_{\rm TF}$, with some empirical values of the total binding energy of atoms is presented in Table 5.1.

We make several comments:

- The quantity $\chi(x)$ in Fig. 5.11 is a universal function of x for all atoms. Since $x = Z^{1/3}r/ba_0$, this implies that the radius of the electron density

[12]It is easiest to start at a very small, but finite x. The calculation shown in Fig. 5.11 starts from $[\chi(x_0), \chi'(x_0)] = [1 + x_0\chi'(x_0), -1.568\cdots]$ with $x_0 = 1 \times 10^{-4}$. The contribution to $C_{\rm TF}$ from $\int_0^{x_0} dx$ was obtained analytically, and the rest of the integral done numerically out to $x = 19.5$, where the remainder is estimated to contribute $< 1\%$.

in Eqs. (5.54) *decreases* as one goes to higher and higher Z — the neutral atoms get *smaller*;

Table 5.1 Total Binding Energy of Atoms.

Atom[a]	$-E^b_{\rm exp}$	$-E_{\rm exp}/Z^{7/3}$	$C_{\rm TF}$
$_{20}$Ca	1.851×10^4 eV	17.0 eV	20.9 eV
$_{26}$Fe	3.463×10^4 eV	17.3 eV	20.9 eV
$_{82}$Pb	$(5.538 \times 10^5$ eV$)^c$	(19.0) eV	20.9 eV

[a] The left subscript on the chemical symbol is Z.
[b] From [Carlson, *et al.* (1970)].
[c] The experimental value is only measured through Fe. This number is from a self-consistent Hartree-Fock calculation.

- The total binding energy of the atom given in Table 5.1 has only been measured experimentally through $_{26}$Fe. The value for $_{82}$Pb is from a self-consistent Hartree-Fock calculation, which agrees with the experimental data in the region of overlap;
- The total binding energy in $_{82}$Pb is greater than the rest energy m_ec^2 of an electron!
- The Thomas-Fermi calculation gives too much binding energy, but comes remarkably close for such a simple approach to a complex many-body system;
- Note that the calculation of $V_{\rm ee}$ incorrectly includes the self-interaction of each electron. One expects this to introduce an error of $O(1/Z)$, and, indeed, the agreement with the Thomas-Fermi calculation improves as Z increases;
- The solution to the Thomas-Fermi equation in Fig. 5.11 is unrealistic at large x. In fact, a decaying analytic solution to the Thomas-Fermi Eq. (5.52) exists

$$\chi(x) \rightarrow \frac{144}{x^3} \qquad ; \; x \rightarrow \infty \tag{5.71}$$

Such a power-law drop off is inconsistent with the exponential fall-off of the bound single-electron wave functions. It has already been pointed out that Thomas-Fermi can be expected to fail at low electron density. A more realistic treatment of the electron density is obtained through the Hartree approximation, which is examined in the next section;
- Numerical integration of Thomas-Fermi theory was used by [Feynman, Metropolis, and Teller (1949)] to obtain an equation of state of atomic matter at high density. This work is remarkable in that it presents one of the first published applications of modern computers to physics.

5.5 Periodic System of the Elements

We next turn to the study of the properties of many-electron atoms, which is the study of the properties of the *elements*.

5.5.1 *Shielded Coulomb Potential*

Start with the nuclear Coulomb potential

$$e\Phi_C(r) = V_{\rm C}(r) = -\frac{Ze^2}{4\pi\varepsilon_0 r} \qquad ; \text{ nuclear Coulomb potential} \tag{5.72}$$

For orientation, the classical newtonian orbits in this potential are sketched in Fig. 4.28, and the quantum mechanical spectrum in Fig. 4.29. Now as we have just learned, the presence of the electrons in the many-electron atom *shields* this potential as one moves out away from the nucleus. In the Thomas-Fermi model, the average potential is given by

$$\begin{aligned} e\Phi &= -\frac{Ze^2}{4\pi\varepsilon_0 r}\chi(x) \qquad ; \text{ Thomas-Fermi potential} \\ x &= \left(\frac{Z^{1/3}}{ba_0}\right) r \end{aligned} \tag{5.73}$$

The Thomas-Fermi shielding function $\chi(x)$ is shown in Fig. 5.11.[13] At the nucleus, one sees the full nuclear Coulomb potential in Eq. (5.72), and it is then reduced by electron screening as one moves away from the nucleus. The situation is illustrated in Fig. 5.12. To examine the single-

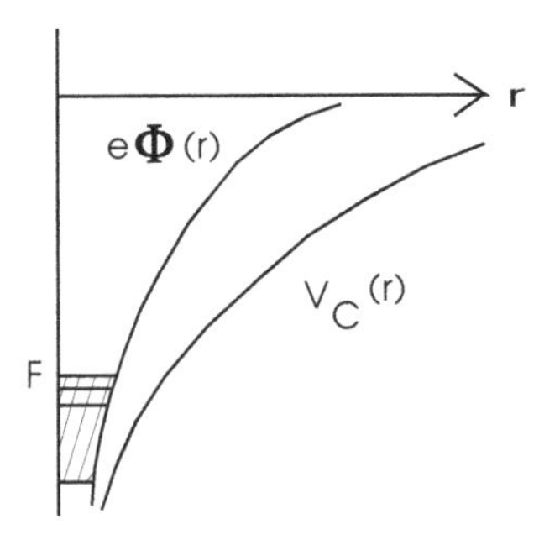

Fig. 5.12 Comparison of point nuclear Coulomb potential $V_{\rm C}(r) = -Ze^2/4\pi\varepsilon_0 r$ with the shielded Thomas-Fermi potential $e\Phi = -(Ze^2/4\pi\varepsilon_0 r)\chi(x)$. Here the single-particle states in the potential $e\Phi$ are filled up to the level F.

[13]The unshielded potential is $\chi(x) = 1$.

particle structure of atoms, one can now imagine solving the single-particle Schrödinger equation numerically in the average potential $e\Phi$. One can then invoke the Pauli exclusion principle and fill these single-particle levels to some Fermi level F.

5.5.2 *Hartree Approximation*

An improved approach to the single-particle structure of atoms is provided by the *Hartree approximation* [Hartree (1928)]. This is still a *mean-field theory.* Here one assumes a set of occupied orbitals, and an electron then moves in the average field created by all the other particles. The Coulomb potential in the Hartree approximation thus takes the form

$$
\begin{aligned}
e\Phi_{\rm H} &= -\frac{Ze^2}{4\pi\varepsilon_0 r} + \frac{e^2}{4\pi\varepsilon_0}\int \frac{n(r')}{|\mathbf{r}-\mathbf{r}'|}d^3r' \qquad ; \text{ Hartree potential} \\
n(r) &= \sum_{\alpha}^{F} |\phi_\alpha(\mathbf{r})|^2
\end{aligned}
\tag{5.74}
$$

We assume *closed shells* in the sense that all the m_l states for a given l are occupied. In this case one can use the relation [Edmonds (1974)]

$$
\sum_{m=-l}^{m=l} |Y_{lm}(\theta,\phi)|^2 = \frac{2l+1}{4\pi} \tag{5.75}
$$

Closed shells give a *spherically symmetric* density distribution. We will thus assume that $n(r)$, and hence $e\Phi_{\rm H}(r)$, are independent of angle.

One can now solve the single-particle Schrödinger equation in this Hartree potential

$$
\left[-\frac{\hbar^2}{2m}\boldsymbol{\nabla}^2 + e\Phi_{\rm H}(r)\right]\phi_\alpha(\mathbf{r}) = \varepsilon_\alpha\phi_\alpha(\mathbf{r}) \tag{5.76}
$$

This presents a *non-linear, self-consistent field problem.* One does not know the Hartree potential until one knows the single-particle wave functions, and one does not know the single-particle wave functions until one knows the Hartree potential. Fortunately, the problem converges well upon iteration. One can, for example, start with the Thomas-Fermi potential and then proceed to modify it upon each iteration.[14]

---

[14]The Hartree approximation, as presented here, again incorrectly includes the self-interaction of each electron. This will again introduce an error of $O(1/Z)$ in the total binding energy. The self-interaction is absent in the improved Hartree-Fock approximation (see Probs. H.3-H.7).

5.5.3 *Structure of the Single-Particle Levels*

The good quantum numbers for a single particle in a spherically symmetric, central potential are

$$\{\bar{n}, l, m_l; s, m_s\} \qquad ; \text{ good quantum numbers} \tag{5.77}$$

We have seen that the level spectrum in the pure Coulomb potential of Eq. (5.72) is that of Fig. 5.13.

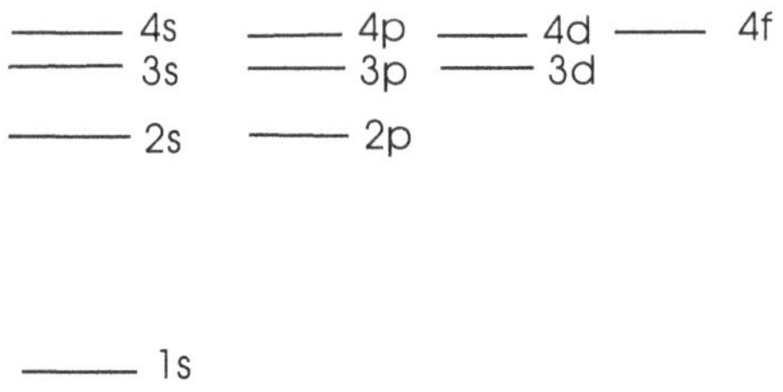

Fig. 5.13 Low-lying single-particle spectrum in the pure Coulomb potential of Eq. (5.72).

This is the "hydrogen-like" spectrum. Consider what happens to this spectrum as one goes over to the potential $e\Phi(r)$ illustrated in Fig. 5.12. We can say quite a bit without actually carrying out the numerical calculation:

- The wave functions that get in *close to the origin* will still feel the very strong, unshielded Coulomb potential. These are the "penetrating orbits";
- The highest-l states for each $\bar{n}$ are the circular orbits. They remain as far away from the nucleus as possible for all time;
- The classical orbits in a $1/r$ potential are sketched in Fig. 4.28. They provide insight into the quantum situation through the *correspondence principle.* In the newtonian case, the center of force is at one focus of the ellipse, and there is a degeneracy reflected in the quantum degeneracy; the energy $E = -Ze^2/8\pi\varepsilon_0 a$ depends only on the semi-major axis a of the ellipse;
- Thus the *penetrating orbits* for each $\bar{n}$ are those of *minimum* l, and the *circular orbits* are those of *maximum* l;
- For each degenerate multiplet of given $\bar{n}$ in Fig. 5.13, therefore, the state with minimum l, that is the s-states ($l = 0$), will be the most tightly bound, and the p-states ($l = 1$) will come next. The circular orbits

with $l = \bar{n} - 1$ will be raised the most by the shielding, and just where they lie with respect to the penetrating orbits of the next multiplet with $\bar{n} + 1$ will depend on the detailed structure of the potential $e\Phi$;

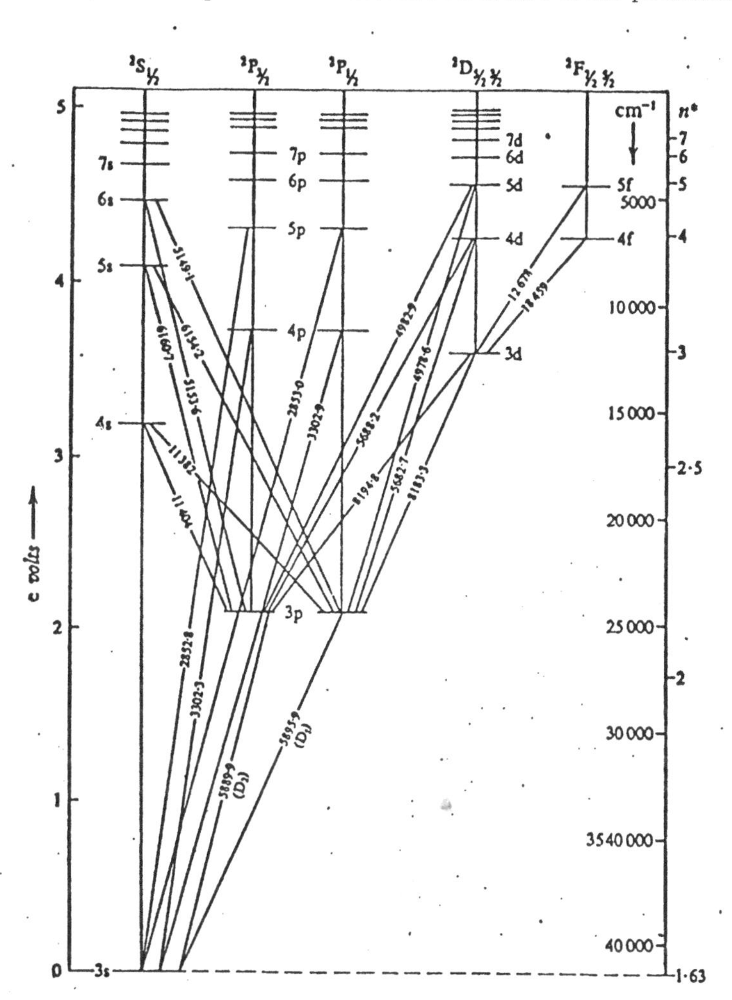

Fig. 5.14 Experimental term diagram for ${}_{11}$Na (from [Kuhn (1969)]). This basically reflects the single-particle level structure of the last valence electron in the spherically symmetric ${}_{11}$Na$^+$ core with electron configuration $(1s)^2(2s)^2(2p)^6$. The left-hand scale is in eV, and the notation at the top is ${}^{2S+1}L_J$ where (S, L, J) are values for the atom. Note the location of the single-particle levels. Note also the fine structure exhibited on the bright yellow "sodium D-lines" involving transitions from the 3p state.

- We note from Fig. 4.28 that the *penetrating orbits are also those that get the farthest away from the nucleus* for each $\bar{n}$![15]

The experimental term diagram for sodium $_{11}$Na, which reflects the single-particle level structure of the last valence electron in a spherically symmetric $_{11}$Na$^+$ core, is shown in Fig. 5.14 (from [Kuhn (1969)]). Notice how all of the features of the previous discussion are manifest here.[16]

The Pauli principle implies that the maximum number of electrons that can be put into each level of given l is $(2l+1)$ for the available m_l values, times 2 for the possible values of m_s

$$\text{max number of electrons in l-shell} = 2(2l+1) \tag{5.78}$$

A simple means for remembering how the single-particle states are actually filled in atoms is presented in Fig. 5.15.

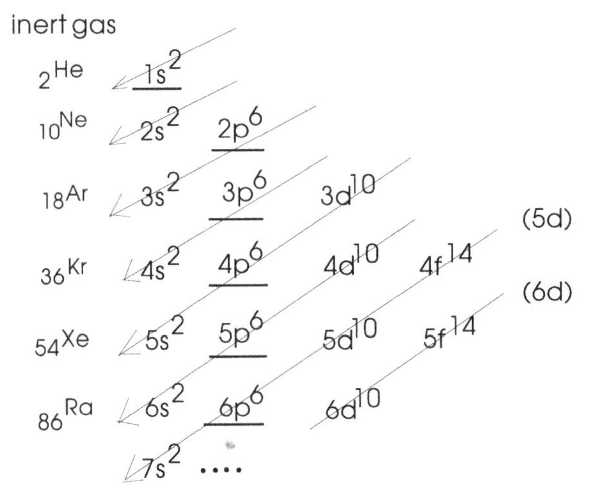

Fig. 5.15 A simple way to remember how the single-particle states are filled in atoms. Start at the top and follow the arrows down. The notation for the levels here is $(\bar{n}l)^n$ where n is the number of electrons that can be accommodated. The locations of the inert gases, identified on the left, are underlined. The electron configuration in $_{35}$Br, for example, is $(1s)^2(2s)^2(2p)^6(3s)^2(3p)^6(4s)^2(3d)^{10}(4p)^5$. In some cases, the $(5d, 6d)$ may be filled before the $(4f, 5f)$.

The radial charge density calculated by Hartree for $_{37}$Rb$^+$, with an electron configuration $(1s)^2(2s)^2(2p)^6(3s)^2(3p)^6(4s)^2(3d)^{10}(4p)^6$, is shown

[15]These correspond to the orbits of the comets in the solar system, while the planets occupy the (essentially) circular orbits.

[16]Atomic data throughout the periodic table can be found in [NIST (2007)].

in Fig. 5.16. Note the change in scale as a function of r/a_0, and how it is the $4p$ and $4s$ wave functions that extend out the farthest in the atom. Note also the oscillations in the density about a smooth (Thomas-Fermi) value.

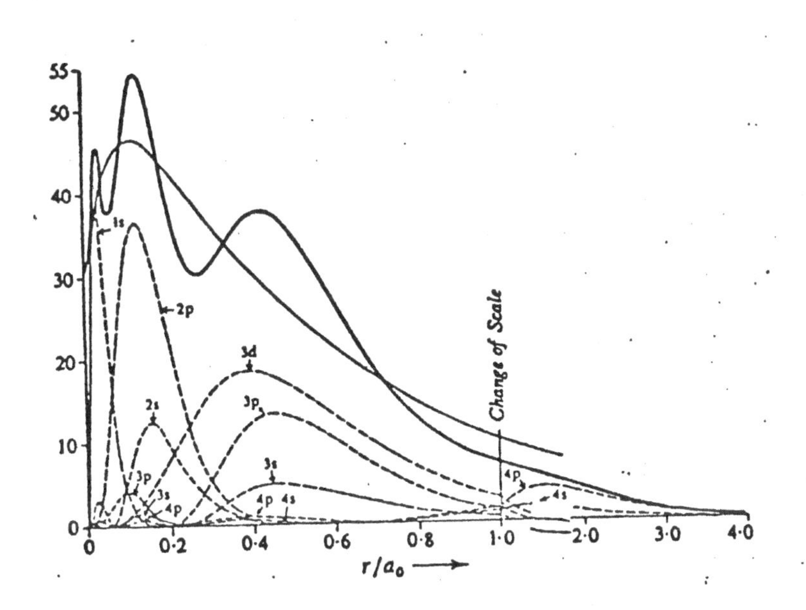

Fig. 5.16 Radial charge density of $_{37}Rb^+$ (D. R. Hartree, ref 8). Note the change of scale as a function of r/a_0, and how it is the $4p$ and $4s$ wave functions that extend out the farthest in the atom. Note also the oscillations in the density about a smooth (Thomas-Fermi) value. What is plotted here, in our previous notation, is $4\pi r^2 n(r)$. From [Kuhn (1969)].

5.5.4 *Chemical Properties of the Elements*

The level spectrum in the potential $e\Phi$ in Fig. 5.12, which reflects that obtained in the Hartree approximation, is sketched in Fig. 5.17. We observe the following:

- Particularly stable are the spherically symmetric atoms with closed s- and p-shells at

$$Z = 2, 10, 18, 36, 54, 86 \qquad ; \text{ inert gases} \tag{5.79}$$

These form the *inert gases*. They are chemically inactive;

- With one additional valence s-electron and an inert-gas core, that electron is only loosely bound and its wave function extends far out from the atom. These elements are very chemically reactive and are electron *donors*. They are known as the *alkali metals*;

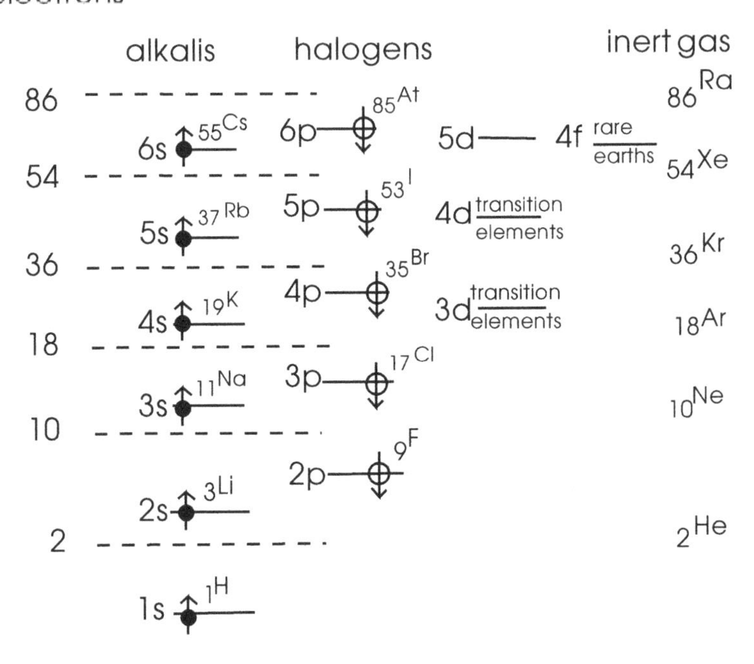

Fig. 5.17 Principal features of the single-particle structure of atoms.

- With one vacancy, or "hole", in the p-shell in an inert-gas atom, adding an additional electron leads to a particularly stable configuration, and the wave function for the state to be occupied again extends far from the atom. These elements can readily accept electrons. They are chemically reactive and are electron *receptors*. They are known as the *halogens*;
- The chemical properties of the remaining elements are governed by those electrons that extend the farthest out from the atomic core. These are the s and p electrons;
- When one is filling the highest l shells for each $\bar{n}$ (the "inner shells"), one is adding electrons to the circular orbits, which changes only the

shielding of the nuclear core and has very little effect on the chemical properties. Thus one has the *transition elements* when the $3d$, $4d$, and $5d$ shells are being filled, and the *rare earths* when the $4f$ shell is being filled.

The "standard periodic table" of the elements obtained by systematically filling the single-particle states is presented in Fig. 5.18.

The behavior of electrons *between* atoms leads to the three basic types of *chemical bonds*:

- Electrons can be *transferred* between atoms. This leads to positively and negatively charged remaining atomic cores held together by electrostatic attraction, and produces an *ionic bond*. This transfer is most readily carried out between a reactive alkali (*e.g.* Na) and reactive halogen (*e.g.* Cl) leading to a very stable ionic solid (*e.g.* NaCl, or *salt*);
- Electrons can be *shared* between atoms. Here a wave function formed from the long-range s and p electron wave functions samples the electrostatic attraction of *both nuclei*, and, in favorable cases, this can lead to stable systems. Since the p-states with $m_l = 0, \pm 1$ are directional, one can get complex stereo configurations (*e.g.* carbon compounds). This is the basis of *organic chemistry*;
- A hydrogen ion H^+ is really an entirely different beast, since it is actually just a tiny proton. The negative electrons can share this charge, leading to systems with *hydrogen bonds*.

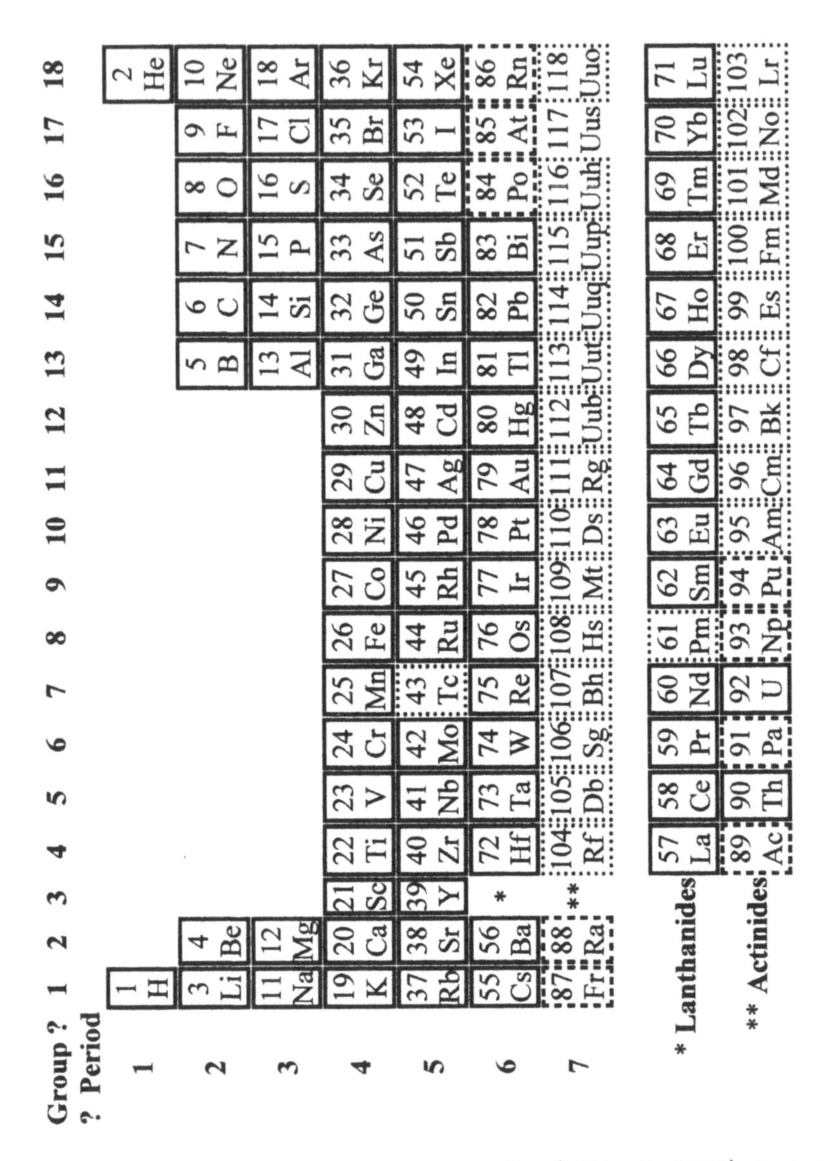

Fig. 5.18 Standard periodic table of the elements from [Wikipedia (2007)]. Go to the website and click on an element. This brings out a wealth of information about each one (*e.g.* 9 pages of detail for sodium). There are several good, alternative websites with similar configurations.

Chapter 6

Nuclear Physics

We turn to the topic of *nuclear physics*.[1] There are three primary forces manifest in nature (in addition to gravity): strong, electromagnetic, and weak. They differ in their strength by several orders of magnitude.[2] The structure of atoms is governed by the *electromagnetic* interaction. The behavior of nuclei is governed by the *strong* interaction. We start the discussion with the definition of a baryon.

6.1 Baryons

To the best of our knowledge, *baryon number* is an exactly conserved quantity, just like electric charge. The nucleons, protons and neutrons, are baryons with baryon number $B = +1$.

$B = 1$. Consider the system with $B = 1$. For the present purposes, this is a single *nucleon* — proton or neutron. The neutron is just slightly heavier than the proton, and the measured mass of these particles is

$$\begin{aligned} m_n c^2 &= 939.6\,\text{MeV} \qquad &; \text{neutron} \\ m_p c^2 &= 938.3\,\text{MeV} \qquad &; \text{proton} \end{aligned} \tag{6.1}$$

Each of these particles has an intrinsic angular momentum, or *spin*, of J=1/2.

They can also be assigned an intrinsic *parity* of $\pi = +1$. Parity represents the behavior of the wave function under spatial reflection. If we return to our discussion of quantum mechanics in one dimension, then if

[1] See [Wong (1999); Preston and Bhaduri (1982); Walecka (2004)].

[2] We shall later become more quantitative here. A major achievement in modern physics is the unification, through the "standard model", of the *electroweak* interactions — see later.

the potential is even under spatial reflections with $V(-x) = V(x)$, the wave function must be either even or odd under these reflections. This was explictly illustrated in the case of the simple harmonic oscillator, where the wave functions alternate in parity as the energy increases (Fig. 6.1). If we denote the intrinsic parity with a superscript, then the nucleons have $J^\pi = \frac{1}{2}^+$. Parity is conserved in the strong interactions (see Prob. 6.5).

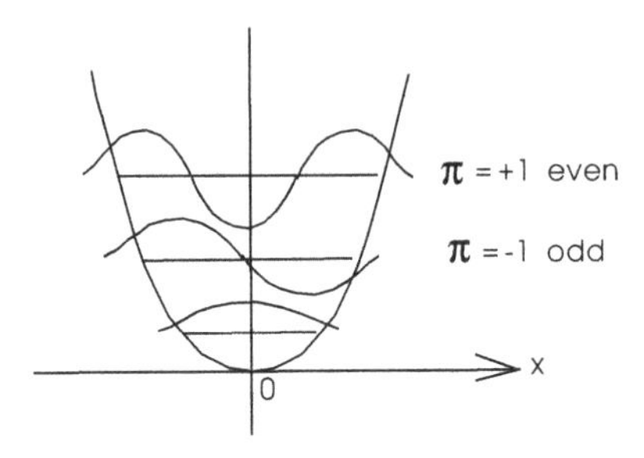

Fig. 6.1 Behavior under spatial reflection, the *parity*, of the eigenstates of the one-dimensional simple harmonic oscillator.

We can thus make an energy-level diagram for the nucleon, and this is shown in Fig. 6.2. Here Z denotes the electric charge, an exactly conserved quantity in all interactions.

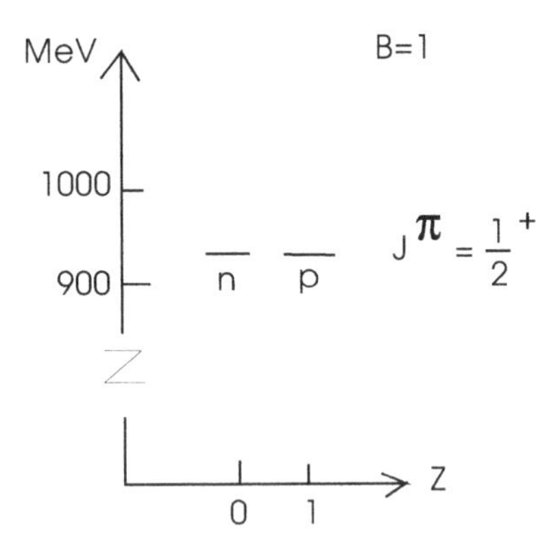

Fig. 6.2 Energy level diagram for the nucleon, a baryon with $B = +1$. Here Z denotes the electric charge.

6.2 β-decay

Nucleons can undergo β-decay through the *weak interaction*, with the emission of an electron and a neutrino. The basic β-decay process is

$$n \rightarrow p + e^- + \bar{\nu}_e \qquad ; \ \beta^-\text{-decay} \tag{6.2}$$

where $\bar{\nu}_e$ is, more specifically, an electron antineutrino (see later). The neutrino is an essentially massless, spin-1/2 particle, which interacts only very weakly with matter. The neutrino was originally introduced by Pauli to explain the apparent lack of energy conservation in the β-decay process. The neutron is sufficiently heavier than the hydrogen atom that the process in Eq. (6.2) can proceed for a free neutron. We denote this process on an energy level diagram as illustrated in Fig. 6.3.

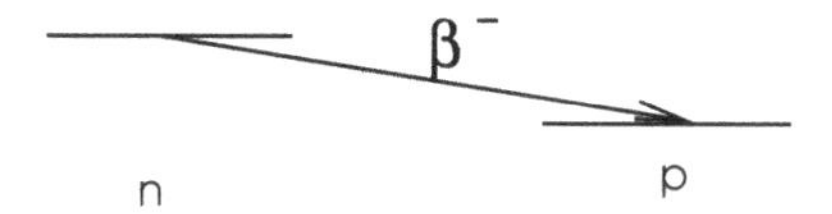

Fig. 6.3 Illustration on the energy-level diagram of the β-decay process for a neutron.

There are additional β-decay processes that can take place in nuclei when allowed by the atomic masses involved. They are

$$\begin{aligned} p &\rightarrow n + e^+ + \nu_e \qquad &&; \ \beta^+\text{-decay} \\ e^- + p &\rightarrow n + \nu_e \qquad &&; \ \text{electron capture} \end{aligned} \tag{6.3}$$

Here e^+ is a *positron*, the antiparticle of the electron, and ν_e is an electron neutrino.

6.3 Mean Life

Suppose one has a sample of N independent decaying systems. The number of decays dN in the time dt will be proportional to the number of systems present, and can be written as

$$dN = -\omega N dt \qquad ; \omega \equiv \text{decay rate} \tag{6.4}$$

Here ω is the constant *decay rate*. This equation can be rearranged to read

$$\frac{dN}{N} = -\omega dt \tag{6.5}$$

This relation can be integrated, with the initial condition $N = N_0$ at $t = 0$, to give

$$\ln\left(\frac{N}{N_0}\right) = -\omega t \tag{6.6}$$

Hence

$$\begin{aligned} N &= N_0 e^{-\omega t} \equiv N_0 e^{-t/\tau} \\ \tau &= \frac{1}{\omega} \qquad ; \text{ mean life} \end{aligned} \tag{6.7}$$

Here τ is the *mean life* of a system.

The *half-life* is defined as the time it takes for the number of systems to be reduced to 1/2 of the original number, so that

$$\frac{N}{N_0} = \frac{1}{2} = e^{-\omega t_{1/2}} \tag{6.8}$$

Therefore

$$\begin{aligned} \ln\frac{1}{2} &= -\omega t_{1/2} = -\frac{t_{1/2}}{\tau} \\ t_{1/2} &= \tau \ln 2 \qquad ; \text{ half-life} \end{aligned} \tag{6.9}$$

The free neutron decays with a mean life of [Particle Data Group (2006)]

$$\tau_n = 885.7 \,\text{sec} \qquad ; \text{ neutron mean life} \tag{6.10}$$

6.4 Deuteron

The nuclear system of two baryons has one bound state, the *deuteron*, composed of a proton and a neutron (pn). Two protons $(p)^2$, or two neutrons $(n)^2$, by themselves are unbound. The quantum numbers of the deuteron are

$$J^\pi = 1^+ \qquad ; \text{ deuteron} \tag{6.11}$$

The level spectrum for $B = 2$ is shown in Fig. 6.4. Here and henceforth we use the following notation for a nucleus

$${}^2_1\text{H} \equiv {}^B_Z[\text{chemical symbol}] \qquad ; \text{ notation for nucleus} \tag{6.12}$$

The *nucleon number* A of a nucleus is the sum of the number of protons (Z) and neutrons (N) of which it is composed. The nucleon number is here identical to the baryon number, and we shall use the terms interchangeably.

$$A \equiv B = N + Z \qquad ; \text{ nucleon number} \tag{6.13}$$

Note that

- Energies in atoms are $O(\text{eV})$;
- Energies in nuclei are $O(\text{MeV}) = O(10^6\,\text{eV})$.

This is reflection of the difference in strength of the strong and electromagnetic interactions.

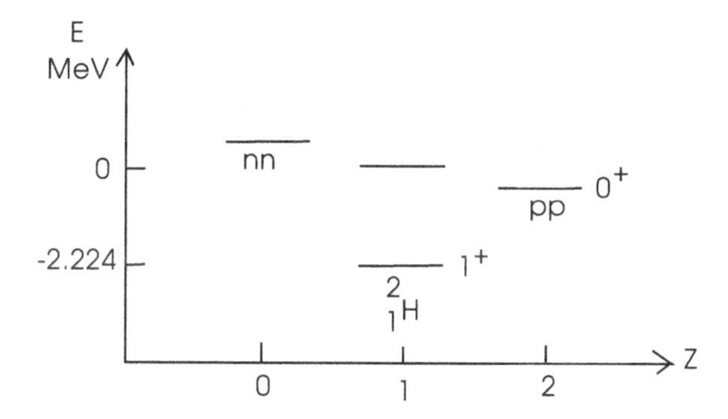

Fig. 6.4 Nuclear level diagram for $B = 2$ (atomic masses—see later). There is one bound state, the deuteron $^2_1$H, with $J^\pi = 1^+$ and a binding energy of 2.224 MeV.

In quantum mechanics, just as in classical mechanics, the two-body problem in a potential that depends only on the relative coordinate $V(r) = V(|\mathbf{r}_2 - \mathbf{r}_1|)$ can always be reduced to a *one-body problem in the relative coordinate* in the center-of-mass system, provided one uses the reduced mass μ defined by[3]

$$\frac{1}{\mu} = \frac{1}{m_1} + \frac{1}{m_2} \qquad ; \text{ reduced mass} \tag{6.14}$$

For the one-body problem in spherical coordinates in quantum mechanics, we know from Eqs. (4.141) and (4.142) that the solution to the time-

[3]See Prob. 6.1. Here the center-of-mass moves freely.

independent Schrödinger equation takes the form

$$\psi(r,\theta,\phi) = \frac{u(r)}{r} Y_{lm}(\theta,\phi) \tag{6.15}$$

where

$$\frac{d^2u(r)}{dr^2} + \left\{ \frac{2\mu}{\hbar^2} [E - V(r)] - \frac{l(l+1)}{r^2} \right\} u(r) = 0 \tag{6.16}$$

For s-waves with $l = 0$, this equation reduces to

$$\frac{d^2u(r)}{dr^2} + \frac{2\mu}{\hbar^2} [E - V(r)]\, u(r) = 0 \quad ; \text{ s-waves} \tag{6.17}$$

The proper boundary condition for s-waves at the origin is, from Eq. (4.144),

$$u(0) = 0 \qquad ; \text{ B. C. at origin} \tag{6.18}$$

The origin acts as a "wall" in the one-dimensional radial equation in spherical coordinates, since $u(r)/r \rightarrow \text{constant}/r$ is too singular as $r \rightarrow 0$ (Prob. 4.27).

We may look for *bound-state* solutions to Eq. (6.17) with $E < 0$

$$E = -|E| \qquad ; \text{ bound state} \tag{6.19}$$

In this case Eq. (6.17) becomes

$$\frac{d^2u(r)}{dr^2} - \frac{2\mu}{\hbar^2} [|E| + V(r)]\, u(r) = 0 \qquad ; \text{ bound state} \tag{6.20}$$

Let us assume a very simple attractive square-well potential

$$\begin{aligned} V(r) &= -V_0 && ; r \leq R \\ &= 0 && ; r > R \end{aligned} \tag{6.21}$$

The energy will always lie above $-V_0$ (or, equivalently, $V_0 - |E| > 0$) since there will always be some additional, positive kinetic energy in the problem. In this case, the solutions to Eq. (6.20) inside and outside of the potential are easily written down (see Fig. 6.5)

$$\begin{aligned} u_I &= \mathcal{A} \sin \kappa r + \mathcal{B} \cos \kappa r && ; \kappa^2 \equiv \frac{2\mu}{\hbar^2}(V_0 - |E|) \\ u_{II} &= \mathcal{C} e^{-\gamma r} + \mathcal{D} e^{\gamma r} && \gamma^2 \equiv \frac{2\mu}{\hbar^2}|E| \end{aligned} \tag{6.22}$$

Here $(\mathcal{A}, \mathcal{B}, \mathcal{C}, \mathcal{D})$ are constants to be determined.

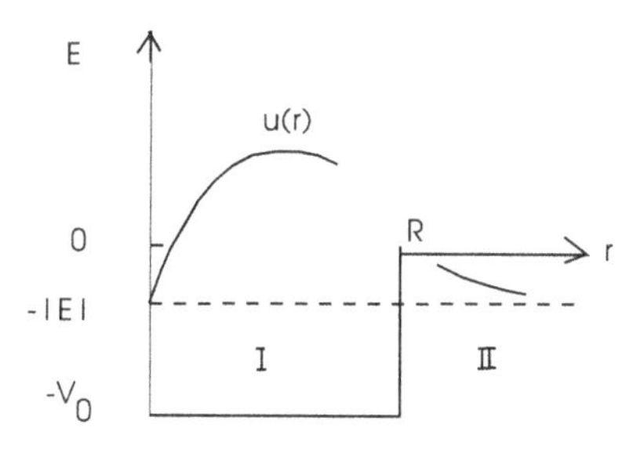

Fig. 6.5 S-wave bound-state wave function $u(r)$ in the attractive square-well potential of Eq. (6.21) plotted relative to the dashed axis. The solutions in regions I ($r \leq R$) and II ($r > R$) are given by Eqs. (6.22) matched to boundary conditions (1)–(3).

We may now impose boundary conditions on the general solution in Eq. (6.22):

(1) The condition $u(0) = 0$ implies $\mathcal{B} = 0$;
(2) The condition of normalizability implies $\mathcal{D} = 0$.
(3) The wave function and its derivative (u, u') must be continuous at $r = R$.[4]

Boundary conditions (3) can now be imposed in a manner that avoids knowledge of the constants (A, C) by matching the *logarithmic derivative* at $r = R$

$$\frac{u'_I}{u_I} = \frac{u'_{II}}{u_{II}} \qquad ; \ r = R \tag{6.23}$$

This leads to the relation

$$\kappa \cot \kappa R = -\gamma \qquad ; \ \text{eigenvalue eqn} \tag{6.24}$$

This is an *eigenvalue equation* for the energy. For only certain values of $|E|$ will it be possible to match a solution that vanishes at the origin to one that decreases exponentially as $r \to \infty$.

Suppose there is just one bound state with vanishingly small binding energy, so that $|E| \to 0$.[5] The solution to the eigenvalue Eq. (6.24) for $\gamma \to 0$ is then

$$\kappa R = \frac{\pi}{2} \qquad ; \ \text{bound state with } |E| = 0 \tag{6.25}$$

[4]Compare Prob. 6.7.
[5]This is to be contrasted to the situation in a Coulomb potential where there will always be an *infinite* number of bound states!

This says that there should be just 1/4 wavelength inside the potential. With the aid of the definition of κ in Eqs. (6.22), Eq. (6.25) can be rearranged to read

$$V_0 R^2 = \frac{\hbar^2 \pi^2}{8\mu} \tag{6.26}$$

For two equal-mass particles, each of mass m, one has $\mu = m/2$. Equation (6.26) then becomes

$$V_0 R^2 = \frac{\hbar^2 \pi^2}{4m} \qquad ; \; |E| = 0 \text{ bound state} \quad \mu = m/2 \tag{6.27}$$

This is an extremely useful relation that fixes the combination $V_0 R^2$ of depth and range for an attractive square-well potential that has just one bound state at zero energy.[6]

Let us put in some numbers. Introduce the unit of a "Fermi" where

$$1\,\text{Fermi} = 1\,\text{F} \equiv 10^{-13}\,\text{cm} = 10^{-15}\,\text{m} \tag{6.28}$$

Now assume that $R = 2\,\text{F}$, which is a typical nuclear distance scale (see later). Equation (6.27) then gives[7]

$$V_0 = 25.5\,\text{MeV} \qquad ; \; R = 2\,\text{F} \tag{6.29}$$

This gives a typical depth for the N-N potential.

We make several comments on these results:

- We shall use the following general labeling of the wave function

$$\psi_{nlm}(\mathbf{r}) = \frac{u_{nl}(r)}{r} Y_{lm}(\theta, \phi) \tag{6.30}$$

  Here n is the number of nodes in the radial wave function $u_{nl}(r)$, including the origin and excluding the point at infinity. Thus the bound-state in Eqs. (6.22)–(6.24) will be denoted by

$$\psi_{100}(\mathbf{r}) = \frac{u_{10}(r)}{r} Y_{00} \tag{6.31}$$

  Let us then examine the behavior of this bound-state wave function $u_{10}(r)$ as the binding energy goes to zero (Fig. 6.6).

[6] The deuteron has only one bound state, and the binding energy is not very large on the nuclear scale [see Eq. (6.29)].

[7] Two convenient numbers to remember in nuclear physics are $\hbar c = 197.3\,\text{MeV-F}$, and $\hbar^2/2m_p = 20.7\,\text{MeV-F}^2$.

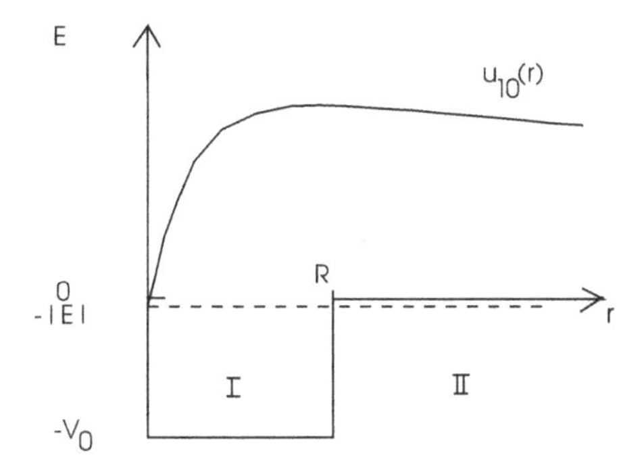

Fig. 6.6 Bound state as $|E| \to 0$ in potential in Fig. 6.5.

- The probability density is given by the square of the wave function in Eq. (6.31), so that[8]

$$|\psi_{100}(\mathbf{r})|^2 4\pi r^2 dr = |u_{10}(r)|^2 dr \tag{6.32}$$

  It is clear from Fig. 6.6 that as $|E| \to 0$, the particle *spends most of its time outside of the potential*, a region where it is forbidden to be classically [see Eq. (4.96)] — this is pure quantum mechanics;
- Most of the normalization integral will now come from region II in Fig. 6.6, and as $|E| \to 0$, *all* of it will come from this region. Thus we can compute the normalization of the wave function in Eq. (6.22) in this limit as

$$\int_0^\infty |\mathcal{C}|^2 e^{-2\gamma r} dr = |\mathcal{C}|^2 \frac{1}{2\gamma} = 1$$
$$\Rightarrow \qquad \mathcal{C} = \sqrt{2\gamma} \tag{6.33}$$

  where we have chosen to make $\mathcal{C}$ real. Hence in this limit, the radial wave function of the bound-state takes the form

$$u_{10}(r) \approx \sqrt{2\gamma}\, e^{-\gamma r} \qquad ; \ |E| \to 0 \tag{6.34}$$

  and this result holds almost everywhere;
- Now note that the size of the probability distribution in Eq. (6.32) has *nothing to do with the range of the potential R, but only depends on the binding energy through* $\gamma = \left(2\mu|E|/\hbar^2\right)^{1/2}$!

[8]Recall $Y_{00} = 1/\sqrt{4\pi}$.

- The *parity* of the wave function is the behavior under spatial reflection, where (see Fig. 4.26)

$$r \to r \qquad ; \; \theta \to \pi - \theta \qquad ; \; \phi \to \phi + 2\pi \tag{6.35}$$

  The spherical harmonics behave the following way under spatial reflection [Edmonds (1974)]

$$Y_{lm}(\pi - \theta, \phi + 2\pi) = (-1)^l Y_{lm}(\theta, \phi) \tag{6.36}$$

  The behavior of the wave function in Eq. (6.30) is therefore

$$\psi_{nlm}(-\mathbf{r}) = (-1)^l \psi_{nlm}(\mathbf{r}) \qquad ; \text{ parity} \tag{6.37}$$

  Thus the wave function in Eq. (6.31) has positive parity, in accord with that of the deuteron;
- The wave function in Eq. (6.31) has angular momentum zero, so where does the angular momentum of the deuteron come from? Each of the nucleons composing it has spin 1/2. These spins must therefore add to give unit total angular momentum

$$\mathbf{S} = \mathbf{s}_p + \mathbf{s}_n \qquad ; \; \mathbf{S}^2 = S(S+1)$$
$$S = 1 \tag{6.38}$$

  We refer to $S = 1$ as the *triplet* state, while $S = 0$ is the *singlet state*;

In this fashion, we have an understanding of the spatial structure and quantum numbers of the ground state of the simplest nucleus $^2_1$H, the deuteron.

6.5 Atomic Masses

Atomic masses are tabulated and readily available to the reader [Lawrence Berkeley Laboratory (2003)]. We make some comments concerning these tables:

- What are tabulated are *atomic masses*. These include the masses of the electrons of the neutral atom, as well as the atomic binding energy;
- Atomic masses are measured in *atomic mass units*, where the mass of the ^{12}C atom is defined to be 12 u

$$Mc^2[^{12}\text{C}] \equiv 12\,\text{u} \qquad ; \; \text{u} = 931.494\,\text{MeV} \tag{6.39}$$

- A most useful quantity is the *mass excess* defined for a nucleus of given (A, Z) and atomic mass M by

$$\Delta \equiv \left[\frac{Mc^2}{\mathrm{u}} - A\right] \mathrm{u} \qquad ; \text{ mass excess} \tag{6.40}$$

Measurements of Δ then ultimately involve measurements of mass ratios to ^{12}C. Values of Δ are tabulated in KeV;

- The *total* energy of a massive, non-relativistic particle is[9]

$$E_i = m_i c^2 + \frac{\mathbf{p}_i^2}{2m_i} \tag{6.41}$$

Energy conservation in the reaction $i \rightarrow f$ then states that

$$\sum_i E_i = \sum_f E_f \qquad ; \text{ energy conservation} \tag{6.42}$$

- The *Q-value* of a nuclear reaction is defined by

$$\begin{aligned} Q &\equiv \sum_f m_f c^2 - \sum_i m_i c^2 \\ &= \Delta_f - \Delta_i \end{aligned} \tag{6.43}$$

The second line follows from the fact that baryon number is conserved in any reaction.

  - A reaction is *exothermic* if the *Q-value is negative.* It will proceed with a release of kinetic energy;
  - A reaction will be *endothermic* if the *Q-value is positive.* It will require an input of kinetic energy to proceed.

Values of Δ for the nucleons and lightest nuclei are given in Table 6.1.

Table 6.1 Atomic mass excess $\Delta = [Mc^2/\mathrm{u} - A]\mathrm{u}$ in MeV.[a]

Atom	^{1}n	^{1}H	^{2}H	^{3}H	^{3}He	^{4}He	^{6}He	^{6}Li
Δ(MeV)[b]	8.071	7.289	13.136	14.950	14.931	2.425	17.594	14.086

[a] Note $Mc^2[^{12}\mathrm{C}] \equiv 12\,\mathrm{u}$, and $\mathrm{u} = 931.494\,\mathrm{MeV}$.
Remember atomic masses include the electrons in the neutral atom and atomic binding.
[b] From [Lawrence Berkeley Laboratory (2003)].

[9] Here we invoke Einstein's celebrated relation between mass and energy (see later). The energy of a massless photon is given by $E_i = p_i c$ [see Eq. (3.50)].

6.6 Light Nuclei

We can make use of this table to calculate the *binding energy* of the lightest nuclear systems[10]

$$\begin{aligned}
\Delta(^1\mathrm{n}) + \Delta(^1\mathrm{H}) - \Delta(^2\mathrm{H}) &= 2.224\,\mathrm{MeV} && ;\ \mathrm{B.\ E.\ of}\ {}^2_1\mathrm{H} \\
\Delta(^2\mathrm{H}) + \Delta(^1\mathrm{n}) - \Delta(^3\mathrm{H}) &= 6.257\,\mathrm{MeV} && ;\ \mathrm{B.\ E.\ of}\ {}^3_1\mathrm{H} \\
\Delta(^3\mathrm{H}) + \Delta(^1\mathrm{p}) - \Delta(^4\mathrm{He}) &= 19.814\,\mathrm{MeV} && ;\ \mathrm{B.\ E.\ of}\ {}^4_2\mathrm{He}
\end{aligned} \tag{6.44}$$

The difference in ground-state energy of some *isobars* (nuclei with the same number of baryons) follows from Table 6.1 according to

$$\begin{aligned}
\Delta(^3\mathrm{H}) - \Delta(^3\mathrm{He}) &= 0.019\,\mathrm{Mev} && ;\ Mc^2[^3\mathrm{H}] - Mc^2[^3\mathrm{He}] \\
\Delta(^6\mathrm{He}) - \Delta(^6\mathrm{Li}) &= 3.508\,\mathrm{Mev} && ;\ Mc^2[^6\mathrm{He}] - Mc^2[^6\mathrm{Li}]
\end{aligned} \tag{6.45}$$

Remember that these are atomic masses.

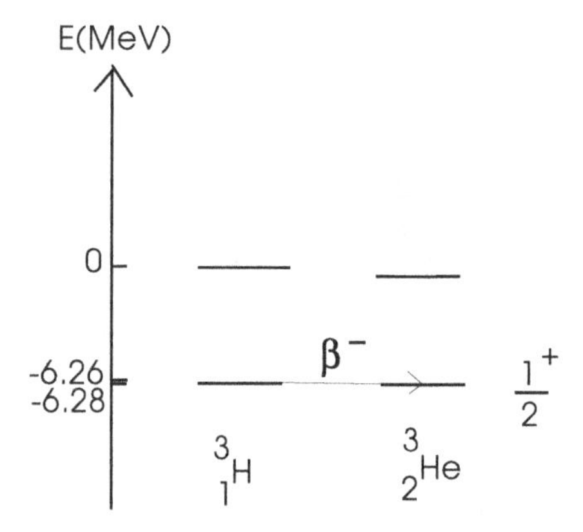

Fig. 6.7 Energy level diagram for $B = 3$ (atomic masses). There are no bound excited states in either ${}^3_1$H (the heavier nucleus) or ${}^3_2$He. Both $(n)^3$ and $(p)^3$ are unbound.

$B = 3$. The energy level diagram for the $B = 3$ system is shown in Fig. (6.7). The ground-state binding energies are calculated above. Neither ${}^3_1$H nor ${}^3_2$He has a bound excited state. Tritium (${}^3_1$H) is sufficiently heavier than ${}^3_2$He that the following nuclear β-decay process takes place

$${}^3_1\mathrm{H} \rightarrow {}^3_2\mathrm{He} + e^- + \bar{\nu}_e \tag{6.46}$$

[10]This is the minimum energy required to cause a given nucleus to disintegrate.

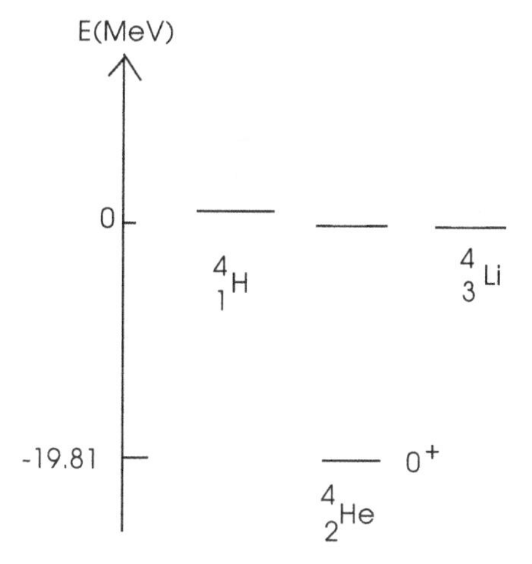

Fig. 6.8 Energy level diagram for $B = 4$ (atomic masses). There is only one stable system, $^4_2$He, and it has no bound excited states.

$B = 4$. The energy level diagram for $B = 4$ is shown in Fig. 6.8. There is only one stable nucleus for this system, $^4_2$He (whose ionized form is the *alpha particle*), and $^4_2$He has no bound excited states.

We make some additional comments on the lightest nuclei:

- The system $B = 5$ is unbound;
- There are only *five stable nuclei* up through $B = 6$, and they are[11]

$$^1_1\text{H},\ ^2_1\text{H},\ ^3_2\text{He},\ ^4_2\text{He},\ ^6_3\text{Li} \qquad ;\ \text{stable nuclei B} \leq 6 \tag{6.47}$$

- The nucleus $^6_3$Li is the first one to have bound excited states;
- The primary nuclear reaction for energy generation in the sun is

$$p + p \rightarrow {}^2_1\text{H} + e^+ + \nu_e \tag{6.48}$$

6.7 Semi-Empirical Mass Formula

The goal of the *semi-empirical mass formula* is to provide a concise description of nuclear binding energies throughout the periodic table. In this approach, due to Weizsäcker (1935), a nucleus composed of B baryons is viewed as an incompressible, uniformly-charged *liquid drop* with a radius R and total charge Ze_p (Fig. 6.9).

[11] To the best of our knowledge, isolated protons and electrons are absolutely stable.

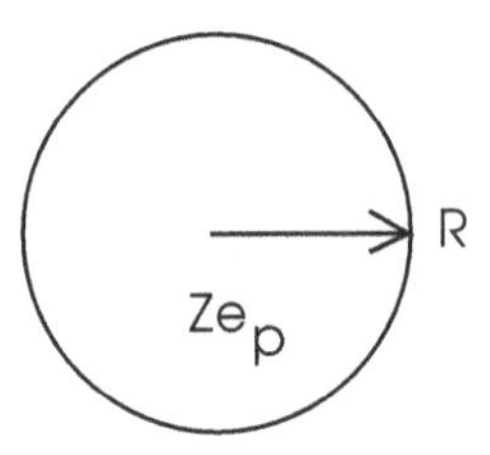

Fig. 6.9 Nucleus of B baryons viewed as a uniformly charged liquid drop with radius R and total charge Ze_p.

One can then identify several contributions to the energy:

6.7.1 *Bulk Properties*

Since the drop is incompressible, its radius must grow as $B^{1/3}$

$$R = r_0 B^{1/3} \qquad ; \; r_0 = \text{constant} \tag{6.49}$$

Here r_0 is a constant that characterizes the equivalent uniform mass distribution of a nucleus. It follows from Eq. (6.49) that the volume of the drop grows with B, as required

$$V = \frac{4\pi}{3} R^3 = \left(\frac{4\pi}{3} r_0^3\right) B \tag{6.50}$$

Just as with the ordinary condensation energy of a drop, there will be one term in the energy of the bound system proportional to the total number of baryons (and hence to the volume)

$$E_b = -a_1 B \qquad ; \; \text{bulk energy} \tag{6.51}$$

Equations (6.49) and (6.51) describe the *bulk properties* of nuclear matter.

6.7.2 *Surface Energy*

Again, as with an ordinary drop, *surface tension* produces a positive *surface energy* proportional to the surface area of the drop [recall Fig. 4.20 and Eq. (4.103)]

$$\begin{aligned} E_s &= +4\pi\sigma R^2 = (4\pi r_0^2) B^{2/3} \\ &\equiv +a_2 B^{2/3} \qquad ; \; \text{surface energy} \end{aligned} \tag{6.52}$$

Here σ is the surface tension.

6.7.3 *Coulomb Energy*

In this picture, there are $Z(Z-1)/2$ pairs of protons distributed uniformly over a sphere of radius R. A simple calculation in electrostatics (Prob. 6.9) then provides the repulsive *Coulomb Energy* in the drop

$$\begin{aligned} E_c &= \frac{6}{5}\frac{e^2}{4\pi\varepsilon_0 R}\frac{1}{2}Z(Z-1) = \frac{3}{5}\frac{e^2}{4\pi\varepsilon_0 r_{0c}}\frac{Z(Z-1)}{B^{1/3}} \\ &\equiv +a_3\frac{Z^2}{B^{1/3}} \qquad ; \text{ Coulomb energy} \end{aligned} \tag{6.53}$$

Here r_{0c} describes the equivalent uniform charge distribution of the nucleus; it may, or may not, be identical to r_0.

Two specific *nuclear effects* are now included:

6.7.4 *Symmetry Energy*

It is observed that nuclei like to have equal numbers of neutrons and protons so that $N = Z$. A *symmetry energy* is then included in the semi-empirical mass formula. A quadratic approximation is made so that the symmetry energy goes as $[(N-Z)/(N+Z)]^2 = (B-2Z)^2/B^2$, and the bulk property of nuclear matter is then invoked so that the whole term is again proportional to B (compare Prob. 6.17)

$$\begin{aligned} E_{sym} &= +a_4\left(\frac{B-2Z}{B}\right)^2 B \\ &= +a_4\frac{(B-2Z)^2}{B} \qquad ; \text{ symmetry energy} \end{aligned} \tag{6.54}$$

6.7.5 *Pairing Energy*

It is observed that nuclei prefer to have even numbers of like particles, protons or neutrons. The isobaric mass surfaces for even B and even-even and odd-odd (N, Z), as well as for odd B, are observed to have the structure illustrated in Fig. 6.10.

There are a few empirical facts of special interest here:

- There are only four stable odd-odd nuclei in nature

$${}^{2}_{1}\text{H},\ {}^{6}_{3}\text{Li},\ {}^{10}_{5}\text{B},\ {}^{14}_{7}\text{N} \qquad ; \text{ stable odd-odd nuclei} \tag{6.55}$$

- There is only one stable odd-B isobar;

- For even B, there may be two or more stable nuclei with even Z and even N

These features are readily understood on the basis of the energy surfaces in Fig. 6.10 and the following theorem (Prob. 6.6)

Of two nuclei with given B, and Z differing by one, at least one is unstable under the β-decay interactions in Eqs. (6.2)–(6.3).

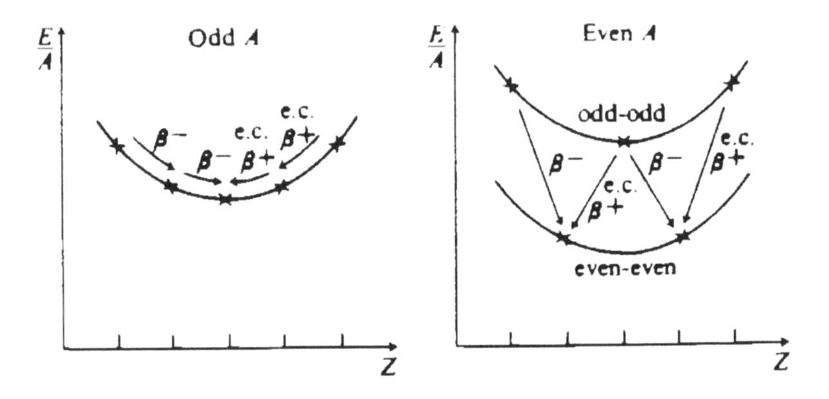

Fig. 6.10 Observed behavior of isobaric mass surfaces for even B and even-even and odd-odd (N, Z), as well as for odd B; here $A \equiv B$. The possible $\beta^{\pm}$ decay and electron capture (e.c.) processes are indicated.

To describe the energy surfaces in Fig. 6.10, one adds a *pairing energy* to the mass formula

$$
\begin{aligned}
E_p = +\lambda\frac{a_5}{B^{3/4}} \qquad ;\lambda &= +1 \qquad ;\ \text{odd-odd} \\
&= 0 \qquad ;\ \text{odd-even} \\
&= -1 \qquad ;\ \text{even-even}
\end{aligned}
\tag{6.56}
$$

The factor of $B^{-3/4}$ is empirical.

6.7.6 *Empirical Fit*

A combination of the above expressions yields the *semi-empirical mass formula*

$$E(B,Z) = -a_1 B + a_2 B^{2/3} + a_3 \frac{Z^2}{B^{1/3}} + a_4 \frac{(B-2Z)^2}{B} + \lambda a_5 \frac{1}{B^{3/4}} \qquad (6.57)$$

This now provides an explicit expression for the ground-state energy of a nucleus with (B, Z), relative to that of B isolated nucleons.

Tabulations of atomic mass excesses $\Delta(B, Z)$ determine total atomic masses. The difference from the sum of the constituent neutron, proton, and electron masses determines total atomic binding energies. These arise from both the nuclear and electron binding energies, with the latter being negligible on the current scale. A fit to measured nuclear binding energies then determines the parameters in the semi-empirical mass formula. One fit by Green is shown in Table 6.2 [Green (1954)].

Table 6.2 One Fit by Green to Observed Nuclear Ground-State Energies [Green (1954)].

Coefficient	a_1	a_2	a_3	a_4	a_5
Fit(MeV)	15.75	17.8	0.710	23.7	34

We make a few observations:

- The semi-empirical mass formula provides a remarkably good fit to the binding energy of observed nuclei throughout the periodic table;
- The repulsive Coulomb energy grows as Z^2 and eventually limits the value of B for which one has bound nuclei;
- The expression in Eq. (6.57) is useful for predicting when the *fission* of a heavy nucleus into two lighter fragments will be an exothermic process, and for computing the energy release in the process (see Prob. 6.11);
- It follows from Eq. (6.57) that there will be a *line of stability* $Z(B)$ for nuclei (see Prob. 6.12);
- When either (N, Z) get too far from the line of stability, nucleons will "drip" from the nucleus . Thus a finite *valley of stability* exists for nuclear systems (see [National Nuclear Data Center (2007)]);
- The empirical value of a_3 allows a determination of r_{0c} (Prob. 6.10)

$$r_{0c} = 1.22\,\mathrm{F} \qquad ; \text{ empirical} \qquad (6.58)$$

This is in excellent agreement with other determinations of r_0, lending credence to this picture of the nucleus.

6.8 Electron Scattering

To learn more about what the nucleus actually looks like, we turn to the topic of electron scattering [Hofstadter (1956); TJNAF (2007)].[12]

6.8.1 *Single-Slit Diffraction*

Consider the single-slit diffraction pattern of Fig. 4.13 and Eq. (4.58), which is reproduced here in Fig. 6.11.

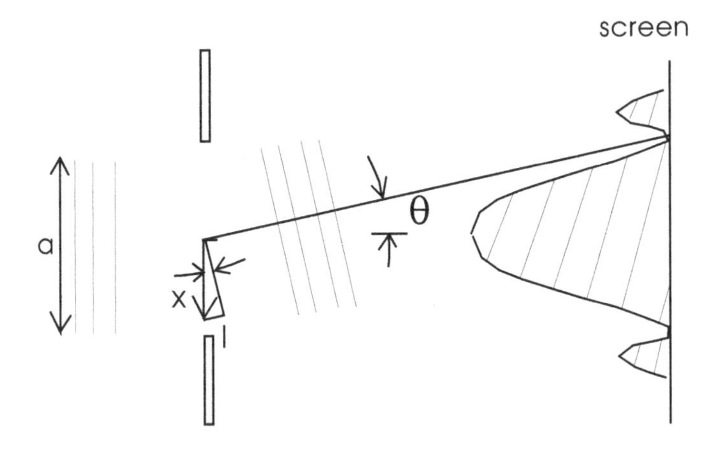

Fig. 6.11 Single-slit diffraction.

Suppose the slit has a width a and x denotes the position along the slit measured from its center. Then the distance l in the figure is given for small angles by $l \approx x\theta$. It follows that the difference in optical pathlength, relative to the central ray, from the point x on the slit to a point at an angle θ on the screen is given for small angles by

$$\Delta = \frac{2\pi}{\lambda}l = \frac{2\pi}{\lambda}x\theta = k\theta x \tag{6.59}$$

Now define the *momentum transfer* by (see Fig. 6.12)

$$\mathbf{q} \equiv \mathbf{k} - \mathbf{k}' \qquad ; \text{ momentum transfer} \tag{6.60}$$

Here $\mathbf{k} = k\,\hat{\mathbf{k}}$ and $\mathbf{k}' = k\,\hat{\mathbf{k}}'$ where $(\hat{\mathbf{k}}, \hat{\mathbf{k}}')$ are unit vectors normal to the wavefront in the incident and outgoing directions, respectively. Then the

[12]For an introduction to electron scattering, see [Walecka (2001)].

difference in optical pathlength relative to the central ray is given for small angles by

$$\Delta \approx qx \qquad ; \text{ difference in optical pathlength} \tag{6.61}$$

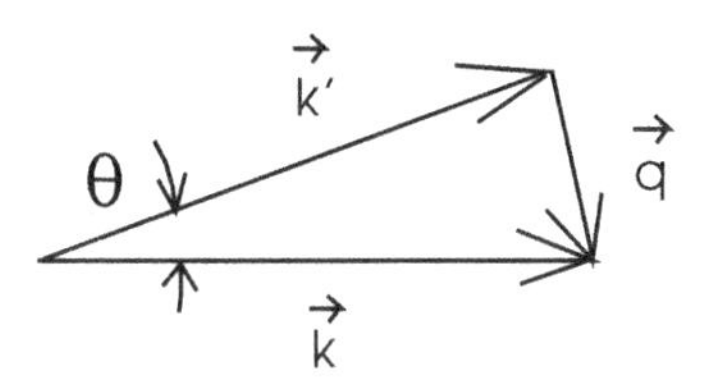

Fig. 6.12 Momentum transfer $\mathbf{q} \equiv \mathbf{k} - \mathbf{k}'$ (in units of $\hbar$). Here $|\mathbf{k}'| = |\mathbf{k}|$, and for small angles, $|\mathbf{q}| = q = k\theta$.

Huygens' principle in optics states that each point on a wavefront acts as a wave source. Thus to get the wave amplitude at an angle θ on the screen, one simply has to add the amplitudes $A\,dx$ from each little element on the slit, with the appropriate phase.[13] Hence

$$\mathcal{A}_{\text{screen}} = A \int_{-a/2}^{a/2} e^{iqx}\,dx \tag{6.62}$$

This is just the *Fourier transform* of the aperture, and the integral is readily done, as previously, to give

$$\mathcal{A}_{\text{screen}} = A \left[\frac{a \sin{(qa/2)}}{qa/2} \right] \tag{6.63}$$

The expression in square brackets is known as the *form factor* $F(q)$.[14] The first minimum of this expression, which can be used to characterize the diffraction pattern, occurs at

$$qa = 2\pi \qquad ; \text{ first diffraction minimum} \tag{6.64}$$

The square of Eq. (6.60) gives

$$q^2 = 4k^2 \sin^2 \frac{\theta}{2} \tag{6.65}$$

[13]Here A is the amplitude of the incident wave, which is assumed to be uniform across the slit.

[14]Here normalized to $F(0) = a$.

Hence Eq. (6.64) can be rewritten as

$$2ka \sin\frac{\theta}{2} = 2\pi \tag{6.66}$$

No matter how small the slit-width a is, one can get this diffraction minimum to appear at a small, but finite angle in the laboratory by using light with large enough k, or equivalently, small enough wavelength. Hence, by making use of light of known wavelength, *one can make a measurement of the width of a microscopic slit by making a measurement of a macroscopic laboratory diffraction pattern!*[15]

6.8.2 *Electron Scattering from a Charge Distribution*

Now consider the scattering of a high-energy electron from the three-dimensional nuclear charge distribution.

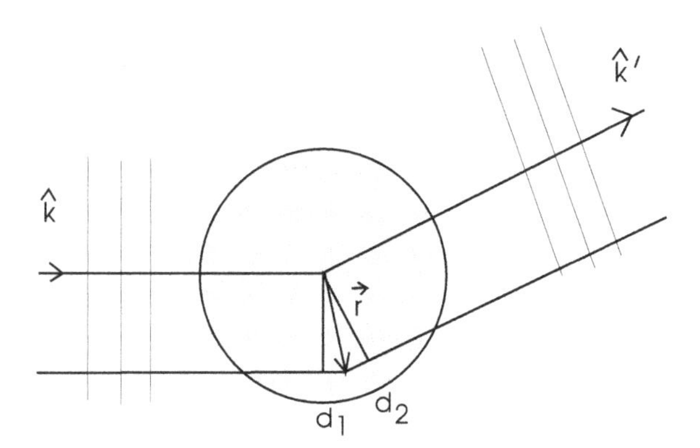

Fig. 6.13 Coulomb scattering from a 3-D charge distribution. Here $\hat{\mathbf{k}}$ and $\hat{\mathbf{k}}'$ are unit vectors in the incident and scattered wave directions.

The difference in optical pathlength relative to the central ray from an arbitrary small element of the charge distribution located at $\mathbf{r}$ is determined from Fig. 6.13 to be

$$\begin{aligned}\Delta &= \frac{2\pi}{\lambda}(d_1 + d_2) = \frac{2\pi}{\lambda}\left(\hat{\mathbf{k}} \cdot \mathbf{r} - \hat{\mathbf{k}}' \cdot \mathbf{r}\right) \\ &= \mathbf{q} \cdot \mathbf{r}\end{aligned} \tag{6.67}$$

The electron scatters from the little element of charge $\rho(\mathbf{r})d^3r$ at each

[15]Compare Prob. 6.16.

point, and again invoking Huygens' principle, the amplitude on a detecting screen is given by

$$\mathcal{A}_{\text{screen}} = A(\mathbf{q}) \int e^{i\mathbf{q}\cdot\mathbf{r}}\, \rho(\mathbf{r})\, d^3r \tag{6.68}$$

where $A(\mathbf{q})$ is the elementary scattering amplitude from a unit charge; it is a known function.

We make several observations:

- The integral measured in Eq. (6.68) is the *form factor* of the charge distribution

$$F(\mathbf{q}) = \int e^{i\mathbf{q}\cdot\mathbf{r}}\, \rho(\mathbf{r})\, d^3r \qquad ; \text{ form factor} \tag{6.69}$$

- The form factor is just the *three-dimensional Fourier transform of the charge distribution with respect to the momentum transfer* $\mathbf{q}$;
- It is clear from the example in Eq. (6.64), and the properties of the Fourier transform, that one can *examine arbitrarily small charge distributions with electron scattering if only the momentum transfer is large enough*;
- For a very high-energy electron with $\varepsilon \gg m_e c^2$, its energy bears the same relation to its wavenumber as that of a photon (see later)

$$\varepsilon = \hbar k c \qquad ; \text{ relativistic electron} \tag{6.70}$$

  Hence, the momentum transfer at a given scattering angle can be made arbitrarily large by simply making the *electron energy high enough*;[16]
- The actual shape of the diffraction pattern depends on the spatial distribution of charge in the nucleus. In principle, through a measurement of the scattering amplitude at all $\mathbf{q}$, one has a measurement of $\rho(\mathbf{r})$ itself through inversion of the three-dimensional Fourier transform;
- In this fashion, through the pioneering experiments of Hofstadter and collaborators [Hofstadter (1956)], we know what the nucleus, as small as it is, actually looks like!

[16]It is ironic that the smaller the target one wants to study, the higher the energy of the electron accelerator one requires, and the larger the spectrometer one needs to see the diffraction pattern.

6.8.3 *Nuclear Charge Distribution*

Although measured in much finer detail for many individual cases, the systematics of the measured nuclear charge distribution throughout the periodic table can be summarized in the following fashion [Hofstadter (1956)] (see Fig. 6.14):

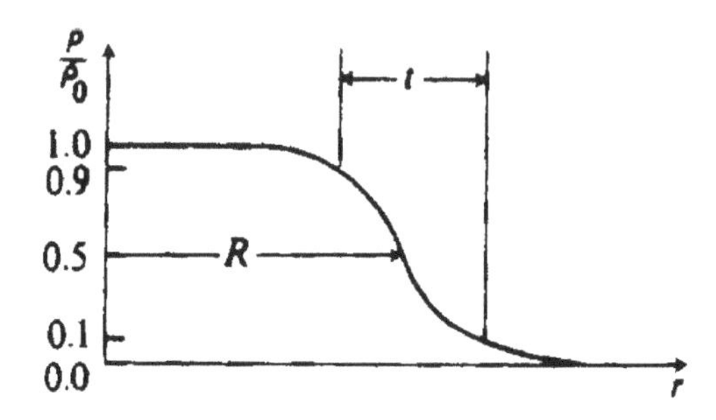

Fig. 6.14 Two-parameter description of systematics of the nuclear charge distribution [Hofstadter (1956)]. Here t is the surface thickness, and $R \equiv R_{1/2}$ is the half-density radius.

(1) The central *nuclear density* is constant

$$\frac{B}{Z}\rho_0 \approx \text{constant} \tag{6.71}$$

Here ρ_0 is the nuclear charge density at the origin.

(2) The *half-density radius* is given by

$$R_{1/2} = 1.07\, B^{1/3}\,\text{F} \tag{6.72}$$

(3) The distance over which the charge density falls from 90% to 10% of its central value, the *surface thickness*, is given by

$$t \approx 2.4\,\text{F} \tag{6.73}$$

6.9 Nuclear Matter

We are now in a position to describe *nuclear matter.* It is approximately the material at the center of a heavy nucleus such as $^{208}_{82}\text{Pb}$, and when composed entirely of neutrons, it is the substance of neutron stars.

- Let the electric charge $e \to 0$ so that there are no Coulomb forces;
- Keep $N = Z$ so there is no symmetry energy;
- Let $B \to \infty$ so that surface effects are negligible with respect to bulk, or volume, effects.

The semi-empirical mass formula of Eq. (6.57) then lets us identify the energy/baryon in this medium

$$\frac{E}{B} = -a_1 = -15.75\,\text{MeV} \qquad ; \text{nuclear matter} \qquad (6.74)$$

Consider a big box of uniform nuclear matter and impose periodic boundary conditions. To a first approximation, one has a *Fermi gas*, and the baryon density in this Fermi gas follows as in Eq. (4.168)

$$\begin{aligned} B &= \frac{gV}{(2\pi)^3}\int_0^{k_\text{F}} d^3k \\ \frac{B}{V} &= g\frac{k_\text{F}^3}{6\pi^2} \end{aligned} \qquad ; \text{baryon density} \qquad (6.75)$$

The degeneracy is $g = 4$, since one has four distinct species $(n\uparrow, n\downarrow, p\uparrow, p\downarrow)$ (see Fig. 6.15)

$$g = 4 \qquad ; \text{nuclear matter} \qquad (6.76)$$

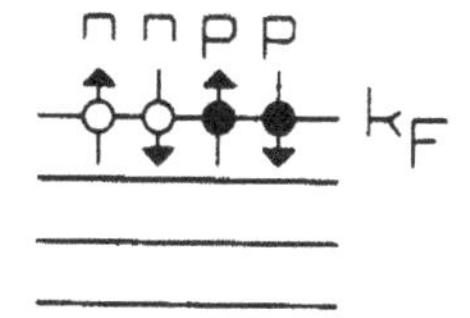

Fig. 6.15 Degeneracy $g = 4$ of nuclear matter with $(n\uparrow, n\downarrow, p\uparrow, p\downarrow)$.

We know the nuclear radius grows as $B^{1/3}$, and the volume as B,

$$R = r_0 B^{1/3} \qquad ; V = \left(\frac{4\pi r_0^3}{3}\right) B \qquad (6.77)$$

Equations (6.75)–(6.77) can then be combined to give

$$\frac{3}{4\pi r_0^3} = \frac{2k_F^3}{3\pi^2}$$

$$\Rightarrow \qquad k_F r_0 = \left(\frac{9\pi}{8}\right)^{1/3} \tag{6.78}$$

The last relation is a very simple and powerful one. It directly relates the radius parameter r_0, or equivalently, the observed baryon density of nuclear matter, to the Fermi wavenumber k_F in this medium.

The appropriate parameter r_0 to be used here is not precisely determined, since there is a surface structure in nuclei (Fig. 6.14). One could use the value r_{0c} determined from the Coulomb energy in Eq. (6.58), or alternatively, the half-density radius parameter determined from electron scattering in Eq. (6.72) could be used. These give the values of k_F shown in Table 6.3.

Table 6.3 Radius parameter and corresponding Fermi wavenumber from two sources.

	Coulomb Energy	Electron Scattering
r_0	1.22 F	1.07 F
k_F	1.25 F^{-1}	1.42 F^{-1}

The observed central nuclear density in $^{208}_{82}$Pb yields a value of $k_F \approx 1.36\,F^{-1}$.

The fact that infinite nuclear matter has a well-defined volume/baryon and energy/baryon is known as the *saturation of nuclear forces.*

6.10 Shell Model

We next turn to a discussion of the *shell model* of the nucleus [Mayer and Jensen (1955)]. We have seen how the semi-empirical mass formula provides a description of the average ground-state energies of nuclei throughout the periodic table. If one looks at detail, there are oscillations about these average energies. There are values of (N, Z) that exhibit *special stability.* These are the "magic numbers"

$$(N, Z) = 2, 8, 20, 28, 50, 82, 126, \cdots \quad ; \text{ magic numbers} \tag{6.79}$$

Nuclei exhibit a *shell structure* similar to that we saw in chapter 5 in the periodic system of elements. Our goal is to understand this structure.

6.10.1 *A Simple Model*

First recall the *one-dimensional simple harmonic oscillator* studied in chapter 4. The energy spectrum is (see Fig. 6.16)

$$E_n = \hbar\omega\left(n + \frac{1}{2}\right) \qquad ; \text{ 1-D simple harmonic oscillator}$$
$$n = 0, 1, 2, \cdots \tag{6.80}$$

The parity alternates between levels; it is even if n is even, and odd if n is odd.

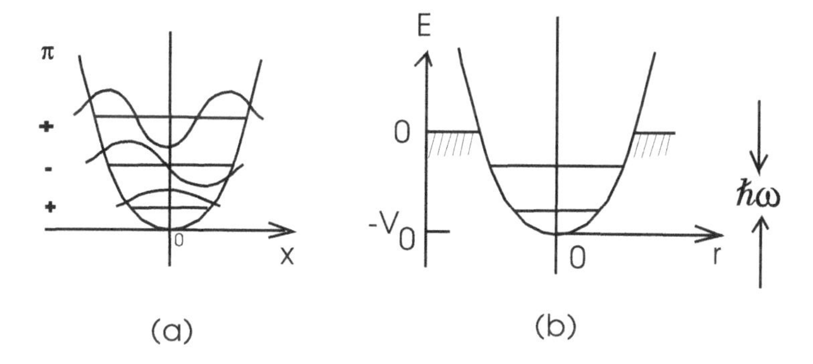

Fig. 6.16 Simple models: (a) One-dimensional simple harmonic oscillator; (b) Three-dimensional simple harmonic oscillator.

Three-Dimensional Simple Harmonic Oscillator. The potential for the three-dimensional simple harmonic oscillator takes the form

$$V = \frac{1}{2}\kappa\left(x^2 + y^2 + z^2\right) - V_0 = \frac{1}{2}\kappa r^2 - V_0 \tag{6.81}$$

The problem *factors* into three, one-dimensional oscillators, and thus we know all the eigenvalues and eigenfunctions. It is, equivalently, a central-force problem with a potential $V(r)$, and it can be solved analytically in spherical coordinates.[17] The eigenvalue spectrum is

$$\begin{aligned}
E_n &= \hbar\omega\left(n_x + n_y + n_z + \frac{3}{2}\right) - V_0 \\
&= \hbar\omega\left(\bar{n} + \frac{3}{2}\right) - V_0 \\
&= \hbar\omega\left[2(n-1) + l + \frac{3}{2}\right] - V_0 \quad ; \text{3-D simple harmonic oscillator} \\
&\qquad\qquad n = 1, 2, 3, \cdots \\
&\qquad\qquad l = 0, 1, 2, \cdots
\end{aligned} \tag{6.82}$$

The spectrum of the three-dimensional oscillator is illustrated in Fig. 6.17.

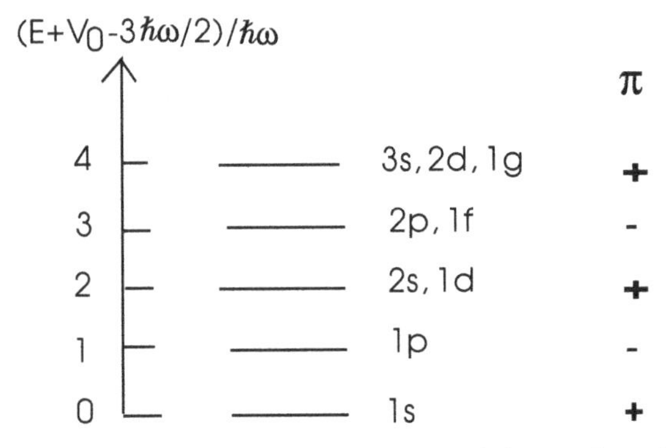

Fig. 6.17 Level structure in the three-dimensional simple harmonic oscillator, with nuclear physics notation.

We make several comments:

- The spacing of the levels is again $\hbar\omega$;
- One can use either (n_x, n_y, n_z) or (n, l, m_l) to characterize the states. The latter set of quantum numbers is more convenient, since it describes states of given angular momentum;
- There is a *dynamical degeneracy* where all levels of given $2(n-1)+l$ have the same energy;
- There is, in addition, the usual angular momentum degeneracy for the $(2l+1)$ values of m_l;
- The parity, which goes as $\pi = (-1)^l$, alternates between levels;

[17]These solutions can be found, for example, in [Walecka (2004)].

- It is customary to use the *nuclear physics notation* and label the levels with (n, l) where n is the number of nodes in the radial wave function $u_{nl}(r)$, including the origin and excluding the point at infinity; in contrast, the *atomic physics notation* employs $(\bar{n}, l)$ where $\bar{n} = 2(n-1)+l$ is again the principal quantum number.

$$\bar{n} = 2(n-1) + l \qquad ; \text{ principal quantum number} \tag{6.83}$$

6.10.2 *More Realistic Model*

The first step in making this model more realistic is to use an improved single-particle potential. Hartree *mean-field theory*, as in chapter 5, would employ

$$\begin{aligned} V_H(r) &= \int V(|\mathbf{r} - \mathbf{r}'|)n(r')\, d^3r' \\ n(r) &= \sum_{\alpha}^{\mathrm{F}} |\phi_\alpha(\mathbf{r})|^2 \end{aligned} \tag{6.84}$$

Here $V(|\mathbf{r} - \mathbf{r}'|)$ is the nucleon-nucleon potential, and n(r) is the baryon density. As in Fig. 6.15, it is assumed that the single-particle levels are filled with $(n\uparrow, n\downarrow, p\uparrow, p\downarrow)$ up to a Fermi level F.[18] The Hartree self-consistent field equations now read

$$\left[-\frac{\hbar^2}{2m}\nabla^2 + V_H(r)\right]\phi_\alpha(\mathbf{r}) = \varepsilon_\alpha \phi_\alpha(\mathbf{r}) \tag{6.85}$$

There are several observations to make:

- As before, if the l-shells are filled, then $n(r)$ will be spherically symmetric;
- This is a self-consistent field problem. The $\phi_\alpha(\mathbf{r})$ are determined from $V_H(r)$, which, in turn, is not known until the $\phi_\alpha(\mathbf{r})$ are found. Fortunately, in applications, the results converge upon iteration;[19]
- The mean field will reflect the nuclear density. This implies that with respect to the simple harmonic oscillator, the nuclear potential will be flattened at the edge of the nucleus as indicated in Fig. 6.18.

[18]Alternatively, one may have separate Fermi levels for the neutrons and protons.

[19]An improvement here is to use the Hartree-Fock equations of Prob. H.3, which correctly incorporate the Pauli principle for the nucleons. The equations are more complicated; however, they also converge upon iteration in most cases.

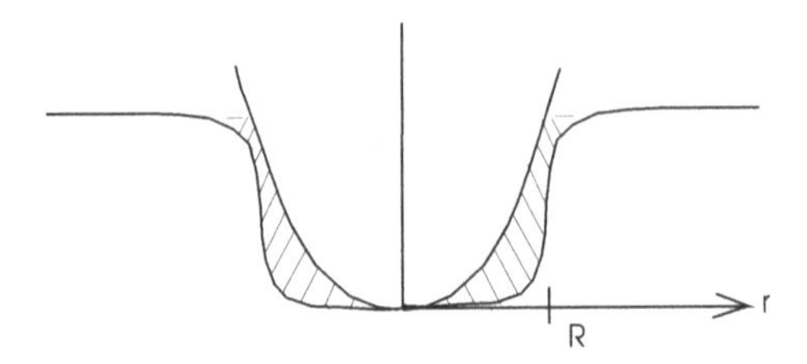

Fig. 6.18 The nuclear mean-field potential reflects the nuclear density, which implies a flattening of the oscillator potential at the edge of the nucleus.

- What happens to the oscillator spectrum when the potential is flattened as in Fig. 6.18? The states with the highest l for each $\bar{n}$ (the "circular orbits") spend most of the time farthest away from the origin. Thus these orbits will be *lowered the most* in energy. One will see a shift in levels corresponding to that in Fig. 6.19;[20]

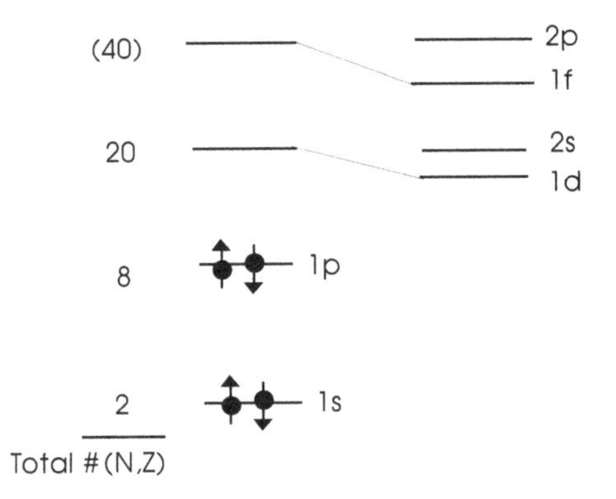

Fig. 6.19 Shift in the first few single-particle levels corresponding to the flattening of the oscillator potential at the edge of the nucleus in Fig. 6.18. See the l.h.s. of Fig. 6.23 for the extension of this figure to the higher oscillator shells.

- If one fills the l-states, then the total number of neutrons or protons that can be put into such a level is

$$\text{number of like nucleons}\,(N,Z) = 2(2l+1) \qquad ;\ \text{l-shell} \qquad (6.86)$$

The factor of 2 comes from the spin (Fig. 6.19), and the $(2l+1)$ comes

[20]An extended view of this spectrum is shown on the l.h.s. of Fig. 6.23.

from the available m_l states;

- The total number of (N, Z) contained in the first four oscillator shells is shown on the l.h.s of Fig. (6.19). The major shells of given $\bar{n}$ are filled at $(N, Z) = 2, 8, 20, 40, \cdots$. We appear to be on the right track in understanding the magic numbers, but we are not yet there.

6.10.3 *Spin-Orbit Interaction*

The major ingredient of the nuclear shell model is a *single-particle spin-orbit force from the strong nuclear interaction*, of opposite sign from the atomic case. There it gave rise to a "fine structure" on the single-particle levels. Here it plays a *dominant role* in the structure of the single-particle states.

Consider the effect of an additional nuclear single-particle spin-orbit interaction of the form

$$H'_{\text{so}} = -\bar{V}_{\text{so}}(r)\,\mathbf{s}\cdot\mathbf{l} \tag{6.87}$$

The vector model of angular momentum gives (see chapter 5)

$$\begin{aligned} \mathbf{j} &= \mathbf{l}+\mathbf{s} \\ \mathbf{l}\cdot\mathbf{s} &= \frac{1}{2}\left[\mathbf{j}^2-\mathbf{l}^2-\mathbf{s}^2\right] = \frac{1}{2}\left[j(j+1)-l(l+1)-s(s+1)\right] \end{aligned} \tag{6.88}$$

The energy shift of the single-particle levels is given by Eq. (5.27)

$$\begin{aligned} \Delta\varepsilon_{j=l+1/2} &= \langle\bar{V}_{\text{so}}(r)\rangle\left[-\frac{1}{2}l\right] \qquad &; j = l+\frac{1}{2}, \quad s=\frac{1}{2} \\ \Delta\varepsilon_{j=l-1/2} &= \langle\bar{V}_{\text{so}}(r)\rangle\left[\frac{1}{2}(l+1)\right] \qquad &; j = l-\frac{1}{2}, \quad s=\frac{1}{2} \end{aligned} \tag{6.89}$$

The *splitting* of these two levels is thus given by

$$\varepsilon_{j=l-1/2}-\varepsilon_{j=l+1/2} = \frac{1}{2}(2l+1)\langle\bar{V}_{\text{so}}(r)\rangle \tag{6.90}$$

We now observe:

- It contrast to the atomic case, here it is the state of highest j, the *stretched state* with $j = l + 1/2$, that is pushed down;
- From the proportionality to l, it is the states with the highest (l, j) that are pushed down the farthest;
- The number of like nucleons (N, Z) that can be placed in a given j-shell is

$$\text{number of like nucleons } (N, Z) = 2j + 1 \qquad ; \text{ j-shell} \qquad (6.91)$$

This is simply the number of m_j states available;

- If the state with highest (l, j) in each major oscillator shell is pushed down far enough in energy, then *the shell model explains the magic numbers*, as illustrated on the r.h.s. of Fig. 6.23 [Mayer and Jensen (1955)].

6.10.4 *Nuclear Spins and Parities*

Consider what happens when one begins to fill a single j-shell with like particle in the shell-model:

(1) *Pairs of like nucleons couple to* $J = 0$. The reason for this is that the interaction between two nucleons is attractive, and this pair state gives the maximum overlap of the wave functions of the two nucleons (see Fig. 6.20). One might expect to get better overlap by putting the two particles into the *same* m_j state, but this is forbidden by the Pauli principle;

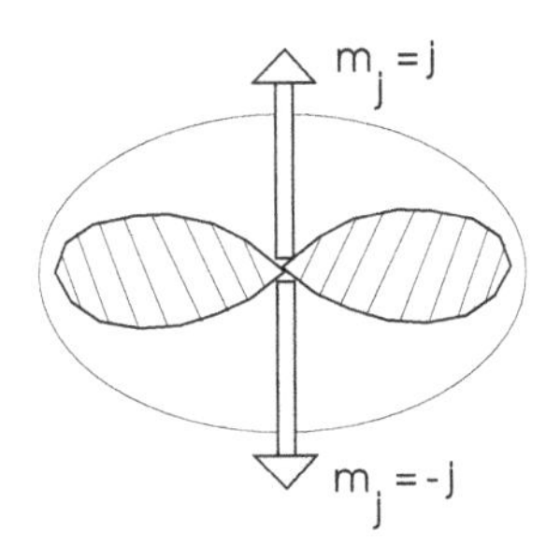

Fig. 6.20 Illustration of why the state with $J = 0$ gives the maximum overlap for a pair of like particles interacting through the attractive nucleon-nucleon interaction.

(2) *All even-even nuclei have ground states with* $J^\pi = 0^+$. The author is unaware of any exceptions to this rule;

(3) *The spin and parity of an odd nucleus then comes from the last valence particle.* This is a very powerful, predictive tool. Consider a few

examples. With the aid of Fig. (6.23), one predicts that

$$ {}^{7}_{3}\mathrm{Li}(J^\pi) = \frac{3}{2}^- \quad ; \; {}^{41}_{20}\mathrm{Ca}(J^\pi) = \frac{7}{2}^- \quad ; \; {}^{209}_{83}\mathrm{Bi}(J^\pi) = \frac{9}{2}^- \tag{6.92} $$

All of these are in accord with the experimental observations [National Nuclear Data Center (2007)].

A *hole* in a j-shell has the spin and parity of the missing state, since if one *adds* a particle and brings in that j^π, one returns to a $J^\pi = 0^+$ ground state. For example

$$ {}^{15}_{8}\mathrm{O}(J^\pi) = \frac{1}{2}^- \quad ; \; {}^{39}_{20}\mathrm{Ca}(J^\pi) = \frac{3}{2}^+ \quad ; \; {}^{65}_{28}\mathrm{Ni}(J^\pi) = \frac{5}{2}^- \tag{6.93} $$

Again, these all agree with experiment.[21]

6.10.5 *Schmidt Lines*

The shell model now allows one to predict nuclear *magnetic dipole moments.* The magnetic moment of the nucleon is

$$ \begin{aligned} \boldsymbol{\mu}_p &= \mu_N(\mathbf{l} + 2\lambda_p \mathbf{s}) \qquad &; \; \lambda_p = 2.793 \\ \boldsymbol{\mu}_n &= \mu_N(2\lambda_n \mathbf{s}) \qquad &; \; \lambda_n = -1.913 \end{aligned} \tag{6.94} $$

Here the nucleon magneton is given by

$$ \mu_N = \frac{|e|\hbar}{2m_p} \tag{6.95} $$

Since the angular momentum of an odd nucleus comes from the last valence particle, one can simply take over the results from the vector model of angular momentum in chapter 5. The total angular momentum of the valence nucleon is

$$ \mathbf{j} = \mathbf{l} + \mathbf{s} \tag{6.96} $$

[21] Sometimes these predicted states, while not the ground states, are found as low-lying isomeric (*i.e.* long-lived) states. Just where the raised state with highest l and $j = l - 1/2$ lies with respect to the other states of the major shell depends on the details of the central potential; in addition, it sometimes pays to promote particles to slightly higher single-particle states with much larger l because of the increased overlap with like nucleons (compare Fig. 6.20) and consequent increased pairing.

As illustrated in Fig. 5.1, in the vector model both $(\mathbf{l}, \mathbf{s})$ precess about $\mathbf{j}$ so that their effective values are

$$\mathbf{l}_{\text{eff}} = (\mathbf{l} \cdot \mathbf{j}) \frac{\mathbf{j}}{\mathbf{j}^2} \qquad ; \ \mathbf{s}_{\text{eff}} = (\mathbf{s} \cdot \mathbf{j}) \frac{\mathbf{j}}{\mathbf{j}^2} \tag{6.97}$$

It follows exactly as in chapter 5 that the effective magnetic moment of the valence nucleon is

$$\boldsymbol{\mu}_{\text{eff}} = \mu_N g_{\text{Landé}} \mathbf{j} \tag{6.98}$$

where the Landé g-factor is again given by[22]

$$g_{\text{Landé}} = \frac{1}{2j(j+1)} \{[j(j+1) + l(l+1) - s(s+1)] + 2\lambda [j(j+1) + s(s+1) - l(l+1)]\} \tag{6.99}$$

The magnetic moment is again obtained by lining the nucleus up as well as possible along the z-axis, so that $m_j = j$

$$\mu = \mu_N (j \, g_{\text{Landé}}) \qquad ; \text{ magnetic moment} \tag{6.100}$$

An evaluation of the above expressions for $s = 1/2$ and $j = l \pm 1/2$ then yields the nuclear magnetic moment as (Prob. 6.18)

$$\begin{aligned} \frac{\mu}{\mu_N} &= j - \frac{1}{2} + \lambda & ; \ j = l + \frac{1}{2} \\ &= j + \frac{j}{j+1}\left(\frac{1}{2} - \lambda\right) & ; \ j = l - \frac{1}{2} \\ & & \text{Schmidt lines} \end{aligned} \tag{6.101}$$

Only the term in λ contributes for the neutron. These are the celebrated *Schmidt lines* for the magnetic dipole moments of odd nuclei. They are drawn in Fig. 6.21. Experimental magnetic moments can be found in [Lawrence Berkeley Laboratory (2005)]. All but the five nuclei indicated in Fig. 6.21 have magnetic moments lying *between the Schmidt lines.*

[22] Note carefully the replacement of g_s by 2λ in Eqs. (6.94) and (6.99).

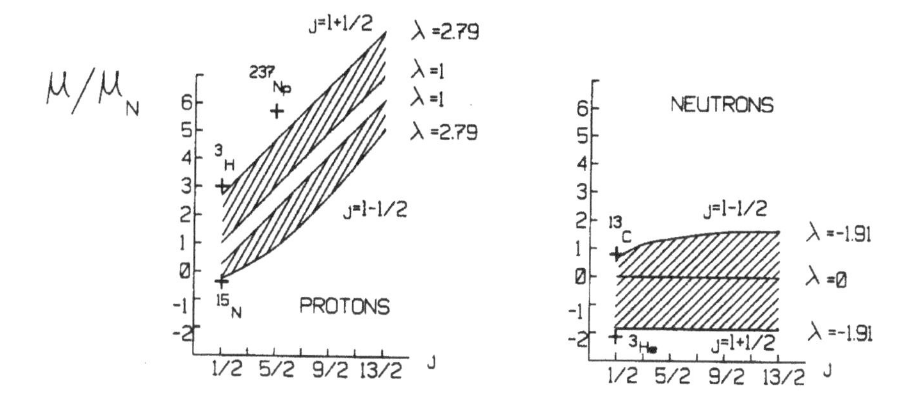

Fig. 6.21 Schmidt lines for nuclear magnetic dipole moments from Eqs. (6.101). Also shown, for comparison purposes, are the lines with *fully quenched* anomalous moments ($\lambda_n = 0, \lambda_p = 1$). All but the indicated five nuclei have magnetic moments lying between the Schmidt lines and the fully quenched values.

6.11 γ-Decay

As in atoms, nuclei can decay with the emission of a photon (Fig. 6.22). The photon energy is given by the Bohr relation

$$h\nu = \hbar kc = \Delta E = E_i - E_f \qquad ; \text{ photon energy} \qquad (6.102)$$

Typically

- Photon energies in atoms are $O(\text{eV})$;
- Photon energies in nuclei are $O(\text{MeV}) = O(10^6\,\text{eV})$.

This is again a reflection of the difference in strength of the strong and electromagnetic interactions. Photons coming from nuclear transitions are referred to as "gammas".

Electromagnetic radiation can carry off angular momentum $J \geq 1$, and when making a nuclear transition $J_i \to J_f$, it follows from our angular momentum rules that J must lie in the interval

$$|J_i - J_f| < J < J_i + J_f \qquad (6.103)$$

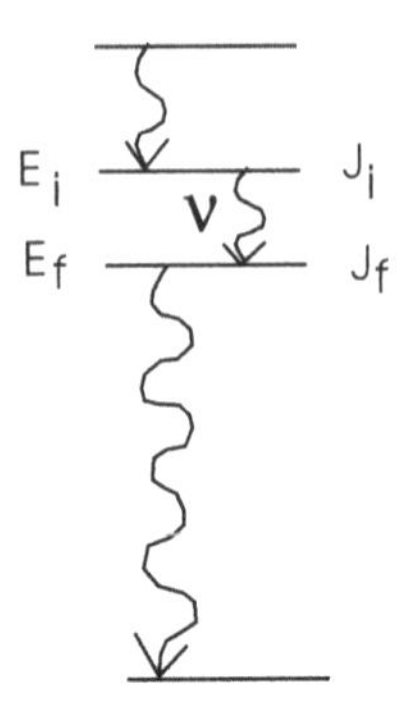

Fig. 6.22 γ-decay.

For given J, the radiation can carry off either parity $\Pi = \pm 1$. The transition multipoles of the radiation field are referred to as

$$\begin{aligned} \Pi &= (-1)^J && ; \text{ electric EJ multipoles} \\ \Pi &= (-1)^{J+1} && ; \text{ magnetic MJ multipoles} \end{aligned} \tag{6.104}$$

The *rate* of gamma emission *increases* with ΔE and *decreases* with J. "Weisskopf estimates" obtained from gross nuclear properties are extremely useful in characterizing these rates [Blatt and Weisskopf (1952)]. The derivation of the general expression for the rate of photon emission from any quantum mechanical system is carried out in Probs. 12.3–12.4. Here we simply make some observations on nuclear γ-transitions:

- As long as the proton and neutron densities remain identical, there is no separation of the center of mass and center of charge (in contrast to the situation in an atom), hence there will be no *electric dipole* (E1) transitions in this case. In fact, the electric dipole strength in nuclei is concentrated in the *giant dipole resonance*, which lies at about 25 to 10 MeV in going from the lightest to the heaviest nucleus, and which evidently involves the motion of the protons against the neutrons;
- Very strong collective quadrupole (E2) transitions involving a deformation of the total nuclear charge Z are observed throughout the periodic table. These are particularly strong in the case of *permanently deformed nuclei* [Bohr and Mottelson (1969); Bohr and Mottelson (1975)];
- The availability in the shell-model of close-lying, excited single-particle states with markedly different j^π leads to *islands of isomerism* in the periodic table;

- Predictions for γ-transitions in the shell model provide a powerful tool for investigating its consequences and validity;[23]
- Inverse transitions from the ground state can be studied in inelastic electron scattering. Here one has the advantage that, in contrast to the situation for a real photon, the momentum transferred by the *virtual photon* exchanged between the electron and nucleus (see later) is only constrained by the inequality $\hbar qc \geq \Delta E$.

[23]The quantum theory of angular momentum [Edmonds (1974)] provides an essential tool in carrying out this analysis.

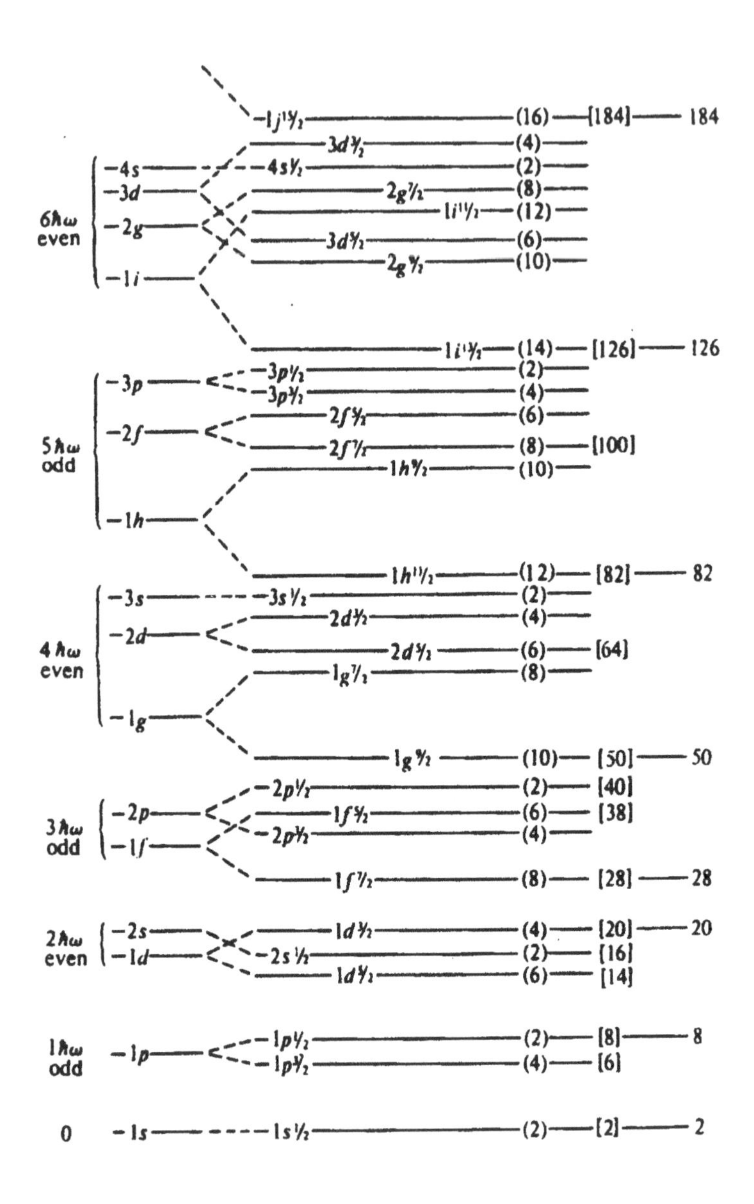

Fig. 6.23 Single-particle level spectrum in the shell model [Mayer and Jensen (1955)]. The figure starts on the l.h.s. with the major oscillator shells, includes the flattening of the nuclear single-particle potential, and then adds the nuclear single-particle spin-orbit potential, leading to the j-shell structure and magic numbers on the r.h.s. (From [Walecka (2004)]).

Chapter 7

Particle Physics

We next turn to the topic of particle physics.[1] To start the discussion, we return to our characterization of the *forces* observed in nature.

7.1 Forces

A principal thrust of modern theoretical physics is to develop a *unified theory* of all the forces, and the *standard model* of the unified electroweak force (see later) is amazingly successful. As manifest in the world around us, however, we observe four forces of very different strength. These forces, and a dimensionless coupling constant characterizing the strength of that force, are as follows:

(1) Strong (nuclear) force with

$$\frac{g_\pi^2}{4\pi\hbar c} = 14.4 \qquad ; \text{ strong} \tag{7.1}$$

Here g_π is the pseudoscalar pion-nucleon coupling constant.

(2) Electromagnetic force with

$$\frac{e^2}{4\pi\varepsilon_0\hbar c} = \frac{1}{137.0} \qquad ; \text{ electromagnetic} \tag{7.2}$$

(3) Weak force with

$$\frac{M_p^2 G_{\rm F}}{\hbar c} = 1.027 \times 10^{-5} \qquad ; \text{ weak} \tag{7.3}$$

Here $M_p = m_p c/\hbar$ is the proton inverse Compton wavelength, and $G_{\rm F}$ is Fermi's weak coupling constant, as determined from μ-decay.

[1]See [Martin (2006); Perkins (2000); Halzen and Martin (1984)].

(4) Gravitation with

$$\frac{m_p^2 G}{\hbar c} = 5.905 \times 10^{-39} \qquad ; \text{ gravitation} \tag{7.4}$$

Here G is Newton's gravitational constant.

The particles in nature interact through these forces.

7.2 Particles

Particles are quanta with well-defined mass and quantum numbers. Particles have antiparticles in which all the quantum numbers are reversed. In some cases, the antiparticle may be identical to the particle.

7.2.1 *Electric Charge*

Electric charge is additively conserved in all interactions.

7.3 Hadrons

Particles that interact through the strong force are known as *hadrons*. There are two types of hadrons.

- *Baryons* have half-integral spin and are *fermions*;
- *Mesons* have integral spin and are *bosons*.

7.3.1 *Baryon Number*

Baryons can be assigned a *baryon number*. To the best of our knowledge, this quantity is additively conserved in all interactions.[2]

7.3.2 *Strangeness*

Hadrons can be assigned a *strangeness*. This quantity is additively conserved in the strong and electromagnetic interactions. Strangeness conservation can be violated in the weak interactions.

[2]Although extensively searched for, proton decay has never been observed.

7.3.3 *Isospin*

We have seen in Fig. 6.2 that the two nucleons, neutron and proton, have almost the same mass. It turns out to be a general result that particles come in (nearly) degenerate multiplets.

Recall from our discussion of angular momentum in quantum mechanics that with a central potential $V(r)$, the eigenstates can be characterized by

$$\begin{aligned} \mathbf{l}^2 &= l(l+1) && \text{; angular momentum} \\ -l \leq m_l &\leq l && \text{; integer steps} \end{aligned} \tag{7.5}$$

We observed that since the potential is invariant under rotations in space, the energy eigenvalues for a given l are independent of m_l.

One can now argue in strict analogy and add an *internal space* similar to spin known as *isospin*. Then, if the potential $V(r)$ is invariant under isospin rotations in this internal space, the energy eigenvalues can be characterized by

$$\begin{aligned} \mathbf{t}^2 &= t(t+1) && \text{; isospin} \\ -t \leq m_t &\leq t && \text{; integer steps} \end{aligned} \tag{7.6}$$

and, again,

> *Since the potential is invariant under rotations in this internal space, the energy eigenvalues for a given t are independent of m_t.*

This leads to multiplets of given total isospin t, with $(2t+1)$ degenerate states.[3]

The nucleon, now denoted by $N(938)$,[4] has isospin $t = 1/2$ and forms an isodoublet

$$(j^\pi, t) = \left(\frac{1}{2}^+, \frac{1}{2}\right) \qquad \text{; nucleon (N)} \tag{7.7}$$

[3]Once there is an additional internal space, there is no particular reason that the symmetry group of the strong interactions should be identical to that of rotations in real space. {The symmetry group of angular momentum in quantum mechanics [Edmonds (1974)] is SU(2) — the special unitary group in two dimensions (see Prob. 7.9).} Indeed, the degree of degeneracy of particle multiplets is a function of the distance at which the multiplets are viewed, and higher symmetry groups, in particular SU(3) and SU(4), play a central role in both particle and nuclear physics (see [Gell-Mann and Ne'eman (1963); Wigner (1937)]).

[4]For its mass in MeV.

The nucleon is observed to have an excited state, the $\Delta(1238)$ with[5]

$$(j^\pi, t) = \left(\frac{3}{2}^+, \frac{3}{2}\right) \qquad ; \text{ delta } (\Delta) \tag{7.8}$$

We emphasize that isospin symmetry is a property of the strong interactions. In the internal isospin space, one has all the analogous consequences of the conservation of angular momentum. Isospin symmetry is violated at the level of the electromagnetic (and weak) interactions.

Particle properties are extensively surveyed and tabulated in [Particle Data Group (2006)], which forms an indispensable resource.[6] The observed low-lying particle spectrum in the baryon sector with $B = 1$ is shown in Fig. 7.1.

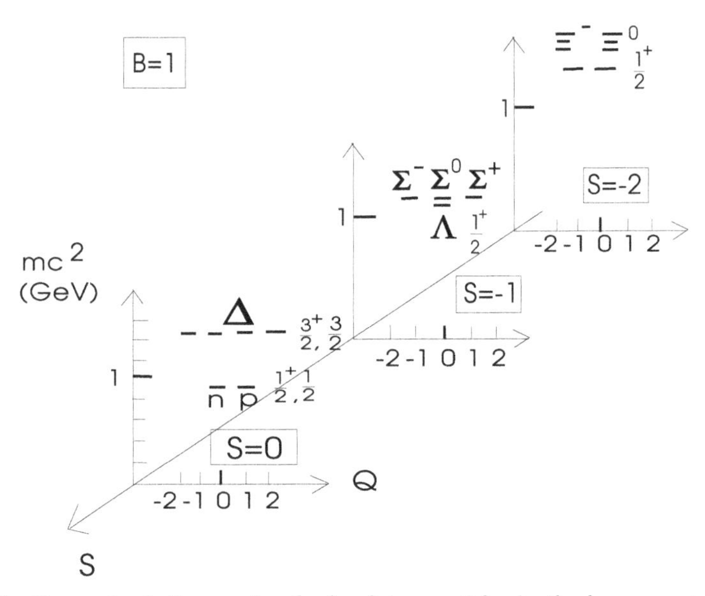

Fig. 7.1 Energy level diagram for the low-lying particles in the baryon sector with $B = 1$. Here Q is the electric charge (in units of $|e|$) and S is the strangeness. In the $S = 0$ sector, the states are explicitly denoted with (j^π, t).

As before, the mass mc^2 of the particles is plotted as ordinate and the electric charge Q (in units of $|e|$) as abscissa; however, there is now a third

[5] This particle is observed as a resonance in pion-nucleon scattering; it has the strong decay $\Delta \rightarrow N + \pi$.

[6] As with all the data references in this book, the data is continually being updated, and the reader is always referred to the latest version of the data source.

dimension, the strangeness S of the multiplets, each member of the multiplet having the same strangeness.

In the strangeness $S = -1$ sector, two low-lying multiplets are observed

$$(j^\pi, t) = \left(\frac{1}{2}^+, 1\right) \qquad ; \text{ sigma } (\Sigma)$$
$$(j^\pi, t) = \left(\frac{1}{2}^+, 0\right) \qquad ; \text{ lambda } (\Lambda) \tag{7.9}$$

In the strangeness $S = -2$ sector there is one low-lying multiplet

$$(j^\pi, t) = \left(\frac{1}{2}^+, \frac{1}{2}\right) \qquad ; \text{ cascade } (\Xi) \tag{7.10}$$

We observe that the mass of the lowest-lying multiplet in each sector increases systematically with $|S|$.

Particularly useful in classifying these particles is the *Gell-Mann–Nishijima relation*, which relates the particle charge Q to the third component of isospin m_t, at a given value of the "hypercharge" $Y \equiv B + S$.

$$Q = m_t + \frac{1}{2}Y \qquad ; \text{ Gell-Mann–Nishijima}$$
$$Y \equiv B + S \qquad ; \text{ hypercharge} \tag{7.11}$$

These energy level diagrams can be extended to the host of excited states, and the reader is referred to [Particle Data Group (2006)] for this data.

As was the case for the nucleon, if allowed by the masses involved, the baryons can decay to each other through the *weak interaction*.[7] Some illustrative weak decay modes of these baryons are

$$\begin{aligned}
\Lambda &\to p + \pi^- \\
&\to p + e^- + \bar{\nu}_e \\
\Sigma^+ &\to p + \pi^0 \\
&\to n + e^+ + \nu_e \\
\Xi^- &\to \Lambda + \pi^- \\
&\to \Sigma^0 + e^- + \bar{\nu}_e
\end{aligned} \tag{7.12}$$

Although some of the above decays involve only hadrons, they violate strangeness conservation, and therefore they only proceed through the weak interaction. All the decay modes, and the branching ratios into them, can be found in [Particle Data Group (2006)].

[7]The decay $\Sigma^0 \to \Lambda + \gamma$ takes place electromagnetically.

The energy spectrum in the low-lying meson sector with $B = 0$ is shown in Fig. 7.2.

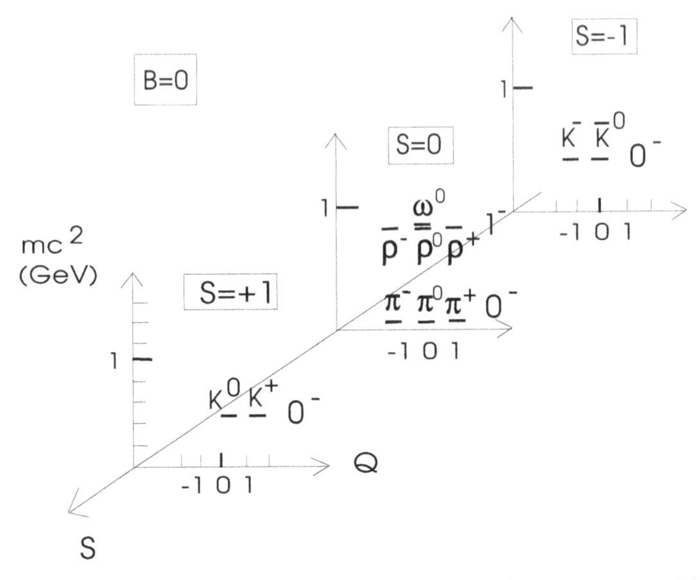

Fig. 7.2 Energy level diagram for the low-lying particles in the meson sector with $B = 0$. Here Q is the electric charge and S is the strangeness.

There are several comments to be made:

- Most striking is the appearance in the $S = 0$ sector, at almost zero energy, of the pseudoscalar *pion* with $(j^\pi, t) = (0^-, 1)$

$$(j^\pi, t) = (0^-, 1) \qquad ; \text{ pion } (\pi) \tag{7.13}$$

- The excited states in the $S = 0$ sector appear as meson resonances in the strong interactions[8]

$$\begin{aligned} (j^\pi, t) &= (1^-, 1) \qquad && ; \text{ rho } (\rho) \\ (j^\pi, t) &= (1^-, 0) \qquad && ; \text{ omega } (\omega) \end{aligned} \tag{7.14}$$

- There is a low-lying multiplet in the $S = +1$ sector

$$(j^\pi, t) = \left(0^-, \frac{1}{2}\right) \qquad ; \text{ kaon } (\mathrm{K}^+, \mathrm{K}^0) \tag{7.15}$$

[8]They have the strong decay modes $\rho \rightarrow \pi + \pi$ and $\omega \rightarrow \pi + \pi + \pi$.

- There is also a low-lying multiplet in the $S = -1$ sector

$$(j^\pi, t) = \left(0^-, \frac{1}{2}\right) \qquad ; \text{ anti-kaon } (\bar{\mathrm{K}}^0, \mathrm{K}^-) \tag{7.16}$$

  Again, the host of excited states in these sectors can be found in [Particle Data Group (2006)].
- The particles in the lowest multiplets in this figure also decay through the weak interactions. For example[9]

$$\begin{aligned} \pi^+ &\rightarrow e^+ + \nu_e \\ \pi^+ &\rightarrow \mu^+ + \nu_\mu \\ K^0 &\rightarrow \pi^+ + \pi^- \\ &\rightarrow \pi^+ + e^- + \bar{\nu}_e \end{aligned} \tag{7.17}$$

  It is an interesting feature of quantum mechanics that the neutral-kaon eigenstates of strangeness produced in the strong interactions are linear combinations of two states (K_L^0, K_S^0) that decay with a definite lifetime through the weak interactions.[10]

For orientation, some representative hadron masses and lifetimes are shown in Table 7.1. All such data may be found in [Particle Data Group (2006)].

Table 7.1 Some selected hadron masses and lifetimes.

Particle	Λ^0	Σ^+	Ξ^0	$\pi^\pm$	ω^0	$K^\pm$
j^π	$1/2^+$	$1/2^+$	$1/2^+$	0^-	1^-	0^-
S	-1	-1	-2	0	0	± 1
mc^2(MeV)	1115.68	1189.37	1314.8	139.57	782.57	493.68
τ(sec) $\times 10^8$	2.63×10^{-2}	0.802×10^{-2}	2.90×10^{-2}	2.603	[a]	1.238

[a] Meson resonance with full width at half maximum of $\Gamma = 8.44$ MeV.

7.3.4 *Charm*

Approximately 1 GeV higher, new types of hadrons are observed. These hadrons can be assigned a new quantum number C, known as *charm*. This quantity is again additively conserved in the strong and electromagnetic

[9] The decay $\pi^0 \rightarrow \gamma + \gamma$ takes place electromagnetically.

[10] An analysis of two particles produced with definite quantum numbers through one interaction, and then mixed by a weaker interaction to produce eigenstates of definite mass (and lifetime), is carried out in the neutrino sector in appendix J.

interactions. Charm conservation can be violated in the weak interactions. The hypercharge now takes the extended form $Y = B + S + C$, and the new Gell-Mann–Nishijima relation reads[11]

$$\begin{aligned} Q &= m_t + \frac{1}{2}Y && ; \text{ Gell-Mann–Nishijima} \\ Y &\equiv B + S + C && ; \text{ hypercharge} \end{aligned} \qquad (7.18)$$

The current results with respect to particle discovery and particle properties can always be found in the latest version of [Particle Data Group (2006)].

7.4 Yukawa Interaction

The electromagnetic force is responsible for the structure of atoms, and the quantum of the electromagnetic interaction is the massless photon. In analogy, the strong force is responsible for the structure of nuclei, and the quantum of that force is the meson. The argument relating the strong interaction between nucleons to the meson was originally presented by Yukawa [Yukawa (1935)].

As motivation, let us return to the Thomas-Fermi theory of the atom. The nuclear Coulomb potential outside a point source of charge Ze_p is given by

$$\Phi_{\text{pt}} = \frac{Ze_p}{4\pi\varepsilon_0 r} \qquad ; \text{ long-range } O(1/r) \qquad (7.19)$$

This electrostatic potential is of long ("infinite") range, going as $O(1/r)$ for large r.

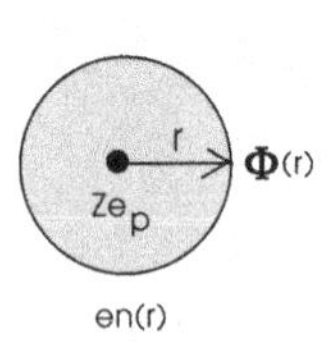

Fig. 7.3 Electron charge distribution $en(r)$ surrounding the nuclear Coulomb point source with charge Ze_p. The electrostatic potential is $\Phi(r)$.

---

[11]The energy level diagram in the strangeness $S = -1$ $(C = 0)$ sector in Fig. 7.1, for example, is mirrored at the higher energy with charm $C = +1$ $(S = 0)$, and appropriately modified charges.

With a spherically symmetric electron charge distribution surrounding that source (Fig. 7.3), the equation for the electrostatic potential inside the charge distribution is modified to take the form in Eq. (5.41)

$$\frac{1}{r}\frac{\partial^2}{\partial r^2}(r\Phi) = \frac{|e|n(r)}{\varepsilon_0} \tag{7.20}$$

This equation describes the shielded Coulomb potential. When expressed in terms the of the electrostatic potential $\Phi(r)$, the precise form of the r.h.s. depends on the specific equation of state employed for the electrons. For a Fermi gas, this becomes the non-linear relation in Eq. (5.42)

$$\frac{1}{r}\frac{\partial^2}{\partial r^2}(r\Phi) = \kappa\Phi^{3/2} \tag{7.21}$$

Now suppose that one has a *strong* point source of strength g, which gives rise to a *meson* field ϕ. Assume that meson field is massive, with a mass m and corresponding inverse Compton wavelength (inverse length)

$$\mu \equiv \frac{mc}{\hbar} \qquad ; \text{ meson mass } m \tag{7.22}$$

It will be assumed that (Φ, ϕ) satisfy analogous equations (compare Fig. 7.4); however, one must incorporate the mass of the meson field.

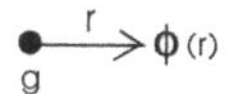

Fig. 7.4 Meson field $\phi(r)$ surrounding the strong nuclear point source with strength g.

One might incorporate that mass by simply using a linear term with the correct dimensions on the r.h.s. of the field equation[12]

$$\frac{1}{r}\frac{\partial^2}{\partial r^2}(r\phi) = \mu^2\phi \qquad ; \text{ meson field equation} \tag{7.23}$$

[12]This modification is justified *a posteriori* by the subsequent argument.

Let us see where this leads. Look for a solution to this equation of the form

$$\phi(r) = \frac{\xi(r)}{r} \tag{7.24}$$

Substitution into Eq. (7.23) then gives

$$\frac{d^2}{dr^2}\xi(r) - \mu^2\xi(r) = 0 \tag{7.25}$$

The solution to this equation that dies out as $r \to \infty$ is $\xi(r) = e^{-\mu r}$. If the source strength g is now matched at the origin as in Eq. (7.19), one has

$$\phi(r) = \frac{g}{4\pi r}e^{-\mu r} \qquad ;\ \text{meson field—point source} \tag{7.26}$$

Since the field has mass, it takes energy to create one of its quanta. One might expect by an uncertainty principle argument that the meson field cannot get very far away from the source, and, indeed, the meson field is now localized about that source.

In the atom, if one brings in another negative charge e, there is an interaction energy of

$$E_{\text{int}} = e\Phi(r) = -|e|\Phi(r) \tag{7.27}$$

In analogy, if one brings in another point source g (*i.e.* another nucleon), one might expect to find an interaction energy of the form[13]

$$E_{\text{int}} = -g\phi(r) = -\frac{g^2}{4\pi r}e^{-\mu r} \tag{7.28}$$

The result is the celebrated *Yukawa interaction*, and the implied Yukawa potential is given by

$$V_{\text{Y}}(r) \equiv -\frac{g^2}{4\pi r}e^{-\mu r} \qquad ;\ \text{Yukawa potential} \tag{7.29}$$

This potential provides a strong, short-range, attractive interaction between two nucleons (Fig. 7.5).

The *range* of the Yukawa potential is related to the Compton wavelength of the meson field by

$$r_{\text{range}} \sim \frac{1}{\mu} = \frac{\hbar}{mc} \tag{7.30}$$

[13]Here we simply assume the minus sign; it is shown in Prob. 7.1 that this is indeed the proper choice.

If one takes m to be the pion mass $m = m_{\pi^\pm}$, then the range is $r_{\text{range}} = 1.41\,\text{F}$, which is an appropriate value for the strong nuclear force.

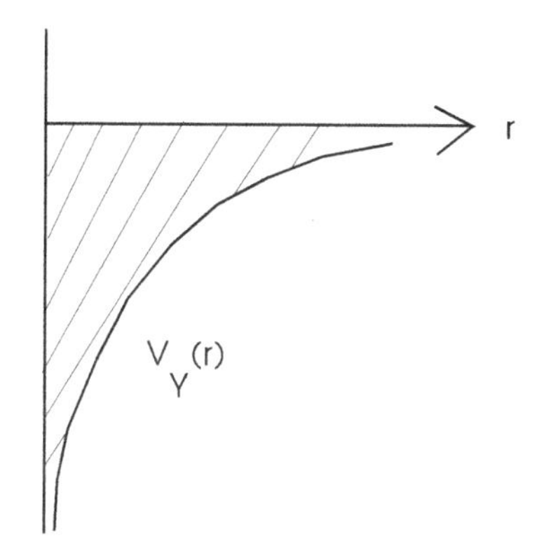

Fig. 7.5 Yukawa interaction potential $V_{\rm Y}(r) = -g^2 e^{-\mu r}/4\pi r$ between two strong nucleon point sources of strength g.

Several comments are relevant:

- The pion was *predicted* by Yukawa on the basis of these arguments:

  Just as there are photons (massless, long-range force) as quanta for the E-M field, there should be pions (massive, short-range force) as quanta for the meson field;

- The muon was discovered in cosmic rays (1937), but the muon proved to be just a heavy electron, with only electromagnetic and weak interactions;
- The strongly-interacting pion was eventually discovered by Powell (1948) using photographic emulsions.

7.5 Leptons

Leptons are spin-1/2 fermions that have only electromagnetic and weak interactions. They are produced in particle decays such as

$$\begin{aligned} n &\rightarrow p + e^- + \bar{\nu}_e \qquad &&; \beta\text{-decay} \\ \pi^+ &\rightarrow \mu^+ + \nu_\mu &&; \pi\text{-decay} \\ \mu^+ &\rightarrow e^+ + \nu_e + \bar{\nu}_\mu &&; \mu\text{-decay} \end{aligned} \tag{7.31}$$

Leptons come in three families (e^-, ν_e), (μ^-, ν_μ), and (τ^-, ν_τ). Each family can be assigned a lepton number (l_e, l_μ, l_τ), which is separately additively conserved in the weak interactions[14]

$$l_i = +1 \qquad ; i = e, \mu, \tau$$
$$\text{conserved lepton number} \tag{7.32}$$

The striking lepton spectrum is shown in Fig. 7.6. The masses of the charged leptons and their lifetimes are given in Table 7.2.

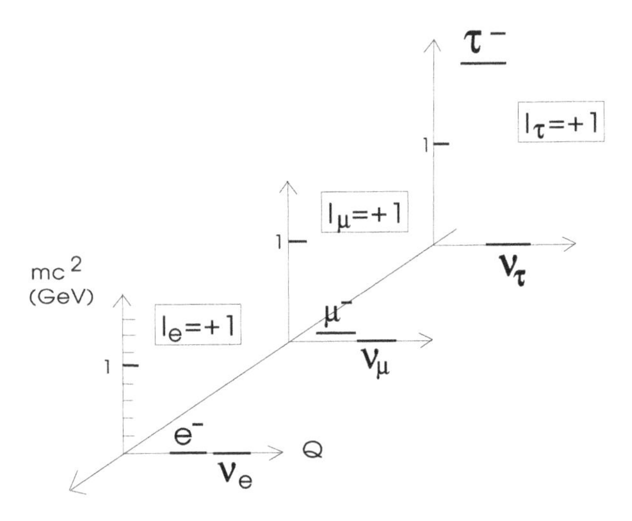

Fig. 7.6 Energy level diagram for the leptons.

Table 7.2 Charged lepton masses and lifetimes.

Particle	e^-	μ^-	τ^-
mc^2(MeV)	0.5110	105.6	1777
τ(sec)	stable	2.197×10^{-6}	0.291×10^{-12}

The situation in the neutrino sector is, in fact, more complicated than this. That situation is currently the subject of intense experimental study. It is now known that there exists an additional contribution to the hamiltonian in the neutrino sector, which, for want of a better word, we shall refer to as the "ultra-weak" hamiltonian $H_{\rm UW}$. This piece gives rise to a

[14]The families of antiparticles $(e^+, \bar{\nu}_e)$, $(\mu^+, \bar{\nu}_\mu)$, and $(\tau^+, \bar{\nu}_\tau)$ then carry separate lepton numbers $l_i = -1$ (see below).

tiny mass for the neutrinos, which allows them to mix through reactions such as

$$\nu_e \rightleftharpoons \nu_\mu \tag{7.33}$$

and the two "flavor" mixing problem is analyzed in appendix J. Such a mixing violates lepton conservation. The origin and detailed structure of $H_{\rm UW}$ is unknown at this writing.[15] The current status can always be determined from the latest version of [Particle Data Group (2006)]. To the best of our knowledge, $H_{\rm UW}$ is only active in the neutrino sector. When we subsequently discuss the "weak interactions" (as, for example, in the extremely successful standard model of electroweak interactions), we shall assume that $H_{\rm UW}$ is absent.

7.6 Antiparticles

Particles all have *antiparticles*, and several comments regarding antiparticles are relevant:

- The quantum numbers $\{Q, B, S, C, l_e, l_\mu, l_\tau\}$ are all reflected for antiparticles;[16]
- The existence of an antiparticle was predicted by Dirac (1926);
- This existence is a consequence of relativistic quantum field theory;
- It follows from very general principles that antiparticles and particles must have the same mass and lifetime;[17]
- The positron e^+, the antiparticle of the electron, was discovered by Anderson (1932);
- Antiparticles can annihilate with particles in reactions such as

$$\begin{aligned} e^+ + e^- &\rightarrow \gamma + \gamma \\ \bar{p} + p &\rightarrow \pi + \pi + \cdots \end{aligned} \tag{7.34}$$

[15]The status at the time of this writing is summarized in chapter 49 of [Walecka (2004)]. Neutrino experiments are notoriously difficult, and the present experimental effort is a herculean one.

[16]They change sign. In some cases, when all these quantum numbers vanish, the antiparticle may be identical to the particle (*e.g.* γ, π^0).

[17]See, for example, [Weyl (1950)].

7.7 Feynman Diagrams

Feynman diagrams provide the *language* of particle physics, and we proceed to describe that language though a series of steps.

7.8 S-matrix

The laboratory is taken to be a huge cubical box of volume $\Omega = L^3$ (Fig.7.7), and periodic boundary conditions are applied so that the single-particle wave functions are those of Eq. (4.118) for spin-zero bosons and Eq. (4.159) for spin-1/2 fermions.

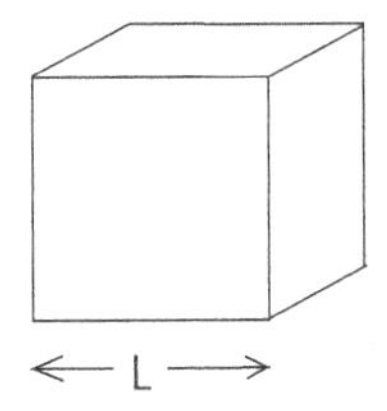

Fig. 7.7 The laboratory is taken to be a big box of volume $\Omega = L^3$ with periodic boundary conditions.

Suppose one wants to describe a general scattering process where two particles are prepared in states of definite momentum,[18] they react, and n particles come out from the reaction process (see Fig. 7.8). The *S-matrix* is defined in the following manner

$$S_{fi} \equiv \text{probability amplitude for process i} \rightarrow \text{f to take place} \qquad (7.35)$$

The general form of the S-matrix will be as follows

$$S_{fi} = -\frac{i}{\Omega^{(n+2)/2}}\left[\Omega\delta_{\mathbf{K}_i,\mathbf{K}_f}\right]\left[\frac{1}{\hbar}\Delta_T(W_f - W_i)\right]T_{fi} \qquad (7.36)$$

Here

- The factor $\Omega^{-(n+2)/2}$ in front comes from the wave functions of the initial and final particles;
- The factor $\left[\Omega\delta_{\mathbf{K}_i,\mathbf{K}_f}\right]$ expresses conservation of the total 3-momentum in the initial and final states. Here

[18]Recall $\mathbf{p} = \hbar\mathbf{k}$; one particle may, of course, be at rest.

$$\mathbf{K}_i = \sum_i \mathbf{k}_i \qquad ; \ \mathbf{K}_f = \sum_f \mathbf{k}'_f \tag{7.37}$$

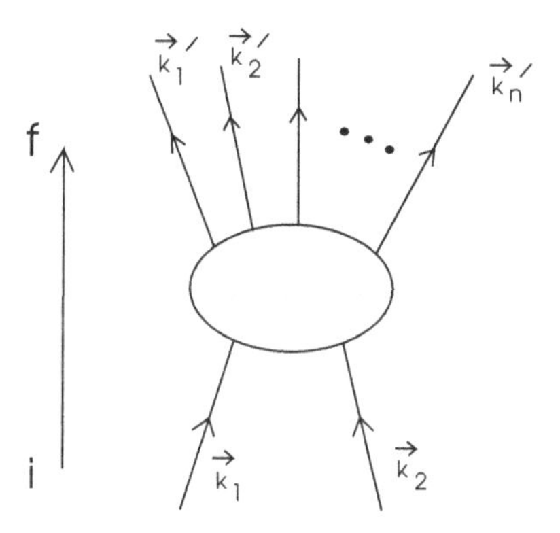

Fig. 7.8 A scattering process describing $2 \rightarrow n$ particles.

- The quantities (W_i, W_f) are the total energies of the initial and final states

$$W_i = \sum_i E_i \qquad ; \ W_f = \sum_f E_f \tag{7.38}$$

T is the time interval over which the interaction is on, and the expression $\Delta_T(W_f - W_i)$ is defined as

$$\begin{aligned} \Delta_T(W_f - W_i) &\equiv \int_{-T/2}^{T/2} dt\, e^{i(W_f - W_i)t/\hbar} \\ &= T\left[\frac{\sin(\omega T/2)}{(\omega T/2)}\right] \qquad ; \ \omega \equiv \frac{W_f - W_i}{\hbar} \end{aligned} \tag{7.39}$$

The integral is done as in appendix I.
- The remaining term T_{fi} in Eq. (7.36) is known as the T-matrix.

7.8.1 *Transition Rate*

The *transition rate* is given by

$$d\omega_{fi} = (\text{probability of the transition i} \rightarrow \text{f})/(\text{time interval}) \tag{7.40}$$

Since the particle states in the big box are spaced very close together (Fig. 4.35), one is *forced* to make a measurement of transitions into a *group* of states. The transition probability will vary smoothly about the central values, and one thus employs

$$\begin{aligned} \frac{|S_{fi}|^2}{T} &= \text{transition rate into one state} \\ dn_f &= \text{number of states observed in detectors} \end{aligned} \tag{7.41}$$

The relevant transition rate is therefore given by

$$d\omega_{fi} = \frac{|S_{fi}|^2}{T} dn_f \qquad ; \text{ transition rate} \tag{7.42}$$

With the aid of the differential form of Eq. (4.166), it follows from the above that

$$d\omega_{fi} = \frac{1}{\hbar^2} \frac{|\Delta_T(W_f - W_i)|^2}{T} \frac{1}{\Omega^n} |T_{fi}|^2 \frac{\Omega d^3k'_1}{(2\pi)^3} \frac{\Omega d^3k'_2}{(2\pi)^3} \cdots \frac{\Omega d^3k'_{n-1}}{(2\pi)^3} \tag{7.43}$$

In this expression:

(1) Overall momentum conservation, expressed through the Kronecker delta $\left[\delta_{\mathbf{K}_i,\mathbf{K}_f}\right]^2 = \delta_{\mathbf{K}_i,\mathbf{K}_f}$, has allowed one (and only one) of the original discrete sums over the wave numbers of the final particles to be carried out; the term dn_f then contains contributions from all but one of the particles in the final state;

(2) The contribution to the transition rate from the term $|\Delta_T|^2/T$ follows from Eq. (7.39) as

$$\frac{|\Delta_T(W_f - W_i)|^2}{T} = T\left[\frac{\sin(\omega T/2)}{(\omega T/2)}\right]^2 \qquad ; \omega \equiv \frac{W_f - W_i}{\hbar} \tag{7.44}$$

The correct prescription is now to let $T \to \infty$. The analysis in appendix I, and the first of Eqs. (7.39), imply that in this limit one arrives at *identical expressions* in both the transition rate and the S-matrix

$$\begin{aligned} \frac{1}{\hbar}\frac{|\Delta_T(W_f - W_i)|^2}{T} &\to 2\pi\,\delta(W_f - W_i) \qquad ; T \to \infty \\ \frac{1}{\hbar}\Delta_T(W_f - W_i) &\to 2\pi\,\delta(W_f - W_i) \end{aligned} \tag{7.45}$$

Here $\delta(W_f - W_i)$ is the Dirac delta function (see Prob. 4.3). This is the form in which energy conservation is actually expressed for the process;[19]

(3) For the *decay* rate of a particle, where there is only one particle in the initial state, the factor in front is replaced by $\Omega^{-n} \to \Omega^{-(n-1)}$.[20]

In summary, if one knows the T-matrix T_{fi}, which is obtained directly from the S-matrix S_{fi} for the process in Fig. (7.8), then one knows the *physical transition rate*

$$d\omega_{fi} = \frac{2\pi}{\hbar}\delta(W_f - W_i)\frac{1}{\Omega^n}|T_{fi}|^2\frac{\Omega d^3k_1'}{(2\pi)^3}\frac{\Omega d^3k_2'}{(2\pi)^3}\cdots\frac{\Omega d^3k_{n-1}'}{(2\pi)^3}$$
$$; \text{ physical transition rate} \qquad (7.46)$$

7.8.2 *Cross Section*

We proceed to define the *cross section* (see Fig. 7.9).

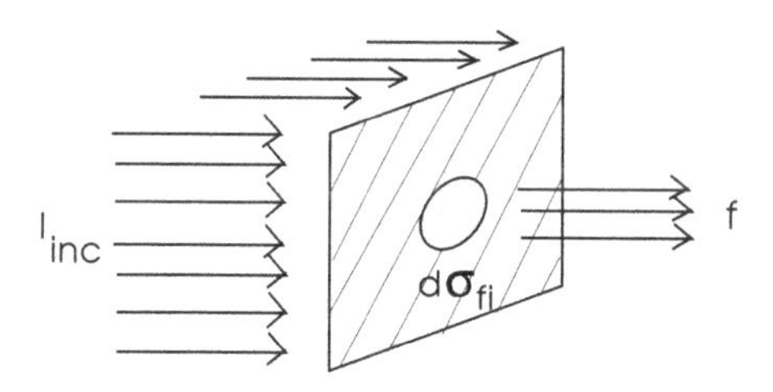

Fig. 7.9 Definition of the cross section.

As previously, we define the incident flux by[21]

$$I_{\text{inc}} \equiv (\text{probability flowing through unit area} \perp \text{beam}) / (\text{time interval})$$
$$; \text{ incident flux} \qquad (7.47)$$

The cross section is then defined by

$$d\sigma_{fi} \equiv \text{little element of transverse area such that if the particle}$$
$$\text{goes through it, an event of type f takes place} \qquad (7.48)$$

---

[19]One can then, in the limit $\Omega \to \infty$, replace $\Omega\delta_{\mathbf{K}_f,\mathbf{K}_i} \to (2\pi)^3\,\delta^{(3)}(\mathbf{K}_f - \mathbf{K}_i)$ in the S-matrix (see Prob. 7.2).

[20]The general expression is $\Omega^{-n} \to \Omega^2/\Omega^{(n_f+n_i)}$.

[21]For a single particle in a box of volume Ω moving with velocity v toward a stationary target, the incident flux is $I_{\text{inc}} = \rho v = (1/\Omega)v$.

It follows that

$$
\begin{aligned}
I_{\text{inc}}\, d\sigma_{fi} &= (\text{probability of transition i} \rightarrow \text{f}) \;/\; (\text{time interval}) \\
&= d\omega_{fi} \qquad\qquad ;\ \text{transition rate}
\end{aligned}
\tag{7.49}
$$

These are the quantities measured in an experiment where one repeats the elementary process over and over:

- $I_{\text{inc}} \rightarrow$ number of incident particles per unit area per unit time;
- $d\omega_{fi} \rightarrow$ number of transitions per unit time.

Thus, through Eqs. (7.46) and (7.49), we have arrived at physics!

7.9 Feynman Diagrams (Continued)

We are now in a position to discuss *Feynman diagrams* and the corresponding set of *Feynman rules.*[22] We shall give the Feynman rules for the quantity $-iT_{fi}/\hbar c$ in various cases. Although their derivation from first principles may require more advanced work,[23] once one has them, they characterize any theory, and a connection with physics is immediately established through the preceeding analysis. The procedure, then, is to :

(1) Draw all possible distinct Feynman diagrams;
(2) Associate factors with each element of the diagram;
(3) Sum the diagrams and calculate

$$
\frac{-iT_{fi}}{\hbar c} \quad \text{from Feynman rules} \tag{7.50}
$$

(4) Make a connection with rates and cross sections through the above;
(5) Check to make sure the artificial quantization volume Ω cancels from any physical result.

In this manner, one can describe any theory.[24]

---

[22] See Feynman's two papers [Feynman, (1949)] in [Schwinger (1958)] — everyone seriously interested in modern physics should read those papers.

[23] See, *e.g.* [Bjorken and Drell (1964); Bjorken and Drell (1965); Cheng and Li (1984)].

[24] Non-relativistic many-body theory is also most concisely formulated in terms of Feynman diagrams and Feynman rules (see [Fetter and Walecka (2003a)]).

7.10 Quantum Electrodynamics (QED)

We give a very brief introduction to *quantum electrodynamics* (QED), probably the most accurate physical theory known.

7.10.1 μ-e Scattering

To lowest order in the fine-structure constant α, there is one Feynman diagram for μ-e scattering. It is shown in Fig. 7.10.

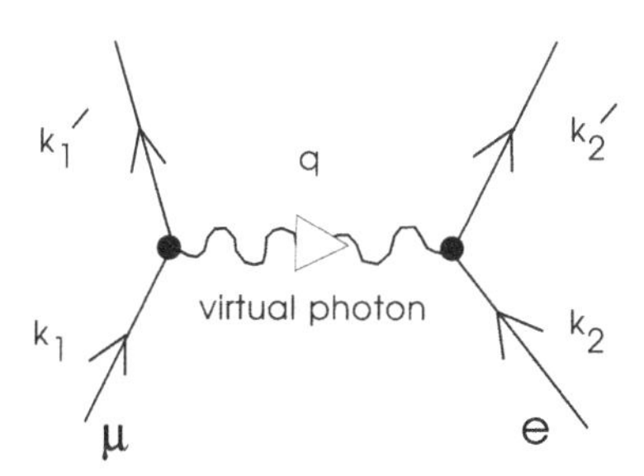

Fig. 7.10 Feynman diagram for μ-e scattering. Here $k = (\mathbf{k}, ik_0)$ represents the four-momentum (next chapter).

To calculate $-iT_{fi}/\hbar c$ in momentum space, the rules are as follows:

- Include a factor of

$$\frac{ie}{\hbar c\sqrt{\varepsilon_0}}j_\gamma \qquad ; \text{E-M current} \tag{7.51}$$

for each of the two vertices. Here j_γ is the Dirac electromagnetic current for each of the spin-1/2 fermions (this will be constructed later);
- Include the following factor for the *virtual photon* propagator

$$\frac{\hbar c}{i}\frac{1}{q^2} \qquad ; \text{ photon propagator} \tag{7.52}$$

Here q^2 is the four-momentum transfer given by

$$\begin{aligned} q^2 &= (\mathbf{k}_1 - \mathbf{k}_1')^2 - (E_1 - E_1')^2/(\hbar c)^2 \\ &= (\mathbf{k}_2 - \mathbf{k}_2')^2 - (E_2 - E_2')^2/(\hbar c)^2 \end{aligned} \tag{7.53}$$

The second line follows from the first by overall energy-momentum conservation.

The T-matrix is then obtained as

$$\frac{-iT_{fi}}{\hbar c} = \left(\frac{ie}{\hbar c\sqrt{\varepsilon_0}}j_\gamma\right)_\mu \left(\frac{\hbar c}{i}\frac{1}{q^2}\right)\left(\frac{ie}{\hbar c\sqrt{\varepsilon_0}}j_\gamma\right)_e \tag{7.54}$$

This expression is exact to order $\alpha = e^2/4\pi\varepsilon_0\hbar c$. It contains the following physics:

- The Coulomb interaction between the two charges;
- The magnetic interaction between the two currents;
- The Lorentz invariance of special relativity (see later).

7.10.2 *Anomalous Magnetic Moment of Electron*

There is a correction to the scattering amplitude for an electron in an external E-M field coming from the emission and absorption of a virtual photon, as illustrated in Fig. 7.11. This leads to a small modification of the g-factor for an electron. This effect was first calculated by Schwinger (1948).[25] The result for the electron magnetic moment in units of the Bohr magneton is

$$\begin{aligned}
\frac{\mu}{\mu_{\rm B}} &= 1 + \frac{\alpha}{2\pi} + \cdots && ;\ \text{Schwinger(1948)} \\
&= 1.001,16\cdots && \text{theory} \\
&= 1.001,16\cdots && \text{experiment}
\end{aligned} \tag{7.55}$$

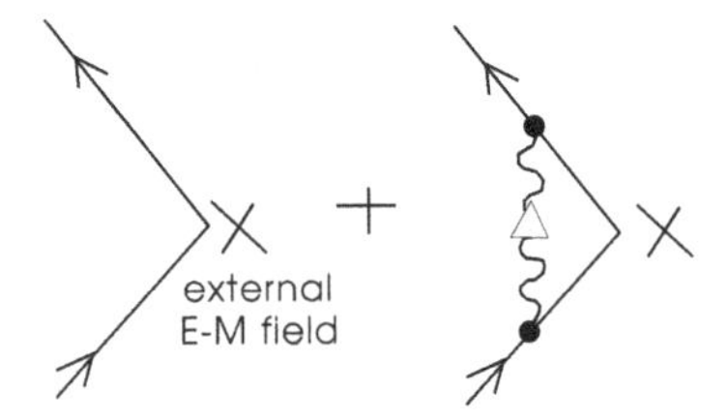

Fig. 7.11 Feynman diagrams for the anomalous magnetic moment of the electron.

A major effort in modern theoretical (and experimental) physics has gone into pushing this comparison as far as possible by including higher-

[25]See [Schwinger (1958)].

order diagrams. The present agreement between theory and experiment is one of the most accurate in physics.

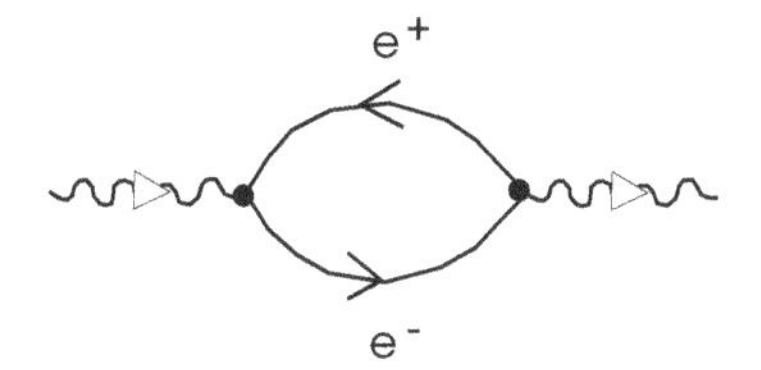

Fig. 7.12 Feynman diagram insertion for vacuum polarization.

One particularly interesting higher-order correction is that coming from *vacuum polarization*. Here the virtual photon creates a virtual (e^-, e^+) pair, that then annihilates back into a virtual photon (Fig. 7.12).

7.11 Quarks

So far, with quantum mechanics as the foundation, we have seen that

- The system of many electrons moving in the average, shielded nuclear Coulomb potential provides a predictive basis for understanding the periodic system of the elements;
- The many-nucleon shell model based on nucleons moving in an average hadronic nuclear potential provides a predictive basis for understanding the main features of nuclear structure.

Hadrons come in multiplets, and quantities such as the nucleon's charge distribution and anomalous magnetic moment indicate that hadrons themselves have internal structure. The question arises, is there an underlying basis that allows one to understand the structure of the hadrons? The answer, due to Gell-Mann (1963) and Zweig (1963), is yes there is. The hadrons appear to be composed of *quarks*.[26]

Quarks are spin-1/2 fermions that have strong, electromagnetic, and weak interactions. They have baryon number 1/3

$$B = \frac{1}{3} \qquad ; \text{ quarks} \tag{7.56}$$

[26]See [Halzen and Martin (1984); Martin (2006)].

The observed baryons with $B = 1$ are composed of triplets of quarks (qqq). Mesons, with $B = 0$, are composed of quark-antiquark pairs $(q\bar{q})$. Thus

$$\begin{aligned} &\text{Baryons are } (qqq) \\ &\text{Mesons are } (q\bar{q}) \end{aligned} \tag{7.57}$$

Quarks are assigned additional quantum numbers $(T, m_t, \mathcal{S}, \mathcal{C})$ for (isospin, third component of isospin, strangeness, charm) that make the quantum numbers for the composite hadrons come out correctly.[27] Quarks, with additional quantum numbers $(\mathcal{B}, \mathcal{T})$, then describe all known hadrons. The electric charge is now given by the extended Gell-Mann–Nishijima relation

$$\begin{aligned} Q &= m_t + \frac{1}{2}Y && \text{; electric charge} \\ Y &\equiv B + \mathcal{S} + \mathcal{C} + \mathcal{B} + \mathcal{T} && \text{; hypercharge} \end{aligned} \tag{7.58}$$

The quantum numbers of the six quarks (u, d, s, c, b, t), referred to as (up, down, strange, charm, bottom, top), are presented in Table 7.3.

Table 7.3 Quarks.

Domain	Quark	B	T	m_t	$\mathcal{S}$	$\mathcal{C}$	$\mathcal{B}$	$\mathcal{T}$	Q
Nuclear	u	1/3	1/2	1/2	0	0	0	0	2/3
	d	1/3	1/2	−1/2	0	0	0	0	−1/3
Extended	s	1/3	0	0	−1	0	0	0	−1/3
	c	1/3	0	0	0	1	0	0	2/3
Full	b	1/3	0	0	0	0	−1	0	−1/3
	t	1/3	0	0	0	0	0	1	2/3

It is important to note that

(1) The quantum numbers (Q, B) are additively conserved in all interactions;
(2) The quantum numbers $(\mathcal{S}, \mathcal{C}, \mathcal{B}, \mathcal{T})$ are additively conserved in the strong and electromagnetic interactions;
(3) The conservation of $(\mathcal{S}, \mathcal{C}, \mathcal{B}, \mathcal{T})$ may be violated in the weak interactions.

The quark picture is very unusual in that, to the best of our knowledge, *free, isolated quarks do not exist in nature.* It is now believed that the underlying strong "color" charge responsible for the interaction between quarks (see later) is completely shielded by strong vacuum polarization,

[27] Henceforth, for consistency, use the symbols $(\mathcal{S}, \mathcal{C})$ for strangeness and charm.

so that only "colorless" composite states can exist asymptotically in the laboratory. Since free quarks do not exist, their masses can only be inferred from other measurements. The masses of the (u, d, s, c, b) quarks inferred by [Gasser and Leutwyler (1982)] are plotted in Fig. 7.13.

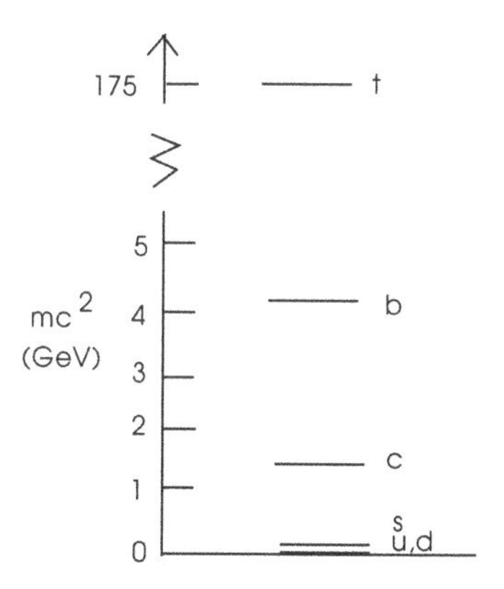

Fig. 7.13 Inferred quark spectrum from [Gasser and Leutwyler (1982)]. Note the change in scale for the subsequently discovered top quark!

The top quark (t), required by compelling theoretical arguments, was discovered at Fermilab (1995) at a mass

$$m_{\text{top}}c^2 = 174 \pm 5\,\text{GeV} \qquad ; \text{ top quark} \qquad (7.59)$$

The spectrum in Fig. (7.13) is bizarre. To the best of the author's knowledge, there is no convincing theoretical argument that gives rise to this spectrum.[28] With this spectrum as a basis, one can focus on various *domains* of physics characterized by the energy scale of interest:

- *Nuclear Domain.* Here one focuses on that subspace of the full theory built on (u, d) quarks. The hadrons built from these quarks describe almost all of nuclear physics. The fact that the hamiltonian describing the strong interaction between these quarks is invariant under "rotations" among them, then leads to the isospin symmetry of nuclear interactions.[29]

---

[28] There's a challenge for bright, young minds!

[29] As pointed out previously, the symmetry group of angular momentum in quantum mechanics [Edmonds (1974)] is not the group $O(3)$ of real rotations in space, but is actually $SU(2)$.

- *Extended Domain.* By enlarging the subspace to include the (s, c) quarks, one has a basis for understanding the strange and charmed hadrons produced at higher laboratory energies;
- *Full Domain.* The extension to the full space with (b, t) quarks is required for many theoretical reasons, and the $(b\bar{b})$ system has now been extensively studied in (e^+, e^-) colliders. While the (b, t) quarks play no apparent role at nuclear energies, their effects become important at very high energies. So far, no additional quarks have been called for.

7.11.1 *Nuclear Domain*

With the empirical observation that quarks are *confined* to the interior of hadrons, we can appeal to the simplest confining model in quantum mechanics, that of the three-dimensional simple harmonic oscillator (Fig. 7.14). From the last chapter, we know the lowest-lying single-particle state in the oscillator will be $(1s)$.

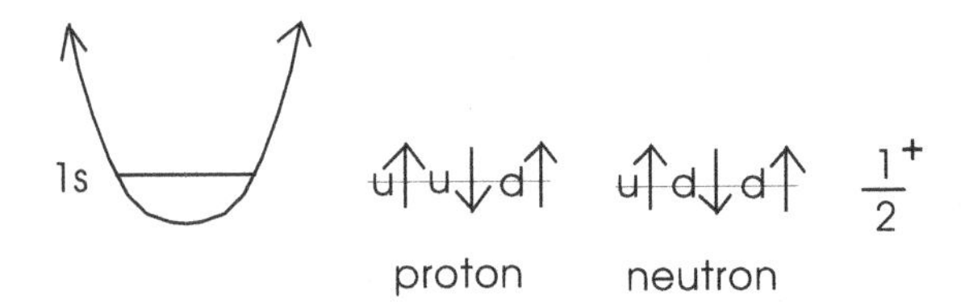

Fig. 7.14 Three-dimensional simple harmonic oscillator as model confining potential, and three-quark $(1s)^3$ ground state in the nuclear domain with $(J^\pi, T) = (1/2^+, 1/2)$.

There are two spin states for each quark $(\uparrow, \downarrow)$,[30] and, with the second observation that the observed hadrons correspond to the multiplets in Eqs. (7.57), one can try building a simple *quark shell model* from $(u\uparrow, u\downarrow, d\uparrow, d\downarrow)$. The proton, with the correct quantum numbers (B, Q), is immediately obtained by placing the non-identical quarks $(u\uparrow, u\downarrow, d\uparrow)$ into the $(1s)$ state, yielding a $(1s)^3$ spatial configuration. The neutron follows analogously by interchanging one $(u \leftrightarrow d)$ to get $(u\uparrow, d\downarrow, d\uparrow)$. Since the total angular momentum of the system can now be taken to be the spin of the last quark, and similarly for the isospin, one obtains $(J^\pi, T) = (1/2^+, 1/2)$ for the nucleon (Fig. 7.14).[31] So far, so good.

[30] More generally, these are *helicity* states.

[31] For clarity, we now use the notation (J^π, T) for the quantum numbers of the composite system, and the quark isospin will be denoted by $\mathbf{t}$.

An obvious excited state of the nucleon is obtained by simply flipping the odd quark spin leading to a state with all the spins aligned, and $J^\pi = 3/2^+$. This indeed corresponds to the first excited state of the nucleon, the $\Delta(1238)$, shown in Fig. 7.1. Furthermore, the $\Delta(1238)^{++}$ component, and then the whole $T = 3/2$ multiplet, are readily obtained from the quark configuration $(u\uparrow, u\uparrow, u\uparrow)$ (Fig. 7.15).

$$u\uparrow \; u\uparrow \; u\uparrow \qquad \frac{3}{2}^+$$

Fig. 7.15 First excited state of the nucleon in the nuclear domain in the $(1s)^3$ configuration in the simple harmonic oscillator potential. Illustrated is that component of the $(J^\pi, T) = (3/2^+, 3/2)$ multiplet with $(M_J, M_T) = (3/2, 3/2)$.

But now one is faced with a seemingly insurmountable problem. Quarks are spin-1/2 fermions, and we know such particles should obey the Pauli exclusion principle. It should not be possible to build a $(1s)^3$ configuration from three identical $(u\uparrow)$ quarks. Indeed, with a $(1s)^3$ configuration, the state in Fig. 7.15 is *totally symmetric* under the interchange of any two quarks! A great deal of time and energy was expended attempting to resolve this issue. The eventual resolution was remarkably simple, and has had the most profound implications for modern physics:

> *Quarks are given an additional, internal degree of freedom that takes three values, say (R, G, B).*

Thus, as with isospin, the quark wave function should actually be written

$$(u\uparrow)_\alpha \qquad ; \; \alpha = R, G, B \quad \text{color} \tag{7.60}$$

This additional degree of freedom is referred to, with some whimsy, as *color.* The quarks in Fig. 7.15 then simply have different colors. The symmetry of the state illustrated in Fig. (7.15) can then be understood if one demands that the color wave function factor be *totally antisymmetric* under the interchange of any two quarks (*i.e.* a color singlet).

A problem here is that one does not see this additional degree of freedom in observed particles, which are *colorless.* The color degree of freedom, like the quarks, is *confined to the interior of hadrons.*

7.11.2 *Some Applications*

Ω^- *Particle.* If either of the (u, d) quarks is replaced with an s quark, the new hadron carries strangeness and lies at a mass $\approx m_s$ above the nucleon, where m_s is the mass of the strange quark in Fig. (7.13). This quark picture can then explain the hadrons in the $S = -1$ sector in Fig. (7.1).[32] With two strange quarks, one has an explanation of the $S = -2$ sector lying at a mass $\approx 2m_s$ in Fig. (7.1).

The quark picture then *predicts* that there should be a hadron composed entirely of three strange quarks (sss), a partner to the one in Fig. 7.15 (see Fig. 7.16), with quantum numbers $(Q, T, \mathcal{S}) = (-1, 0, -3)$ at $\approx 3m_s$ above the Δ. This striking prediction was subsequently verified by the experimental discovery of the Ω^- particle at Brookhaven (1964), providing firm support for the underlying quark structure.

$$s\uparrow s\uparrow s\uparrow \qquad \frac{3}{2}^+$$

Fig. 7.16 A predicted low-lying state of three strange quarks, a partner to the one in Fig. 7.15, in the extended domain in the simple harmonic oscillator potential. This prediction was subsequently verified by the experimental discovery of the Ω^- particle.

Mesons. Mesons are composed of $(q\bar{q})$ in this picture, and since the additive quantum numbers are simply reversed in the antiquark, one can just read off the quantum numbers of the composites from Table 7.3. Some examples are given in Fig. 7.17.

The spins of the quark and antiquark in the meson can be combined using our previous results from the theory of angular momentum in quantum mechanics[33]

$$\begin{aligned} \mathbf{S} &= \mathbf{s}_1 + \mathbf{s}_2 \\ \mathbf{S}^2 &= S(S+1) \qquad ; S = 0, 1 \end{aligned} \tag{7.61}$$

An exactly analogous argument can be made for isospin

$$\begin{aligned} \mathbf{T} &= \mathbf{t}_1 + \mathbf{t}_2 \\ \mathbf{T}^2 &= T(T+1) \qquad ; T = 0, 1 \end{aligned} \tag{7.62}$$

[32]See Prob. 7.5.

[33]Recall that $|s_1 - s_2| \leq S \leq s_1 + s_2$ in integer steps; here $s_1 = s_2 = 1/2$.

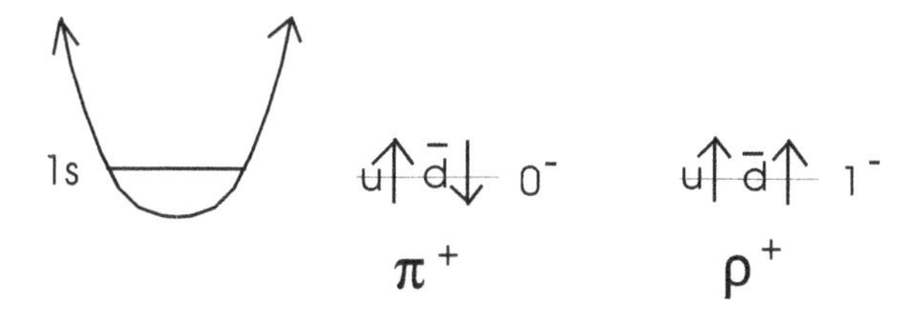

Fig. 7.17 Some low-lying $(q\bar{q})$ meson states in the nuclear domain in the simple harmonic oscillator potential with a $(1s)^2$ spatial configuration.

Thus in the nuclear domain, this quark picture predicts the low-lying mesons shown in Table 7.4.

Table 7.4 Spin and isospin of the low-lying $(q\bar{q})$ states in the nuclear domain.

S	0	0	1	1
T	1	0	0	1
particles	π^+, π^0, π^-	η^0	ω^0	ρ^+, ρ^0, ρ^-

It is a consequence of relativistic quantum field theory that, just as with all the other quantum numbers, the intrinsic parity of the antiquark is opposite to that of the quark, and thus the parity of a bound state with given l for the quark-antiquark system is

$$\Pi = (-1)^{l+1} \qquad ; \text{ parity of } (q\bar{q}) \tag{7.63}$$

This accords precisely with the parities of the observed mesons in Table 7.4, all of which are odd with $l = 0$.

In fact, since the quantum numbers reverse for antiparticles, one gets the same predictions for the quantum numbers of the mesons in a picture of mesons where they are composed of nucleons and antinucleons $(N\bar{N})$. This is the celebrated Fermi-Yang model of mesons, which appeared long before the quark model.

Fig. 7.18 Predicted meson state with one strange antiquark in the extended domain, based on Fig. 7.17.

Again, replacing a (u, d) quark with an s quark gives a strange meson at a higher mass. One example of a predicted meson, which appears in nature with all the correct quantum numbers, is shown in Fig. 7.18.

Wave Functions The detailed construction of the wave functions for the (qqq) baryons and the $(q\bar{q})$ mesons is more complicated, and depends on the dynamical model of the hadrons. One can work at various levels here:

- In the non-relativistic quark model, the quarks are assumed to have effective masses $m_q^\star \approx m_p/3$ where m_p is the nucleon mass, and to move non-relativistically in a confining potential such as the simple harmonic oscillator. The spatial ground state of the nucleon is the totally symmetric $(1s)^3$ state. One then constructs totally symmetric products of spin and isospin wave functions with given (S, T) for the three quarks. Various properties of the nucleon now follow from quantum mechanics. The model is readily extended to the excited states of the nucleon, and to strange and charmed hadrons. This model has had striking success (see [Bhaduri (1988)]);
- A more realistic dynamical model is provided by the "M.I.T. Bag", where the (u, d) quarks are light and move relativistically inside a confining bubble hollowed out in the vacuum. This model, too, has had many successes ([Bhaduri (1988)]);
- At the deepest level, one now solves the relativistic quantum field theory QCD (see next section) with a computer on a finite space-time lattice [Wilson (1974)], and then proceeds to calculate hadronic properties directly from the underlying theory. Enough evidence has now accumulated that this is the correct way to go, and this *lattice gauge theory* will form the basis for modern hadronic particle theory for the foreseeable future. Convincing evidence already exists from such calculations, for example, that quarks are indeed confined to the interior of hadrons.

7.12 Quantum Chromodynamics (QCD)

We previously briefly discussed *quantum electrodynamics* (QED), the theory of the electromagnetic interactions. Figure (7.10) shows the Feynman diagram for (e, μ) scattering, and the accompanying Feynman rules are given in the text. The scattering takes place through the exchange of a virtual photon with a propagator $(\hbar c/iq^2)$ where q^2 is the four-momentum transfer. The photon couples to the electromagnetic current j_γ through a factor $(iej_\gamma/\hbar c\sqrt{\varepsilon_0})$ at each vertex.

The theory of the strong interactions binding quarks into the observed hadrons is *quantum chromodynamics* (QCD). In this theory, virtual *gluons* are exchanged, which couple to the *color current* of the quark. The lowest-order Feynman diagram for quark-quark scattering in QCD is given in Fig. (7.19).

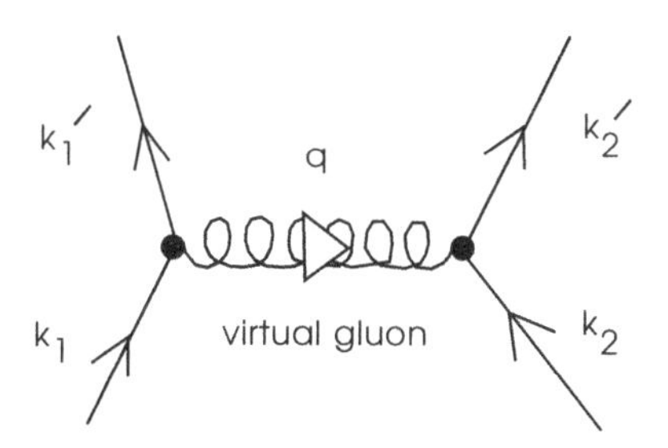

Fig. 7.19 Feynman diagram for quark-quark scattering through one-gluon exchange.

The accompanying Feynman rules for $(-iT_{fi}/\hbar c)$ are:

- Include a factor of

$$\frac{ig}{\hbar c}j_C \qquad ;\text{color current} \tag{7.64}$$

for each of the two vertices. Here j_C is the Dirac color current for each of the spin-1/2 fermions (constructed later), and g is the strong color charge;

- Include the following factor for the *virtual gluon* propagator

$$\frac{\hbar c}{i}\frac{1}{q^2} \qquad ;\text{ gluon propagator} \tag{7.65}$$

Here q^2 is the four-momentum transfer given by

$$\begin{aligned} q^2 &= (\mathbf{k}_1 - \mathbf{k}_1')^2 - (E_1 - E_1')^2/(\hbar c)^2 \\ &= (\mathbf{k}_2 - \mathbf{k}_2')^2 - (E_2 - E_2')^2/(\hbar c)^2 \end{aligned} \tag{7.66}$$

The second line again follows from the first by overall energy-momentum conservation.

QCD has additional, non-linear, self-couplings of the gluons that are not present for photons in QED. There are both three-gluon and four-gluon vertices as shown in Fig. (7.20).

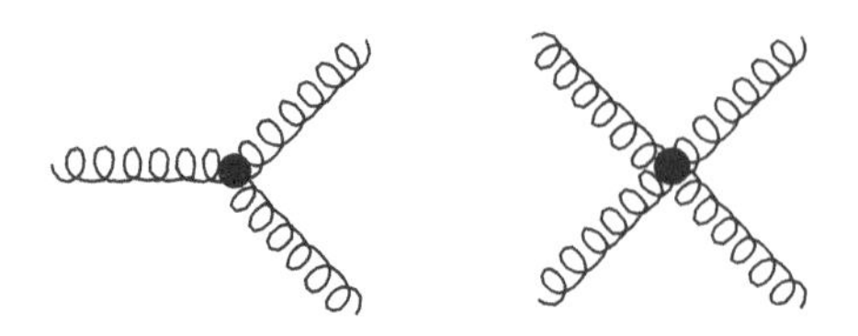

Fig. 7.20 Three and four gluon self-couplings in QCD.

As consequences of the non-linear gluon couplings of QCD:

(1) Color is confined to the interior of hadrons:
   - The interaction is *strong* at large distances;
   - There is complete *shielding* of the strong color charge at large distances due to strong vacuum polarization;
   - Lattice gauge theory indicates that confinement is indeed a dynamical consequence of QCD.

(2) At small distances, the interaction gets *weak*. As a statement on the strong coupling constant measured at various q^2 in Fig. 7.19, as modified by higher-order corrections to the process, one has

$$g(q^2) \to 0 \qquad ; \; q^2 \to \infty \qquad \text{asymptotic freedom} \tag{7.67}$$

This most remarkable property of QCD, known as *asymptotic freedom*, was discovered by [Gross and Wilczek (1973); Politzer (1973)]. It allows one to test the theory by doing perturbation theory in g at high momentum transfer. It also makes lattice gauge theory possible, in that it allows one to obtain the true continuum theory in the limit as the lattice spacing goes to zero.

In QED, the polarization of the vacuum illustrated in Fig. 7.12 *shields* a bare point charge, so that $e(q^2)$ *grows* as $q^2 \to \infty$, which probes shorter and shorter distances. The vacuum acts like a *dielectric* medium in QED. In contrast, a small color charge surrounds itself with color charge so that the opposite situation holds in QCD, as seen in Eq. (7.67). In QCD, the vacuum acts like a *paramagnetic* medium, where a small magnetic moment can induce a large magnetization.

Dynamic evidence for a point-like substructure in the nucleon was first obtained from deep inelastic (e, e') experiments at SLAC. As the four-momentum transfer q^2 and energy loss in the laboratory frame ν both become large, the observed response surfaces become functions of only the *ratio* of these two quantities $x = q^2/2M\nu$ [Friedman and Kendall (1972)]. This is directly interpreted in terms of the *quark-parton model* of Feynman and Bjorken and Paschos [Bjorken and Paschos (1969)] (see Prob. 9.14).[34]

7.13 Standard Model of Electroweak Interactions

The unified theory of electroweak interactions developed by [Salam and Ward (1964); Weinberg (1967); Glashow, Iliopoulos, and Maiani (1970)] represents one of the great achievements of modern physics. Its phenomenological consequences for most of the weak interactions can be summarized by the Feynman diagram in Fig. 7.21.

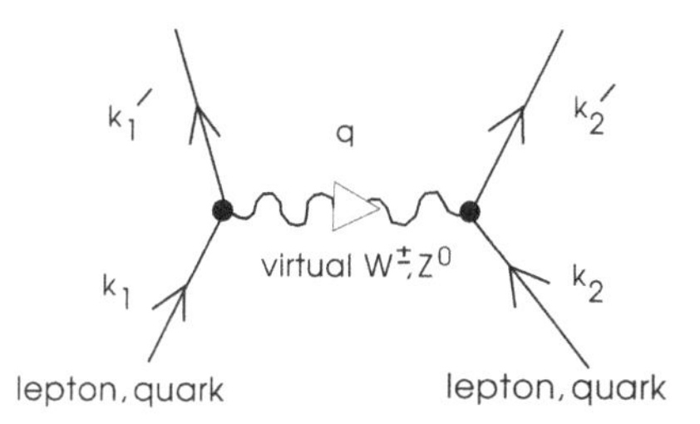

Fig. 7.21 Exchange of virtual massive weak vector bosons ($W^{\pm}, Z^0$) in the standard model of the electroweak interactions.

Leptons and quarks exchange virtual heavy, *weak vector bosons*, and the Feynman rules for $(-iT_{fi}/\hbar c)$ that accompany this diagram are:

- Include a factor of

$$\frac{ig_W}{\hbar c} j_W \qquad \text{; weak current} \tag{7.68}$$

  for each of the two vertices. Here j_W is the appropriate Dirac weak charged or neutral current for each of the spin-1/2 fermions (see later), and g_W is the weak charge;
- Include the following factor for the *virtual vector boson* propagator

$$\frac{\hbar c}{i} \frac{1}{q^2 + M_W^2} \qquad \text{; vector boson propagator (see later)} \tag{7.69}$$

[34]Introductions to QCD and lattice gauge theory, as well as to electron scattering and the standard model of electroweak interactions, can be found in [Walecka (2004)].

Here $M_W = m_W c/\hbar$ is the appropriate vector boson inverse Compton wavelength, and q^2 is the four-momentum transfer given by

$$\begin{aligned} q^2 &= (\mathbf{k}_1 - \mathbf{k}_1')^2 - (E_1 - E_1')^2/(\hbar c)^2 \\ &= (\mathbf{k}_2 - \mathbf{k}_2')^2 - (E_2 - E_2')^2/(\hbar c)^2 \end{aligned} \qquad (7.70)$$

The weak coupling constant g_W is related to the Fermi constant $G_{\rm F}$ of Eq. (7.3) by

$$\frac{g_W^2}{M_W^2} = \frac{G_{\rm F}}{\sqrt{2}} \qquad (7.71)$$

The Heisenberg uncertainty principle states that

$$\Delta p \Delta x \sim \hbar \qquad (7.72)$$

If a virtual particle of mass m is emitted, then one can estimate $\Delta p \sim mc$, and hence

$$\Delta x \sim \frac{\hbar}{mc} \qquad ; \text{ if } \Delta p \sim mc \qquad (7.73)$$

Hence the interaction is of very *short range* if the virtual quantum transmitting that interaction is very heavy.[35]

After years of searching, the weak vector bosons $(W^{\pm}, Z^0)$ were discovered at CERN (1983) with masses

$$\begin{aligned} m_W c^2 &= 80.42\,\text{GeV} \\ m_Z c^2 &= 91.19\,\text{GeV} \end{aligned} \qquad (7.74)$$

The Large Hadron Collider (LHC) under construction at CERN will study the collisions of $p + p$, as well as various heavy ions, in the C-M system.[36] It is the world's highest energy particle accelerator, and it is designed to explore the frontiers of particle physics [LHC (2008)]. The LHC is scheduled to begin operation in 2008.

[35] We have already seen one example of this relationship in the range of the nuclear force arising from the exchange of a massive meson.

[36] See next chapter.

Chapter 8

Special Relativity

In addition to quantum mechanics, the second pillar of modern physics is the theory of special relativity due to Einstein (1905). Although now verified by countless experiments, the original foundation lay in the experiment of Michelson and Morley (1881).

8.1 Michelson-Morley Experiment

Consider a fluid at rest. Suppose a sound signal is sent out from the origin O and moves with velocity c_{sound} along the x-axis (Fig. 8.1).

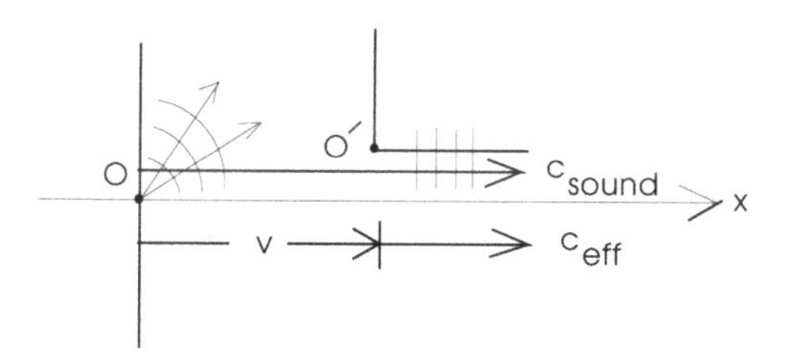

Fig. 8.1 Sound signal sent out in a fluid at rest and viewed from a frame moving with a velocity v along the x-axis relative to the stationary fluid.

If the sound wave is viewed from a frame whose origin O' moves with a velocity v along the x-axis relative to the stationary fluid, then classical physics tells us that in the moving frame the sound wave appears to have a velocity

$$c_{\text{eff}} = c_{\text{sound}} - v \tag{8.1}$$

Quite generally, these velocities simply add as classical vectors

$$\mathbf{c}_{\text{eff}} = \mathbf{c}_{\text{sound}} - \mathbf{v} \tag{8.2}$$

Consider Maxwell's equations for light in vacuum (chapter 2). As we have seen, there is a solution describing a plane wave traveling in the x-direction (Fig. 8.2).

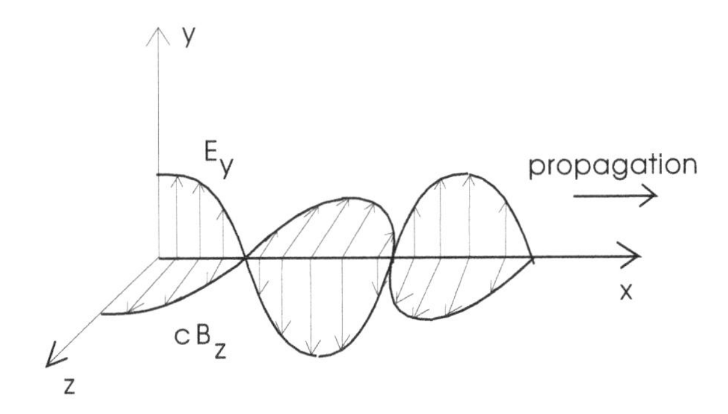

Fig. 8.2 Field configuration in Eqs. (2.94) describing an electromagnetic plane wave traveling to the right along the x-axis with velocity c.

The fields (E_y, cB_z) in this plane wave satisfy the one-dimensional wave equation

$$\frac{\partial^2 \phi}{\partial x^2} = \frac{1}{c^2}\frac{\partial^2 \phi}{\partial t^2} \qquad ; \ \phi = (E_y, cB_z)$$
$$c = \frac{1}{\sqrt{\mu_0 \varepsilon_0}} \qquad ; \text{ velocity of light} \tag{8.3}$$

Now one can ask the question, is there an "ether", in which light has a velocity c, corresponding to a "fluid at rest" for light? If so, then on classical grounds, one would expect to measure a light velocity $\mathbf{c}_{\text{eff}} = \mathbf{c} - \mathbf{v}$ when moving with velocity $\mathbf{v}$ with respect to that ether.

To search for such an effect, Michelson and Morley employed a Michelson interferometer. In this instrument, a light beam from a source is split into two beams that then travel down two perpendicular arms. The beams are reflected from mirrors and then recombined to form an interference pattern on a screen, which depends on the difference in optical pathlength down the two arms (Fig. 8.3).

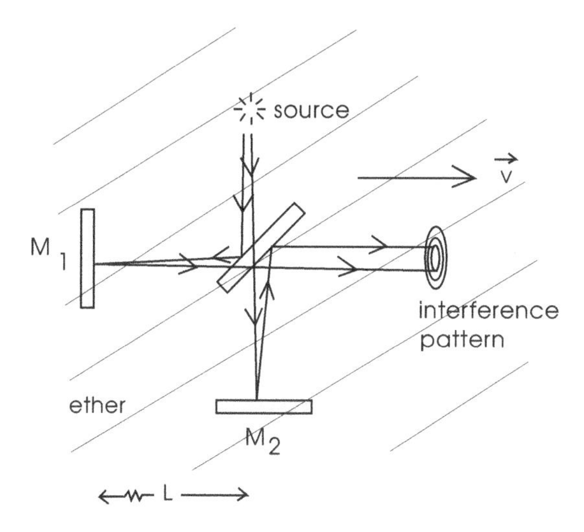

Fig. 8.3 Schematic of Michelson-Morley experiment. A light beam from a source is split, travels to two mirrors (M_1, M_2), and is then recombined to form an interference pattern. The apparatus travels with various velocities $\mathbf{v}$ relative to an assumed background ether where light has a velocity c, and one searches for shifts in the fringes of the interference pattern as the apparatus is rotated with respect to $\mathbf{v}$.

Let us analyze the situation, assuming the presence of the ether and the classical addition of velocities. The difference in optical pathlength can be written as

$$\Delta = 2\pi \left[\frac{c(t_I - t_{II})}{\lambda} \right] \tag{8.4}$$

Here $c/\lambda = \nu$ is the frequency of the light, and (t_I, t_{II}) are the times that the light signal takes to traverse the two arms of the interferometer.

The l.h.s. of Fig. 8.4 shows the configuration in arm I where the velocity $\mathbf{v}$ of the mirror M_1 is parallel to the arm. If L is the length of the arms, then the time t_I to go down and back in arm I is given by

$$t_I = \frac{L}{c - v} + \frac{L}{c + v} \tag{8.5}$$

In arm II, which is perpendicular to $\mathbf{v}$, the light must travel with the velocity vectors shown on the r.h.s of Fig. 8.4 in order that it be reflected back into the interferometer from mirror M_2. Thus, with the classical addition of velocities, the time t_{II} is given by

$$t_{II} = 2\frac{L}{\sqrt{c^2 - v^2}} \tag{8.6}$$

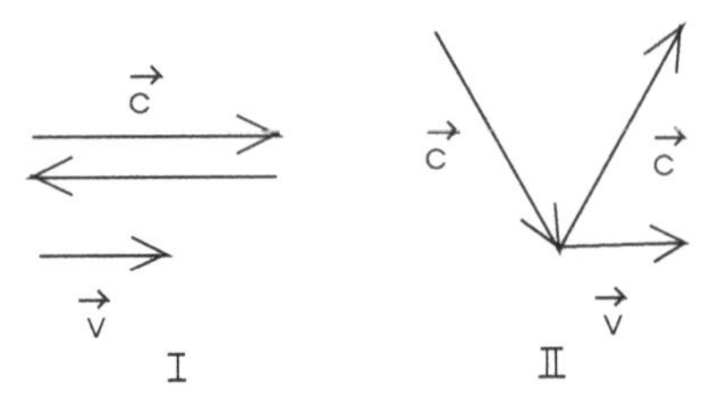

Fig. 8.4 Orientation of the two arms of the interferometer with respect to **v**, and classical addition of velocities. Here I denotes the configuration of (**v**, **c**) in the first arm where the mirror M_1 moves with velocity **v** parallel to the arm, while II then denotes the corresponding configuration of the velocities (**v**, **c**) in the second arm, where the velocity **v** is perpendicular to the arm.

The difference in optical pathlength is thus given by

$$\begin{aligned} \Delta &= \frac{2\pi L}{\lambda}\left[\frac{1}{1-v/c} + \frac{1}{1+v/c} - \frac{2}{\sqrt{1-v^2/c^2}}\right] \\ &= \frac{2\pi L}{\lambda}\left(\frac{v}{c}\right)^2 \qquad ;\ \text{to } O\left[\left(\frac{v}{c}\right)^2\right] \end{aligned} \tag{8.7}$$

The second line is obtained through an expansion in v/c (Prob. 8.1).

The whole apparatus is assumed to be moving with velocity v with respect to the ether, and there is an interference pattern produced because the optical pathlength from M_1 is a little longer. Suppose the apparatus is now *rotated* with respect to **v** so that M_2 occupies the position formerly occupied by $M1$. During that rotation, the interference pattern should go through a cycle of reversals.[1] In their experiment, Michelson and Morley:

- Varied the velocity **v** with which their interferometer moved with respect to the assumed ether;
- Made use of the fact that the earth's rotation can be used to reverse that velocity every 12 hours to give them a periodic signal;
- Counted the number of interference fringes moving past a fixed point as the apparatus was rotated.

[1] For example, suppose the difference in optical pathlength is π. Then the center spot should go from dark to light to dark under 1/4 turn.

No effect was seen! It was a conclusion of the Michelson-Morley experiment that

> *There is no underlying ether with respect to which the velocity of light is c.*

It was the eventual conclusion of the Michelson-Morley experiment that

> *The velocity of light takes the same value in any inertial frame.*[2]

The reader should really let this result sink in, because it is in complete variance with how one adds velocities in classical physics in going from one inertial frame to another.

8.2 Lorentz Transformation

Lorentz observed that there is a *mathematical transformation* that leaves the form of the wave equation for light in Eq. (8.3) unchanged (leaves it "invariant"). That transformation is

$$\begin{aligned} x &= \frac{(x' + Vt')}{\sqrt{1 - V^2/c^2}} \qquad ; \text{Lorentz transformation} \\ t &= \frac{(t' + Vx'/c^2)}{\sqrt{1 - V^2/c^2}} \end{aligned} \tag{8.8}$$

where V is simply some constant. Equations (8.8) are readily inverted, and the result is obtained by merely changing the sign of $V \leftrightarrow -V$

$$\begin{aligned} x' &= \frac{(x - Vt)}{\sqrt{1 - V^2/c^2}} \qquad ; \text{Lorentz transformation} \\ t' &= \frac{(t - Vx/c^2)}{\sqrt{1 - V^2/c^2}} \end{aligned} \tag{8.9}$$

It is readily verified that this transformation leaves the following quadratic form invariant[3]

$$x^2 - c^2t^2 = x'^2 - c^2t'^2 \qquad ; \text{invariant} \tag{8.10}$$

[2] Just go to the rest frame of the interferometer—there is no effect.

[3] The demonstration of Eqs. (8.9) and (8.10) is left as an exercise (Prob. 8.1).

Let us verify Lorentz's result. Write $\phi[x(x',t'),t(x',t')]$ and use the chain rule for differentiation of an implicit function twice. This gives

$$\frac{\partial\phi}{\partial x'} = \frac{\partial\phi}{\partial x}\frac{\partial x}{\partial x'} + \frac{\partial\phi}{\partial t}\frac{\partial t}{\partial x'} = \frac{1}{\sqrt{1-V^2/c^2}}\left[\frac{\partial\phi}{\partial x} + \frac{V}{c^2}\frac{\partial\phi}{\partial t}\right]$$

$$\frac{\partial^2\phi}{\partial x'^2} = \frac{1}{1-V^2/c^2}\left\{\frac{\partial}{\partial x}\left[\frac{\partial\phi}{\partial x} + \frac{V}{c^2}\frac{\partial\phi}{\partial t}\right] + \frac{V}{c^2}\frac{\partial}{\partial t}\left[\frac{\partial\phi}{\partial x} + \frac{V}{c^2}\frac{\partial\phi}{\partial t}\right]\right\} \quad (8.11)$$

In a similar fashion, one obtains

$$\frac{\partial\phi}{\partial t'} = \frac{\partial\phi}{\partial x}\frac{\partial x}{\partial t'} + \frac{\partial\phi}{\partial t}\frac{\partial t}{\partial t'} = \frac{1}{\sqrt{1-V^2/c^2}}\left[V\frac{\partial\phi}{\partial x} + \frac{\partial\phi}{\partial t}\right]$$

$$\frac{\partial^2\phi}{\partial t'^2} = \frac{1}{1-V^2/c^2}\left\{V\frac{\partial}{\partial x}\left[V\frac{\partial\phi}{\partial x} + \frac{\partial\phi}{\partial t}\right] + \frac{\partial}{\partial t}\left[V\frac{\partial\phi}{\partial x} + \frac{\partial\phi}{\partial t}\right]\right\} \quad (8.12)$$

Now take Eq. (8.11) and subtract $(1/c^2)$ times Eq. (8.12) to arrive at

$$\frac{\partial^2\phi}{\partial x'^2} - \frac{1}{c^2}\frac{\partial^2\phi}{\partial t'^2} = \frac{\partial^2\phi}{\partial x^2} - \frac{1}{c^2}\frac{\partial^2\phi}{\partial t^2} \quad ;\ \text{Lorentz} \quad (8.13)$$

This is the result of Lorentz.

8.3 Einstein's Theory

The Michelson-Morley experiment, and the Lorentz transformation, provide the basis for Einstein's theory of special relativity [Einstein (1905)]. Recall our definition of the *primary inertial frame* as one that is at rest with respect to the fixed stars. We observed that any frame moving with constant velocity relative to the fixed stars is again inertial. Einstein assumed that

- The laws of physics are the same in any inertial frame;
- The speed of light, which provides a limiting velocity on the propagation of signals, is the same in any inertial frame.

As a consequence, there is no experiment that can be done in a closed inertial frame that tells how fast one is moving with respect to the fixed stars. Einstein further assumed, and this was his real stroke of genius, that

- The Lorentz transformation is a *physical transformation* that relates the coordinates (x,t) measured in one inertial frame to those (x',t') measured in a second inertial frame moving with velocity $v \equiv V$ relative to the first.

Time is no longer absolute, as it is in newtonian physics, but is now relative, and varies from frame to frame. This revolutionizes our notion of space-time.

Let us see how this works. The Lorentz transformation now provides the physical relation between the spatial coordinates and the time in two inertial frames moving with a constant relative velocity $v \equiv V$. Let an "event" denote a process that occurs at a given point in space-time (a particle decay, the emission of a light signal, the coincidence of two points, *etc.*). Suppose there are two observers, one in an inertial "unprimed" frame, and a second in another inertial "primed" frame moving with velocity v along the x-axis relative to the first. Suppose the observers agree to start their clocks at a first event that occurs when the origins of the two coordinate systems coincide (Fig. 8.5).

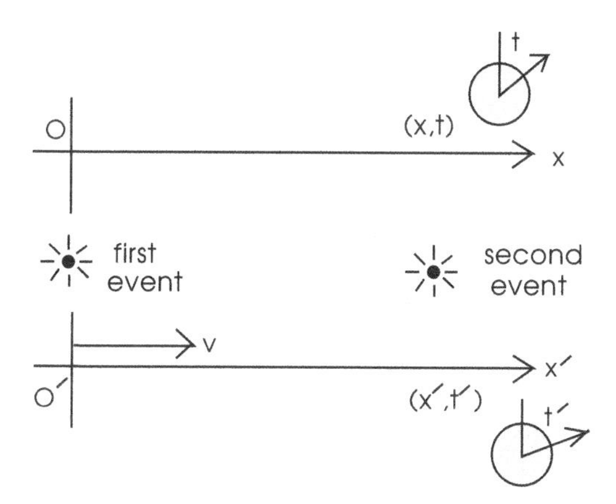

Fig. 8.5 Two events in special relativity as viewed in two different inertial frames. The primed frame is moving with constant velocity v along the x-axis relative to the unprimed frame. The clocks in the two frames start with the first event, which occurs when the origins coincide. The second event appears to occur at (x', t') in the primed frame and (x, t) in the unprimed frame.

Let a second event occur along the x-axis. The first observer says the second event occurs at (x, t) and the second observer says it occurs at (x', t'). According to Einstein, these coordinates are related by the following *Lorentz transformation*

$$x' = \frac{(x - vt)}{\sqrt{1 - v^2/c^2}} \qquad ; \text{ Lorentz transformation}$$

$$t' = \frac{(t - vx/c^2)}{\sqrt{1 - v^2/c^2}} \tag{8.14}$$

Note that in the first frame, all observers agree on the time t as in a usual laboratory experiment, and in the second frame, all observers agree on the time t', but now the *time is relative between the frames.* This is a major break with newtonian mechanics where time is absolute between all inertial frames.

As before, Eqs. (8.14) are readily inverted, and the result is obtained by merely changing the sign of the relative velocity $v \leftrightarrow -v$

$$x = \frac{(x' + vt')}{\sqrt{1 - v^2/c^2}} \qquad ; \text{ Lorentz transformation}$$

$$t = \frac{(t' + vx'/c^2)}{\sqrt{1 - v^2/c^2}} \tag{8.15}$$

It is now a consequence of Eq. (8.10) that observers will measure the *same velocity of light* c in the two inertial frames. Consider the situation in Fig. (8.6) where the first event at the origins, which coincide, denotes the emission of a light signal, and the second event marks its detection.

Fig. 8.6 Two events in special relativity as viewed in two different inertial frames. The first denotes the emission of a light signal at the origins, which coincide, and the second the detection of the light signal, which occurs at the point (x, t) in the first frame and (x', t') in the second.

The first observer measures the speed of light as $x^2/t^2 = c^2$, and thus from Eq. (8.10)

$$x^2 - c^2t^2 = x'^2 - c^2t'^2 = 0 \tag{8.16}$$

The second observer measures the speed of light as x'^2/t'^2 and gets the same value $x'^2/t'^2 = c^2$. Hence *both observers measure the same speed of light.*

It further follows from Eq. (8.13) that

$$\frac{\partial^2 \phi}{\partial x'^2} - \frac{1}{c^2}\frac{\partial^2 \phi}{\partial t'^2} = \frac{\partial^2 \phi}{\partial x^2} - \frac{1}{c^2}\frac{\partial^2 \phi}{\partial t^2} \tag{8.17}$$

Therefore *both observers write the same wave equation for light, with the same c.* Hence the physical consequences will be identical in the two inertial frames.[4]

8.4 Time Dilation

Consider a particle at rest at the origin O' of a frame moving with velocity v along the x-axis, as observed in the inertial laboratory frame. As above, it is assumed that the clocks in the two frames start when the origins (O, O') coincide (Fig. 8.7)

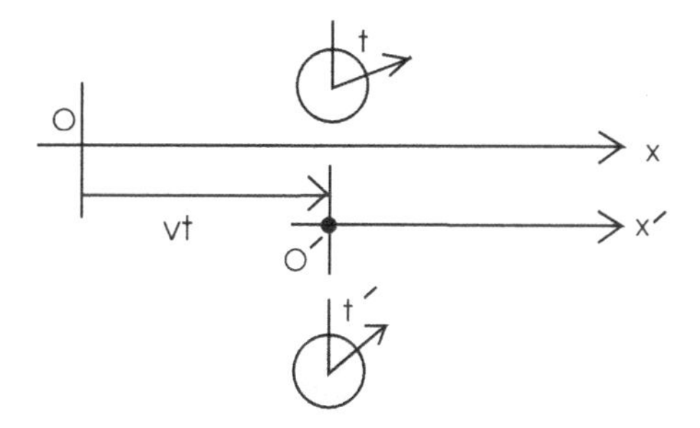

Fig. 8.7 Particle at rest at the origin O' of a frame moving with velocity v in the x-direction as observed in an inertial laboratory frame. As above, it is assumed that the clocks in the two frames start when the origins coincide.

The coordinates of the particle in its rest frame are then

$$\begin{aligned} x' &= 0 \\ t' &\equiv \tau \qquad ; \text{ proper time} \end{aligned} \tag{8.18}$$

The time in the particle's rest frame is known as the *proper time*. The

[4]They will, in fact, write the same set of Maxwell's equations (see Prob. 8.8).

coordinates of the particle in the laboratory frame are evidently

$$x = vt$$
$$t = t \tag{8.19}$$

The Lorentz transformation in Eqs. (8.15) provides the relation between the two sets of coordinates[5]

$$t = \frac{\tau}{\sqrt{1 - v^2/c^2}} \quad ; \text{ time dilation} \tag{8.20}$$

This remarkable relation states that the *laboratory clock runs faster than the clock in the rest frame of the particle*! At complete odds with the newtonian concept of time, this *time dilation* has now been confirmed in many ways, for example:

- Cosmic-ray muons, created in the upper atmosphere, decay very quickly in their rest frame

$$\mu^- \rightarrow e^- + \bar{\nu}_e + \nu_\mu$$
$$\tau = 2.197 \times 10^{-6}\,\text{sec} \tag{8.21}$$

  If moving with $v \approx c$, however, they live long enough to be detected at the surface of the earth;
- Muons in a storage ring with $v \approx c$ live long enough so that very accurate measurements can be made of their magnetic moment.

8.5 Lorentz Contraction

Consider a meter stick of length d in its rest frame lying parallel to the x-axis and moving with velocity v in the x-direction. Assume that in the rest frame, one end of the stick lies at the origin O' and the other at $x' = d$ (Fig. 8.8). Assume again that the clocks start when the origins (O, O') coincide. In the laboratory frame, the length of the stick is measured by locating the two ends *simultaneously*. Thus, in the laboratory frame, the length of the stick is measured at $t = 0$. The coordinates thus satisfy the relations

$$x' = d$$
$$t = 0 \tag{8.22}$$

[5]Note that both of Eqs. (8.15) lead to the same result.

The Lorentz transformation again provides the relation between the coordinates in the two inertial frames. The first of Eqs. (8.14) states that

$$x = d\sqrt{1 - \frac{v^2}{c^2}} \qquad ; \text{ Lorentz contraction} \tag{8.23}$$

The meter stick appears *shorter in the laboratory frame*! This *Lorentz contraction* has also been repeatedly verified over the years.

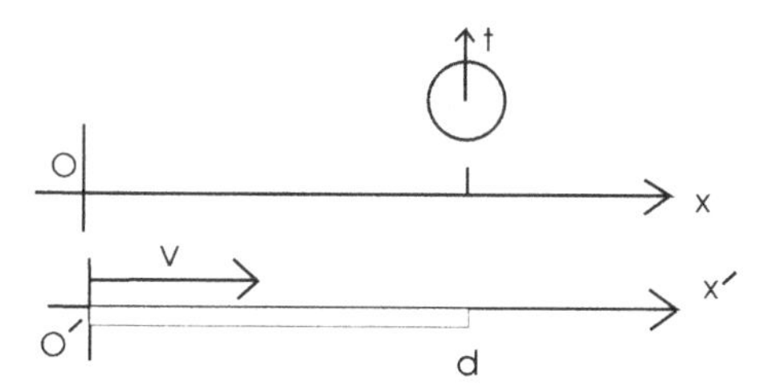

Fig. 8.8 Length of a meter stick lying along the x-axis and moving with velocity v in the x direction. In its rest frame, the length of the stick is d. It is again assumed that the clocks in the two frames start when the origins coincide.

As one interesting example, the Stanford Linear Accelerator (SLAC) is two miles long. To the electrons moving down it with a velocity close the speed of light, its length appears to be less than one meter.[6] This plays a crucial role in the ability to steer those electrons down the accelerator pipe.

8.6 Transverse Dimension

It is easy to argue that a transverse dimension should be unchanged under a Lorentz transformation. Consider a meter stick of length d oriented in the transverse dimension, and a slit of exactly size d, both measured in the inertial laboratory frame. Now let the meter stick move with velocity v toward the slit (Fig. 8.9). Suppose the meter stick grew in the transverse direction because of its motion. Then it would not pass though the slit. Now reverse the roles of the slit and the meter stick. Let the slit move toward the stationary meter stick. It will also grow, and hence pass over the meter stick. The situations are identical since v is simply the relative velocity of the slit and meter stick. One therefore has a logical contradiction, and the

---

[6]See Prob. 8.5.

meter stick can neither grow nor shrink in the transverse dimension because of its motion.

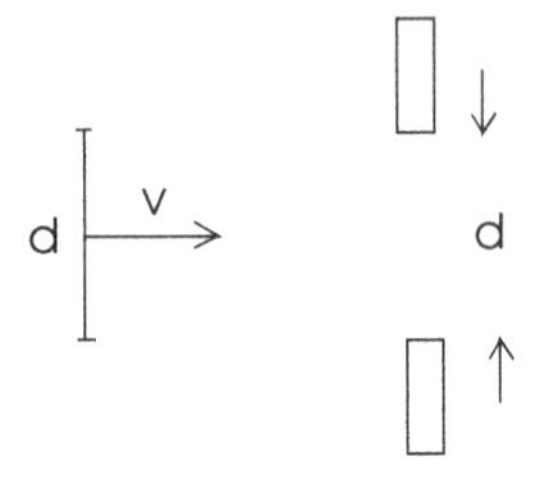

Fig. 8.9 Behavior of a transverse dimension under a Lorentz transformation.

Thus one can simply include the unchanged transverse dimensions in the three-dimensional Lorentz transformation (see Fig. 8.10).

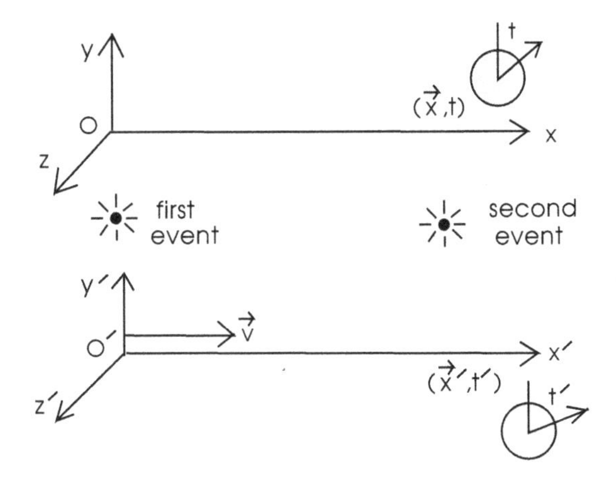

Fig. 8.10 Two events in special relativity as viewed in two different three-dimensional inertial frames. The observers in the two frames agree to start their clocks when the origins (O, O') coincide. The first observer says the second event then takes place at $(\mathbf{x}, t)$ and the second observer, in the frame moving with velocity $\mathbf{v}$ in the x-direction, says it takes place at $(\mathbf{x}', t')$ where the space-time coordinates are related through the Lorentz transformation in Eq. (8.24).

In three dimensions, for a second inertial "primed" frame moving with velocity v along the x-axis relative to an inertial "unprimed" frame, and with clocks started at the first event that marks the coincidence of the origins, one has

$$
\begin{aligned}
y' &= y \qquad\qquad ; \text{ Lorentz transformation in 3-D} \\
z' &= z \\
x' &= \frac{(x - vt)}{\sqrt{1 - v^2/c^2}} \\
t' &= \frac{(t - vx/c^2)}{\sqrt{1 - v^2/c^2}}
\end{aligned}
\tag{8.24}
$$

It is readily verified that the following quadratic form is left invariant under this transformation[7]

$$
\mathbf{x}^2 - c^2t^2 = \mathbf{x}'^2 - c^2t'^2 \tag{8.25}
$$

This again ensures that observers in the two frames will measure the same velocity of light.[8]

8.7 Minkowski Space

A Lorentz transformation is a rule by which the coordinates $(\mathbf{x}, t)$ of a event as viewed in the first inertial frame are related to the coordinates $(\mathbf{x}', t')$ of the event as viewed in a second inertial frame moving with velocity $\mathbf{v}$ relative to the first. The *interval* between the two events $(1, 2)$ in space-time is defined by the following

$$
s_{21}^2 = [x_{(2)} - x_{(1)}]^2 \equiv [\mathbf{x}_{(2)} - \mathbf{x}_{(1)}]^2 - c^2[t_{(2)} - t_{(1)}]^2 \quad ; \text{ interval} \tag{8.26}
$$

A Lorentz transformation between two inertial frames can then be characterized, quite generally, as a coordinate transformation that leaves the

---

[7]See Prob. 8.1(d).

[8]There is an apparent contradiction in special relativity known as the "twin paradox". If time goes slower in a moving frame, and velocity is relative, which twin will actually age relative to the other? The nicest resolution of this apparent paradox known to the author was that presented by Prof. Phil Morse in a colloquium at M.I.T. more than 50 years ago. He solved the simple harmonic oscillator problem in general relativity, where one learns how to deal with *accelerated frames*, and compared the ages of the two twins, one in the inertial laboratory frame, and one going out and back. Indeed, the laboratory twin ages more. The resolution of this apparent paradox is that one has to compare ages for the twins at rest at the same point in space, and an *accelerated frame* is involved in bringing one of the twins back to her sister. If done correctly, one obtains the same answer in special relativity.

interval invariant

$$\begin{aligned} s_{21}^2 = s'^2_{21} &= [x'_{(2)} - x'_{(1)}]^2 \\ &= [\mathbf{x}'_{(2)} - \mathbf{x}'_{(1)}]^2 - c^2[t'_{(2)} - t'_{(1)}]^2 \qquad ; \text{ invariant} \end{aligned} \tag{8.27}$$

Consider the *null interval* between two events (Fig. 8.11)

$$s_{21}^2 = 0 \qquad ; \text{ null interval} \tag{8.28}$$

These represent events that can be connected with a light signal, and invariance of the interval implies that observers in all inertial frames will measure the *same velocity of light c*.

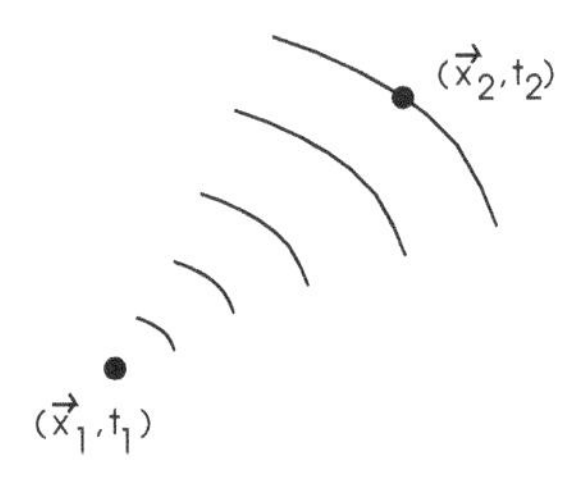

Fig. 8.11 Light travels on the null interval $s_{21}^2 = (\mathbf{x}_2 - \mathbf{x}_1)^2 - c^2(t_2 - t_1)^2 = 0$.

To examine the implied coordinate transformations, introduce the following notation

$$\begin{aligned} x_4 &\equiv ict \\ x_\mu &= (\mathbf{x}, x_4) = (x_1, x_2, x_3, x_4) \end{aligned} \tag{8.29}$$

The interval then takes the form

$$\begin{aligned} s_{21}^2 &= [x_{(2)} - x_{(1)}]^2 \\ &= \sum_{\mu=1}^{4} [x_{(2)} - x_{(1)}]_\mu [x_{(2)} - x_{(1)}]_\mu = \sum_{\mu=1}^{4} [x_{(2)} - x_{(1)}]_\mu^2 \end{aligned} \tag{8.30}$$

The quantity $[x_{(2)} - x_{(1)}]_\mu$ forms a *four-vector* connecting the events (1) and (2) in space-time. This four-vector has an imaginary fourth component and lies in a four-dimensional *Minkowski space.* The interval in Eq. (8.30) is the square of the length of the four-vector. Lorentz transformations are then *rotations and translations* that keep the length of four-vectors constant in the four-dimensional Minkowski space

$$x'_\mu = \sum_{\nu=1}^{4} a_{\mu\nu}(v)x_\nu + \alpha_\mu \qquad ;\text{Lorentz transformation} \tag{8.31}$$

Here α_μ is a translation, which cancels in four-vectors. For rotations, the coefficients $a_{\mu\nu}(v)$ must satisfy the following orthogonality conditions

$$\sum_{\mu=1}^{4} a_{\nu\mu}a_{\rho\mu} = \sum_{\mu=1}^{4} a_{\mu\nu}a_{\mu\rho} = \delta_{\nu\rho} \tag{8.32}$$

where $\delta_{\nu\rho}$ is the Kronecker delta

$$\begin{aligned} \delta_{\nu\rho} &= 1 \qquad ;\ \nu = \rho \\ &= 0 \qquad ;\ \nu \neq \rho \end{aligned} \tag{8.33}$$

One can now explicitly verify that the transformation in Eq. (8.31) leaves the interval in Eq. (8.30) invariant

$$\begin{aligned} s'^2_{21} &= \sum_{\mu=1}^{4}\sum_{\nu=1}^{4}\sum_{\rho=1}^{4} a_{\mu\nu}a_{\mu\rho}[x_{(2)} - x_{(1)}]_\nu[x_{(2)} - x_{(1)}]_\rho \\ &= \sum_{\nu=1}^{4}[x_{(2)} - x_{(1)}]_\nu[x_{(2)} - x_{(1)}]_\nu = s^2_{21} \end{aligned} \tag{8.34}$$

At this point, it is convenient to introduce the *convention* that repeated Greek indices are summed from one to four, thus[9]

$$v_\mu v_\mu \equiv \sum_{\mu=1}^{4} v_\mu v_\mu = v_\mu^2 \qquad ;\ \text{convention} \tag{8.35}$$

With this shorthand, the interval in Eq. (8.30) is written as

$$s^2_{21} = [x_{(2)} - x_{(1)}]^2 = [x_{(2)} - x_{(1)}]_\mu[x_{(2)} - x_{(1)}]_\mu = [x_{(2)} - x_{(1)}]^2_\mu \tag{8.36}$$

and the Lorentz transformation in Eq. (8.31) takes the form

$$\begin{aligned} x'_\mu &= a_{\mu\nu}(v)x_\nu + \alpha_\mu \qquad ;\text{Lorentz transformation} \\ \text{or;} \qquad \underline{x}' &= \underline{a}\,\underline{x} + \underline{\alpha} \end{aligned} \tag{8.37}$$

[9]Readers should spend a little time familiarizing themselves with this (standard) convention, since it will henceforth be employed without comment.

Here the second line is simply a rewriting of the first in matrix notation (see below). The conditions that the Lorentz transformation coefficients form an orthogonal matrix in Eqs. (8.32) can be written in matrix notation as

$$\underline{a}^T \underline{a} = \underline{a}\,\underline{a}^T = \underline{1} \tag{8.38}$$

where the transpose is defined by $[\underline{a}^T]_{\rho\mu} \equiv a_{\mu\rho}$.[10] It is evident from Eqs. (8.38) that the transpose of $\underline{a}$ provides the inverse of that matrix

$$\underline{a}^T = \underline{a}^{-1} \qquad ; \text{ orthogonal matrix} \tag{8.39}$$

Hence $\underline{a}$ is an *orthogonal* matrix.

As an example, we explicitly construct the Lorentz transformation matrix for the configuration illustrated in Fig. 8.12.

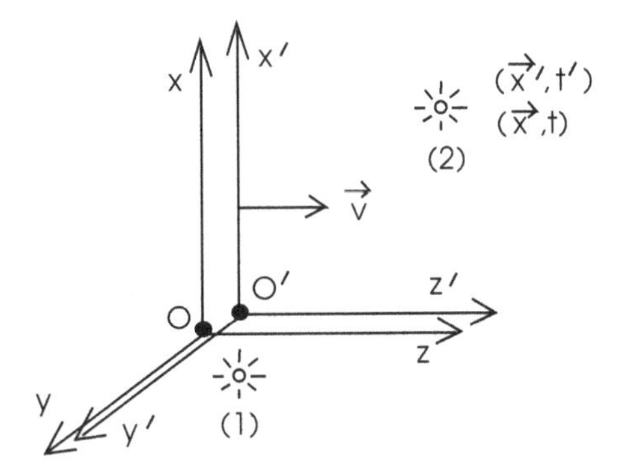

Fig. 8.12 Configuration for a Lorentz transformation along the z-axis. Here it is assumed there is a first event that occurs when the origins (O, O') coincide. The second event is then located at $(\mathbf{x}, t)$ in the first inertial frame and $(\mathbf{x}', t')$ in the second inertial frame moving with velocity $\mathbf{v}$ in the z-direction with respect to the first. The position four-vector in space-time is now defined by $x_\mu = (\mathbf{x}, ict)$ and $x'_\mu = (\mathbf{x}', ict')$.

The observers in the two inertial frames agree to start their clocks at the first event that occurs when the origins (O, O') coincide. This implies that the first space-time point is the origin in both frames $x_{(1)} = x'_{(1)} = 0$, and hence *all intervals are now with respect to the origin.*

We may write the position four-vector $x_\mu = [x_{(2)} - x_{(1)}]_\mu$ of the second event in the two frames as column matrices, where the designation (2) is

[10] The transpose of matrix is obtained by interchanging its rows and columns.

now suppressed

$$\underline{x} = \begin{pmatrix} x \\ y \\ z \\ ict \end{pmatrix} \qquad ; \; \underline{x}' = \begin{pmatrix} x' \\ y' \\ z' \\ ict' \end{pmatrix} \tag{8.40}$$

Under the Lorentz transformation in Eq. (8.37), the four-vector x_μ transforms according to

$$\begin{aligned} x'_\mu &= a_{\mu\nu} x_\nu \qquad ; \text{ four-vector} \\ \text{or;} \qquad \underline{x}' &= \underline{a}\,\underline{x} \end{aligned} \tag{8.41}$$

One says that this four-vector transforms *homogeneously* under the Lorentz transformation.[11]

It follows from Eqs. (8.14) that for the Lorentz transformation in the z-direction, the matrix $\underline{a}$ is given by

$$\underline{a} = \begin{bmatrix} 1 & 0 & 0 & 0 \\ 0 & 1 & 0 & 0 \\ 0 & 0 & \frac{1}{\sqrt{1-v^2/c^2}} & \frac{iv/c}{\sqrt{1-v^2/c^2}} \\ 0 & 0 & \frac{-iv/c}{\sqrt{1-v^2/c^2}} & \frac{1}{\sqrt{1-v^2/c^2}} \end{bmatrix} \tag{8.42}$$

Here the individual elements of the matrix are obtained as $a_{\mu\nu} = [\underline{a}]_{\mu\nu}$ where μ denotes the row and ν denotes the column. It is now readily verified by matrix multiplication that the second relation in Eqs. (8.41) generates the correct Lorentz transformation for the configuration shown in Fig. 8.12. It is also readily verified by matrix multiplication that Eqs. (8.38) are satisfied.[12]

The relation in Eq. (8.41) can be inverted though the use of the orthogonality relations in Eqs. (8.32) to give

$$\begin{aligned} a_{\nu\mu} x'_\nu &= a_{\nu\mu} a_{\nu\sigma} x_\sigma = x_\mu \\ \text{or;} \qquad \underline{x} &= \underline{a}^T \, \underline{x}' \end{aligned} \tag{8.43}$$

[11]The translation α_μ cancels in the four-vector, which is defined as the *difference* in position $x_\mu = [x_{(2)} - x_{(1)}]_\mu$.

[12]See Prob. 8.3.

8.8 Four-Vectors

A four-vector is defined in general by how it transforms under a Lorentz transformation.

If $u'_\mu = a_{\mu\nu}(v)u_\nu$, *then* u_μ *is a four-vector.*

This transformation property has two important applications:

- It allows one to relate components of the four-vector u_μ in one inertial frame to those in another inertial frame;
- It allows one to construct *invariants* that are unchanged from frame to frame, for if u_μ and w_μ are two four-vectors, then their scalar product $u \cdot w \equiv u_\mu w_\mu$ is unchanged under a Lorentz transformation[13]

$$u' \cdot v' \equiv u'_\mu w'_\mu = u_\mu w_\mu = u \cdot v \qquad ; \text{ invariant} \tag{8.44}$$

We proceed to give some examples of four-vectors:

(1) It is clear from the above that the position with respect to the origin, the event (1) in the previous discussion, is a four-vector.

$$x_\mu = (\mathbf{x}, ict) \qquad ; \text{ position w.r.t. origin} \tag{8.45}$$

(2) The *differential displacement* is similarly a four-vector, since from Eq. (8.37)

$$dx'_\mu = a_{\mu\nu} dx_\nu \qquad ; \text{ differential} \tag{8.46}$$

This relation can be inverted exactly as in Eq. (8.43)

$$dx_\nu = a_{\mu\nu} dx'_\mu \tag{8.47}$$

(3) The *gradient* with respect to x_μ is a four-vector.[14]

$$\frac{\partial}{\partial x_\mu} = \left(\boldsymbol{\nabla}, \frac{1}{ic}\frac{\partial}{\partial t}\right) \qquad ; \text{ gradient} \tag{8.48}$$

The proof of this statement follows from the chain rule of differentiation

$$\frac{\partial}{\partial x'_\mu}\phi\left[x_\nu(x'_\mu)\right] = \frac{\partial \phi}{\partial x_\nu}\frac{\partial x_\nu}{\partial x'_\mu} = a_{\mu\nu}\frac{\partial \phi}{\partial x_\nu} \tag{8.49}$$

---

[13] This follows directly from the previous discussion—see Prob. 8.4.

[14] Note carefully where the i appears in this relation.

The last equality follows from Eq. (8.47)

$$\frac{\partial x_\nu}{\partial x'_\mu} = a_{\mu\nu} \tag{8.50}$$

(4) If τ denotes the time in the rest frame of a particle (the "proper time"), and x_μ denotes the position of the particle in an inertial frame, then the *four-velocity* of a particle is defined by

$$u_\mu \equiv \frac{dx_\mu}{d\tau} \qquad ; \text{ four-velocity}$$
$$\tau \equiv \text{proper time} \tag{8.51}$$

The four-velocity forms a four-vector. To establish this result, recall that the invariant interval between the position of the origin O' at two different space-time points, expressed in the particle's rest frame, is given by

$$\begin{aligned} s_{21}^2 &= [\mathbf{x}'_{(2)} - \mathbf{x}'_{(1)}]^2 - c^2[t'_{(2)} - t'_{(1)}]^2 = -c^2[t'_{(2)} - t'_{(1)}]^2 \\ &= -c^2\tau^2 \qquad ; \text{ proper time} \end{aligned} \tag{8.52}$$

Thus the proper time is an invariant. Equation (8.46) establishes that dx_μ is a four-vector. The four-velocity defined in Eq. (8.51) is therefore also a four-vector.

The relation between an infinitesimal element of time in the inertial laboratory frame and in the particle's rest frame follows from Eq. (8.20)

$$dt = \frac{d\tau}{\sqrt{1 - v^2/c^2}} \tag{8.53}$$

With a second use of the chain rule, the four-velocity can thus be rewritten in terms of laboratory quantities according to

$$u_\mu = \frac{dx_\mu}{dt}\frac{dt}{d\tau} = (\mathbf{v}, ic)\frac{1}{\sqrt{1 - v^2/c^2}} \qquad ; \text{ four-velocity} \tag{8.54}$$

Note that

$$u_\mu^2 = -c^2 \tag{8.55}$$

(5) If m_0 denotes the constant *rest mass* of a particle, then the *four-momentum* defined by $m_0 u_\mu$ is evidently also a four-vector

$$p_\mu \equiv m_0 u_\mu \qquad ; \text{ four-momentum} \qquad (8.56)$$

With the aid of Eq. (8.54), the four-momentum can be written in terms of laboratory quantities as

$$\begin{aligned} p_\mu &= \left(\frac{m_0 \mathbf{v}}{\sqrt{1 - v^2/c^2}}, \frac{i}{c} \frac{m_0 c^2}{\sqrt{1 - v^2/c^2}} \right) \\ &\equiv \left(\mathbf{p}, \frac{i}{c} E \right) \end{aligned} \qquad (8.57)$$

The relativistic momentum $\mathbf{p}$ and relativistic energy E of a particle are then identified as

$$\begin{aligned} \mathbf{p} &= \frac{m_0 \mathbf{v}}{\sqrt{1 - v^2/c^2}} \equiv m\mathbf{v} \qquad &; \text{ relativistic momentum} \\ E &= \frac{m_0 c^2}{\sqrt{1 - v^2/c^2}} \equiv mc^2 \qquad &; \text{ relativistic energy} \end{aligned} \qquad (8.58)$$

These quantities are evidently related by

$$\begin{aligned} p_\mu^2 &= \mathbf{p}^2 - \left(\frac{E}{c} \right)^2 = -(m_0 c)^2 \\ \text{or;} \qquad E^2 &= \mathbf{p}^2 c^2 + m_0^2 c^4 \qquad ; \text{ particle} \end{aligned} \qquad (8.59)$$

Note the following important limits of Eq. (8.59):

- Particle at rest, where $\mathbf{p} = 0$

$$E = m_0 c^2 \qquad ; \mathbf{p} = 0 \qquad (8.60)$$

This is Einstein's celebrated relation between mass and energy;

- Non-relativistic limit (NRL), where $\mathbf{p}/m_0 c \to 0$

$$E = m_0 c^2 + \frac{\mathbf{p}^2}{2m_0} \qquad ; \frac{\mathbf{p}}{m_0 c} \to 0 \qquad (8.61)$$

- Extreme relativistic limit (ERL), where $m_0 \to 0$

$$E = pc \qquad ; m_0 \to 0 \qquad (8.62)$$

All of these relations have been previously employed in this book.

8.9 Some Applications

It was observed in Eq. (8.44) that the scalar product of two four-vectors is invariant under Lorentz transformations

$$u' \cdot v' = u'_\mu v'_\mu = u_\mu v_\mu = u \cdot v \tag{8.63}$$

We have seen several examples of this

$$\begin{aligned} u^2 &= u_\mu u_\mu = -c^2 && \text{; four-velocity} \\ p^2 &= p_\mu p_\mu = -(m_0 c)^2 && \text{; four-momentum} \\ \frac{\partial}{\partial x_\mu}\frac{\partial}{\partial x_\mu} &= \nabla^2 - \frac{1}{c^2}\frac{\partial^2}{\partial t^2} && \text{; wave operator} \end{aligned} \tag{8.64}$$

In going from one inertial frame to another, one can either relate components by explicitly carrying out the appropriate Lorentz transformation, or one can construct invariants, which can then be expressed in either frame. The latter procedure provides a very simple and convenient method for relating quantities in the two inertial frames. We give some examples.

8.9.1 *Relativistic Kinematics*

Suppose one has a reaction involving a particle 1, with mass m_1 and energy E_L, incident on a target particle 2, with mass m_2, initially at rest in the laboratory frame. In order to determine what masses are accessible in the reaction, it is most convenient to go to the center-of-momentum (C-M) frame, where the total energy is available to produce the new state at rest. This involves a Lorentz transformation along the incident direction from the inertial laboratory frame to the inertial C-M frame.

We label the kinematic variables in the two frames as shown in Fig. 8.13. In the laboratory frame, the incident and target four-momenta are given by

$$\begin{aligned} p_1 &= \left(\mathbf{p}_L, \frac{i}{c}E_L\right) && \text{; laboratory frame} \\ p_2 &= \left(\mathbf{0}, \frac{i}{c}m_2c^2\right) && \end{aligned} \tag{8.65}$$

In the C-M frame, these four-momenta take the form

$$p_1 = \left(\mathbf{p}, \frac{i}{c}E_1\right) \qquad ; \text{ C-M frame}$$

$$p_2 = \left(-\mathbf{p}, \frac{i}{c}E_2\right) \tag{8.66}$$

The total energy in the C-M frame will be denoted by

$$W = E_1 + E_2 \qquad ; \text{ total energy in C-M frame} \tag{8.67}$$

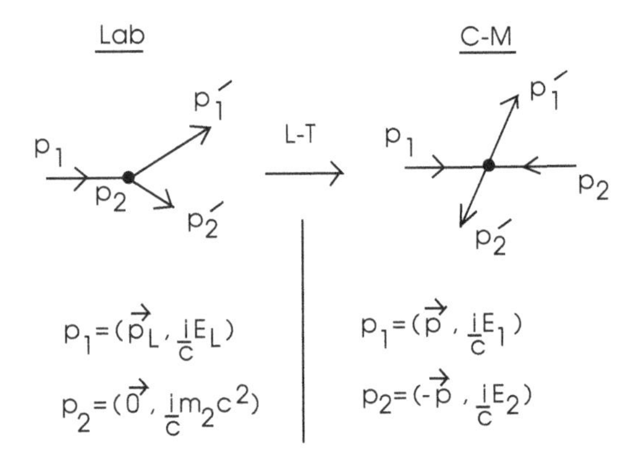

Fig. 8.13 Kinematic variables in the laboratory and center-of-momentum (C-M) inertial frames. Here p denotes the four-momentum, and the same symbols are used for these in the two different frames. They are then expressed in terms of the appropriate three-momenta and energies in the two frames.

Now construct the following *Lorentz invariant* from the four-vectors (p_1, p_2)

$$s \equiv -\left(p_1 + p_2\right)^2 \qquad ; \text{ Lorentz invariant} \tag{8.68}$$

If s is evaluated in the C-M frame using Eqs. (8.66), one has

$$s = -\left(\mathbf{0}^2 - \frac{1}{c^2}W^2\right) = \frac{1}{c^2}W^2 \qquad ; \text{ C-M frame} \tag{8.69}$$

Thus sc^2 is just the square of the total energy in the C-M frame.

If s is evaluated in the laboratory frame using Eqs. (8.65), one finds

$$
\begin{aligned}
s &= -\left[\mathbf{p}_L^2 - \frac{1}{c^2}\left(E_L + m_2c^2\right)^2\right] \\
&= -\left[-(m_1c)^2 - \frac{2E_Lm_2c^2}{c^2} - \frac{(m_2c^2)^2}{c^2}\right] \qquad ; \text{ lab frame} \qquad (8.70)
\end{aligned}
$$

Here Eq. (8.59) has been used for the incident particle 1 in the second line.

The two expressions for s can now equated since it is a Lorentz invariant

$$
-\left[-(m_1c)^2 - \frac{2E_Lm_2c^2}{c^2} - \frac{(m_2c^2)^2}{c^2}\right] = \frac{1}{c^2}W^2 \qquad (8.71)
$$

This expression can be solved for E_L to give

$$
E_L = \frac{1}{2m_2c^2}\left[W^2 - (m_1c^2)^2 - (m_2c^2)^2\right] \qquad ; \text{ lab energy} \qquad (8.72)
$$

The magnitude of the momentum $\mathbf{p}$ in the C-M frame follows from W

$$
W = \left(\mathbf{p}^2c^2 + m_1^2c^4\right)^{1/2} + \left(\mathbf{p}^2c^2 + m_2^2c^4\right)^{1/2} \qquad (8.73)
$$

The threshold laboratory energy E_L^{thresh} required to just create a state of mass M in the C-M system is therefore[15]

$$
E_L^{\text{thresh}} = \frac{1}{2m_2c^2}\left[\left(Mc^2\right)^2 - (m_1c^2)^2 - (m_2c^2)^2\right] \qquad ; \text{ threshold} \quad W = Mc^2 \qquad (8.74)
$$

As an example, consider the reaction

$$
e^+ + e^- \rightarrow \pi^+ + \pi^- \qquad (8.75)
$$

The threshold energy for a laboratory positron beam on electrons at rest (say in an atom) is

$$
\begin{aligned}
E_L^{\text{thresh}} &= \frac{1}{2 \times 0.511\,\text{MeV}}\left[(2 \times 139.6\,\text{MeV})^2 - 2\left(0.511\,\text{MeV}\right)^2\right] \\
&= 7.627 \times 10^4\,\text{MeV} \qquad (8.76)
\end{aligned}
$$

In contrast, if one had *colliding* positron-electron beams, where the laboratory frame is also the C-M system, then the corresponding threshold is

[15] In this case the created state is at rest, and there is no excess kinetic energy in the C-M system, as there must be in the laboratory frame in order to conserve momentum.

just

$$E_{CM}^{\text{thresh}} = 139.6\,\text{MeV} \tag{8.77}$$

This shows the great advantage of constructing storage rings for colliding-beam experiments in the search for interesting higher mass states.[16]

As a second example, consider the relation between the scattering angle θ_{CM} in the C-M system and that in the laboratory frame θ_L for an arbitrary elastic scattering process (see Fig. 8.14).

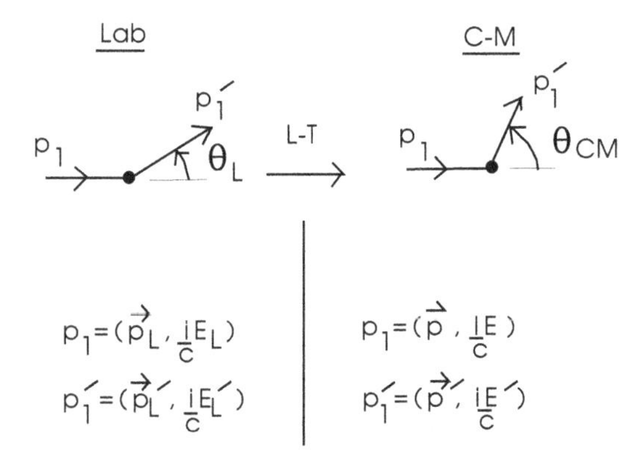

Fig. 8.14 Kinematic variables in the laboratory and center-of-momentum (C-M) in the case of elastic scattering. Here θ_L is the laboratory scattering angle and θ_{CM} is the corresponding angle in the C-M system. The notation is the same as in Fig. 8.13. For elastic scattering in the C-M system, $|\mathbf{p}'| = |\mathbf{p}|$ and $E' = E$.

Write the four-momenta of the incident particle in the laboratory frame before and after the scattering as

$$\begin{aligned} p_1 &= \left(\mathbf{p}_L, \frac{i}{c}E_L\right) \qquad ; \text{ lab frame} \\ p_1' &= \left(\mathbf{p}_L', \frac{i}{c}E_L'\right) \end{aligned} \tag{8.78}$$

[16]The first convincing evidence for the existence of the c-quark came from the creation of the ψ bound state of a $(c\bar{c})$ pair in the colliding $e^+ + e^-$ experiments carried out at SLAC by Burt Richter and collaborators (1974).

In the same fashion, write these four-vectors in the C-M frame as

$$p_1 = \left(\mathbf{p}, \frac{i}{c}E\right) \qquad ; \text{ C-M frame}$$
$$p'_1 = \left(\mathbf{p}', \frac{i}{c}E\right) \tag{8.79}$$

Note the energy of the particle is unchanged in the C-M system for elastic scattering, and $|\mathbf{p}'| = |\mathbf{p}|$.

Now introduce the *invariant momentum transfer* as

$$t \equiv (p'_1 - p_1)^2 \qquad ; \text{ invariant} \tag{8.80}$$

It follows from Eqs. (8.79) that in the C-M sytem this expression takes the form

$$t = (\mathbf{p}' - \mathbf{p})^2 = 2\mathbf{p}^2(1 - \cos\theta_{CM}) \tag{8.81}$$

When evaluated in the laboratory frame using Eqs. (8.78), the invariant momentum transfer becomes

$$\begin{aligned} t &= (\mathbf{p}'_L - \mathbf{p}_L)^2 - \frac{1}{c^2}(E'_L - E_L)^2 \\ &= \frac{2E_L E'_L}{c^2} - 2p_L p'_L \cos\theta_L - 2(m_1 c)^2 \end{aligned} \tag{8.82}$$

The two expressions for t can now be equated since it is a Lorentz invariant, and one finds

$$E_L E'_L - (m_1 c^2)^2 - p_L p'_L c^2 \cos\theta_L = \mathbf{p}^2 c^2 (1 - \cos\theta_{CM}) \tag{8.83}$$

To be of use, one must find another relation that determines E'_L. This can be done with the aid of overall energy-momentum conservation, which can be written as a relation between four-vectors as

$$p_1 + p_2 = p'_1 + p'_2 \qquad ; \text{ energy-momentum conservation four-vectors} \tag{8.84}$$

This allows Eq. (8.80) to be re-expressed as

$$t = (p'_1 - p_1)^2 = (p_2 - p'_2)^2 \tag{8.85}$$

Since the initial four-momentum of the target in the laboratory frame is $p_2 = (\mathbf{0}, im_2c)$, the last expression is evaluated in that frame as

$$\begin{aligned} t &= -2(m_2c)^2 - 2p_2 \cdot p_2' = -2(m_2c)^2 + \frac{2}{c^2}(m_2c^2)E_2' \\ &= \frac{2}{c^2}m_2c^2\left(E_2' - m_2c^2\right) = \frac{2}{c^2}m_2c^2\left(E_L - E_L'\right) \end{aligned} \tag{8.86}$$

Here E_2' is the final energy of the target in the laboratory frame. Energy conservation has been used once more in obtaining the final expression. Hence, from Eq. (8.81), one has

$$m_2c^2\left(E_L - E_L'\right) = \mathbf{p}^2c^2(1 - \cos\theta_{CM}) \tag{8.87}$$

We now note:

- If the l.h.s. of Eqs. (8.83) and (8.87) are equated, one obtains a relation that determines the final laboratory energy of the scattered particle E_L' in terms of its initial energy and the scattering angle $(E_L, \cos\theta_L)$;
- Equations (8.73) and (8.72) provide the C-M momentum $\mathbf{p}^2$ in terms of the laboratory energy (E_L);
- Then either of Eqs. (8.83) or (8.87) determines the C-M scattering angle $\cos\theta_{CM}$ in terms of the incident energy and laboratory scattering angle $(E_L, \cos\theta_L)$;

As one application of these results, consider the non-relativistic limit (NRL) for the elastic scattering of equal-mass particles. In this case these relations give [see Prob. 8.12(a)]

$$\begin{aligned} 4\mathbf{p}^2 &= \mathbf{p}_L^2 && ;\ \text{NRL} \\ p_L' &= p_L\cos\theta_L && m_1 = m_2 \\ \cos\theta_{CM} &= \cos(2\theta_L) \end{aligned} \tag{8.88}$$

As a second application, consider the extreme-relativistic limit (ERL) for the elastic scattering of equal-mass particles. In this case these relations yield [see Prob. 8.12(b)]

$$\begin{aligned} 2\mathbf{p}^2 &= (mc)p_L && ;\ \text{ERL} \\ \frac{1}{p_L'} - \frac{1}{p_L} &= \frac{1}{mc}(1 - \cos\theta_L) && m_1 = m_2 \equiv m \\ \frac{\sin(\theta_L/2)}{\sin(\theta_{CM}/2)} &= \left(\frac{mc}{2p_L'}\right)^{1/2} \end{aligned} \tag{8.89}$$

8.9.2 *White Dwarf Stars*

The topic of white dwarf stars presents a nice extension of the previous discussions of quantum statistics and Thomas-Fermi theory.[17] As preparation, we need the equation of state of a *relativistic* degenerate Fermi gas. The extreme relativistic limit (ERL) of the relation between a particle's energy and momentum is that given in Eq. (8.62)

$$E = pc = \hbar kc \qquad ; \text{ ERL} \tag{8.90}$$

This expression is exactly the same as for the photon [see Eq. (3.50)]. The ground state of the relativistic, non-interacting Fermi gas is obtained by filling the single-particle levels with energy $E = \hbar kc$ up to a maximum wavenumber $k_{\rm F}$ (Fig. 8.15). The spin degeneracy is g, and for electrons $g = 2$.

With a big box of volume V and periodic boundary conditions, the total number of filled levels follows from Eq. (4.167) as

$$N = \frac{gV}{(2\pi)^3}\int_0^{k_{\rm F}} d^3k = \frac{gV}{6\pi^2}k_{\rm F}^3 \tag{8.91}$$

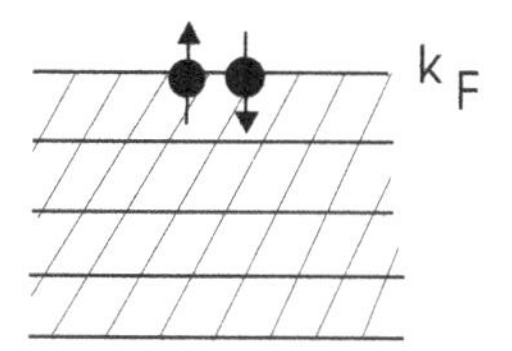

Fig. 8.15 Ground state of a relativistic, non-interacting Fermi gas. The single-particle levels with energy $E = \hbar kc$ are filled to a maximum wavenumber $k_{\rm F}$. The spin degeneracy is g, and for electrons $g = 2$.

The total energy follows in a similar fashion as

$$E = \frac{gV}{(2\pi)^3}\int_0^{k_{\rm F}} (\hbar kc)\, d^3k = \frac{gV}{6\pi^2}\left(\frac{3}{4}\hbar k_{\rm F}c\right)k_{\rm F}^3 \tag{8.92}$$

Hence

$$\frac{E}{N} = \frac{3}{4}\hbar k_{\rm F}c = \frac{3}{4}\hbar c\left(\frac{6\pi^2 n}{g}\right)^{1/3} \qquad ; \text{ relativistic F-G} \tag{8.93}$$

[17] See, for example, [Lifshitz and Landau (1984)].

Here n is the particle density

$$n = \frac{N}{V} \qquad ; \text{ particle density} \tag{8.94}$$

The pressure is then given by Eq. (4.173)

$$P = -\left(\frac{\partial E}{\partial V}\right)_N \tag{8.95}$$

Therefore

$$P = \frac{1}{4}\hbar c \left(\frac{6\pi^2}{g}\right)^{1/3} n^{4/3} \qquad ; \text{ relativistic F-G} \quad T = 0 \tag{8.96}$$

A white dwarf star has the following properties:

- Totally ionized helium atoms $^4\text{He}^{++}$ (α-particles) provide an inert, positive, uniformly-charged background in which the electrons move;
- The *mass* of the star comes from the $^4\text{He}^{++}$;
- The electrons provide the repulsive *pressure* that keeps the star from collapsing under the gravitational attraction.

Since the entire medium is neutral, the α-particle number density n_α is related to the electron density n_e by

$$n_\alpha = \frac{1}{2}n_e \qquad ; \text{ neutrality} \tag{8.97}$$

Since the mass comes from the α-particles, the mass density ρ is given by

$$\rho = m_{\text{He}}n_\alpha = 4m_p n_\alpha = 2m_p n_e \qquad ; \text{ mass density} \tag{8.98}$$

We denote the gravitational force per unit mass (the gravitational *field*) in the medium by $\boldsymbol{\mathcal{F}}_{\text{grav}}$.[18] Then, in a local-density approximation, the condition of *hydrostatic equilibrium* is that the gravitational attraction toward the center of the star should just be compensated by the repulsive pressure force arising from the degenerate electron gas (see Fig. 8.16)

$$\boldsymbol{\mathcal{F}}_{\text{grav}} = \frac{1}{\rho}\boldsymbol{\nabla}P \qquad ; \text{ hydrostatic equilibrium} \tag{8.99}$$

[18]Analogous to the electric field $\boldsymbol{\mathcal{E}}$ in electrodynamics.

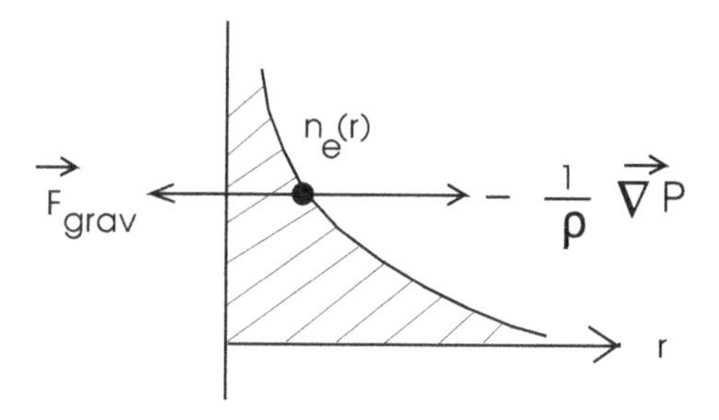

Fig. 8.16 In analogy to the Thomas-Fermi theory of atoms, in white dwarf stars the attractive (gravitational force)/(unit mass) is balanced by the repulsive pressure force arising from the relativistic, degenerate Fermi gas of electrons.

The gravitational force points in the radial direction, and exactly as in Thomas-Fermi theory, Gauss' theorem allows one to calculate the gravitational field at a radius r arising from a spherically symmetric mass distribution [compare Eq. (5.39)]

$$(\mathcal{F}_{\text{grav}})_r = -\frac{4\pi G}{r^2}\int_0^r \rho(r')r'^2\,dr' \tag{8.100}$$

Here G is Newton's gravitational constant

$$G = 6.673 \times 10^{-11}\,\text{m}^3/\text{kg-s}^2 \qquad ;\ \text{Newton's constant} \tag{8.101}$$

Thus the condition of hydrostatic equilibrium in the star becomes

$$\frac{1}{\rho}\frac{dP}{dr} = -\frac{4\pi G}{r^2}\int_0^r \rho(r')r'^2\,dr' \tag{8.102}$$

There are some *boundary conditions* that go along with this equation:

- The mass density clearly vanishes at the surface of the star, which lies at a radius R

$$\rho(R) = 0 \qquad ;\ \text{surface} \tag{8.103}$$

- At the origin, the gravitational force will vanish, while the mass density will be finite. Hence, from Eq. (8.102), the slope of the pressure will vanish there[19]

$$\frac{dP(0)}{dr} = 0 \qquad ;\ \rho(0) \neq 0 \tag{8.104}$$

[19]There is no *cusp* in the pressure at the origin.

The equation of state to be employed follows from Eqs. (8.96) and (8.98), and it takes the form

$$P = \eta\rho^{4/3} \qquad \text{; equation of state}$$
$$\eta = \frac{\hbar c}{4}\left(\frac{6\pi^2}{g}\right)^{1/3}\frac{1}{(2m_p)^{4/3}} \tag{8.105}$$

Insertion into Eq. (8.102) gives

$$\frac{4\eta}{3}\frac{1}{\rho^{2/3}}\frac{d\rho}{dr} = 4\eta\frac{d}{dr}\rho^{1/3} = -\frac{4\pi G}{r^2}\int_0^r \rho(r')r'^2\,dr' \tag{8.106}$$

Introduce the definitions

$$\xi = \frac{r}{R} \qquad \text{; } R = \text{ star radius}$$
$$\bar{\lambda}^2 = \frac{4\pi G R^2}{4\eta} \tag{8.107}$$

Equation (8.106) then takes the form

$$\xi^2\frac{d}{d\xi}\rho^{1/3} = -\bar{\lambda}^2\int_0^{\xi}\rho(\xi')\xi'^2\,d\xi' \tag{8.108}$$

This can now be converted into dimensionless form with the introduction of[20]

$$\bar{\lambda}^3\rho \equiv f^3 \qquad \text{; dimensionless} \tag{8.109}$$

Differentiation with respect to ξ then recasts the condition of hydrostatic equilibrium in Eq. (8.108) into the following

$$\frac{1}{\xi^2}\frac{d}{d\xi}\left(\xi^2\frac{d}{d\xi}f\right) = -f^3 \qquad \text{; N-L diff eqn} \tag{8.110}$$

This dimensionless, non-linear differential equation for the density is to be solved with the two boundary conditions in Eqs. (8.103) and (8.104), which can be written as[21]

$$f(1) = 0 \qquad ;\ f'(0) = 0 \qquad \text{; B.C.} \tag{8.111}$$

---

[20] The dimensions of $\eta/\hbar c$ are $[M^{-4/3}]$, of $G/\hbar c$ are $[M^{-2}]$, and of $\bar{\lambda}^3$ are $[M^{-1}L^3]$.
[21] See Prob. 8.15.

The total mass of the star is obtained by integrating the mass density out to the star's radius

$$\begin{aligned} M &= 4\pi \int_0^R \rho(r) r^2 \, dr = 4\pi R^3 \int_0^1 \rho(\xi) \xi^2 \, d\xi \\ &= \frac{4\pi R^3}{\bar{\lambda}^3} \int_0^1 f^3 \xi^2 \, d\xi \end{aligned} \tag{8.112}$$

The differential expression on the l.h.s. of Eq. (8.110) can now be substituted for f^3 in the integral to give

$$\int_0^1 f^3 \xi^2 \, d\xi = -\int_0^1 \frac{d}{d\xi}\left(\xi^2 \frac{d}{d\xi} f\right) d\xi = -f'(1) \tag{8.113}$$

Hence

$$M = \frac{4\pi R^3}{\bar{\lambda}^3}[-f'(1)] = 4\pi R^3 \left[\frac{4\eta}{4\pi G R^2}\right]^{3/2} [-f'(1)] \tag{8.114}$$

Note that the radius of the star R cancels in this expression,[22] and it becomes

$$M = -4\pi \left(\frac{\hbar c}{4\pi G}\right)^{3/2} \left(\frac{6\pi^2}{g}\right)^{1/2} \frac{1}{(2m_p)^2} f'(1) \tag{8.115}$$

The problem has thus been reduced to integrating the nonlinear differential Eq. (8.110) subject to the two boundary conditions in Eqs. (8.111), and determining the resulting $f'(1)$, which then provides the mass of the star through Eq. (8.115).

The results obtained from numerical integration of Eq. (8.110) using Mathcad11 and the Runge-Kutta algorithm are shown in Fig. 8.17. The differential equation was integrated in from $\xi = 1$, and the initial slope was adjusted until the curve became flat at the origin. The values obtained (rounded to three decimal places) are

$$\begin{aligned} f'(1) &= -2.018 \qquad ; \text{ numerical integration} \\ f(0) &= 6.897 \end{aligned} \tag{8.116}$$

The total mass of the white dwarf can now be calculated from Eq. (8.115), using $g = 2$ and the following numerical values

$$\begin{aligned} G &= 6.67 \times 10^{-11}\,\text{Nm}^2/\text{kg}^2 \qquad ; \; c = 3.00 \times 10^8\,\text{m/s} \\ m_p &= 1.67 \times 10^{-27}\,\text{kg} \qquad ; \; \hbar = 1.05 \times 10^{-34}\,\text{J-s} \end{aligned} \tag{8.117}$$

[22]See Prob. 8.16.

The result is

$$M = 2.85 \times 10^{30}\,\text{kg} \tag{8.118}$$

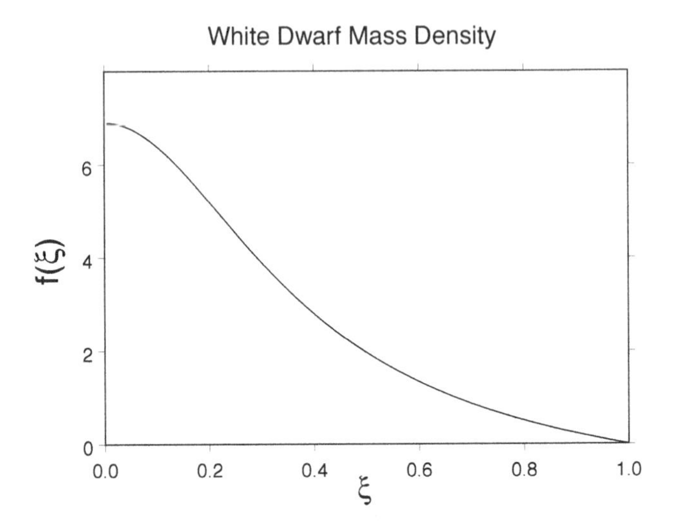

Fig. 8.17 Dimensionless white dwarf mass density $f(\xi) = \bar{\lambda}^3\rho$ versus $\xi = r/R$. Equations (8.107) and (8.105) give $\bar{\lambda}^3 = (2m_p)^2(4\pi G/\hbar c)^{3/2}(g/6\pi^2)^{1/2}R^3$.

It is instructive to write the total mass in terms of the mass of the sun

$$M_\odot = 1.99 \times 10^{30}\,\text{kg} \qquad ; \text{ solar mass} \tag{8.119}$$

Then from Eq. (8.118)

$$\frac{M}{M_\odot} = 1.43 \qquad ; \text{ Chandrasekhar limit} \tag{8.120}$$

This is known as the *Chandrasekhar limit* for the mass of a white dwarf. This is the largest mass that can be supported against the gravitational attraction by the Fermi pressure of a cold, fully-relativistic, electron gas.

If the star has a larger mass than the Chandrasekhar limit, it collapses down to densities where the *nuclear force* comes into play, with resulting supernovae and neutron stars. Eventually, if the mass is large enough, black holes are formed (see later).[23]

[23]The corresponding calculation of the properties of a white dwarf star in the NRL, appropriate to small mass stars, and the interpolation between the NRL and ERL are discussed in Probs. 8.16 and 8.18.

Chapter 9

Relativistic Quantum Mechanics

The Schrödinger equation, when written in terms of local relativistic quantum fields, and allowing for the creation and destruction of particles (see later), provides a successful description of the microscopic world. The Schrödinger equation for a single non-relativistic particle, however, is *inherently non-relativistic.* It is based on the non-relativistic dispersion relation [see Eq. (4.20)]

$$\begin{aligned} \hbar\omega_k &= \frac{\hbar^2\mathbf{k}^2}{2m_0} && ; \text{ N-R dispersion relation} \\ &\rightarrow \frac{-\hbar^2\boldsymbol{\nabla}^2}{2m_0} && \end{aligned} \tag{9.1}$$

The use of the relativistic relation between energy and momentum as a starting point

$$\begin{aligned} \hbar\omega_k &= \left(m_0^2c^4 + \hbar^2c^2\mathbf{k}^2\right)^{1/2} && ; \text{ relativistic} \\ &\rightarrow \left(m_0^2c^4 - \hbar^2c^2\boldsymbol{\nabla}^2\right)^{1/2} && \end{aligned} \tag{9.2}$$

immediately gets one into serious difficulties with defining the square-root operator, or equivalently, with an infinite power series in $\boldsymbol{\nabla}^2$.

The first successful union of quantum mechanics and special relativity for a single-particle was achieved by Dirac, and here we give the lovely historical argument [Dirac (1926)].

9.1 The Dirac Equation

One wants the theory to possess the following features:

(1) A positive-definite probability density

$$\rho = \Psi^\star \Psi \geq 0 \qquad ; \text{ probability density} \tag{9.3}$$

(2) A Schrödinger equation that is first-order in the time derivative

$$i\hbar \frac{\partial \Psi}{\partial t} = H\Psi \qquad ; \text{ Schrödinger equation} \tag{9.4}$$

(3) A continuity equation

$$\frac{\partial \rho}{\partial t} + \boldsymbol{\nabla} \cdot \mathbf{s} = 0 \qquad ; \text{ continuity equation} \tag{9.5}$$

As before, this will provide a basis for the interpretation of the theory and ensure that, for a localized particle,

$$\frac{d}{dt}\int \rho(\mathbf{x}, t) d^3x = 0 \tag{9.6}$$

(4) The correct relativistic relation between energy and momentum

$$E^2 = \mathbf{p}^2c^2 + m_0^2c^4 \qquad ; \text{ relativistic relation} \tag{9.7}$$

Now we do know of a theory that is Lorentz covariant and involves only first-order time derivatives, and that is the set of Maxwell's equations in electrodynamics [Eqs. (2.90) and Prob. 8.8]. Here one has a set of eight *coupled* equations for the components of the electric and magnetic fields $(\mathbf{E}, \mathbf{B})$. Dirac argued by analogy. He introduced a wave function Ψ that had a set of n components

$$\psi_\sigma \qquad ; \sigma = 1, 2, \cdots, n \quad \text{components of } \Psi \tag{9.8}$$

with a corresponding positive-definite probability density defined by

$$\rho \equiv \sum_{\sigma=1}^{n} \psi_\sigma^\star \psi_\sigma \tag{9.9}$$

To satisfy Lorentz covariance, one expects to have to treat space and time on an equal footing, and to satisfy the second requirement above, they must then occur linearly. Thus Dirac assumed an equation of motion of the form

$$i\hbar\frac{\partial\psi_\sigma}{\partial(ct)} = \frac{\hbar}{i}\sum_{k=1}^{3}\sum_{\rho=1}^{n}\alpha^k_{\sigma\rho}\frac{\partial\psi_\rho}{\partial x_k} + m_0c\sum_{\rho=1}^{n}\beta_{\sigma\rho}\psi_\rho \quad ; \text{ Dirac eqn}$$

$$\sigma = 1,\cdots,n \qquad (9.10)$$

Here $(\alpha^k_{\sigma\rho}, \beta_{\sigma\rho})$ are simply constants that couple the various components of the wave function. One now has a set of n coupled, linear, partial differential equations. In order to satisfy the third requirement above, we need to investigate $\partial\rho/\partial t$. To this end, consider

$$i\hbar\frac{\partial}{\partial(ct)}\sum_\sigma\psi^\star_\sigma\psi_\sigma = \sum_\sigma\psi^\star_\sigma\left[\frac{\hbar}{i}\sum_{k=1}^{3}\sum_\rho\alpha^k_{\sigma\rho}\frac{\partial\psi_\rho}{\partial x_k} + m_0c\sum_\rho\beta_{\sigma\rho}\psi_\rho\right] +$$
$$\sum_\sigma\psi_\sigma\left[\frac{\hbar}{i}\sum_{k=1}^{3}\sum_\rho\alpha^{k\star}_{\sigma\rho}\frac{\partial\psi^\star_\rho}{\partial x_k} - m_0c\sum_\rho\beta^\star_{\sigma\rho}\psi^\star_\rho\right] \qquad (9.11)$$

where the second line is obtained from the complex conjugate of Eq. (9.10). Now interchange dummy summation indices $\sigma \rightleftharpoons \rho$ in the second term on the r.h.s., and add it to the first term. The result can be written as

$$i\hbar\frac{\partial}{\partial(ct)}\sum_\sigma\psi^\star_\sigma\psi_\sigma = \frac{\hbar}{i}\sum_{k=1}^{3}\frac{\partial}{\partial x_k}\left(\sum_\sigma\sum_\rho\psi^\star_\sigma\alpha^k_{\sigma\rho}\psi_\rho\right)$$
$$-\frac{\hbar}{i}\sum_{k=1}^{3}\sum_\sigma\sum_\rho\left(\frac{\partial\psi^\star_\sigma}{\partial x_k}\right)\left(\alpha^k_{\sigma\rho} - \alpha^{k\star}_{\rho\sigma}\right)\psi_\rho$$
$$+m_0c\sum_\sigma\sum_\rho\psi^\star_\sigma\left(\beta_{\sigma\rho} - \beta^\star_{\rho\sigma}\right)\psi_\rho \qquad (9.12)$$

To have a continuity equation, the r.h.s. should be the divergence of some quantity. It will be a divergence if the last two terms are absent. They will disappear if the numerical coefficients in Eq. (9.10) are required to satisfy the relations

$$\beta_{\sigma\rho} = \beta^\star_{\rho\sigma}$$
$$\alpha^k_{\sigma\rho} = \alpha^{k\star}_{\rho\sigma} \qquad (9.13)$$

These requirements can be rewritten in *matrix notation* as[1]

$$\begin{aligned} \beta &= \beta^\dagger \\ \alpha^k &= \alpha^{k\dagger} \qquad ; \; k = 1, 2, 3 \end{aligned} \tag{9.14}$$

Here we have used the fact that the complex conjugate transpose of a matrix is the *hermitian adjoint*

$$\left[\underline{m}^\dagger\right]_{\sigma\rho} \equiv m^\star_{\rho\sigma} \qquad ; \text{ hermitian adjoint} \tag{9.15}$$

If one defines the "vector" $\boldsymbol{\alpha} \equiv (\alpha_1, \alpha_2, \alpha_3)$, then Eqs. (9.14) take the form

$$\begin{aligned} \beta &= \beta^\dagger \qquad && ; \text{ hermitian} \\ \boldsymbol{\alpha} &= \boldsymbol{\alpha}^\dagger && \boldsymbol{\alpha} = (\alpha_1, \alpha_2, \alpha_3) \end{aligned} \tag{9.16}$$

Such matrices are said to be *hermitian.*

The previous results can also be rewritten in matrix notation as follows. Introduce the column vector

$$\Psi = \begin{pmatrix} \psi_1 \\ \psi_2 \\ \vdots \\ \psi_n \end{pmatrix} \qquad ; \text{ Dirac wave function} \tag{9.17}$$

Then the probability density in Eq. (9.9) takes the form (see Prob. B.6)

$$\rho = \Psi^\dagger \Psi \qquad ; \text{ probability density} \tag{9.18}$$

The Dirac Eq. (9.10) becomes

$$\begin{aligned} i\hbar \frac{\partial \Psi}{\partial t} &= H\Psi \qquad && ; \text{ Dirac equation} \\ H &\equiv c\boldsymbol{\alpha} \cdot \mathbf{p} + \beta m_0 c^2 && ; \; \mathbf{p} = \frac{\hbar}{i} \boldsymbol{\nabla} \end{aligned} \tag{9.19}$$

It follows from Eqs. (9.12) and (9.13) that the continuity equation can be written in matrix notation as

$$\begin{aligned} \frac{\partial \rho}{\partial t} + \boldsymbol{\nabla} \cdot \mathbf{s} &= 0 \qquad && ; \text{ continuity equation} \\ \rho &= \Psi^\dagger \Psi && ; \text{ probability density} \\ \mathbf{s} &= c\,\Psi^\dagger \boldsymbol{\alpha} \Psi && ; \text{ probability current} \end{aligned} \tag{9.20}$$

---

[1]We forgo the underlining of $(\boldsymbol{\alpha}, \beta)$, and later, of Ψ; furthermore $\alpha^k \equiv \alpha_k$.

It remains to satisfy point (4) and obtain the correct relativistic relation between energy and momentum. As with the Schrödinger equation, one looks for stationary-state solutions to the Dirac equation and converts it to time-independent form

$$\Psi = \psi(\mathbf{x})\, e^{-iEt/\hbar}$$
$$H\psi = (c\boldsymbol{\alpha}\cdot\mathbf{p} + \beta m_0 c^2)\psi = E\psi \tag{9.21}$$

If H is applied to both sides once again, one obtains

$$H^2\psi = E^2\psi = (c^2\mathbf{p}^2 + m_0^2 c^4)\psi \tag{9.22}$$

The r.h.s. is the required result. The quantity H^2 on the l.h.s. is now obtained through matrix multiplication[2]

$$H^2 = \left(c\boldsymbol{\alpha}\cdot\mathbf{p} + \beta m_0 c^2\right)\left(c\boldsymbol{\alpha}\cdot\mathbf{p} + \beta m_0 c^2\right) \tag{9.23}$$

Since matrices do not commute, one must keep careful track of the order of the factors in this expression. Thus

$$H^2 = c^2 \sum_{k=1}^{3}\sum_{l=1}^{3} \alpha^k \alpha^l p_k p_l + m_0 c^3 \sum_{k=1}^{3}\left(\alpha^k \beta + \beta \alpha^k\right) p_k + m_0^2 c^4 \beta^2 \tag{9.24}$$

Since the components of p_k do commute with each other, a change of dummy summation variables allows the first term on the r.h.s. to be rewritten as

$$c^2 \sum_k \sum_l \alpha^k \alpha^l p_k p_l = c^2 \sum_k \sum_l \frac{1}{2}\left(\alpha^k \alpha^l + \alpha^l \alpha^k\right) p_k p_l \tag{9.25}$$

The required expression for H^2 in Eq. (9.22) is then reproduced, provided the following relations are imposed on the Dirac matrices $(\boldsymbol{\alpha}, \beta)$

$$\begin{aligned} \beta\alpha^k + \alpha^k\beta &= 0 \qquad ;\ \text{anti-commute} \\ \alpha^k\alpha^l + \alpha^l\alpha^k &= 2\delta_{kl} \qquad (\boldsymbol{\alpha},\beta)\ \text{n} \times \text{n matrices} \\ \beta^2 &= 1 \end{aligned} \tag{9.26}$$

Thus (α^k, β) must be hermitian, *anti-commuting*, $n\times n$ matrices that satisfy the last two conditions.[3]

---

[2]Note that one has effectively taken $\sqrt{c^2\mathbf{p}^2 + m_0^2 c^4}$ through the clever use of matrices!

[3]The unit matrix is again suppressed on the r.h.s. of the last two relations.

The smallest dimension with which one can satisfy the relations in Eqs. (9.26) is $n = 4$. The *standard representation* of the Dirac matrices can then be exhibited in 2×2 form as (see appendix B)

$$\boldsymbol{\alpha} = \begin{pmatrix} 0 & \boldsymbol{\sigma} \\ \boldsymbol{\sigma} & 0 \end{pmatrix} \quad ; \beta = \begin{pmatrix} 1 & 0 \\ 0 & -1 \end{pmatrix} \qquad \text{; standard representation} \quad 2 \times 2 \text{ form} \tag{9.27}$$

Here $\boldsymbol{\sigma} = (\sigma_x, \sigma_y, \sigma_z)$ are the *Pauli matrices* given by[4]

$$\sigma_x = \begin{pmatrix} 0 & 1 \\ 1 & 0 \end{pmatrix} \quad ; \sigma_y = \begin{pmatrix} 0 & -i \\ i & 0 \end{pmatrix} \quad ; \sigma_z = \begin{pmatrix} 1 & 0 \\ 0 & -1 \end{pmatrix} \text{ ; Pauli matrices} \tag{9.28}$$

The properties of the Pauli matrices are summarized in Eqs. (B.13).

For a free particle in a cubical box of volume Ω with periodic boundary conditions, one can again look for solutions to the Dirac equation of the form in Eq. (4.159)

$$\psi = \frac{1}{\sqrt{\Omega}} e^{i\mathbf{k}\cdot\mathbf{x}} u(\mathbf{k})$$
$$\mathbf{p} = \hbar\mathbf{k} \qquad \text{; eigenvalue} \tag{9.29}$$

Equation (9.22) then becomes[5]

$$E_k^2 = (\hbar\mathbf{k}c)^2 + m_0^2c^4 \qquad \text{; eigenvalue} \tag{9.30}$$

9.1.1 *Non-Relativistic Reduction*

In order to relate this discussion to that of the one-particle Schrödinger equation, consider a non-relativistic reduction of the Dirac Equation. Write the Dirac wave function in the following two-component form

$$\psi = \begin{pmatrix} \phi \\ \chi \end{pmatrix} \qquad \text{; two-component form} \tag{9.31}$$

Now use the standard representation of the Dirac matrices in Eqs. (9.27), and substitute Eq. (9.31) into the last of Eqs. (9.21). The stationary-state Dirac equation then takes the form

$$\left[\begin{pmatrix} 0 & c\boldsymbol{\sigma}\cdot\mathbf{p} \\ c\boldsymbol{\sigma}\cdot\mathbf{p} & 0 \end{pmatrix} + m_0c^2 \begin{pmatrix} 1 & 0 \\ 0 & -1 \end{pmatrix}\right] \begin{pmatrix} \phi \\ \chi \end{pmatrix} = E \begin{pmatrix} \phi \\ \chi \end{pmatrix} \tag{9.32}$$

[4]Note that in this discussion, $k = (1, 2, 3)$ is the same as $k = (x, y, z)$.

[5]See Prob. 9.6 for the construction of the Dirac spinor $u(\mathbf{k})$ for a free particle.

The upper and lower components of this matrix relation are

$$\begin{aligned} c\boldsymbol{\sigma}\cdot\mathbf{p}\chi + m_0c^2\phi &= E\phi \\ c\boldsymbol{\sigma}\cdot\mathbf{p}\phi - m_0c^2\chi &= E\chi \end{aligned} \tag{9.33}$$

We remind the reader that each of these equations is itself a two-component relation, the $\boldsymbol{\sigma}$ are the Pauli matrices, $\mathbf{p} = (\hbar/i)\boldsymbol{\nabla}$, and E is the eigenvalue. Equations (9.33) are still exact.

Consider the positive energy eigenvalue with [compare Eq. (9.30)]

$$E = +\sqrt{(m_0c^2)^2 + \cdots} \qquad ; \text{ positive-energy solution} \tag{9.34}$$

The second of Eqs. (9.33) can be written as

$$\chi = \frac{c\boldsymbol{\sigma}\cdot\mathbf{p}}{E + m_0c^2}\,\phi \tag{9.35}$$

This term is now of $O(\mathbf{p}/m_0c)$, and hence it is small in the NRL. Substitution of Eq. (9.35) into the first of Eqs. (9.33) gives

$$\frac{(c\boldsymbol{\sigma}\cdot\mathbf{p})^2}{E + m_0c^2}\,\phi = (E - m_0c^2)\phi \tag{9.36}$$

Write out the numerator on the l.h.s.

$$(\boldsymbol{\sigma}\cdot\mathbf{p})^2 = \sum_{i=1}^{3}\sum_{j=1}^{3}\sigma_i\sigma_j p_i p_j = \sum_{i=1}^{3}\sum_{j=1}^{3}\frac{1}{2}\left(\sigma_i\sigma_j + \sigma_j\sigma_i\right)p_i p_j \tag{9.37}$$

The last equality comes from a change of dummy indices $i \rightleftharpoons j$ and the fact that the p_i commute. The properties of the Pauli matrices in Eq. (B.13) reduce this expression to

$$(\boldsymbol{\sigma}\cdot\mathbf{p})^2 = \mathbf{p}^2 \tag{9.38}$$

Now in the NRL

$$\begin{aligned} E - m_0c^2 &\equiv \varepsilon \qquad\qquad ; \text{ NRL-eigenvalue} \\ E + m_0c^2 &= 2m_0c^2 + \varepsilon \approx 2m_0c^2 \end{aligned} \tag{9.39}$$

Thus in the NRL, the positive-energy, stationary-state Dirac Eq. (9.33) for the upper components ϕ of the Dirac wave function reduces to the free-particle Schrödinger equation[6]

$$\frac{\mathbf{p}^2}{2m_0}\,\phi = -\frac{\hbar^2\nabla^2}{2m_0}\,\phi = \varepsilon\phi \qquad ; \text{ Schrödinger equation} \tag{9.40}$$

[6]The lower components are then given in the NRL by $\chi = (\boldsymbol{\sigma}\cdot\mathbf{p}/2m_0c)\phi$.

9.1.2 *Electromagnetic Current*

The continuity equation in the Dirac case is given by Eq. (9.20)

$$\begin{aligned}
\frac{\partial \rho}{\partial t} + \boldsymbol{\nabla}\cdot \mathbf{s} = 0 \qquad & ;\ \text{continuity equation} \\
\rho = \Psi^\dagger \Psi \qquad & ;\ \text{probability density} \\
\mathbf{s} = c\,\Psi^\dagger \boldsymbol{\alpha} \Psi \qquad & ;\ \text{probability current}
\end{aligned} \tag{9.41}$$

The electromagnetic current four-vector for a Dirac particle is obtained from these quantities as [see Probs. 8.4(c) and 8.8]

$$ej_\mu = e\left(\frac{1}{c}\mathbf{s}, i\rho\right) \qquad ;\ \text{electromagnetic current} \tag{9.42}$$

where the dependence on the electromagnetic charge e is now made explicit.

At this point, it is convenient to define a new set of Dirac matrices

$$\begin{aligned}
\boldsymbol{\gamma} &\equiv i\boldsymbol{\alpha}\beta = -i\beta\boldsymbol{\alpha} \\
\gamma_4 &\equiv \beta
\end{aligned} \tag{9.43}$$

and a new adjoint wave function defined by (see Prob. B.6)

$$\overline{\Psi} = \Psi^\dagger \gamma_4 \tag{9.44}$$

Then

$$\begin{aligned}
ej_\mu &= ie\left(\overline{\Psi}\,\boldsymbol{\gamma}\Psi,\ \overline{\Psi}\,\gamma_4\Psi\right) \qquad && ;\ \text{electromagnetic current} \\
ej_\mu &= ie\,\overline{\Psi}\,\gamma_\mu\Psi && ;\ \gamma_\mu \equiv (\boldsymbol{\gamma}, \gamma_4)
\end{aligned} \tag{9.45}$$

9.1.3 *Covariant Form*

The Dirac equation can be written in Lorentz covariant form as follows. Start from

$$\left(c\boldsymbol{\alpha}\cdot\mathbf{p} + \beta m_0 c^2\right)\Psi = i\hbar\frac{\partial \Psi}{\partial t} \tag{9.46}$$

with $\mathbf{p} = (\hbar/i)\boldsymbol{\nabla}$. Now multiply this relation on the left by the matrix β, then by $1/\hbar c$, and then use the definitions in Eqs. (9.43)

$$\begin{aligned}
(\boldsymbol{\gamma}\cdot\boldsymbol{\nabla} + M)\,\Psi = -\gamma_4\frac{\partial}{\partial x_4}\Psi \qquad & ;\ x_4 = ict \\
& M \equiv \frac{m_0 c}{\hbar}
\end{aligned} \tag{9.47}$$

Then, taking the last term to the l.h.s.,

$$\left(\gamma_\mu \frac{\partial}{\partial x_\mu} + M\right)\Psi = 0 \qquad ; \text{ covariant form}$$

$$M = m_0 c/\hbar \qquad (9.48)$$

This expression now *looks* covariant, as the first term appears to be the scalar product of two four-vectors γ_μ and $\partial/\partial x_\mu$. In addition, the current in the second of Eqs. (9.45) now *looks* like a four-vector.

9.1.4 *Dirac Hole Theory*

We have to face the problem of the *negative-energy* solutions to the Dirac equation, for example, those with eigenvalue $-\sqrt{(\hbar\mathbf{k}c)^2 + (m_0c^2)^2}$ in Eq. (9.30). Within the principles of quantum mechanics, under some perturbation, an isolated particle can simply keep falling down into these levels without end. There is no ground state for a free particle!

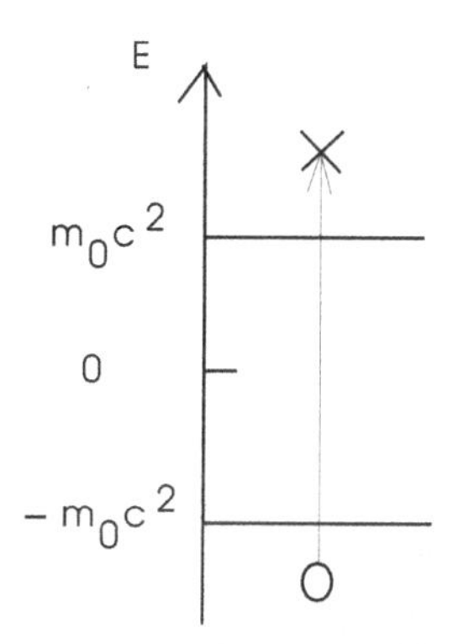

Fig. 9.1 Particle promoted to a positive-energy state, leaving a hole in the filled negative-energy Dirac sea.

Dirac came up with an extremely clever solution to this problem by invoking the Pauli exclusion principle. We shall see below that the Dirac equation describes a particle of spin-1/2, which is then a fermion. Dirac assumed that *all the negative-energy states are already filled with identical fermions.* Thus they are unavailable to the particle in a positive-energy state. Everything is then measured with respect to this filled negative-energy sea, the new vacuum.

This picture has some immediate consequences, for example:

- Within this picture, one of the negative-energy fermions can be promoted to a positive-energy state leaving a *hole* in the negative-energy sea (Fig. 9.1). The hole must have just the opposite properties of the particle, since if it is filled with a particle, one returns to the vacuum. The hole is thus just an *antiparticle*. Based on this picture, Dirac *predicted* the existence of antiparticles before they were discovered!
- Within this picture, the vacuum has dynamics. For example, the charge in the vacuum can be rearranged by the presence of another charge, and the vacuum is *polarizable*.

Both of these observations imply that, from the outset, one is faced with a *many-body problem* in relativistic quantum mechanics.

9.1.5 *Electromagnetic Interactions*

Consider a Dirac particle in an electromagnetic field. Such fields can be described in terms of potentials $(\mathbf{A}, \Phi)$ according to (see Prob. 8.9)

$$\begin{aligned} \mathbf{B} &= \boldsymbol{\nabla} \times \mathbf{A} & &; \text{ electromagnetic potentials} \\ \mathbf{E} &= -\boldsymbol{\nabla}\Phi - \frac{\partial \mathbf{A}}{\partial t} & & \end{aligned} \tag{9.49}$$

In quantum mechanics, as well as in classical mechanics, these fields can be incorporated by making the following replacements

$$\begin{aligned} \mathbf{p} &\rightarrow \mathbf{p} - e\mathbf{A} & &; \text{ incorporate E-M field} \\ H &\rightarrow H + e\Phi & & \end{aligned} \tag{9.50}$$

Thus the Dirac equation in the presence of an electromagnetic field becomes[7]

$$\left[c\boldsymbol{\alpha} \cdot (\mathbf{p} - e\mathbf{A}) + \beta m_0 c^2 + e\Phi\right] \Psi = i\hbar \frac{\partial \Psi}{\partial t} \qquad ; \text{ Dirac eqn} \tag{9.51}$$

If the arguments in Eqs. (9.31)–(9.40) are repeated in the presence of such fields, the Dirac equation leads to the following remarkable results:

(1) When applied to an electron in a static magnetic field described by $\mathbf{A}(\mathbf{x})$, one finds a magnetic moment of (see Prob. 9.2)

$$\boldsymbol{\mu}_{\text{el}} = \frac{e\hbar}{2m_e} 2\mathbf{S} \qquad ; \mathbf{S} = \frac{1}{2}\boldsymbol{\sigma} \tag{9.52}$$

[7]See Prob. 9.4.

Thus

- Since $\mathbf{S}$ is a spin angular momentum [see Eqs. (B.14)], one concludes that the Dirac equation describes a *particle of spin-1/2*;
- The electron is predicted to have a *g-factor of* $g_s = 2$, in accord with observation (see Table 4.1).

(2) In a static, central, electric field described by $\Phi(r)$, the electron experience a *spin-orbit interaction* given by (see Prob. 9.3)

$$V_{\rm SO} = e\left(\frac{\hbar}{2m_e c}\right)^2 \frac{1}{r}\left(\frac{d\Phi}{dr}\right) 2\mathbf{S}\cdot\mathbf{l} \tag{9.53}$$

This is again in accord with the experimental observation (see Prob. 5.1).

9.2 Quantum Electrodynamics (QED)

In chapter 7 we gave a brief introduction to *quantum electrodynamics* (QED). Let us review those arguments. In lowest order, there is one Feynman diagram for $\mu^- + e^- \to \mu^- + e^-$ scattering. It is shown in Fig. 9.2.

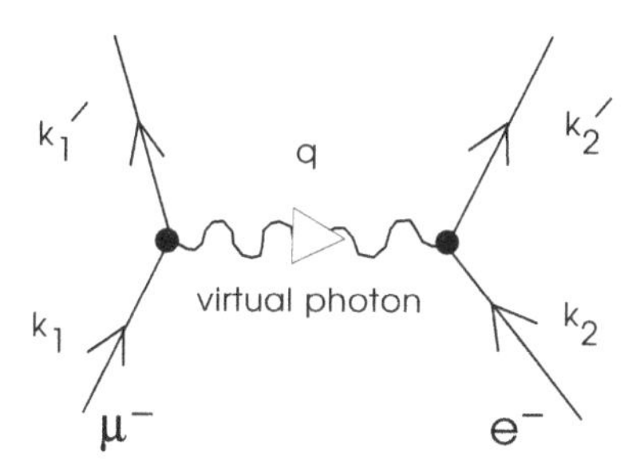

Fig. 9.2 Feynman diagram for $\mu^- + e^- \to \mu^- + e^-$ scattering.

To calculate $-iT_{fi}/\hbar c$ in momentum space, the rules are as follows:

- Include a factor of

$$\frac{ie}{\hbar c\sqrt{\varepsilon_0}}\,j^\gamma \qquad ;\text{E-M current} \tag{9.54}$$

for each of the two vertices. Here j^γ is the Dirac electromagnetic current for each of the spin-1/2 fermions;

- Include the following factor for the *virtual photon* propagator

$$\frac{\hbar c}{i}\frac{1}{q^2} \qquad ; \text{ photon propagator} \tag{9.55}$$

Here q^2 is the four-momentum transfer given by

$$\begin{aligned} q^2 &= (\mathbf{k}_1 - \mathbf{k}_1')^2 - (E_1 - E_1')^2/(\hbar c)^2 \\ &= (\mathbf{k}_2 - \mathbf{k}_2')^2 - (E_2 - E_2')^2/(\hbar c)^2 \end{aligned} \tag{9.56}$$

The second line follows from the first by overall energy-momentum conservation.

We are now in a position to *explicitly construct* the current j^γ. The Dirac electromagnetic current four-vector is given by Eq. (9.45)

$$ej_\mu = ie\,\overline{\Psi}\gamma_\mu\Psi \tag{9.57}$$

The space-time behavior of the Dirac wave functions has been used in generating the form of the S-matrix in Eq. (7.36), and what is left in the T-matrix is the Dirac spinor $u(\mathbf{k})$ in Eq. (9.29). Thus the current j^γ to be used in the T-matrix for the Feynman diagram in Fig. 9.2 for (μ^-, e^-) is

$$ej^\gamma = ie\,\bar{u}(\mathbf{k}')\gamma_\mu u(\mathbf{k}) \tag{9.58}$$

Hence, combining these results, the T-matrix for the process in Fig. 9.2 is

$$\frac{-iT_{fi}}{\hbar c} = \left[-\frac{e}{\hbar c\sqrt{\varepsilon_0}}\bar{u}(\mathbf{k}_1')\gamma_\mu u(\mathbf{k}_1)\right]_{\mu^-}\frac{\hbar c}{i}\frac{1}{q^2}\left[-\frac{e}{\hbar c\sqrt{\varepsilon_0}}\bar{u}(\mathbf{k}_2')\gamma_\mu u(\mathbf{k}_2)\right]_{e^-} \tag{9.59}$$

The free-particle spinors are constructed in Prob. 9.6, and this expression is now completely explicit.[8] It is exact to order $\alpha = e^2/4\pi\varepsilon_0\hbar c$. Through Eqs. (7.49) and (7.46), it gives a cross section for the process that is exact to order α^2.

Consider next the following process, in lowest order,

$$e^- + e^+ \rightarrow \mu^- + \mu^+ \tag{9.60}$$

There is again just the one Feynman diagram shown in Fig. 9.3, and all the factors are the same as above! The only modifications are:

(1) The statement of energy-momentum conservation is now

$$k_1 + k_2 = q = k_1' + k_2' \tag{9.61}$$

[8] Additional gauge-dependent terms in the photon propagator do not contribute here due to the conservation of the Dirac current [*i.e.* $q_\mu\bar{u}(\mathbf{k}')\gamma_\mu u(\mathbf{k}) = 0$].

(2) One needs to use the negative-energy solutions of Prob. 9.6(b) for the antiparticles e^+ and μ^+, and hence[9]

$$\frac{-iT_{fi}}{\hbar c} = \left[-\frac{e}{\hbar c\sqrt{\varepsilon_0}}\bar{u}(\mathbf{k}_1')\gamma_\mu v(-\mathbf{k}_2')\right]_\mu \frac{\hbar c}{i}\frac{1}{q^2}\left[-\frac{e}{\hbar c\sqrt{\varepsilon_0}}\bar{v}(-\mathbf{k}_2)\gamma_\mu u(\mathbf{k}_1)\right]_e \tag{9.62}$$

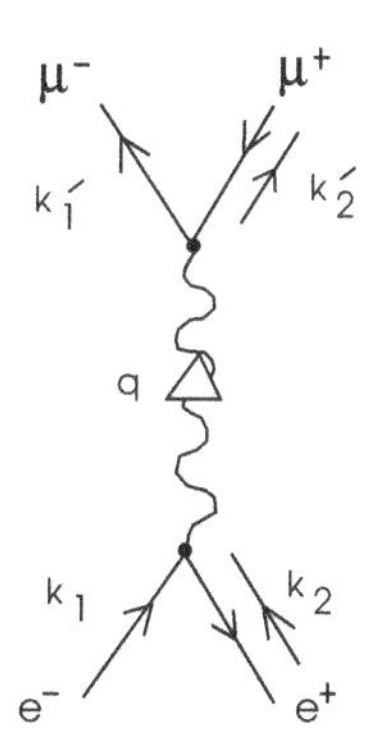

Fig. 9.3 Feynman diagram for $e^- + e^+ \rightarrow \mu^- + \mu^+$.

This again provides a cross section that is exact to order α^2 for the process in Eq. (9.60).

Consider the production of a *quark pair* through the same process

$$e^- + e^+ \rightarrow q + \bar{q} \tag{9.63}$$

The analysis is exactly the same, provided one uses the correct charge e_q for the quark in the first factor in Eq. (9.62).[10]

There are now higher-order *virtual processes* that can be analyzed with a systematic application of Feynman diagrams and Feynman rules (see [Bjorken and Drell (1964); Bjorken and Drell (1965)]). We give some examples in Fig. 9.4.

- The vacuum polarization illustrated in Fig. 9.4(a) has already been touched on. It arises from virtual pair production and shields the charge in QED. There is a modification of Coulomb's law at short distances arising from vacuum polarization. The graph in Fig. 9.4(a) gives a correction of $O(\alpha)$ to the cross section for μ-e scattering;

[9] Recall Dirac hole theory.
[10] See Table 7.3 for $Q = e_q/e_p$.

- The vertex insertion illustrated in Fig. 9.4(b) arises from the emission and absorption of virtual photons during the scattering process. It provides a correction of $O(\alpha)$ to the cross section for μ-e scattering. This vertex correction is also responsible for the leading contribution to the anomalous magnetic moment of the electron (see chapter 7);
- The insertion illustrated in Fig. 9.4(c) represents the proper relativistic, quantum mechanical, electromagnetic self-energy. This also provides a correction of $O(\alpha)$ to the cross section for μ-e scattering;

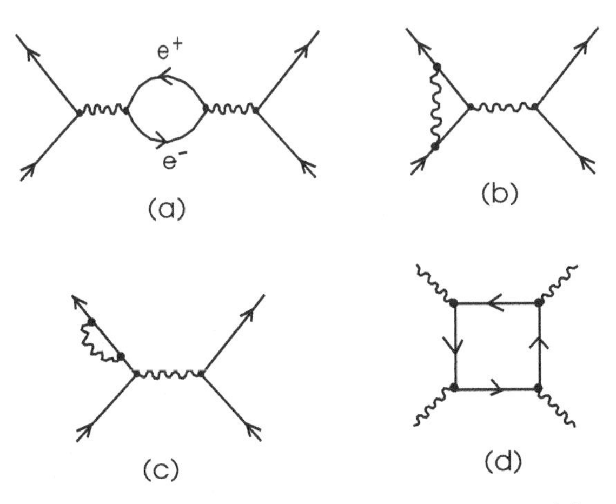

Fig. 9.4 Some virtual processes in QED: (a) vacuum polarization; (b) vertex correction (on each vertex); (c) self-energy insertion (on each leg); (d) scattering of light by light.

Each of the above virtual contributions involves an infinite integral over intermediate four-momenta. If one identifies, and inputs, the experimental values of the mass and charge $(m_{\rm exp}, e_{\rm exp})$, then the resulting expressions for other experimental quantities, such as cross sections, are finite with measurable higher-order corrections. It is an amazing result that this *renormalization theory*, due to Feynman, Schwinger, Tomonaga, Dyson, Ward and others (1948), can be systematically carried out to all orders in α in QED.[11] The demonstration of this result, and the initiation of the calculation of the higher-order corrections, which have all now been accurately verified by experiment, is one of the great achievements of modern physics.

- The diagram in Fig. 9.4(d), arising from virtual pair production, gives

[11]See the papers in [Schwinger (1958)].

rise to the scattering of light by light. The net contribution from this and related diagrams is finite in QED.

9.3 Weak Interactions

Consider the following weak process

$$\nu_\mu + e^- \rightarrow \nu_e + \mu^- \tag{9.64}$$

To lowest order, there is only the one Feynman diagram shown in Fig. 9.5; it involves the exchange of a heavy, charged, weak vector boson W^-.

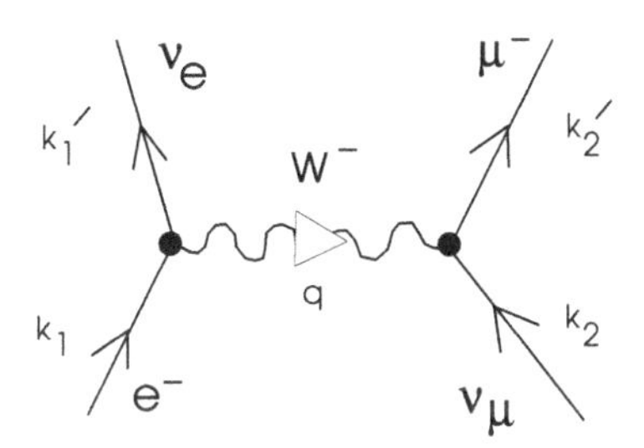

Fig. 9.5 Lowest-order Feynman diagram for the process $\nu_\mu + e^- \rightarrow \nu_e + \mu^-$ through the exchange of a massive, charged, weak vector boson.

The modifications of the above analysis of μ-e scattering are as follows:

(1) One must use the proper weak interaction vertex, and we are now in a position to again *explicitly construct* that vertex. In the standard model, it is given by

$$\frac{ig_W}{\hbar c} j_W \rightarrow -\frac{g_W}{\hbar c}\bar{u}(\mathbf{k}')\gamma_\mu(1+\gamma_5)u(\mathbf{k}) \qquad ; \text{ weak vertex} \tag{9.65}$$

Here the matrix γ_5 is defined and discussed below;

(2) The correct propagator must be employed for the massive vector boson

$$\frac{\hbar c}{i}\frac{1}{q^2+M_W^2}\left(\delta_{\mu\nu}+\frac{q_\mu q_\nu}{M_W^2}\right) \approx \frac{\hbar c}{i}\frac{1}{q^2+M_W^2}\delta_{\mu\nu} \quad ; \text{ vector meson} \tag{9.66}$$

The last expression neglects $q_\mu q_\nu / M_W^2$ applied to the lepton vertices.

We proceed to discuss these expressions:

(1) In the effective low-energy theory where $q^2 \ll M_W^2$, one has a point four-fermion coupling with the Fermi coupling constant of Eqs. (7.71) and (7.3)[12]

$$\frac{g_W^2}{M_W^2} = \frac{G_F}{\sqrt{2}} \qquad ; \frac{M_p^2 G_{\rm F}}{\hbar c} = 1.027 \times 10^{-5} \tag{9.67}$$

(2) We claim that

If the Dirac particles are massless, then these weak interactions involve only the left-handed particles and right-handed antiparticles.

Let us establish this result. The matrix γ_5 appearing in Eq. (9.65) is defined by

$$\begin{aligned} \gamma_5 &\equiv \gamma_1\gamma_2\gamma_3\gamma_4 = \begin{pmatrix} 0 & -1 \\ -1 & 0 \end{pmatrix} \\ \gamma_5^2 &= 1 \end{aligned} \tag{9.68}$$

where the last expression in the first line is the standard representation for γ_5 in 2×2 form (see Prob. 9.8). The standard representation for $\boldsymbol{\alpha}$ is given in Eqs. (9.27). It follows that

$$\gamma_5 \boldsymbol{\alpha} = \boldsymbol{\alpha}\gamma_5 = \begin{pmatrix} -\boldsymbol{\sigma} & 0 \\ 0 & -\boldsymbol{\sigma} \end{pmatrix} \equiv -\boldsymbol{\Sigma} \tag{9.69}$$

Here $(1/2)\boldsymbol{\Sigma}$ is a 4×4 matrix satisfying the angular momentum commutation relations of Eqs. (B.14), and thus the full Dirac spin operator is (note Prob. 9.7)

$$\mathbf{S} = \frac{1}{2}\boldsymbol{\Sigma} \qquad ; \text{ Dirac spin} \tag{9.70}$$

Now a free, massless Dirac particle satisfies the following stationary-state Dirac equation

$$\begin{aligned} c\boldsymbol{\alpha} \cdot \mathbf{p}\,\psi &= +E\psi \\ \text{or;} \qquad \boldsymbol{\alpha} \cdot \mathbf{k}\,\psi &= k\,\psi \qquad ; \text{ eigenvalue } \mathbf{p} = \hbar\mathbf{k} \end{aligned} \tag{9.71}$$

Multiply this equation on the left by $(1 + \gamma_5)/2$, and use the fact that γ_5 and $\boldsymbol{\alpha}$ commute [see Eq. (9.69)]. Thus

$$\boldsymbol{\alpha} \cdot \mathbf{k}\,\phi = k\,\phi \qquad ; \phi \equiv \frac{1}{2}(1 + \gamma_5)\psi \tag{9.72}$$

---

[12] Compare Probs. 7.7 and 7.8.

Multiply this equation on the left by γ_5, use Eq. (9.69), and use $\gamma_5^2 = 1$. This gives

$$-\boldsymbol{\Sigma} \cdot \mathbf{k}\, \phi = k\, \phi \tag{9.73}$$

Hence

$$\boldsymbol{\Sigma} \cdot \hat{\mathbf{k}}\, \phi = -\phi \qquad ; \; \hat{\mathbf{k}} \equiv \frac{\mathbf{k}}{k} \tag{9.74}$$

The component of (twice) the spin along the direction of motion is $\boldsymbol{\Sigma} \cdot \hat{\mathbf{k}}$, and thus the state $\phi = (1/2)(1 + \gamma_5)\psi$ appearing in the weak-interaction vertex has *negative helicity* as advertised. This is illustrated in Fig. 9.6.[13]

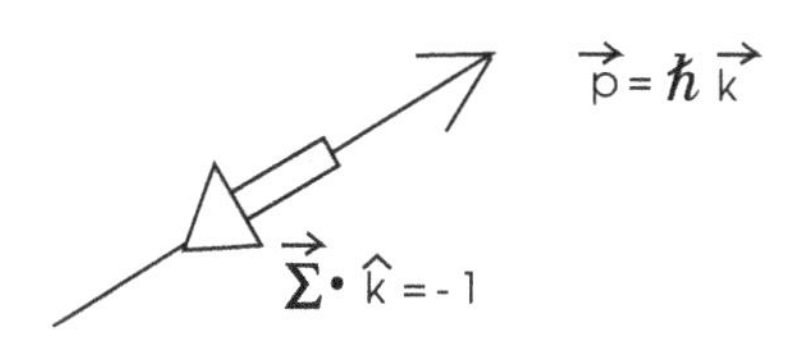

Fig. 9.6 Particles couple through $\phi = [(1 + \gamma_5)/2]\psi$ and, if massless, have negative helicity in the weak interactions. Here $\hat{\mathbf{k}} = \mathbf{k}/k$ is a unit vector in the direction of $\mathbf{k}$.

It is also true that while the γ_μ in the vertex in Eq. (9.65) represents an ordinary polar vector, the $\gamma_\mu\gamma_5$ represents an *axial vector*, and hence the current has a mixed behavior under spatial reflections (see Prob. 9.10). Thus the weak interactions *do not conserve parity*, as first proposed by Lee and Yang [Lee and Yang (1956)].

(3) For the quarks the weak vertex is somewhat more involved, as the weak (ud) vertex gets multiplied by $\cos\theta_{\rm C}$ and the weak (us) vertex by $\sin\theta_{\rm C}$, where $\theta_{\rm C}$ is the Cabibbo angle (see Fig. 9.7)

$$\theta_{\rm C} = 0.23 \qquad ; \text{ Cabibbo angle} \tag{9.75}$$

The weak coupling is the same for all colors of quarks.

(4) The coupling of the leptons and quarks to the heavy weak neutral boson in the standard model is more complicated.[14]

[13]Since $\bar{\psi}\gamma_\mu(1+\gamma_5)\psi = 2\bar{\phi}\gamma_\mu\phi$ (see Prob. 9.9), this analysis holds for both the final and initial particles. We leave the corresponding demonstration of the helicity of massless antiparticles as an exercise for the reader.

[14]These couplings are given, for example, in [Walecka (2004)].

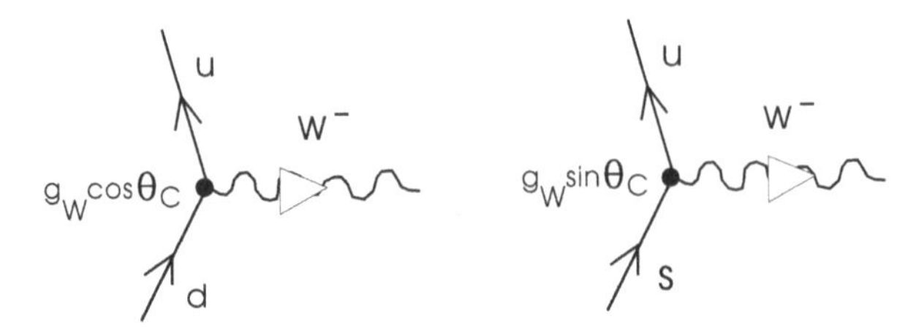

Fig. 9.7 Modification of the weak vertex for the quarks by the Cabibbo angle.

9.4 Quantum Chromodynamics (QCD)

We previously introduced quantum chromodynamics (QCD), and the lowest-order Feynman diagram for quark-quark scattering is repeated here in Fig. 9.8. The vertices involve the strong color current j_C, and we are now in a position to *explicity* construct that current. The Dirac wave functions now have three additional components denoting the color of the Dirac quarks

$$\psi = \begin{pmatrix} \psi_R \\ \psi_G \\ \psi_B \end{pmatrix} \qquad ; \text{ quark color components} \tag{9.76}$$

We write the strong vertex out for all the possible processes. Any individual process is then governed by the appropriate term in the following expression for j_C. The strong vertex is given by

$$\frac{ig_S}{\hbar c} j_C = -\frac{g_S}{\hbar c}\bar{\psi}\gamma_\mu \frac{1}{2}\lambda^a \psi \tag{9.77}$$

The λ are the 3×3 hermitian, traceless, Gell-Mann matrices, the analogues of the 2×2 Pauli matrices (see appendix B). The matrices $(\lambda^a)_{ij}$ for $a = 1, \cdots, 8$ are given in order by[15]

$$\begin{pmatrix} & 1 & \\ 1 & & \\ & & \end{pmatrix} \begin{pmatrix} & -i & \\ i & & \\ & & \end{pmatrix} \begin{pmatrix} 1 & & \\ & -1 & \\ & & \end{pmatrix} \begin{pmatrix} & & 1 \\ & & \\ 1 & & \end{pmatrix} \begin{pmatrix} & & -i \\ & & \\ i & & \end{pmatrix}$$
$$\begin{pmatrix} & & \\ & & 1 \\ & 1 & \end{pmatrix} \begin{pmatrix} & & \\ & & -i \\ & i & \end{pmatrix} \begin{pmatrix} 1/\sqrt{3} & & \\ & 1/\sqrt{3} & \\ & & -2/\sqrt{3} \end{pmatrix} \tag{9.78}$$

[15]We again suppress the underlining of these matrices.

The vertex then involves the coupling to eight corresponding *gluons* $G^{(a)}$ with $a = 1, \cdots, 8$.

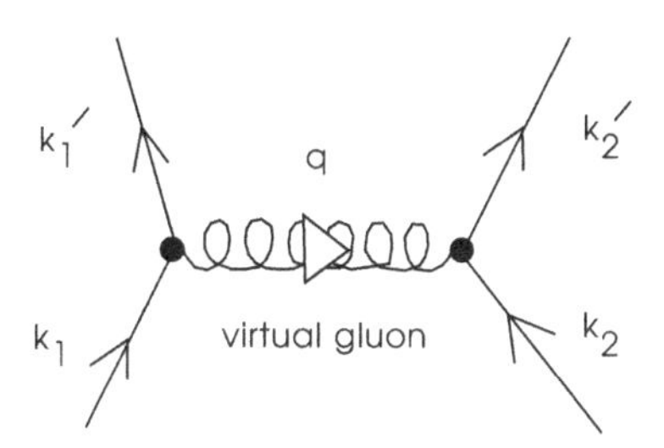

Fig. 9.8 Feynman diagram for quark-quark scattering through one-gluon exchange.

This notation in Eq. (9.77) is very compact, for each of the ψ_C in ψ contains all the *flavors* of quarks $(u, d, s, c, \cdots)_C$, for example

$$\psi_R = \begin{pmatrix} u_R \\ d_R \\ s_R \\ c_R \\ \vdots \end{pmatrix} \qquad ; \text{ flavors for each color} \tag{9.79}$$

Each of these, in turn, is a *Dirac wave function* for that flavor of quark, for example

$$u_R = \begin{pmatrix} u_1 \\ u_2 \\ u_3 \\ u_4 \end{pmatrix}_R \qquad ; \text{ Dirac wave function} \tag{9.80}$$

The matrix γ_μ in Eq. (9.77) then sits between the various Dirac wave functions. Everything in Eq. (9.77) is the unit matrix with respect to flavor; the *flavor* of the quarks does not change in the strong interactions.

To be concrete, let us explicitly exhibit all the indices on the quark wave functions

$$\begin{aligned} \psi_{i,l,\alpha} \qquad & ; \ i = R, G, B && ; \text{ color} \\ & \quad\ l = u, d, s, c, \cdots && ; \text{ flavor} \\ & \quad\ \alpha = 1, \cdots, 4 && ; \text{ Dirac} \end{aligned} \tag{9.81}$$

The strong color vertex in Eq. (9.77) then has the form

$$\bar{\psi}\gamma_\mu \frac{1}{2}\lambda^a \psi = \sum_{(i,j)=R,G,B} \; \sum_{(l,m)=u,d,s,c,\cdots} \; \sum_{(\alpha,\beta)=1,\cdots 4} \bar{\psi}_{i,l,\alpha} \left(\gamma_\mu\right)_{\alpha\beta} \left(\frac{1}{2}\lambda^a\right)_{ij} (\underline{1})_{lm}\, \psi_{j,m,\beta} \tag{9.82}$$

The basic process of the quark coupling to a gluon is illustrated in Fig. 9.9.

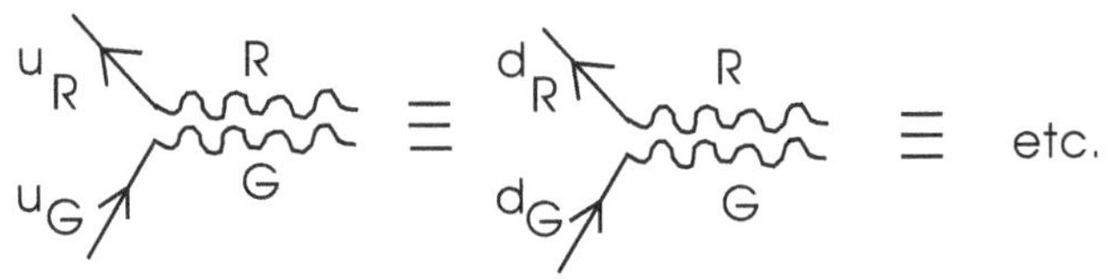

Fig. 9.9 Basic quark-quark-gluon vertex in QCD, involving, for example, $\lambda^{(1)}$. The color of the quark changes, and the gluon $G^{(1)}$ then effectively carries off the two colors (the gluons are "dichromatic"). The interaction is the same for all flavors of quarks.

Everything is now explicit, and one can proceed to the cross section for any one of the lowest-order processes in QED, QCD, or the standard model of the weak interactions.[16]

In summary:

- The electroweak interactions couple to the quarks. That coupling involves the *flavor* of the quarks. The electroweak couplings are *independent of the color* of the quarks; they are color-blind. The gluons are absolutely neutral in the electroweak interactions;
- The strong interactions couple to *color*. They are the same for all flavors of quarks, and thus are *independent of flavor*. The gluons carry color and have the elementary, strong, non-linear couplings illustrated in Fig. 7.20. It is these strong non-linear gluon couplings that give QCD its unusual properties of confinement and asymptotic freedom.

[16]The cross section, as a little element of transverse area $d\sigma_\perp$, must be invariant under Lorentz transformations along the incident direction. The results in Probs. 9.11–9.13 can be used to explicitly exhibit the Lorentz invariance of the cross section in Eqs. (7.49) and (7.46).

Chapter 10

General Relativity

Eleven years after his theory of special relativity revolutionized our notion of space-time [Einstein (1905)], Einstein's theory of general relativity completely changed our concept of how mass and energy determine the *structure* of that space-time [Einstein (1916)]. These works form two of the most influential papers in the history of physics.

We introduce general relativity by considering the motion of a point particle of mass m constrained to move without friction on a surface of arbitrary shape.[1]

10.1 Motion on a Two-Dimensional Surface

Consider a two-dimensional surface of arbitrary shape (Fig. 10.1). Introduce a coordinate grid (q^1, q^2) on that surface that defines the position of the particle. This can evidently be done in a wide variety of ways. All we ask of the coordinates is that they be linearly independent.

We will assume that at the point (q^1, q^2) the surface has a unique *tangent plane* (Fig. 10.2). To first order in the infinitesimals (dq^1, dq^2), the corresponding physical displacement in the surface $d\mathbf{s}$ is identical to that in the tangent plane, and it can be written as

$$d\mathbf{s} = \mathbf{e}_1 dq^1 + \mathbf{e}_2 dq^2 \qquad ; \text{ line element} \tag{10.1}$$

The vectors $(\mathbf{e}_1, \mathbf{e}_2)$, which generate the physical displacements corresponding the coordinate changes (dq^1, dq^2), are known as the *basis vectors*, and Eq. (10.1) defines the *line element*. It gives a physical displacement in the tangent plane, which to first order in infinitesimals is identical to that in

[1] For an introduction to general relativity, see [Walecka (2007)].

the surface.

The square of this physical displacement $(d\mathbf{s})^2 = d\mathbf{s} \cdot d\mathbf{s}$ is known as the *interval*. It is given by

$$(d\mathbf{s})^2 = \sum_{i=1}^{2}\sum_{j=1}^{2} g_{ij}(q)dq^i dq^j \qquad ; \text{ interval}$$
$$g_{ij}(q) \equiv \mathbf{e}_i \cdot \mathbf{e}_j \qquad ; \text{ metric} \qquad (10.2)$$

The second line defines the *metric* $g_{ij}(q)$, which depends both on the choice of coordinates and location on the surface.

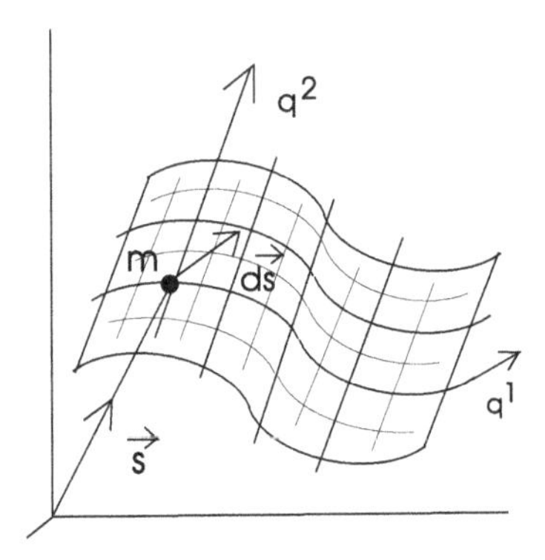

Fig. 10.1 Particle of mass m constrained to move without friction on a two-dimensional surface of arbitrary shape. Generalized coordinates (q^1, q^2) locate the position of the particle on the surface. Here $d\mathbf{s}$ is the line element.

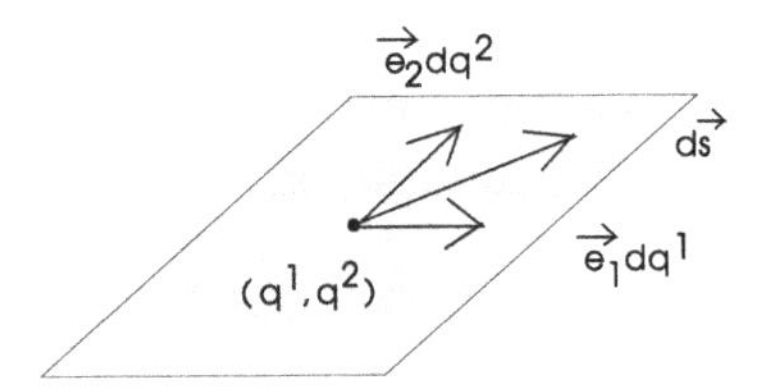

Fig. 10.2 Tangent plane to the surface in Fig. 10.1 at the point (q^1, q^2). Here $(\mathbf{e}_1, \mathbf{e}_2)$ are the basis vectors.

It is convenient at this point to introduce the convention that repeated upper and lower indices are summed over the dimension of the problem, here from 1 to 2. With this convention, Eq. (10.2) takes the form

$$(d\mathbf{s})^2 = g_{ij}(q)dq^i dq^j \qquad ; \text{ summation convention} \qquad (10.3)$$

Although the coordinate system is arbitrary, the displacements in space represented by the line element and interval are *physical quantities.*[2]

Consider the path of a particle with mass m moving without friction between two points (1) and (2) on this surface (Fig.10.3). Imagine confining a string to the surface and pulling it tight to minimize the distance between the two points. The string then defines a curve in the surface known as the *geodesic*. It follows from Newton's laws that

The trajectory of the particle between the two points on the surface forms a geodesic.

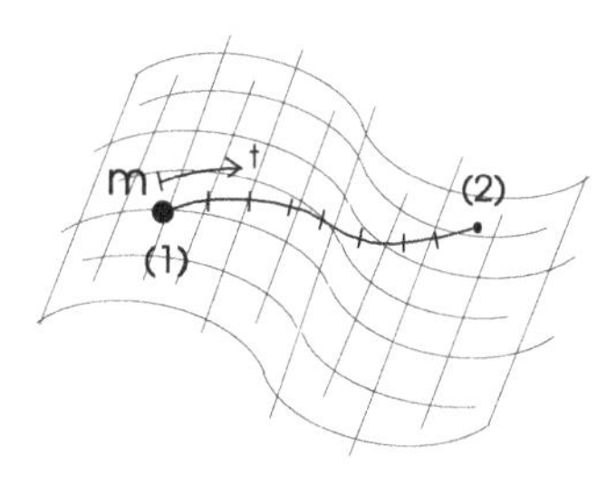

Fig. 10.3 Particle trajectory between points (1) and (2) on the surface. Here t is the time, which parameterizes the distance along the path. Since the energy $E = T = mv^2/2$ is conserved, the magnitude of the velocity along the path is constant, and the particle covers equal distances in equal times.

Thus the trajectory is such that it minimizes the distance between the two points[3]

$$\text{minimize} \int_1^2 \sqrt{(d\mathbf{s})^2} \Rightarrow \text{ path} \qquad ; \text{ geodesic} \qquad (10.4)$$

There are no forces in the surface, and thus we know from elementary physics that the energy of the particle is conserved

$$E = T = \frac{1}{2}m\mathbf{v}^2 = \text{constant} \qquad ; \text{ energy} \qquad (10.5)$$

[2]With another choice of coordinates such that $dq^i = a^i{}_j d\xi^j$, one has $(d\mathbf{s})^2 = g_{ij}dq^i dq^j = \bar{g}_{ij}d\xi^i d\xi^j$. Here $\bar{g}_{ij}$ is the metric with the new coordinates. This relation defines the tensor transformation law.

[3]The definition of a geodesic is actually more general than this. It is a curve that minimizes, or *maximizes* this integral (or, more generally, makes it *stationary* under variations about the actual path).

This implies that the magnitude of the velocity $v = \sqrt{\mathbf{v}^2}$ is constant along the path, and thus the particle covers equal distances in equal times. One can then use the time t along the actual path to parameterize the distance along the path, and the integral in Eq. (10.4) can be rewritten as

$$\int_1^2 \sqrt{(d\mathbf{s})^2} = \int_{t_1}^{t_2} dt \sqrt{\left(\frac{d\mathbf{s}}{dt}\right)^2} \tag{10.6}$$

It turns out that minimizing this express is the same as minimizing

$$\frac{m}{2}\int_{t_1}^{t_2} dt \left(\frac{d\mathbf{s}}{dt}\right)^2 = \int_{t_1}^{t_2} T\, dt \tag{10.7}$$

Thus the path can equally well be determined from the relation

$$\text{minimize} \int_{t_1}^{t_2} T\, dt \Rightarrow \text{path} \qquad ;\ \text{Hamilton's principle} \tag{10.8}$$

where T is the kinetic energy. This is *Hamilton's principle.* It is fully equivalent to Newton's laws, and can be considered to be the basic principle of classical mechanics.[4] The great advantage of Hamilton's principle is that it can be implemented with *any* choice of underlying coordinate system. With our previous choice of coordinates (q^1, q^2), and with the definition $\dot{q} \equiv dq(t)/dt$, the kinetic energy T appearing in Hamilton's principle takes the form

$$T(q^1, q^2; \dot{q}^1, \dot{q}^2) = \frac{m}{2}\left(\frac{d\mathbf{s}}{dt}\right)^2 = \frac{1}{2} m g_{ij}(q) \frac{dq^i}{dt}\frac{dq^j}{dt} \tag{10.9}$$

10.2 Equivalence Principle

Consider a particle of mass m moving in the central gravitational field of a heavy mass M (Fig. 10.4).

[4]The proof that the trajectory is a geodesic lies in first developing lagrangian mechanics from Hamilton's principle (see Probs. 10.4–10.5), and then carrying out the argument in the opposite direction. The step from Eq. (10.7) to (10.6) rests on the fact that $(d\mathbf{s}/dt)^2 = \mathbf{v}^2$ is constant along the actual path.

In the presence of an additional potential, Hamilton's equations read (see Prob. 10.4)

$$\text{minimize} \int_{t_1}^{t_2} L\, dt \Rightarrow \text{path} \qquad ;\ \text{Hamilton's principle}$$
$$L = T - V \qquad ;\ \text{lagrangian}$$

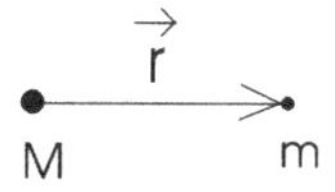

Fig. 10.4 Particle of mass m in central gravitational field of heavy mass M.

Newton's second law, together with his law of gravitation, give the equation of motion of the particle as

$$m_i \frac{d^2\mathbf{r}}{dt^2} = -m_g MG \frac{\mathbf{r}}{r^3} \tag{10.10}$$

Here we are careful to distinguish the inertial mass m_i of the particle, the one entering into the second law and governing the acceleration with respect to the fixed stars, from the gravitational mass m_g, the one entering into his universal law of gravitation. The *equivalence principle* says that these two masses are, in fact, *identical*

$$m_i \equiv m_g \qquad ; \text{ equivalence principle} \tag{10.11}$$

The particle mass then *cancels* in Eq. (10.10),[5] and it becomes

$$\frac{d^2\mathbf{r}}{dt^2} = -MG \frac{\mathbf{r}}{r^3} \tag{10.12}$$

This result implies that all particles move the same way in this gravitational field, a fact confirmed long ago by Galileo in his famous experiment dropping various objects from the leaning tower of Pisa. The path of the particles is determined solely by the *geometry* of the gravitational field arising from the source mass M. It was Einstein's genius to realize the full implications of this result.

At the surface of the earth, which for a short time can be assumed to form an inertial frame, Eq. (10.12) becomes

$$\frac{d^2\mathbf{r}}{dt^2} = \mathbf{g} \qquad ; \text{ surface of earth} \tag{10.13}$$

where $g = 9.80\,\text{m/s}^2$ is the familiar acceleration of gravity. Now consider a second primed frame at the surface of the earth that is accelerating downward with respect to the inertial frame as illustrated in Fig. 10.5. Here

$$\mathbf{z}' = \mathbf{z} - \mathbf{l} \tag{10.14}$$

[5]A cancelation that is usually just blithely carried out in freshman physics.

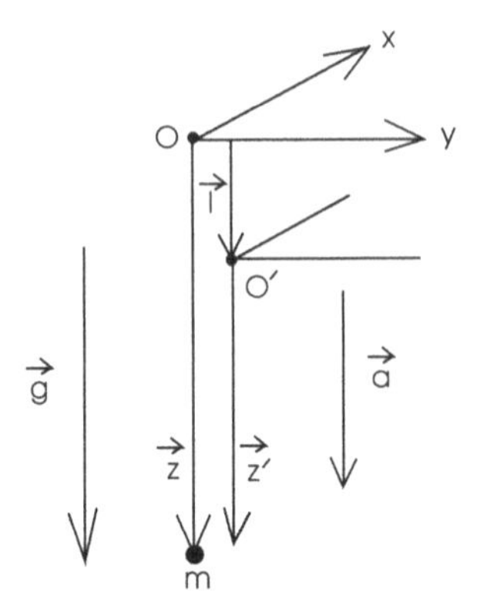

Fig. 10.5 Primed frame at the surface of the earth accelerating downward with respect to the inertial frame. Here $\mathbf{l}$ is the downward displacement of the origin O'.

Two differentiations with respect to time now give

$$\frac{d^2\mathbf{z}'}{dt^2} = \frac{d^2\mathbf{z}}{dt^2} - \frac{d^2\mathbf{l}}{dt^2} \equiv \frac{d^2\mathbf{z}}{dt^2} - \mathbf{a} \tag{10.15}$$

where $\mathbf{a} = d^2\mathbf{l}/dt^2$ is the *acceleration* of the origin O' as observed in the inertial frame. With the invocation of the equivalence principle $m_i = m_g \equiv m$, Newton's second law in the inertial frame, in the presence of the earth's gravitational field, reads

$$m\frac{d^2\mathbf{z}}{dt^2} = m\mathbf{g} + \delta\mathcal{F}_{\rm app} \qquad ; \text{ Newton's second law} \tag{10.16}$$

Here $\delta\mathcal{F}_{\rm app}$ is any additional applied force. Substitution of Eq. (10.15) gives the corresponding equation of motion in the primed, accelerating frame

$$m\frac{d^2\mathbf{z}'}{dt^2} = m(\mathbf{g} - \mathbf{a}) + \delta\mathcal{F}_{\rm app} \tag{10.17}$$

The effect of the gravitational field is *exactly canceled* if[6]

$$\mathbf{a} = \mathbf{g} \qquad ; \text{ cancels gravity} \tag{10.18}$$

The equation of motion in the accelerated frame is then exactly the same as Newton's law in an inertial frame, only the effect of gravity has been removed completely.

How does one achieve the condition in Eq. (10.18)? Just let the primed frame *fall freely* in the gravitational field!

[6]Note this cancelation can only take place *locally*, since $\mathbf{g}$ varies from point to point.

10.3 Local Freely Falling Frame (LF^3)

We thus introduce the concept of the *local freely falling frame* (LF^3). It is a frame that is just held and let go at a given point in a gravitational field. In the inertial system the (LF^3) has no velocity $\mathbf{v}$, rather it has an acceleration $\mathbf{a}$, and *gravity is absent in the* (LF^3).

Two basic concepts of special and general relativity, which we now state relative to the primary inertial frame, can then be summarized as follows (see Fig. 10.6):

(1) No experiment that we can do in a closed laboratory will tell us whether or not we are moving with constant velocity $\mathbf{v}$ with respect to the fixed stars;
(2) No experiment that we can do in a *small* closed laboratory will distinguish between

- An acceleration $\mathbf{a}$ with respect to the fixed stars;
- A *local* gravitational field with $\mathbf{g} = -\mathbf{a}$.

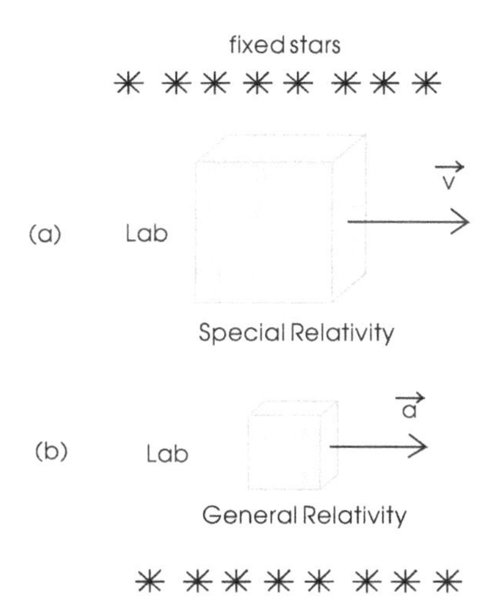

Fig. 10.6 (a) A global closed lab moving with constant velocity $\mathbf{v}$ with respect to the fixed stars in special relativity; (b) A local closed lab moving with acceleration $\mathbf{a}$ with respect to the fixed stars in general relativity.

10.4 Special Relativity

Let us return to special relativity and rewrite the previous results in a different form. We live in a four-dimensional space-time with coordinates

$$x^{\mu} = (\mathbf{x}, ct) \qquad ; \text{No "}i\text{" now} \tag{10.19}$$

Physics lies in the *interval* given by

$$(d\mathbf{s})^2 = g_{\mu\nu} dx^{\mu} dx^{\nu} \qquad ; \text{physical interval} \tag{10.20}$$

Here we use the convention that *repeated upper and lower Greek indices are summed from 1 to 4.*

The space we live in is a *Minkowski space.* This fact is incorporated by introducing a metric in a cartesian basis of the form[7]

$$g_{\mu\nu} = [\underline{g}]_{\mu\nu}$$

$$\underline{g} = \begin{bmatrix} 1 & 0 & 0 & 0 \\ 0 & 1 & 0 & 0 \\ 0 & 0 & 1 & 0 \\ 0 & 0 & 0 & -1 \end{bmatrix} \qquad ; \text{metric in cartesian basis} \tag{10.21}$$

Again, as discussed in chapter 8, it was Einstein's genius to give a physical interpretation to the Lorentz transformations between inertial frames moving with a relative velocity $\mathbf{v}$ (see Fig. 8.5).[8]

10.5 Einstein's Theory of General Relativity

The Minkowski space of special relativity is not a nice, simple euclidian space, but rather a space with the metric of Eq. (10.21). Einstein asked the question, could this space be even more complicated? Could it also be *curved*? Could the presence of mass and energy determine a curved space such that the geodesics in this space are just the newtonian orbits? If so, one would have a unified description of both Newton's second law and his

[7]Note that the nature of the Minkowski space is now entirely contained in the metric, and no longer in the complexity of the coordinates. Just where one now puts the minus signs in the metric is a matter of convention.

[8]In special relativity, we could talk *globally* about the coordinate transformation between inertial frames.

universal law of gravitation, two cornerstones of physics. It is just this problem that is solved by Einstein's theory of general relativity.

We present Einstein's theory through a set of three assumptions:

(1) We live in a four-dimensional riemannian space (see Fig. 10.7).[9] The physical interval in this space is given by

$$(d\mathbf{s})^2 = g_{\mu\nu} dx^\mu dx^\nu \qquad ; \text{ physical interval} \qquad (10.22)$$

Here $g_{\mu\nu}$ is the metric, which in free space in cartesian coordinates has the form in Eq. (10.21).

- In the local freely falling frame (LF^3) there is no gravity, and one has just special relativity with the metric in cartesian coordinates of Eq. (10.21).

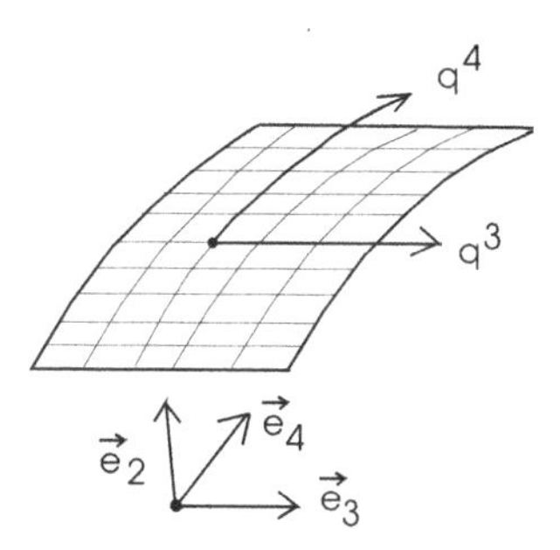

Fig. 10.7 Sketch of the four-dimensional riemannian space in which we live. The surface suppresses the two dimensions (q^1, q^2), and the set of basis vectors suppresses $\mathbf{e}_1$.

(2) Particles move along the geodesics in this space. The geodesics are determined by[10]

$$\text{minimize} \int_1^2 d\tau \left(\frac{d\mathbf{s}}{d\tau}\right)^2 \Rightarrow \text{path} \qquad ; \text{ geodesics} \qquad (10.23)$$

Here $d\tau$ is an element of proper time along the actual path (Fig. 10.8)

$$(d\tau)^2 \equiv -\frac{1}{c^2}(d\mathbf{s})^2 \qquad ; \text{ along actual path} \quad \text{proper time} \qquad (10.24)$$

[9]The previous two-dimensional surface of arbitrary shape forms a two-dimensional riemannian space — a most useful analogy.

[10]More generally, the geodesics are determined by making this integral stationary.

(3) The structure of the space is given by the Einstein field equations

$$G^{\mu\nu} = \kappa T^{\mu\nu} \qquad \text{; field equations} \tag{10.25}$$

In this expression:

- $G^{\mu\nu}$ is the Einstein field tensor. It is a second-order, non-linear, differential form in the metric;
- $T^{\mu\nu}$ is the energy-momentum tensor for the matter fields present;
- $\kappa = 8\pi G/c^4$ is the constant of proportionality.

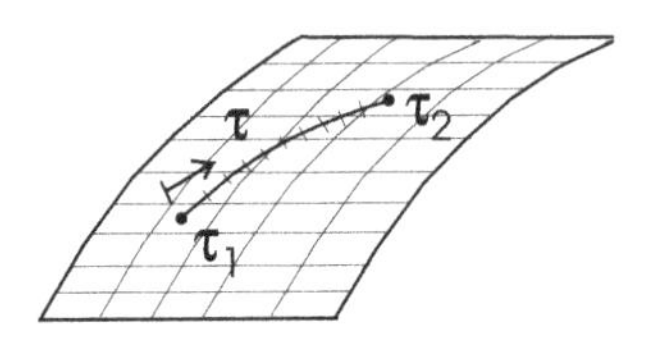

Fig. 10.8 Particle trajectory in the space. Here τ is the proper time.

10.6 Schwarzschild Solution

Schwarzschild found a solution to the Einstein field equations outside of a spherically symmetric source of mass M [Schwarzschild (1916)]. Although the details of his solution go beyond the scope of this book,[11] several properties of that solution are of interest to us here. The Schwarzschild solution for the interval in spherical coordinates is given by

$$(d\mathbf{s})^2 = \frac{(dr)^2}{(1 - 2MG/c^2r)} + r^2(d\theta)^2 + r^2 \sin^2\theta\,(d\phi)^2 - \left(1 - \frac{2MG}{c^2r}\right)(cdt)^2$$
$$\text{; Schwarzschild solution} \tag{10.26}$$

The Schwarzschild metric is immediately identified from this interval.

10.6.1 *Interpretation*

For the interpretation of the Schwarzschild solution, we introduce three different frames of reference (see Fig. 10.9):

I. *Global Inertial Laboratory Frame.* In the global laboratory system, the coordinates are (r, θ, ϕ, ct). All observers in this frame agree on the values

[11]They can be found in [Walecka (2007)].

of these coordinates. The physical interval in terms of these coordinates is given by the Schwarzschild metric;

II. *Far-Away Frame.* This system is sufficiently far away from the source so that $2MG/c^2r \to 0$. In this case, one has just free flat space with the metric (in a cartesian basis) of Eq. (10.21) and interval $(d\mathbf{s})^2 = (d\vec{x})^2 - c^2(dt)^2$. *The coordinate t is the time in this frame.*

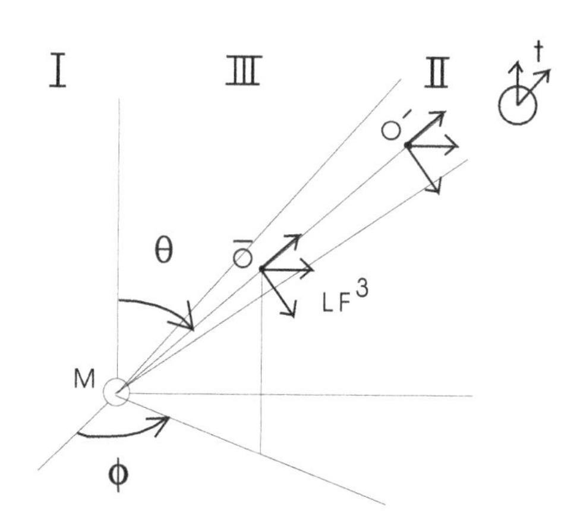

Fig. 10.9 Three reference frames introduced to obtain a physical interpretation of the implications of the Schwarzschild metric. Frame I is the global inertial laboratory system with coordinates (r, θ, ϕ, ct). Frame II is one very far away from the source, and frame III is the local freely falling frame (LF^3).

III. *Local Freely Falling Frame* (LF^3). This is a system that is just held and let go in the gravitational field. The LF^3 has no *velocity* in the global laboratory frame, but it does have a finite *acceleration.* In this system, an observer sees just free flat space and special relativity, with a metric given in cartesian coordinates by Eq. (10.21); there is no gravity in the LF^3. In this frame, we define (see Fig. 10.10):

- $d\bar{t}$ is an element of time in the LF^3;
- $d\bar{l}$ is an element of length in the radial direction $\hat{\mathbf{e}}_r$ in the LF^3.

The interval corresponding to the displacements $(d\bar{l}, d\bar{t})$ in the LF^3 is then

$$(d\mathbf{s})^2 = (d\bar{l}\,)^2 - c^2(d\bar{t}\,)^2 \qquad ; \text{ interval in } LF^3 \qquad (10.27)$$

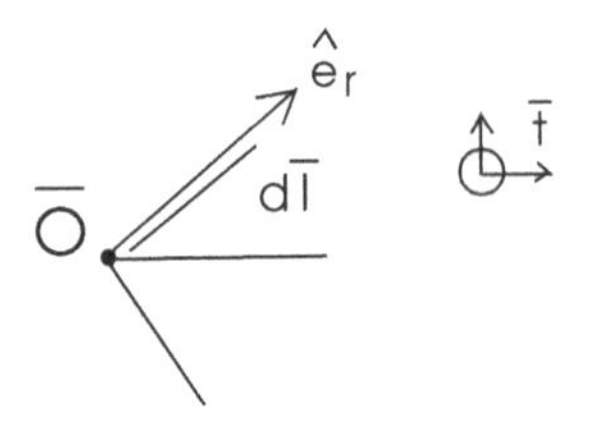

Fig. 10.10 Reference frame III, the local freely falling frame LF^3.

10.6.2 *Some Applications*

Suppose an observer sits at the origin $\bar{O}$ in the LF^3. The physical interval for this observer is given by

$$(d\mathrm{s})^2 = -c^2(d\bar{t}\,)^2 \qquad ;\ \text{interval in frame III} \qquad (10.28)$$

Here $d\bar{t}$ is an element of *proper time*. What does an observer in the global laboratory frame see for this same physical interval? Since the LF^3 has no velocity in frame I, $(dr, d\theta, d\phi) = (\dot{r}, \dot{\theta}, \dot{\phi})dt$ all vanish. Thus the observer in frame I says the physical interval is given by

$$(d\mathrm{s})^2 = -\left(1 - \frac{2MG}{c^2 r}\right) c^2(dt)^2 \qquad ;\ \text{interval in frame I} \quad (10.29)$$

Since it is the same physical interval, these expressions can be equated, with the result that

$$dt = \frac{d\bar{t}}{\sqrt{1 - 2MG/c^2 r}} \qquad ;\ \text{time dilation} \qquad (10.30)$$

There is a *time dilation* in the global laboratory frame depending on the radial position in the gravitational field. Here the time dilation comes from the relative *acceleration* of the LF^3, rather than the relative *velocity* of special relativity.

Evidently the relevant gravitational parameter is the *Schwarzschild radius* given by

$$R_{\mathrm{S}} \equiv \frac{2MG}{c^2} \qquad ;\ \text{Schwarzschild radius} \qquad (10.31)$$

Equation (10.30) then reads

$$dt = \frac{d\bar{t}}{\sqrt{1 - R_{\mathrm{S}}/r}} \qquad ;\ \text{time dilation} \qquad (10.32)$$

As a second application, consider a measurement in the LF^3 of the length $d\bar{l}$ of a standard meter stick oriented in the radial direction $\hat{\mathbf{e}}_r$. Since the length measurement is carried out at a given time in that frame $d\bar{t} = 0$, and the physical interval becomes

$$(ds)^2 = (d\bar{l}\,)^2 \qquad ; \text{ interval in frame III} \tag{10.33}$$

The corresponding length dl in the global laboratory frame is also measured in one time in that frame so that $dt = 0$. The physical interval frame I is thus given by

$$(ds)^2 = \frac{(dl)^2}{(1 - R_{\rm S}/r)} \qquad ; \text{ interval in frame I} \tag{10.34}$$

Since it is the same physical interval, there is a *radial length contraction* given by

$$dl = d\bar{l}\sqrt{1 - R_{\rm S}/r} \qquad ; \text{ radial length contraction} \tag{10.35}$$

Once again, this arises from the relative *acceleration* of the LF^3, rather than from the relative *velocity* of special relativity.

10.6.3 *Schwarzschild Radius*

The Schwarzschild solution is singular at the Schwarzschild radius defined in Eq. (10.31). Let us put in some numbers for this quantity. We use

$$\begin{aligned} c &= 3.00 \times 10^8\,\text{m/s} &&; \ M_\odot = 1.99 \times 10^{30}\,\text{kg} \\ G &= 6.67 \times 10^{-11}\,\text{m}^3/\text{kg-s}^2 &&; \ M_e = 5.98 \times 10^{24}\,\text{kg} \end{aligned} \tag{10.36}$$

This gives for the sun and the earth, together with the observed radii,

$$\begin{aligned} &\text{sun;} && R_\odot^{\rm S} = 2.95 \times 10^3\,\text{m} &&; \ R_\odot = 6.96 \times 10^8\,\text{m} \\ &\text{earth;} && R_e^{\rm S} = 8.86 \times 10^{-3}\,\text{m} &&; \ R_e = 6.38 \times 10^6\,\text{m} \end{aligned} \tag{10.37}$$

Even for the largest mass object in our solar system, the sun, the Schwarzschild radius is a tiny fraction of the actual radius.

We note that one can actually *get to the singularity if all the mass M of an object lies inside of the Schwarzschild radius* $R_{\rm S}$. Such objects do exist in our universe as *black holes.* At radial distances beyond the Schwarzschild radius, the gravitational field of a black hole is described by the Schwarzschild metric; however, light emitted at the Schwarzschild radius is trapped by gravity and can never get out — hence the name.[12]

Take the following values of the coordinates in the Schwarzschild metric

$$\begin{aligned} t &= \text{constant} \\ \theta &= \text{constant} = \pi/2 \end{aligned} \tag{10.38}$$

The Schwarzschild interval in Eq. (10.26) then reduces to the following

$$(d\mathbf{s})^2 = \frac{(dr)^2}{1 - 2MG/c^2r} + r^2\,(d\phi)^2 \tag{10.39}$$

This interval describes a two-dimensional surface, and for visualization, we can return to our paradigm of a point particle moving without friction on such a surface. This surface is plotted as a function of (r, ϕ) in Fig. 10.11.

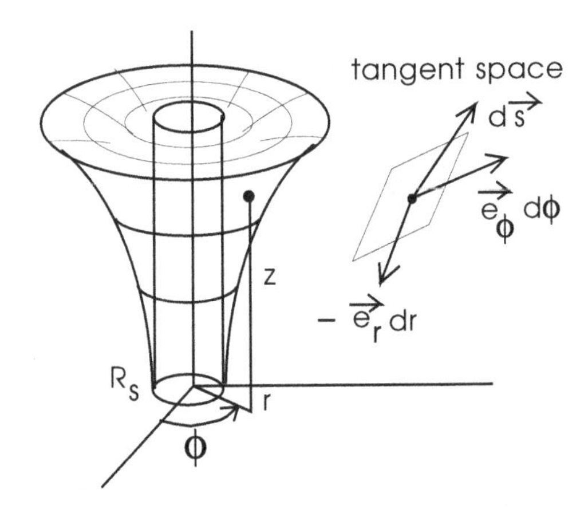

Fig. 10.11 The two-dimensional riemannian surface defined by the Schwarzschild metric with the two coordinates in Eq. (10.38) held constant, as viewed in three-dimensional euclidian space. The generalized coordinates used to locate a point on the surface are (r, ϕ). The tangent space at the point (r, ϕ) is indicated, as well as the basis vectors and line element at that point. The surface is actually described by $z = 2[R_S(r - R_S)]^{1/2}$.

If one is confined to the surface, there is no ambiguity about the angular coordinate ϕ, and the radial coordinate can be measured by making a complete tour about the z-axis and taking $r \equiv (\text{circumference})/2\pi$.

Suppose we introduce someone we will call the "record keeper". He or she simply records and plots the coordinates as they are reported. All

[12] What happens *inside* of the Schwarzschild radius for a black hole is anybody's guess.

the record keeper sees is a point in the space of the coordinates (r, ϕ) as indicated in Fig. 10.12.

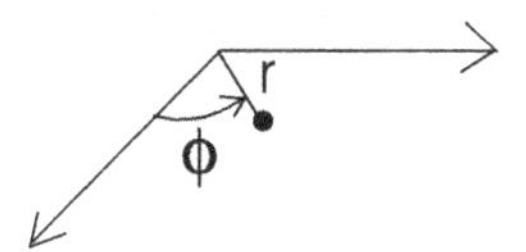

Fig. 10.12 Location on the surface as reported to, and plotted by, the "record keeper" in the space of the coordinates (r, ϕ).

It is the *metric* that allows the record keeper to go from these coordinates to what is actually happening as the particle moves along the surface. For example, as the particle moves in radially toward the Schwarzschild radius, a very small change in the coordinate r corresponds to a very large physical displacement on the surface.

10.6.4 *Motion of a Point Mass*

A point mass moves along a geodesic in general relativity.

$$\text{minimize} \quad m_0 \int_1^2 d\tau \left(\frac{d\mathbf{s}}{d\tau}\right)^2 \Rightarrow \text{path}$$

$$(d\tau)^2 \equiv -\frac{1}{c^2}(d\mathbf{s})^2 \quad ; \text{ along actual path}$$

$$\text{proper time} \tag{10.40}$$

Suppose that instead of parameterizing the geodesic curve with the proper time along the actual path, one parameterizes it with the coordinate t appearing in the Schwarzschild metric, which is the time in the global inertial laboratory frame. Everything else in the integrand is then allowed to vary about the actual path in order to find the minimum of the integral. In this case

$$(d\tau)^2 \equiv -\frac{1}{c^2}(d\mathbf{s})^2 \qquad ; \text{ everywhere now} \tag{10.41}$$

A simple change of variables in the integral in Eq. (10.40) then gives

$$\text{minimize } (-m_0c^2)\int_1^2 \frac{d\tau}{dt}\,dt \Rightarrow \text{ path} \qquad ;\text{ geodesic}$$
$$(d\tau)^2 \equiv -\frac{1}{c^2}(d\mathbf{s})^2 \qquad ;\text{ everywhere} \tag{10.42}$$

This result can be rewritten as

$$\text{minimize } \int_1^2 L(r,\theta,\phi;\,\dot{r},\dot{\theta},\dot{\phi})\,dt \Rightarrow \text{ path} \qquad ;\text{ geodesic}$$
$$L(r,\theta,\phi;\,\dot{r},\dot{\theta},\dot{\phi}) = -m_0c^2\frac{d\tau}{dt} \tag{10.43}$$

This is now in the form of Hamilton's principle in classical mechanics,[13] and with the Schwarzschild interval of Eq. (10.26), the lagrangian is given by

$$L(r,\theta,\phi;\,\dot{r},\dot{\theta},\dot{\phi}) =$$
$$-m_0c^2\left[\left(1-\frac{R_S}{r}\right)-\frac{1}{c^2}\left(\frac{1}{1-R_S/r}\,\dot{r}^2+r^2\dot{\theta}^2+r^2\sin^2\theta\,\dot{\phi}^2\right)\right]^{1/2} \tag{10.44}$$

The problem is now reduced to one in ordinary classical mechanics.

To make a connection with something familiar, one can go to the *newtonian limit* by letting $c^2 \to \infty$. In this case, an expansion of Eq. (10.44) gives[14]

$$L = -m_0c^2 + \frac{1}{2}m_0\mathbf{v}^2 - \left(-\frac{m_0MG}{r}\right) + O\left(\frac{1}{c^2}\right)$$
$$= \text{constant} + T - V + O\left(\frac{1}{c^2}\right) \tag{10.45}$$

The first term is a constant, which does not affect the equation of motion. The second term is the usual non-relativistic kinetic energy. The third term is minus the newtonian gravitational potential energy, which allows an identification of the mass M of the source and the gravitational constant G.

---

[13]See Prob. 10.4.

[14]Recall $R_S = 2MG/c^2$; we identify the usual velocity as $\mathbf{v}^2 = \dot{r}^2 + r^2\dot{\theta}^2 + r^2\sin^2\theta\,\dot{\phi}^2$.

10.7 Cosmology

The Einstein field equations can be solved relatively easily in the case of a uniform mass density ρ that extends throughout all space (see Fig. 10.13). Rather surprisingly, this turns out to provide a good approximation to what is actually observed in the visible universe.

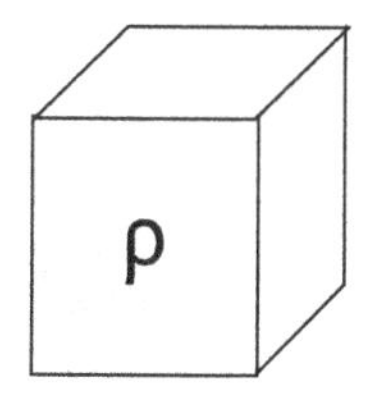

Fig. 10.13 Model of a finite, uniform mass density throughout the universe. The mass density is a function of time, and the value required at the present time is $\approx (10\,\text{H-atoms})/\text{m}^3$.

The mass density is a function of time, and at the present time the required mass density is approximately

$$\begin{aligned} \rho &\approx 2 \times 10^{-29}\,\text{gm/cm}^3 && ;\ \text{at present time} \\ m_p &= 1.67 \times 10^{-24}\,\text{gm} && ;\ \text{proton mass} \end{aligned} \tag{10.46}$$

The second line reminds the reader of the proton mass. A good way to remember the required mass density is that it is approximately ten H-atoms per cubic meter.

10.7.1 *Robertson-Walker Metric* ($k = 0$)

The solution to the Einstein field equations that we shall focus on for this uniform mass cosmology is known as the Robertson-Walker metric with $k = 0$. This is just one of the possible solutions with a uniform mass density — the simplest one. The situation is illustrated in Fig. 10.14. The coordinates are (q^1, q^2, q^3, ct), and it is assumed that the spatial coordinates form an orthogonal system. The physical interval and corresponding metric

$g_{\mu\nu} = [\,\underline{g}\,]_{\mu\nu}$ are given by

$$(d\mathbf{s})^2 = \Lambda^2(t)(d\mathbf{q})^2 - c^2(dt)^2 \qquad ; \text{ physical interval}$$

$$\underline{g} = \begin{bmatrix} \Lambda^2(t) & 0 & 0 & 0 \\ 0 & \Lambda^2(t) & 0 & 0 \\ 0 & 0 & \Lambda^2(t) & 0 \\ 0 & 0 & 0 & -1 \end{bmatrix} \quad ; \text{ Robertson-Walker } (k=0) \qquad (10.47)$$

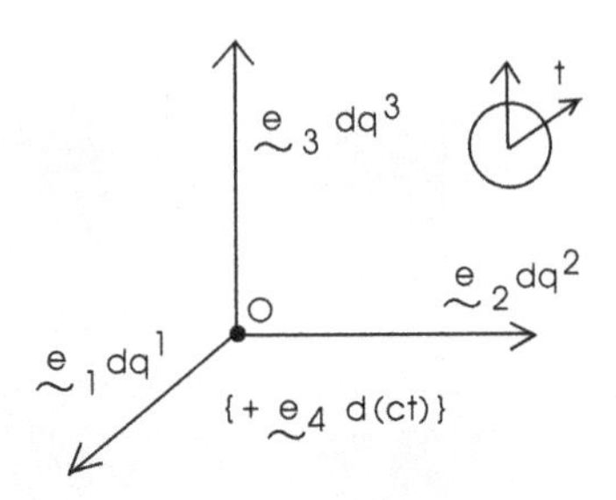

Fig. 10.14 Illustration of physical spatial displacements in the Robertson-Walker metric with $k = 0$ for the cosmology where there is a uniform mass density throughout all space. The fourth coordinate displacement $\mathbf{e}_4\, d(ct)$ is suppressed.

The Einstein field equations in this case reduce to the following[15]

$$h(t) \equiv \frac{1}{\Lambda(t)}\frac{d\Lambda(t)}{dt} = \frac{2}{3(t-t_0)}$$
$$t - t_0 = \frac{2}{3h(t)} = \left[\frac{1}{6\pi\rho(t)G}\right]^{1/2} \qquad (10.48)$$

Here t_0 is simply an integration constant, and $t - t_0$ is given the grand name of "age of the universe".

Let us put in some numbers for the present time $t = t_p$ using

$$\begin{aligned} G &= 6.67 \times 10^{-11}\,\text{m}^3/\text{kg-s}^2 && ; \ 1\,\text{yr} = 3.16 \times 10^7\,\text{s} \\ c &= 3.00 \times 10^8\,\text{m/s} && ; \ 1\,\text{light-yr} = 9.46 \times 10^{15}\,\text{m} \\ \rho(t_p) &= 2 \times 10^{-26}\,\text{kg/m}^3 && ; \ 1\,\text{pc} = 3.26\,\text{light-yr} \\ & && ; \ 1\,\text{Mpc} = 3.08 \times 10^{19}\,\text{km} \end{aligned} \qquad (10.49)$$

[15]The details can again be found in [Walecka (2007)].

The last two relations on the right define useful units of length known as the "parsec" (pc) and "megaparsec" (Mpc) where 1 Mpc $= 10^6$ pc. these numbers give

$$
\begin{aligned}
t_p - t_0 &= 0.63 \times 10^{10}\,\text{yr} \\
h(t_p) &= 3.34 \times 10^{-18}\,\text{s}^{-1} = 103\,(\text{km/s})/\text{Mpc}
\end{aligned}
\tag{10.50}
$$

We shall see shortly why the last, rather convoluted, way of writing s^{-1} provides a useful expression.

10.7.2 *Interpretation*

To provide an interpretation of the Robertson-Walker metric, we return to our paradigm of motion on an arbitrary two-dimensional surface. Suppose that now the surface is a flat *rubber sheet* that stretches uniformly in all directions as a function of time (see Fig. 10.15).

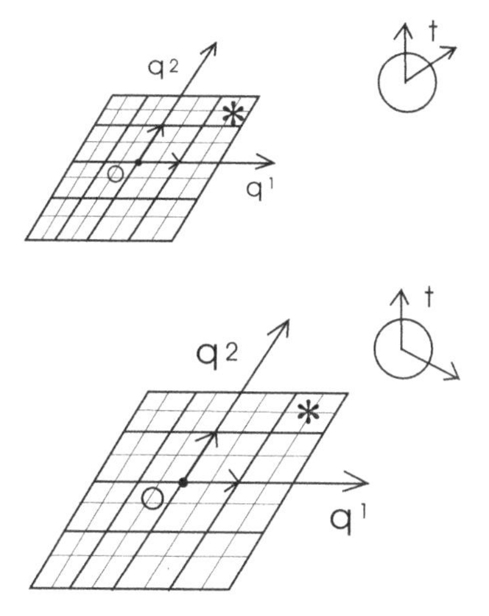

Fig. 10.15 Two-dimensional rubber sheet analogy for the Robertson-Walker metric with $k = 0$. Shown is the configuration at two times, the bottom one later than the top. The stars mark two events that occur at a fixed coordinate position (q^1, q^2).

At a given time t_1, the physical interval on the sheet is given by

$$
(d\mathbf{s})^2 = [\Lambda(t_1)]^2\,(d\mathbf{q})^2 \qquad ;\ \text{interval at } t_1 \tag{10.51}
$$

Where $d\mathbf{q}$ is the coordinate displacement. Suppose that the coordinates

$\mathbf{q} = (q^1, q^2)$ *do not change with time.* In this case, the two points stay fixed on the record keeper's screen, and only the clock moves; nothing appears to be happening (Fig. 10.16). On the contrary, the *physical distance* between the two events marked on the record keeper's screen is changing, and it is the metric that describes how the physical separation between the events increases with time.

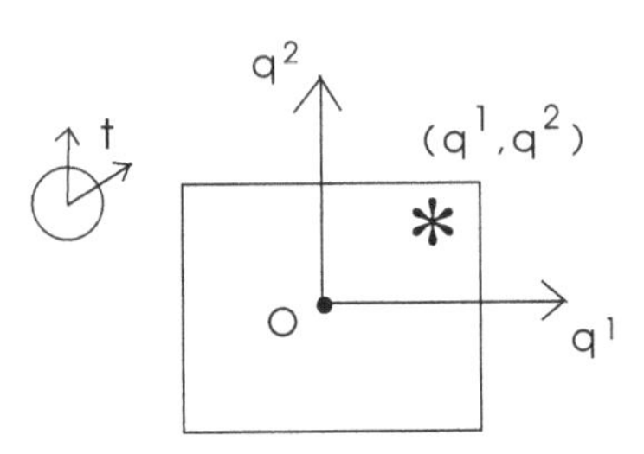

Fig. 10.16 Record keeper's screen corresponding to the situation in Fig. 10.15. Only the clock moves. It is the metric that describes how the physical separation between the events increases with time.

Since the Robertson-Walker metric is the same everywhere in space,[16] one can talk about a finite coordinate separation of the two points on the record keeper's screen, and a corresponding finite coordinate displacement $(\mathbf{q}^2)^{1/2}$. The physical distances l between the two events at the time t_1 and at the later time t_2 are therefore given by

$$\begin{aligned} l_1 &= l(t_1) = \Lambda(t_1)(\mathbf{q}^2)^{1/2} \\ l_2 &= l(t_2) = \Lambda(t_2)(\mathbf{q}^2)^{1/2} \end{aligned} \tag{10.52}$$

Now form the ratio $(l_2 - l_1)/(t_2 - t_1)$, and then substitute the first of Eqs. (10.52), to obtain

$$\begin{aligned} \frac{l_2 - l_1}{t_2 - t_1} &= \left[\frac{\Lambda(t_2) - \Lambda(t_1)}{t_2 - t_1}\right](\mathbf{q}^2)^{1/2} \\ &= \left[\frac{\Lambda(t_2) - \Lambda(t_1)}{t_2 - t_1}\right]\left[\frac{l_1}{\Lambda(t_1)}\right] \end{aligned} \tag{10.53}$$

We may define a *relative velocity* v_r of the two events by

$$l_2 = l_1 + v_r(t_2 - t_1) \tag{10.54}$$

[16]It is *global.*

As $t_2 \to t_1$, and with the definition $l_1 \equiv l$, this relation becomes [recall Eq. (10.48)]

$$v_r = \frac{d\Lambda}{dt}\frac{l}{\Lambda} = h(t)l \qquad ; \text{ relative velocity} \qquad (10.55)$$

We know from freshman physics that light emitted from a source moving away from us is *Doppler shifted.* That shift is given to order v/c by[17]

$$\frac{\lambda}{\lambda_0} = 1 + \frac{v}{c} \qquad ; \text{ Doppler shift} \qquad (10.56)$$

Substitution of Eq. (10.55) then gives

$$\frac{\lambda - \lambda_0}{\lambda_0} = \frac{1}{c}h(t)l \qquad ; \text{ cosmological redshift} \qquad (10.57)$$

This is known as the *cosmological redshift.* The quantity $h(t_p)$ is known as *Hubble's constant*[18]

$$h(t_p) \equiv H_0 \qquad ; \text{ Hubble's constant} \qquad (10.58)$$

If Hubble's constant is determined by some means, then the cosmological redshift in Eq. (10.57) allows one to measure the *distance l of the emitting star.* This plays an important role in astronomy.

10.7.3 *Horizon*

Since the time coordinate is not modified in the metric in Eq. (10.47), and since light always follows the null interval $(d\mathbf{s})^2 = 0$, it follows that the physical velocity of light in the global inertial laboratory frame is still c in this cosmology. Thus there is a finite physical distance $c(t_p - t_0)$ that light traveling for the current age of the universe can have traveled. This distance marks the location of our *horizon* (see Fig. 10.17).

$$D_{\text{H}} = c(t_p - t_0) \qquad ; \text{ distance to horizon} \qquad (10.59)$$

We make several comments:

[17] The full expression for the Doppler shift in special relativity is derived in Prob. 10.1. The present derivation of the cosmological redshift is correct through $O(v/c)$; note that there are other sources of the redshift of light (for example, the gravitational redshift of light propagating away from a massive object).

[18] The best current measured value of the Hubble constant is $H_0 = 72 \pm 8\,(\text{km/s})/\text{Mpc}$ (see [Walecka (2007)]). This gives $H_0^{-1} = 13.5 \times 10^9\,\text{yr}$, a number which readers will often see quoted.

- This is as far as we can see at the present time;[19]
- This light was radiated at the birth of the universe — it is "old light";
- This light will be strongly redshifted when it gets to us;
- The value of Hubble's constant in Eq. (10.50) gives a distance to the horizon of

$$D_{\rm H} = 1.94 \times 10^3\,{\rm Mpc} \tag{10.60}$$

- It is clear from the rubber sheet analogy that the universe looks the same to all observers in this cosmology.

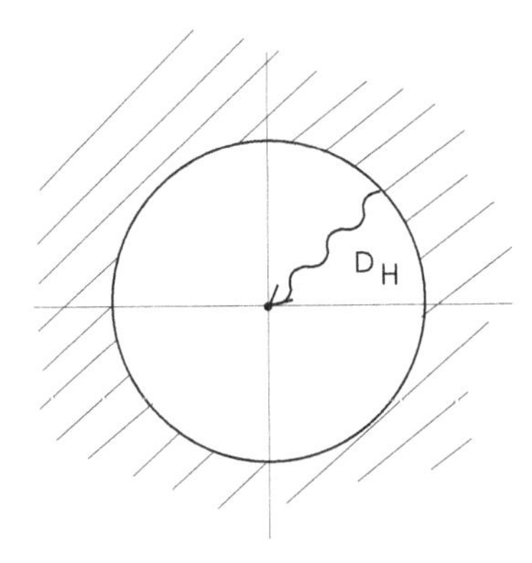

Fig. 10.17 Our horizon lies at a physical spatial distance $D_{\rm H} = c(t_p - t_0)$. This is the distance that light that was emitted at the birth of the universe has traveled to get to us. It is as far as we can see at the present time. Here $D_{\rm H} \approx 2000\,{\rm Mpc}$.

[19]The instantaneous velocity of the horizon in this cosmology is $v_r/c = 2/3$ for all times (see Prob. 10.7).

Chapter 11

Quantum Fluids

Quantum fluids refers to macroscopic many-body systems whose behavior reflects the underlying quantum mechanics. We focus here on two such systems — superfluid ^{4}He and the electron fluid in a superconducting metal. We start the discussion with ^{4}He.[1]

11.1 Superfluid ^{4}He

The phase diagram in (P, T) for ^{4}He is sketched in Fig. 11.1.

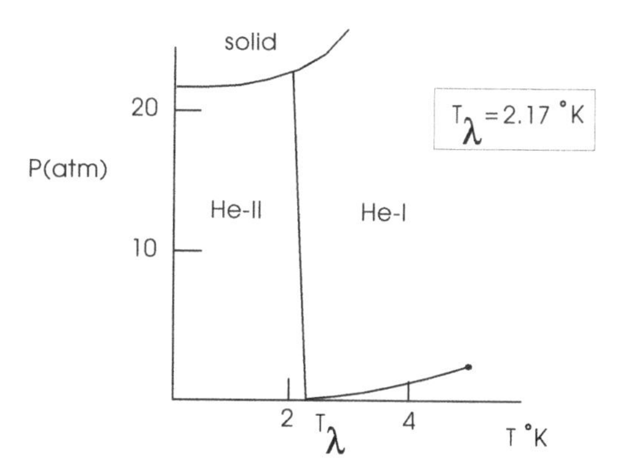

Fig. 11.1 Sketch of phase diagram for liquid ^{4}He.

This has several interesting features:

- At high enough pressure, He forms a solid. At low enough pressure, He

[1] Most of the material in this chapter is taken from [Fetter and Walecka (2003a)].

is a gas. There is an ordinary gas-liquid phase transition, and a critical point. Below $\sim 20\,\text{atm}$, He remains a liquid down to $T = 0$;

- There is a phase transition from the ordinary liquid He-I to a second phase He-II, which exhibits many fascinating properties. At zero pressure, this phase transition takes place at a temperature $T_\lambda = 2.17\,^\circ\text{K}$;
- He-II exhibits superfluid flow with *vanishing viscosity*. It behaves as a mixture of two interpenetrating, incompressible fluids of constant total density, one which is a normal fluid and one the superfluid. There is a smooth transition from all normal fluid to essentially all superfluid as the temperature is decreased from $T = T_\lambda$ to $T = 0$;
- The heat capacity exhibits a logarithmic singularity at T_λ (Fig. 11.2).

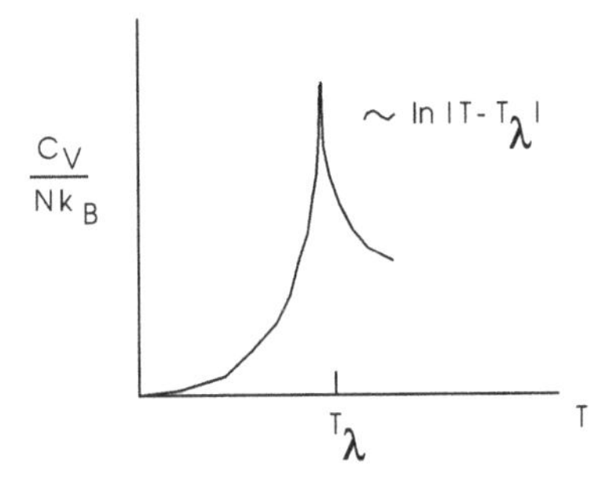

Fig. 11.2 Sketch of heat capacity for liquid ^{4}He.

Atomic ^{4}He consists of a $(1s)^2$ electronic configuration about a $^4_2$He nucleus. The nucleus, in the shell model, consists of $(n\uparrow n\downarrow p\uparrow p\downarrow)$ in a $(1s)^4$ configuration (see Fig. 11.3).

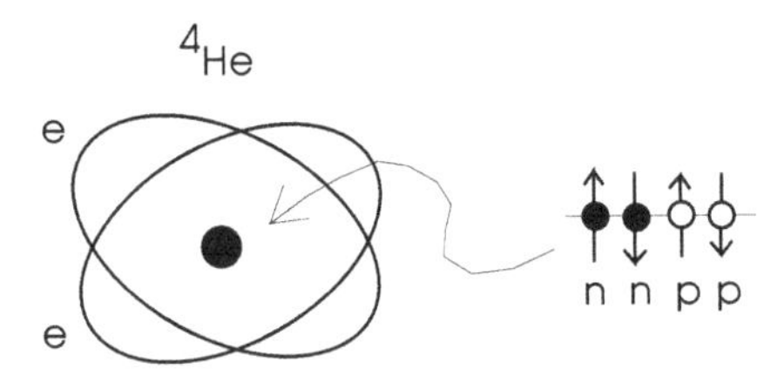

Fig. 11.3 Electron and nucleon configuration of ^{4}He.

Since it consists of six fermions, a ^{4}He atom behaves as a *boson*, and since all spins are paired, it forms a *spin-zero* boson with $J^\pi = 0^+$. With two tightly-bound $1s$ electrons, the atom is very small and the interatomic forces are very weak outside that core. ^{4}He is indeed chemically inactive; it forms an *inert gas* (chapter 5). It is the lightest of the inert gases, and it is in liquid He that one expects quantum effects to become most important.[2]

As a first approximation, one can consider ^{4}He to be a collection of non-interacting point bosons with the quantum statistics discussed at the end of chapter 4. At low temperature, the particles will be condensed into the single-particle ground state (the "Bose condensate"), while a certain fraction of the particles will be thermally excited into single-particle excited states. At $T = 0$ they will all be in the condensate, which forms the superfluid. The excitations at $T \neq 0$ can then be identified with the normal fluid. As previously discussed, there is a phase transition in the non-interacting Bose system which is not dissimilar to that discussed above.

11.1.1 *Hartree Approximation*

The many-particle ground-state wave function for a collection of N non-interacting spin-zero bosons is of the form (see appendix H)

$$\Phi(\mathbf{x}_1, \mathbf{x}_2, \cdots, \mathbf{x}_N) = \phi_0(\mathbf{x}_1)\phi_0(\mathbf{x}_2)\cdots\phi_0(\mathbf{x}_N) \tag{11.1}$$

Here $\phi_0(\mathbf{x})$ is the lowest-energy single-particle state for the problem at hand [see, for example, Eqs. (4.115) or (4.118)]. The wave function in Eq. (11.1) is totally symmetric under the interchange of any two particles. Suppose the particles now interact through a two-particle potential of the form $V(\mathbf{x}_i - \mathbf{x}_j)$. The effects of this potential can be most simply included by appealing to the Hartree approximation discussed in chapter 5 and Probs. H.3–H.7. Here the form of the wave function in Eq. (11.1) is maintained, while the effect on a particle of the mean field created by all the other particles is taken into account.

Since all the particles occupy the same single-particle state in Eq. (11.1), the particle density $n(\mathbf{x})$ and corresponding particle current $n(\mathbf{x})\mathbf{v}(\mathbf{x})$ for the many-body system are just N times the probability density and probability current for one particle. Hence

$$\begin{aligned} n(\mathbf{x}) &= N|\phi_0(\mathbf{x})|^2 && ;\ \text{particle density} \\ n(\mathbf{x})\mathbf{v}(\mathbf{x}) &= N\frac{\hbar}{2im}\left[\phi_0^\star \boldsymbol{\nabla}\phi_0 - (\boldsymbol{\nabla}\phi_0)^\star\phi_0\right] && ;\ \text{particle current} \end{aligned} \tag{11.2}$$

---

[2]Recall Prob. 4.12.

The Hartree potential is the direct analogue of that given in Eq. (5.74)

$$V_H(\mathbf{x}) = \int d^3y\, V(\mathbf{x}-\mathbf{y})n(\mathbf{y}) \qquad ;\ \text{Hartree potential} \quad (11.3)$$

The Hartree equation is then that of Eq. (5.76)

$$\left[-\frac{\hbar^2\nabla^2}{2m} + V_H(\mathbf{x})\right]\phi_0(\mathbf{x}) = \varepsilon_0\phi_0(\mathbf{x}) \qquad ;\ \text{Hartree equation} \quad (11.4)$$

Here ε_0 is the lowest single-particle eigenvalue for the problem at hand. We note that this equation is applicable to both uniform and non-uniform systems.

It is convenient at this point to define a new single-particle wave function that scales out the factor of N

$$\psi_0(\mathbf{x}) \equiv \sqrt{N}\,\phi_0(\mathbf{x}) \qquad (11.5)$$

The Hartree equation then takes the form

$$\left[-\frac{\hbar^2\nabla^2}{2m} + \int d^3y\, V(\mathbf{x}-\mathbf{y})|\psi_0(\mathbf{y})|^2\right]\psi_0(\mathbf{x}) = \varepsilon_0\psi_0(\mathbf{x}) \quad ;\ \text{Hartree} \quad (11.6)$$

This is a non-linear, integro-differential equation that can be solved by iteration in the general case, or analytically in the case of a uniform system (Prob. H.6). The corresponding particle density and particle current are

$$\begin{aligned} n(\mathbf{x}) &= |\psi_0(\mathbf{x})|^2 && ;\ \text{particle density} \\ n(\mathbf{x})\mathbf{v}(\mathbf{x}) &= \frac{\hbar}{2im}\left[\psi_0^\star\boldsymbol{\nabla}\psi_0 - (\boldsymbol{\nabla}\psi_0)^\star\psi_0\right] && ;\ \text{particle current} \quad (11.7) \end{aligned}$$

11.1.2 *Velocity Field*

The single-particle wave function in Eq. (11.5) can be parameterized in terms of its modulus and phase as

$$\psi_0(\mathbf{x}) \equiv |\psi_0(\mathbf{x})|e^{i\phi(\mathbf{x})} \qquad (11.8)$$

A simple calculation then shows that the particle density and current in Eqs. (11.7) take the form

$$\begin{aligned} n(\mathbf{x}) &= |\psi_0(\mathbf{x})|^2 \\ \mathbf{v}(\mathbf{x}) &= \frac{\hbar}{m}\boldsymbol{\nabla}\phi(\mathbf{x}) \end{aligned} \qquad (11.9)$$

The second equation is particularly interesting. The velocity field comes from the gradient of the phase of the single-particle wave function. It follows immediately that the velocity field is *irrotational* with vanishing curl [see Prob. F.3(a)].

$$\nabla \times \mathbf{v}(\mathbf{x}) = 0 \qquad ; \text{ irrotational flow} \tag{11.10}$$

Stokes theorem (appendix F) then implies that the *circulation* around any closed curve C in a plane lying entirely within the fluid [Fig. 11.4(a)] *vanishes*

$$\oint_C \mathbf{v} \cdot d\mathbf{l} = 0 \qquad ; \text{ circulation in fluid} \tag{11.11}$$

Note the remarkable fact that we have obtained properties of the macroscopic fluid flow of the condensed Bose system from the single-particle wave function!

11.1.3 *Quantized Circulation*

While the circulation vanishes in the bulk fluid, it is possible to set superfluid ^{4}He into rotation and create a *vortex* where the superfluid component vanishes at its center. This creates a hole in the superfluid. Suppose one has a rectilinear vortex and the integration contour in Eq. (11.11) lies completely in the fluid, but now goes *around the vortex.* In this case, one can obtain a finite circulation, and we are in a position to say something about the allowed values of this circulation. The current in Eq. (11.9) comes from the phase of the single-particle wave function, and it will be assumed that for a Bose system this wave function is *single-valued throughout the fluid.* Suppose the integration contour in Eq. (11.11) is a circle that lies in a plane transverse to the vortex and centered on it as illustrated in Fig. 11.4(b). The integral then just gives the difference in phase between the initial and final points on the circle[3]

$$\oint_C \mathbf{v} \cdot d\mathbf{l} = \oint \frac{\hbar}{m} \nabla\phi \cdot d\mathbf{l} = \frac{\hbar}{m} \oint d\phi = \frac{\hbar}{m}\left[\phi(2\pi) - \phi(0)\right] \tag{11.12}$$

Since we come back where we started when we go around the circle, and since the wave function must be single-valued, the difference in phase must

[3]For the second step, see Prob. F.3(b).

be an integral number of 2π

$$\oint_C \mathbf{v}\cdot d\mathbf{l} = \frac{\hbar}{m}(2\pi\nu) = \frac{h}{m}\nu \qquad ; \nu = 0, 1, 2, \cdots$$

quantized cirulation (11.13)

The circulation around the vortex must be *quantized* in units of h/m. For ^{4}He the unit of circulation has the value

$$\frac{h}{m_{\mathrm{He}}} = 0.997 \times 10^{-3}\,\mathrm{cm}^2/\mathrm{s} \tag{11.14}$$

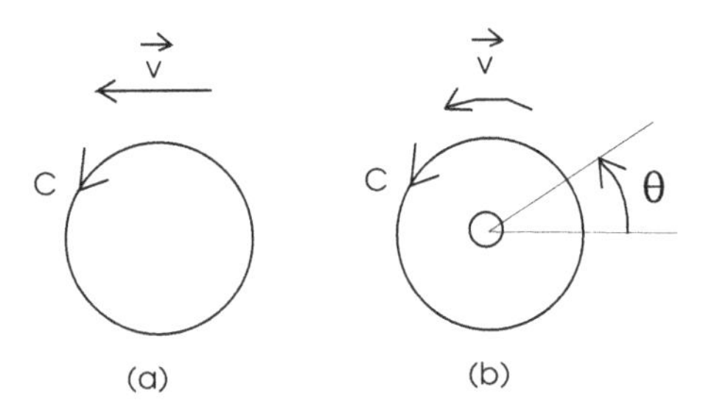

Fig. 11.4 Circulation of superfluid in ^{4}He: (a) Bulk fluid; (b) Around a vortex which is transverse to page.

11.1.4 *Gross-Pitaevskii Equation*

Consider the following modifications of the Hartree equation:

(1) First, model the force between the Bose particles by a repulsive delta-function potential[4]

$$V(\mathbf{x}-\mathbf{y}) = g\delta^{(3)}(\mathbf{x}-\mathbf{y}) \qquad ; g > 0 \tag{11.15}$$

(2) Now note that for a uniform system with particle density n, it follows from Prob. H.5 that the total ground-state energy density is given by

$$\frac{E}{V} = \frac{1}{2}n\varepsilon_0 = \frac{1}{2}n^2\int d^3y\, V(\mathbf{y}) \tag{11.16}$$

[4]This models the repulsive van der Waals force between the atoms in liquid ^{4}He.

The chemical potential is then given by[5]

$$\mu \equiv \left(\frac{\partial E}{\partial N}\right)_{S,V} = \varepsilon_0 \tag{11.17}$$

The Hartree equation will be generalized by replacing ε_0 by μ, an experimental observable, on the r.h.s. As with finite temperature, the interactions take some of the particles out of the Hartree condensate and distribute them over the other Hartree single-particle states. The number of particles in the condensate is not a conserved quantity, and the use of the chemical potential allows one to take this into account. Correspondingly, the particle density n will be replaced by the condensate density n_0.

These steps lead to the *Gross-Pitaevskii equation*

$$\left[-\frac{\hbar^2\nabla^2}{2m} + g|\psi_0(\mathbf{x})|^2\right]\psi_0(\mathbf{x}) = \mu\,\psi_0(\mathbf{x}) \qquad ; \text{ Gross-Pitaevskii}$$
$$|\psi_0(\mathbf{x})| = \sqrt{n_0(\mathbf{x})} \tag{11.18}$$

This is now a *local* non-linear differential equation that is readily integrated with existing numerical programs.[6]

11.1.5 *Vortex*

Consider the solution to the Gross-Pitaevskii equation in the case of a rectilinear vortex as illustrated in Fig. 11.4(b). Look for a solution of the form of

$$\psi_0(r,\theta) = \sqrt{n_0}\, f(r)e^{i\theta} \tag{11.19}$$

where n_0 is the condensate density in the bulk fluid, and $f(r)$ is real. Then from the second of Eqs. (11.9) [see Prob. F.3(c) for the second step]

$$\mathbf{v}(r,\theta) = \frac{\hbar}{m}\boldsymbol{\nabla}\theta = \frac{\hbar}{mr}\hat{\mathbf{e}}_\theta \tag{11.20}$$

The quantities involved here are illustrated in Fig. 11.5.

---

[5]The system is in the volume V and in its ground state ($S = 0$).

[6]The Gross-Pitaevskii equation is derived with more sophistication in [Fetter and Walecka (2003a)]. For the present purposes, it will be assumed that any excitations out of the condensate are small, so that $T \to 0$ and $n_0 \to n$.

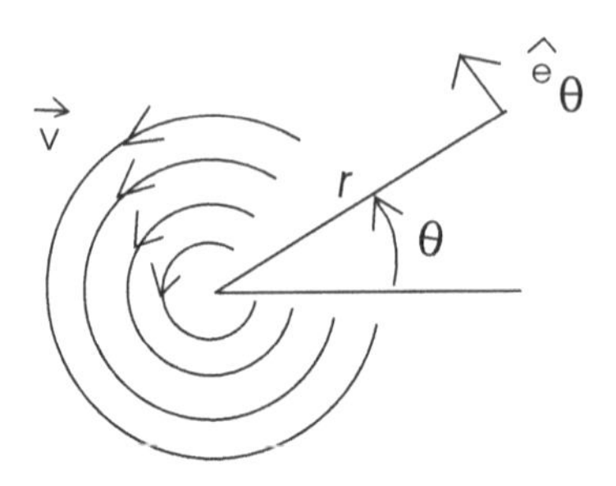

Fig. 11.5 Vortex velocity field in Eq. (11.20).

Since $d\mathbf{l} = \hat{\mathbf{e}}_\theta \, r d\theta$, the circulation about the origin is immediately evaluated as

$$\oint_C \mathbf{v} \cdot d\mathbf{l} = \int_0^{2\pi} \frac{\hbar}{mr} r d\theta = \frac{2\pi\hbar}{m} = \frac{h}{m} \tag{11.21}$$

The velocity field in Eq. (11.20) represents a vortex with *one unit of circulation*. Note the magnitude of the velocity falls off as $1/r$.

With the assumed form of the solution in Eq. (11.19), and the laplacian in polar coordinates of Prob. 4.26(a),[7] the Gross-Pitaevskii Eq. (11.18) becomes

$$\frac{\hbar^2}{2m}\left(\frac{1}{r}\frac{d}{dr}r\frac{d}{dr} - \frac{1}{r^2}\right) f + \mu f - n_0 g f^3 = 0 \tag{11.22}$$

where an overall factor of $\sqrt{n_0}\, e^{i\theta}$ has been canceled.

Far away from the vortex, one should have $|\psi_0|^2 \to n_0$, and hence there is a boundary condition that goes with this equation

$$f(r) \to 1 \qquad ; \; r \to \infty \tag{11.23}$$

It follows from Eq. (11.22) that the chemical potential is thus given by

$$\mu = g n_0 \tag{11.24}$$

This is the energy required to insert one boson into the medium. Division by this factor reduces Eq. (11.22) to

$$\frac{\hbar^2}{2mgn_0}\left(\frac{1}{r}\frac{d}{dr}r\frac{d}{dr} - \frac{1}{r^2}\right) f + f - f^3 = 0 \tag{11.25}$$

[7] Here $(\rho, \phi) \to (r, \theta)$.

This equation is reduced to dimensionless form with the introduction of the new variable

$$\zeta \equiv \frac{r}{\xi} \qquad ; \xi \equiv \left(\frac{\hbar^2}{2mgn_0}\right)^{1/2} \tag{11.26}$$

Thus the Gross-Pitaevskii equation in this case is reduced to the following non-linear differential equation and boundary condition

$$\frac{d^2f}{d\zeta^2} + \frac{1}{\zeta}\frac{df}{d\zeta} - \frac{1}{\zeta^2}f + f - f^3 = 0$$

$$f(\zeta) \to 1 \qquad ; \zeta \to \infty \tag{11.27}$$

We make the following observations:

(1) As $\zeta \to 0$, the angular momentum barrier, given by the third term in this equation, dominates. It is readily verified that the regular solution takes the following form at the origin

$$f = C\zeta \qquad ; \zeta \to 0 \tag{11.28}$$

where C is a constant.

(2) As $\zeta \to \infty$, a power series solution in $1/\zeta^2$ gives

$$f = 1 - \frac{1}{2\zeta^2} + \cdots \qquad ; \zeta \to \infty \tag{11.29}$$

(3) Numerical integration is readily carried out for all values of ζ, and the result obtained using the Runge-Kutta algorithm in Mathcad11 are shown in Fig. 11.6.

We observe that the superfluid is excluded from the vortex core, as advertised, and the size of the core is $\zeta \sim 1$, or

$$r_{\text{core}} \sim \xi \qquad ; \text{ core size} \tag{11.30}$$

The observed core size in He-II is ~ 1 Å.[8]

This solves the Gross-Pitaevskii equation in the case of one particular non-uniform system — a vortex. That equation is also applicable to cold, isolated, laser-trapped Bose systems, whose experimental study provides one of the more fascinating aspects of modern physics (see [Colorado (2007)]).

[8]Note Prob. 11.2.

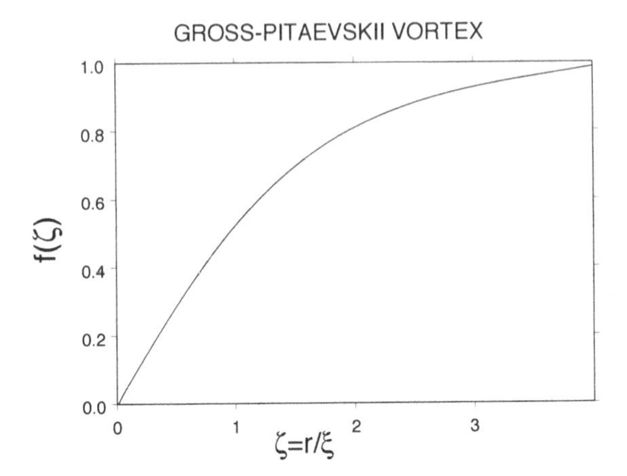

Fig. 11.6 Numerical integration of the Gross-Pitaevskii Eqs. (11.27) and (11.28) for a vortex with unit circulation.

11.1.6 *Superfluidity*

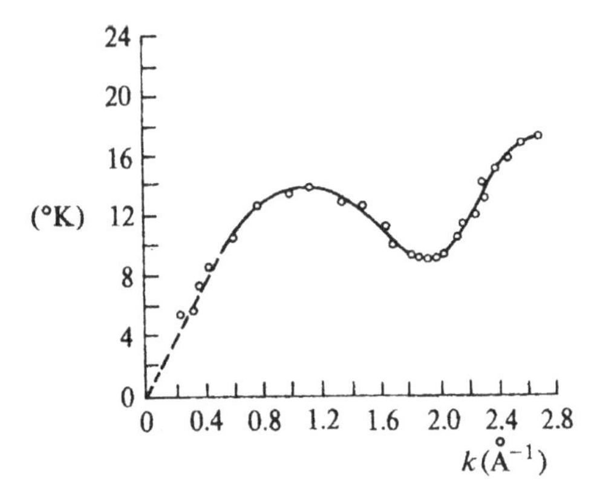

Fig. 11.7 Experimental low-temperature quasiparticle spectrum $\varepsilon_k(k)/k_{\rm B}$ in He-II. From [Fetter and Walecka (2003a)].

There are various types of excitations in He-II. They have a dispersion relation $\varepsilon_k(k)$ that is experimentally accessible through specific heat measurements,[9] or more directly, through inelastic neutron scattering where the response of the fluid to a specified amount of momentum and energy

[9] Recall the Debye theory of the heat capacity.

transfer $(\hbar\mathbf{k}, \varepsilon_k)$ is examined. Landau called the quanta of these excitations "quasiparticles" [Landau (1941)]. The measured low-temperature quasiparticle spectrum in He-II is shown in Fig. 11.7. At long wavelength, one sees phonons, the quanta of the sound waves in the fluid,[10] with $\varepsilon_k = \hbar k c_{\rm sound}$. At higher k, the spectrum is more complicated, exhibiting a dip (corresponding to "rotons"); ε_k then continues to grow with k.

Landau gave a simple argument based on conservation of energy and momentum that greatly helps one understand superfluidity [Landau (1941)]. Suppose one has a heavy object of mass M moving with velocity $\mathbf{v}$ through the uniform Bose fluid. Consider a collision process that creates an excitation in the fluid as illustrated in Fig. 11.8.

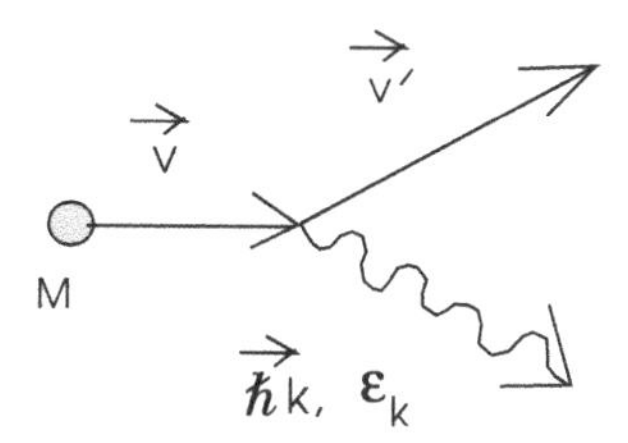

Fig. 11.8 Creation of an excitation in the Bose fluid by a heavy object moving through it.

Energy and momentum will be conserved in the collision, and hence

$$
\begin{aligned}
\frac{1}{2}M\mathbf{v}^2 &= \frac{1}{2}M\mathbf{v}'^2 + \varepsilon_k && ;\ \text{energy} \\
M\mathbf{v} &= M\mathbf{v}' + \hbar\mathbf{k} && ;\ \text{momentum}
\end{aligned}
\tag{11.31}
$$

Substitution of the second relation in the first gives

$$
\begin{aligned}
\frac{1}{2}M\mathbf{v}^2 &= \frac{1}{2}M\left(\mathbf{v}^2 - \frac{2\hbar\mathbf{k}\cdot\mathbf{v}}{M} + \frac{\hbar^2\mathbf{k}^2}{M^2}\right) + \varepsilon_k \\
&\approx \frac{1}{2}M\left(\mathbf{v}^2 - \frac{2\hbar\mathbf{k}\cdot\mathbf{v}}{M}\right) + \varepsilon_k
\end{aligned}
\tag{11.32}
$$

where the second line neglects $\hbar^2\mathbf{k}^2/M^2$. A cancelation of terms then leads to the relation

$$
\varepsilon_k - \hbar\mathbf{k}\cdot\mathbf{v} = 0 \quad ;\ \text{energy-momentum conservation} \tag{11.33}
$$

[10]See chapter 12.

Suppose $\varepsilon_k > \hbar kv$, then this condition *cannot be satisfied.* This leads to a *critical velocity* given by

$$v_{\text{critical}} = \left(\frac{\varepsilon_k}{\hbar k}\right)_{\min} \qquad ; \text{ critical velocity} \qquad (11.34)$$

If the velocity v of the mass M moving through the Bose fluid is *less* that this, then it *cannot create excitations of the fluid,* and hence *there will be no viscosity.* For a fluid moving in a tube, just go to the rest frame of the fluid. The wall is then the heavy moving object (Fig. 11.9). It is clear from Fig. 11.7 that in He-II the quantity $(\varepsilon_k/\hbar k)$ has a minimum reached in the roton part of the curve.

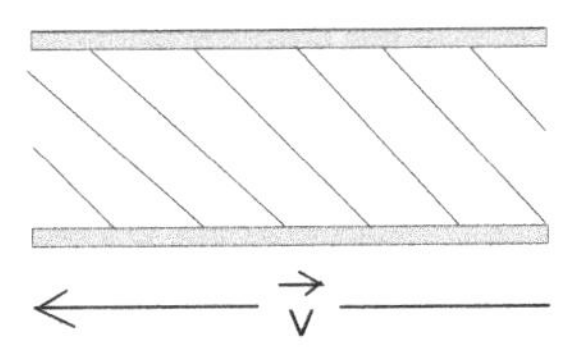

Fig. 11.9 Superfluid moving through a heavy tube.

For collisions with a particle in a free Fermi gas, one has an energy transfer of

$$\varepsilon_k = \frac{(\mathbf{p} + \hbar\mathbf{k})^2}{2m} - \frac{\mathbf{p}^2}{2m} \approx \hbar\mathbf{k}\cdot\mathbf{v}_0 = \hbar k v_0 \cos\theta_{\mathbf{pk}} \qquad (11.35)$$

Hence

$$\left(\frac{\varepsilon_k}{\hbar k}\right)_{\min} = 0 \qquad ; \text{ free Fermi gas} \qquad (11.36)$$

Hence, by this argument, a *free Fermi gas will not exhibit superfluidity.*[11] For a Fermi gas to exhibit superfluidity, there must be some modification of the excitation spectrum by the interactions, for example, by the introduction of an *energy gap* Δ so that the energy transfer satisfies $\varepsilon_k \geq \Delta$.

11.2 Superconductivity

We next turn to the topic of superconductors, which have many remarkable properties (see [Superconductors (2007)]).

[11]Neither will a *free* Bose gas. It is the interactions between the bosons that give rise to the spectrum in Fig. 11.7 (compare Prob. 11.2).

11.2.1 *Experimental Properties*

When many metallic elements or compounds are cooled to within a few degrees of absolute zero, they lose all traces of electrical resistivity at a definite temperature T_c. The following phenomena are then observed:

1). *Infinite conductivity.* Ohm's law states that in a material the electromagnetic current is proportional to the electric field

$$\mathbf{j} = \sigma \mathbf{E} \qquad ; \text{ Ohm's law} \tag{11.37}$$

where σ is the electrical conductivity. In a perfect superconductor, the vanishing resistivity implies $\sigma \to \infty$, and with an infinite conductivity, *there will be no electric field in a superconductor.* It follows from the third of Maxwell's Eqs. (2.90) that when $\mathbf{E} = 0$

$$\frac{\partial \mathbf{B}}{\partial t} = 0 \tag{11.38}$$

Hence the magnetic field $\mathbf{B}$ is constant in a superconductor.

2). *Meissner effect.* It is a remarkable fact that when a material makes the superconducting transition in an applied magnetic field, or alternatively, when a sufficiently weak magnetic field is applied to a superconductor, the magnetic field is *excluded* from the bulk material. Hence, in fact, $\mathbf{B} = 0$ in the bulk superconductor. This is the *Meissner effect.* The last of Maxwell's Eqs. (2.90) implies that the current also vanishes in the bulk superconductor

$$\mu_0 \mathbf{j} = \boldsymbol{\nabla} \times \mathbf{B} = 0 \tag{11.39}$$

The magnetic field can penetrate into the *surface* of a superconductor; just how far it gets inside depends on the characteristic "penetration depth" of the material. There is a corresponding surface current, which also vanishes exponentially as one goes into the bulk superconductor.[12]

3). *Critical field.* At strong enough applied magnetic fields, the superconducting material is driven normal. There is a characteristic *critical field* at which this transition occurs, the critical field decreasing with temperature and vanishing at T_c.

4). *Flux quantization.* There are materials known as type-II superconductors where *magnetic flux tubes* first penetrate the bulk superconductor, before it is driven normal by a further increase in the applied magnetic field. The interior metal in these flux tubes is in the normal state, and the

[12]See Prob. 11.4.

field is maintained by a persistent supercurrent at their boundary. It is observed that the magnetic flux in these tubes is *quantized* in units of $h/2e$.

5). *Specific heat.* The specific heat of a typical superconductor is sketched in Fig. 11.10. It is discontinuous at $T = T_c$, with no latent heat for the transition. In the superconducting state, the specific heat C_s vanishes exponentially as $T \to 0$

$$C_s \propto \exp\left\{-\frac{\Delta_0}{k_B T}\right\} \tag{11.40}$$

This dependence indicates the existence of a gap in the energy spectrum separating the excited states from the ground state by an energy Δ_0.[13]

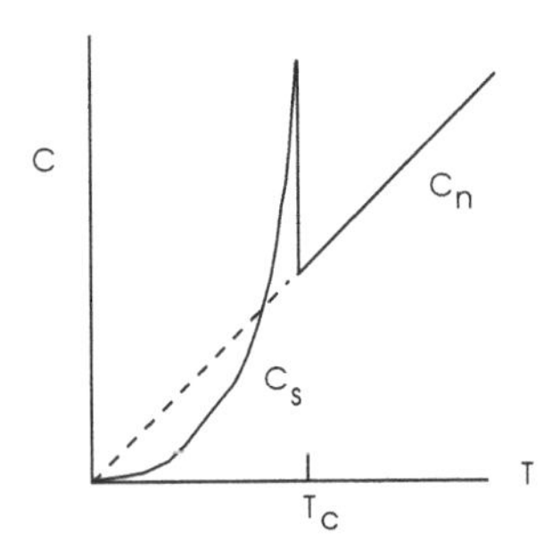

Fig. 11.10 Schematic drawing of the specific heat C_s of a superconductor. $C_n \propto T$ is the electronic specific heat of the normal metal, which dominates at low T.

6). *Isotope effect.* The transition temperature in isotopically pure metals typically varies with the ionic mass according to

$$T_c \propto M^{-1/2} \tag{11.41}$$

11.2.2 *Some Observations*

The simplest picture of a metal is a non-interacting Fermi gas of negatively charged electrons moving in a much heavier, uniform, positive ionic background. Although this picture does allow one to understand many metallic properties, it provides no understanding of the remarkable properties of superconductors.

The vanishing resistivity appears to be analogous to the vanishing viscosity of a superfluid. One way of understanding the specific heat results is

[13]Typically, Δ_0 is somewhat less that $2k_B T_c$. There are also "gapless" superconductors.

that the electrons form *bound pairs* in the medium, with a binding energy of Δ_0. Indeed, if the electrons were to form bound pairs, then the electron pair could be viewed as a *boson* and all our results on superfluidity applied to superconductors! In particular, one could understand the vanishing resistivity in terms of superfluid flow. The magnetic flux tube in a type-II superconductor looks intriguingly like a vortex in He-II, and magnetic flux quantization could then be understood in analogy to the quantization of circulation in a vortex (see below).

An important clue here is given by the isotope effect. This result indicates the importance of the interaction between the electrons and phonons, the quanta of the sound waves in the underlying ionic lattice. This attractive electron-phonon interaction indeed provides a mechanism for the formation of *bound Cooper pairs* of electrons lying close to the Fermi surface. It is to this topic that we turn next.

11.2.3 *Cooper Pairs*

Start with the above picture of a metal as a non-interacting Fermi gas of negatively charged electrons moving in a much heavier, uniform, positive ionic background. It is important to realize that this system is neutral and hence has *no Coulomb energy*. The negative exchange term in the electron-electron interaction energy, discussed in appendix H, then provides the binding energy of the metal. At this point, the many-body wave function, at least for those electrons near the Fermi surface is well described by that of a free Fermi gas.

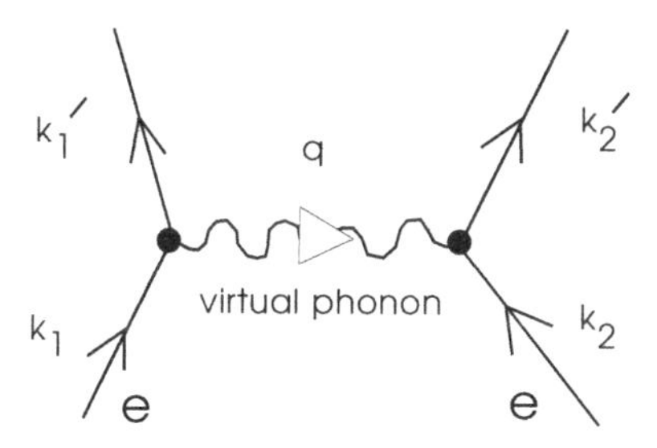

Fig. 11.11 Two electrons near the Fermi surface interacting through the exchange of a phonon.

Now there is an additional interaction between electrons arising from the exchange of a *phonon*, the quantum of the sound waves in the underlying ionic lattice, as illustrated in Fig. 11.11. The modes of oscillation of the lattice were discussed in the Debye theory of the specific heat in chapter 3, where it was observed that there is a cut-off to the frequency of these modes at $\omega_D = 2\pi\nu_{\rm max}$.

It is shown in [Fetter and Walecka (2003a)] that for electron energies ϵ_k within a band of thickness $\hbar\omega_D$ about the Fermi surface, the equivalent potential arising from the phonon exchange in Fig. 11.11 is attractive and takes the form

$$V_{\rm eq}(\mathbf{x}) = -\gamma^2\delta^{(3)}(\mathbf{x}) \qquad ; \text{ phonon-exchange} \quad |\epsilon_k - \epsilon_{k_{\rm F}}| < \hbar\omega_D \tag{11.42}$$

The Pauli principle implies that it is only electrons with opposite spins that can exist in relative s-states and feel the effect of this contact potential.

Let us then investigate the model quantum mechanical problem of two particles close to the Fermi surface, with spins opposed,[14] interacting through a short-range, attractive interaction. To simplify the discussion, assume that the center of momentum of the pair is at rest in the laboratory system, so that the initial momenta of the pair (in units of $\hbar$) are $(\mathbf{k}_2, \mathbf{k}_1) = (\mathbf{k}, -\mathbf{k})$. The relative coordinate and relative momentum of Prob. 6.1 are then given by

$$\mathbf{x} = \mathbf{x}_2 - \mathbf{x}_1 \qquad ; \quad \frac{m\mathbf{k}_2 - m\mathbf{k}_1}{2m} = \mathbf{k} \tag{11.43}$$

The situation is illustrated in Fig. 11.12.

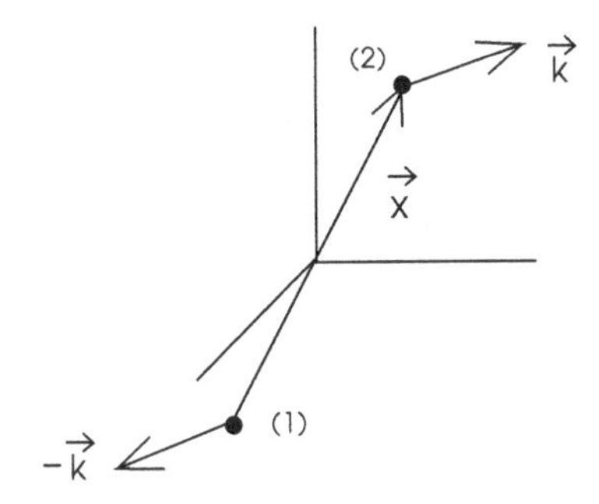

Fig. 11.12 Two electrons near the Fermi surface, with initial momenta (in units of $\hbar$) of $(\mathbf{k}_2, \mathbf{k}_1) = (\mathbf{k}, -\mathbf{k})$ and relative coordinate $\mathbf{x}$, which interact through a short-range, attractive interaction.

[14]We now suppress the spin wave functions in the subsequent arguments.

The two-body Schrödinger equation for the pair, in the relative coordinate $\mathbf{x}$, and with a potential $V(x)$, is derived in Prob. 6.1. It takes the form

$$(\nabla^2 + \kappa^2)\psi(\mathbf{x}) = \lambda v(x)\psi(\mathbf{x}) \qquad ;\ \text{Schrödinger eqn} \qquad (11.44)$$

Here we have defined[15]

$$\kappa^2 \equiv \frac{2\mu E}{\hbar^2} \qquad ;\ \lambda v \equiv \frac{2\mu V}{\hbar^2} \qquad ;\ \frac{1}{\mu} = \frac{1}{m} + \frac{1}{m} \qquad (11.45)$$

We work in a big cubical box of volume $\Omega = L^3$ with periodic boundary conditions, where the unperturbed wave functions are

$$\phi_{\mathbf{k}} = \frac{1}{\sqrt{\Omega}} e^{i\mathbf{k}\cdot\mathbf{x}} \qquad ;\ \mathbf{k} = \frac{2\pi}{L}(n_1, n_2, n_3)$$
$$n_i = 0, \pm 1, \pm 2, \cdots \qquad (11.46)$$

These wave functions are both orthonormal and complete[16]

$$\int_\Omega d^3x\, \phi^\star_{\mathbf{k}_1}(\mathbf{x})\phi_{\mathbf{k}_2}(\mathbf{x}) = \delta_{\mathbf{k}_1\mathbf{k}_2} \qquad ;\ \text{orthonormal}$$
$$\sum_{\mathbf{k}} \phi_{\mathbf{k}}(\mathbf{x})\phi^\star_{\mathbf{k}}(\mathbf{y}) = \delta^{(3)}(\mathbf{x} - \mathbf{y}) \qquad ;\ \text{complete} \qquad (11.47)$$

The Schrödinger Eq. (11.44) can be rewritten in the following form

$$\psi(\mathbf{x}) = \sum_{\mathbf{t}} \phi_{\mathbf{t}}(\mathbf{x}) \frac{1}{\kappa^2 - t^2} \int_\Omega d^3y\, \phi^\star_{\mathbf{t}}(\mathbf{y})\lambda v(y)\psi(\mathbf{y}) \qquad (11.48)$$

The proof lies in simply operating on this expression with $(\nabla^2 + \kappa^2)$, which eliminates the denominator on the r.h.s., and then using the completeness relation in Eqs. (11.47). This is now an *integral equation* for $\psi(\mathbf{x})$, which is complicated by the presence of the eigenvalue κ^2 in the denominator on the r.h.s.

Let us choose a somewhat different *normalization* of the wave function ψ,[17] which we now refer to as $\psi_{\mathbf{k}}$,

$$\int_\Omega d^3x\, \phi^\star_{\mathbf{k}}(\mathbf{x})\psi_{\mathbf{k}}(\mathbf{x}) = 1 \qquad ;\ \text{choice of norm} \qquad (11.49)$$

---

[15]It is convenient to explicitly include a factor of λ characterizing the sign and strength of the potential.

[16]The completeness relation is proven in the limit $L \to \infty$ in Probs. 4.3 and 4.24. The proof for finite L proceeds in exactly the same fashion.

[17]Recall that we are free to choose any normalization of the wave function, as long as expectation values are computed as in Prob. 4.5.

Equation (11.48) can then be written as a pair of equations as follows

$$\psi_{\mathbf{k}}(\mathbf{x}) = \phi_{\mathbf{k}}(\mathbf{x}) + \sum_{\mathbf{t}\neq\mathbf{k}} \phi_{\mathbf{t}}(\mathbf{x})\frac{1}{\kappa^2 - t^2}\int d^3y\,\phi^\star_{\mathbf{t}}(\mathbf{y})\lambda v(y)\psi_{\mathbf{k}}(\mathbf{y}) \quad ; \text{ S-eqn}$$

$$\kappa^2 - k^2 = \int_\Omega d^3x\,\phi^\star_{\mathbf{k}}(\mathbf{x})\lambda v(x)\psi_{\mathbf{k}}(\mathbf{x}) \qquad (11.50)$$

These equations represent an exact rewriting of the Schrödinger equation, where the wave function is normalized according to Eq. (11.49). The first is now an *inhomogeneous* integral equation, which has the great advantage that

$$\psi_{\mathbf{k}}(\mathbf{x}) \to \phi_{\mathbf{k}}(\mathbf{x}) \qquad ; \text{ as } \lambda \to 0 \qquad (11.51)$$

Equations (11.50) refer to an isolated pair of particles. We now modify those equations to take into account one crucial *many-body* aspect of the problem. The pair interacts in the presence of a Fermi gas of other electrons, and the Pauli principle states that two fermions cannot occupy the same state. Consequently, the only states that can be admixed into the wave function for the pair must lie *outside* of the Fermi sphere (Fig. 11.13).

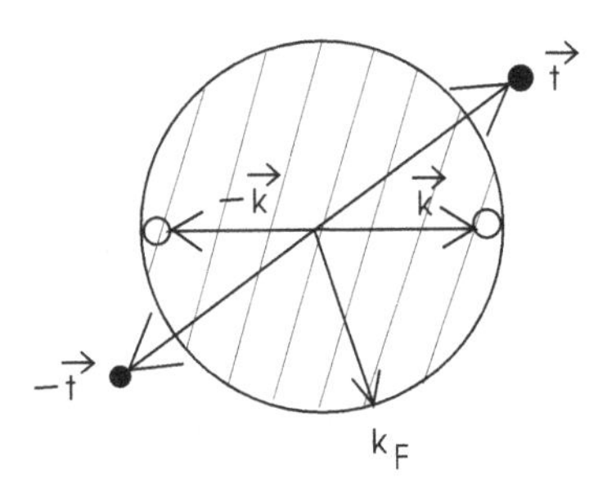

Fig. 11.13 Two electrons start with initial momenta (in units of $\hbar$) of $(\mathbf{k}_2, \mathbf{k}_1) = (\mathbf{k}, -\mathbf{k})$ just below the Fermi surface. Since the other levels below k_F are already filled with electrons, the Pauli principle implies that only states with $(\mathbf{t}, -\mathbf{t})$ *above* the Fermi surface can be admixed into the wave function of the pair.

Thus the restriction $|\mathbf{t}| > k_F$ will be put on the sum in the first of Eqs. (11.50), and they take the form

$$\psi_{\mathbf{k}}(\mathbf{x}) = \phi_{\mathbf{k}}(\mathbf{x}) + \sum_{|\mathbf{t}|>k_F} \phi_{\mathbf{t}}(\mathbf{x})\frac{1}{\kappa^2 - t^2}\int_\Omega d^3y\,\phi^\star_{\mathbf{t}}(\mathbf{y})\lambda v(y)\psi_{\mathbf{k}}(\mathbf{y}) \quad ; \text{ B-G eqn}$$

$$\kappa^2 - k^2 = \int_\Omega d^3x\,\phi^\star_{\mathbf{k}}(\mathbf{x})\lambda v(x)\psi_{\mathbf{k}}(\mathbf{x}) \qquad (11.52)$$

This is known as the *Bethe-Goldstone equation* [Bethe and Goldstone (1957)]. It allows a pair of particles in the Fermi sea, through an interaction between them, to be virtually excited into any of the unfilled states above the Fermi level.

The Bethe-Goldstone equation can be solved numerically as it stands; however, to gain insight, we introduce a particular type of two-body potential that will allow us to obtain an analytic solution. First, generalize the notion of a potential from local to *non-local*, through the replacement

$$\int d^3x\, e^{-i\mathbf{k}\cdot\mathbf{x}}\, v(x)\psi(\mathbf{x}) \rightarrow \int d^3x\, d^3y\; e^{-i\mathbf{k}\cdot\mathbf{x}}\, v(\mathbf{x},\mathbf{y})\psi(\mathbf{y}) \quad ; \text{ non-local} \tag{11.53}$$

One then recovers a local potential with the replacement

$$v(\mathbf{x},\mathbf{y}) \rightarrow v(|\mathbf{x}|)\delta^{(3)}(\mathbf{x}-\mathbf{y}) \qquad ; \text{ local} \tag{11.54}$$

Now introduce the notion of a *separable* potential[18]

$$\begin{aligned} v(\mathbf{x},\mathbf{y}) &= u(|\mathbf{x}|)u(|\mathbf{y}|) \qquad ; \text{ separable} \\ \tilde{u}(\mathbf{k}) &= \tilde{u}(k) \equiv \int d^3x\, e^{-i\mathbf{k}\cdot\mathbf{x}}\, u(x) \end{aligned} \tag{11.55}$$

We note that the electron-electron potential arising from phonon exchange in Eq. (11.42) happens to be both *local and separable*, since the identification

$$v(\mathbf{x},\mathbf{y}) = \delta^{(3)}(\mathbf{x})\delta^{(3)}(\mathbf{y}) \tag{11.56}$$

reproduces results obtained with the local potential $v(x) = \delta^{(3)}(\mathbf{x})$.

Substitution of Eqs. (11.55) into the second of Eqs. (11.52), together with the identification of the wave function in Eq. (11.46), gives

$$\kappa^2 - k^2 = \frac{\lambda}{\sqrt{\Omega}}\tilde{u}(k)\int d^3y\, u(y)\psi_{\mathbf{k}}(\mathbf{y}) \tag{11.57}$$

Now take $\int_\Omega d^3x\ u(x)$ of the first of Eqs. (11.52), and then multiply by $\lambda\tilde{u}(k)/\sqrt{\Omega}$ to obtain

$$\kappa^2 - k^2 = \frac{\lambda}{\Omega}|\tilde{u}(k)|^2 + \frac{\lambda}{\Omega}\sum_{t>k_{\rm F}} \tilde{u}^\star(t)\frac{1}{\kappa^2 - t^2}\tilde{u}(t)(\kappa^2 - k^2) \tag{11.58}$$

[18]Here $u(x)$ is real.

In the large volume limit one can replace $\sum_{\mathbf{t}} \to \Omega(2\pi)^{-3}\int d^3t$, and the above result is then rearranged to give

$$f(\kappa^2) \equiv \frac{1}{\Omega}\frac{|\tilde{u}(k)|^2}{\kappa^2 - k^2} + \int_{k_F}^{\infty}\frac{d^3t}{(2\pi)^3}\frac{|\tilde{u}(t)|^2}{\kappa^2 - t^2} = \frac{1}{\lambda} \qquad ; \text{ eigenvalue eqn} \quad (11.59)$$

This is now an *eigenvalue* equation for κ^2, which gives the energy shift of the interacting pair from the non-interacting value $E_0(k) = \hbar^2k^2/2\mu$ as

$$\Delta E(k) = \frac{\hbar^2}{2\mu}(\kappa^2 - k^2) \qquad ; \text{ energy shift} \quad (11.60)$$

A graphical solution to the eigenvalue equation is sketched in Fig. 11.14.

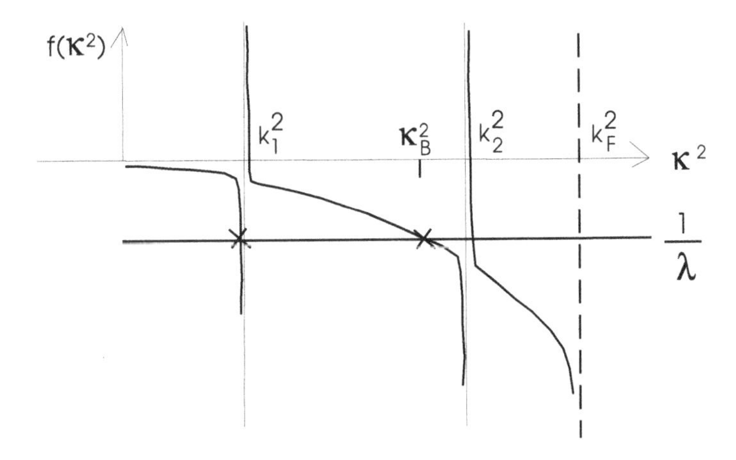

Fig. 11.14 Graphical solution of the eigenvalue equation $f(\kappa^2) = 1/\lambda$ [Eq. (11.59)] for two different values (k_1^2, k_2^2) of the initial k^2, and an attractive interaction with $\lambda < 0$. Both the ordinary eigenvalue in the case $k_1^2 < \kappa_B^2$, and the extraordinary eigenvalue in the case $k_2^2 > \kappa_B^2$, are identified with crosses.

Let us discuss these results:

(1) For a repulsive potential with $\lambda > 0$, the lowest eigenvalue is always given by

$$\kappa^2 - k^2 \approx \frac{\lambda}{\Omega}|\tilde{u}(k)|^2 \qquad ; \lambda > 0 \qquad (11.61)$$

This is the lowest-order energy shift for the pair,[19] and the result is exact as $\lambda \to 0$. It is important to note that this shift is proportional

[19]See Prob. 11.6.

to $1/\Omega$. This factor reflects the uniform relative probability density d^3x/Ω for free particles in a large box of volume Ω, with the wave function in Eq. (11.46). As a consequence of the factor $1/\Omega$, this term only plays a role in $f(\kappa^2)$ for κ^2 very, very near to k^2.

(2) For $\lambda < 0$, the lowest eigenvalue κ^2 is again given by the ordinary solution (the solution for $k^2 = k_1^2$ in Fig. 11.14) until $k^2 > \kappa_B^2$, where κ_B^2 is the *extraordinary* eigenvalue defined by

$$\int_{k_\mathrm{F}}^{\infty} \frac{d^3t}{(2\pi)^3} \frac{|\tilde{u}(t)|^2}{\kappa_B^2 - t^2} = \frac{1}{\lambda} \qquad \text{; extraordinary eigenvalue} \quad (11.62)$$

(3) For all $k^2 > \kappa_B^2$, *the lowest eigenvalue is still given by* κ_B^2! (See the solution for $k^2 = k_2^2$ in Fig. 11.14.)

(4) If $1/|\lambda|$ is such that $\kappa_B^2 \to k_\mathrm{F}^2$, the integrand in Eq. (11.62) becomes singular at the lower limit. A partial integration allows one to identify the singular part of the integral,[20] and retaining just this part, the eigenvalue equation becomes

$$\frac{k_\mathrm{F}}{2\pi^2}|\tilde{u}(k_\mathrm{F})|^2 \frac{1}{2} \ln \frac{k_\mathrm{F}^2}{k_\mathrm{F}^2 - \kappa_B^2} \approx \frac{1}{|\lambda|} \qquad ;\kappa_B^2 \to k_\mathrm{F}^2 \quad (11.63)$$

Write the extraordinary eigenvalue as

$$\kappa_B^2 \equiv k_\mathrm{F}^2 - \frac{2\mu}{\hbar^2}\Delta \quad (11.64)$$

Then the energy shift of the interacting pair for $k_\mathrm{F}^2 > k^2 > \kappa_B^2$ is given by

$$\Delta E(k) = -\Delta + \frac{\hbar^2}{2\mu}(k_\mathrm{F}^2 - k^2) \qquad \text{; energy shift} \quad (11.65)$$

Substitution of Eq. (11.64) into (11.63) gives

$$\ln\left(\frac{\hbar^2 k_\mathrm{F}^2}{2\mu\Delta}\right) = \frac{4\pi^2}{k_\mathrm{F}|\lambda||\tilde{u}(k_\mathrm{F})|^2}$$

$$\Rightarrow \qquad \Delta = \frac{\hbar^2 k_\mathrm{F}^2}{m} \exp\left\{-\frac{4\pi^2}{k_\mathrm{F}|\lambda||\tilde{u}(k_\mathrm{F})|^2}\right\} \quad (11.66)$$

(5) The energy shift of a pair at the Fermi surface is now given by

$$\Delta E(k_\mathrm{F}) = -\Delta \quad (11.67)$$

Where Δ, given by Eq. (11.66), has the following properties:

[20]See Prob. 11.5.

- It is *independent of* Ω;
- It thus represents the binding energy of a *bound-state pair*;
- It goes as $\sim \exp\{1/\lambda \cdots\}$ and thus has an *essential singularity* in the coupling constant λ as $\lambda \to 0^-$. One can never obtain the result in Eq. (11.66) from a perturbation series in powers of λ.

These two-body bound states arising from a weak attractive interaction between particles at the Fermi surface of a Fermi gas were first found by Cooper [Cooper (1956)], and they are known as *Cooper pairs.* With the attractive electron-phonon interaction, all particles within a energy distance of $\sim \hbar\omega_D$ of the Fermi surface in a metal form such pairs. The construction of the many-body wave function for a system with many bound fermion pairs near the Fermi surface clearly presents a formidable problem. The solution to the problem in the case of superconductors was given by Bardeen, Cooper, and Schrieffer, whose successful theory now provides the foundation for our understanding of superconductivity [Bardeen, Cooper, and Schrieffer (1957)].[21]

11.2.4 *Flux Quantization*

Let us use the previous results to make a very simple model of superconductors. The electrical conductivity comes from the electrons near the Fermi surface. Assume that near the Fermi surface one has $N_{\rm C}$ Cooper pairs of electrons, with binding energy Δ, which behave as spin-zero bosons. The bosons have mass $m_b \approx 2m$ and charge $e_b = 2e$. The previous superfluid analysis can now be taken over intact. The bosons condense into the single-particle state $\varphi_0(\mathbf{x})$.[22]

Suppose one has a time-independent one-body Schrödinger equation, probability density, and probability current of the form

$$
\begin{aligned}
\frac{\mathbf{p}^2}{2m}\varphi &= \varepsilon\varphi \\
\rho &= |\varphi|^2 \\
\mathbf{s} &= \frac{1}{2m}\left[\varphi^\star(\mathbf{p}\varphi) + (\mathbf{p}\varphi)^\star\varphi\right]
\end{aligned}
\tag{11.68}
$$

where $\mathbf{p} = (\hbar/i)\boldsymbol{\nabla}$ is the canonical momentum. Now add a magnetic field

[21] The BCS theory is developed in detail in [Fetter and Walecka (2003a)].

[22] If we return to the argument on superfluidity, then in this simple picture the heavy moving object must first break up a pair at the Fermi surface, with C-M momentum $\mathbf{K} = 0$, in order to excite the system; this costs an energy $\varepsilon_k = \Delta$.

$\mathbf{B}(\mathbf{x}) = \nabla \times \mathbf{A}(\mathbf{x})$. The proper way to do this is to make the following gauge-invariant replacement in these expressions[23]

$$\mathbf{p} \rightarrow \mathbf{p} - e\mathbf{A} \tag{11.69}$$

The time-independent one-body Schrödinger equation in a magnetic field, and the accompanying probability density and probability current, then take the form[24]

$$\begin{aligned}
\frac{(\mathbf{p} - e\mathbf{A})^2}{2m}\varphi &= \varepsilon\varphi \\
\rho &= |\varphi|^2 \\
\mathbf{s} &= \frac{1}{2m}\left\{\varphi^\star[(\mathbf{p} - e\mathbf{A})\varphi] + [(\mathbf{p} - e\mathbf{A})\varphi]^\star\varphi\right\}
\end{aligned} \tag{11.70}$$

In the many-body problem, define a new single-particle wave function for the equivalent bosons by

$$\psi_0(\mathbf{x}) \equiv \sqrt{N_{\rm C}}\, \varphi_0(\mathbf{x}) \tag{11.71}$$

The many-body particle density and particle current then follow as in Eqs. (11.7)

$$\begin{aligned}
n(\mathbf{x}) &= |\psi_0(\mathbf{x})|^2 \\
n(\mathbf{x})\mathbf{v}(\mathbf{x}) &= \frac{\hbar}{2im_b}\left[\psi_0^\star \boldsymbol{\nabla}\psi_0 - (\boldsymbol{\nabla}\psi_0)^\star\psi_0\right] - \frac{e_b\mathbf{A}(\mathbf{x})}{m_b}|\psi_0(\mathbf{x})|^2
\end{aligned} \tag{11.72}$$

Now parameterize ψ_0 as in Eq. (11.8)

$$\psi_0(\mathbf{x}) \equiv |\psi_0(\mathbf{x})|e^{i\phi(\mathbf{x})} \tag{11.73}$$

It follows that

$$\begin{aligned}
n(\mathbf{x}) &= |\psi_0(\mathbf{x})|^2 \\
\mathbf{v}(\mathbf{x}) &= \frac{\hbar}{m_b}\boldsymbol{\nabla}\phi(\mathbf{x}) - \frac{e_b}{m_b}\mathbf{A}(\mathbf{x})
\end{aligned} \tag{11.74}$$

Suppose that instead of a vortex in a superfluid in Fig. 11.4(b), one has a flux tube in a type-II superconductor. The electromagnetic current vanishes exponentially as one goes into the bulk superconductor, and hence

[23] See Eq. (9.50) and Prob. 9.4.
[24] See Prob. 11.7.

so must the velocity field. Thus far enough away from the flux tube in Fig. 11.5 one has

$$\oint_C \mathbf{v} \cdot d\mathbf{l} = 0 \qquad ; r \rightarrow \infty \tag{11.75}$$

Hence, from Eq. (11.74)

$$\hbar \oint_C \boldsymbol{\nabla}\phi \cdot d\mathbf{l} = e_b \oint_C \mathbf{A} \cdot d\mathbf{l} \tag{11.76}$$

The term on the l.h.s. is evaluated exactly in Eqs. (11.12)–(11.13), and Stokes' theorem can be used to transform the term on the r.h.s. into the magnetic flux through the surface S bounded by the curve C[25]

$$\oint_C \mathbf{A} \cdot d\mathbf{l} = \int_S \boldsymbol{\nabla} \times \mathbf{A} \cdot d\mathbf{S} = \int_S \mathbf{B} \cdot d\mathbf{S} \tag{11.77}$$

A combination of these results yields

$$\begin{aligned} \int_S \mathbf{B} \cdot d\mathbf{S} = \frac{h}{e_b}\nu \qquad & ; \text{flux quantization} \\ & \nu = 0, 1, 2, \cdots \end{aligned} \tag{11.78}$$

This is *flux quantization.* Notice that $N_{\rm C}$ and m_b have dropped out of this result, and it only depends on fundamental constants. Note also that this argument holds any time one has a bulk superconductor (for example a ring) surrounding a tube of magnetic flux. The unit of quantized flux in SI units is given by [see Eqs. (2.80)]

$$\begin{aligned} \frac{h}{2|e|} &= 2.068 \times 10^{-15}\,\frac{\rm Js}{\rm C} \qquad ; \text{flux quantum} \\ &= 2.068 \times 10^{-15}\,{\rm Tm}^2 \end{aligned} \tag{11.79}$$

When many flux tubes penetrate type-II superconductors, they form regular arrays, and lovely pictures exist that exhibit these arrays (see, for example, [CNST (2007)]).

As one large-scale application of quantum fluids, CEBAF (the Continuous Electron Beam Accelerator Facility), constructed to study the structure of nuclei and hadrons, has accelerating microwave cavities made of superconducting niobium, which are cooled in a bath of superfluid helium (see [TJNAF (2007)]).

---

[25] To avoid confusion, we here denote the surface with S.

Chapter 12

Quantum Fields

We conclude with an introduction to the quantum theory of fields, which underlies most of modern theoretical physics.

12.1 String

To introduce quantum field theory, we return to the starting point of this book, which was the classical theory of the transverse planar motion of a string. Let us go back and compute the total energy of the moving string.

12.1.1 *Energy*

Assume again that the string has a constant mass density ρ and tension τ, and consider the element of string in Fig. 2.9. If the length of that element in the horizontal equilibrium configuration is dx, then the length ds of the stretched element is (see Prob. 10.2)

$$ds = dx\left\{1 + \left[\frac{\partial q(x,t)}{\partial x}\right]^2\right\}^{1/2} \approx dx\left\{1 + \frac{1}{2}\left[\frac{\partial q(x,t)}{\partial x}\right]^2\right\} \tag{12.1}$$

The last expression holds for small amplitude oscillations. The potential energy stored in the element of string comes from the work done by stretching the string against the constant tension τ

$$dW = \tau(ds - dx) \approx \frac{\tau}{2}\left[\frac{\partial q(x,t)}{\partial x}\right]^2 dx \tag{12.2}$$

The potential energy in the string of length L is obtained by summing this over all elements[1]

$$V = \frac{\tau}{2}\int_0^L \left[\frac{\partial q(x,t)}{\partial x}\right]^2 dx \qquad ; \text{ potential energy} \tag{12.3}$$

The kinetic energy is obtained by summing the kinetic energy of each element of string due to its transverse motion

$$T = \frac{\sigma}{2}\int_0^L \left[\frac{\partial q(x,t)}{\partial t}\right]^2 dx \qquad ; \text{ kinetic energy} \tag{12.4}$$

The total energy of the string is the sum these two contributions

$$E = T + V = \frac{\sigma}{2}\int_0^L \left[\frac{\partial q(x,t)}{\partial t}\right]^2 dx + \frac{\tau}{2}\int_0^L \left[\frac{\partial q(x,t)}{\partial x}\right]^2 dx \tag{12.5}$$

With the introduction of the wave velocity $c^2 = \tau/\sigma$, this expression is[2]

$$E = \frac{\sigma}{2}\int_0^L dx \left\{\left[\frac{\partial q(x,t)}{\partial t}\right]^2 + c^2\left[\frac{\partial q(x,t)}{\partial x}\right]^2\right\} \qquad ; \text{ energy}$$

$$c^2 = \tau/\sigma \tag{12.6}$$

12.1.2 *Normal Modes*

We will consider the configuration where the string of length L circles a cylinder and is free to move without friction in the transverse direction (Fig. 12.1 and Prob. 2.4). In this case, one has periodic boundary conditions

$$q(x+L,t) = q(x,t) \qquad ; \text{ p.b.c.} \tag{12.7}$$

The normal-modes are given by

$$q_k(x,t) = \frac{1}{\sqrt{L}}e^{i(kx-\omega_k t)} \qquad ; \text{ normal modes}$$

$$\omega_k = |k|c \qquad k_p = \frac{2\pi p}{L} \quad ; p = 0, \pm 1, \pm 2, \cdots \tag{12.8}$$

The wavenumber k goes over a discrete, infinite set; there is one for each value of the integer p. For ease of notation, we henceforth suppress the subscript p on the wavenumber k.

---

[1]We now write equalities.

[2]For uniformity, we again use the symbol c for the velocity of wave propagation in the string; however, this should not here be confused with the speed of light.

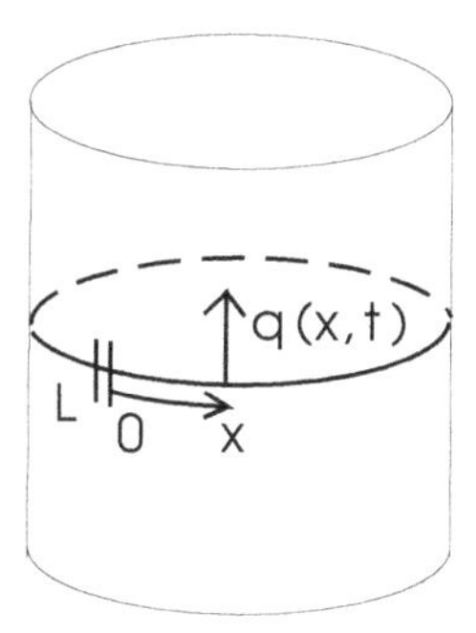

Fig. 12.1 String of length L stretched around a cylinder free to oscillate without friction in a direction parallel to the axis of the cylinder. Illustration of periodic boundary conditions.

The normal-modes satisfy the orthonormality relation

$$\int_0^L dx\, q_k^\star(x,t) q_{k'}(x,t) = \delta_{kk'} \qquad ;\ \text{orthonormal} \qquad (12.9)$$

Let us expand the general solution to the wave equation in terms of the normal-modes

$$q(x,t) = \sum_k \left(\frac{\hbar}{2\omega_k \sigma L}\right)^{1/2} \left[a_k e^{i(kx-\omega_k t)} + a_k^\star e^{-i(kx-\omega_k t)}\right]$$

$$;\ \text{general solution} \qquad (12.10)$$

Here the $\{a_k\}$ are the normal-mode amplitudes; they form an infinite set of complex constants. It is readily verified that the general solution to the wave equation for the string can be written in this form:

- This expression is *real*;
- Since each normal mode satisfies the wave equation, so does the sum over modes

$$\Box\, q(x,t) = 0 \qquad ;\ \text{wave equation} \qquad (12.11)$$

- This solution obeys periodic boundary conditions

$$q(x+L,t) = q(x,t) \qquad ;\ \text{p.b.c.} \qquad (12.12)$$

- It has in it enough flexibility to match the initial conditions (see chap. 2)

$$q(x,0) = \sum_k \left(\frac{\hbar}{2\omega_k \sigma L}\right)^{1/2} \left[a_k e^{ikx} + a_k^\star e^{-ikx}\right] \qquad ;\ \text{initial cond.}$$

$$\left[\frac{\partial q(x,t)}{\partial t}\right]_{t=0} = \frac{1}{i}\sum_k \left(\frac{\hbar\omega_k}{2\sigma L}\right)^{1/2} \left[a_k e^{ikx} - a_k^\star e^{-ikx}\right] \qquad (12.13)$$

Let us compute the energy E contained in the general solution $q(x,t)$ by substituting Eq. (12.10) into Eq. (12.6). The kinetic energy contribution is

$$\begin{aligned} T = \frac{\sigma}{2}\sum_k \sum_{k'} \int_0^L dx\, \frac{\hbar}{2\sigma L(\omega_k \omega_{k'})^{1/2}} \times \\ \left\{-i\omega_k \left[a_k e^{i(kx-\omega_k t)} - a_k^\star e^{-i(kx-\omega_k t)}\right]\right\} \times \\ \left\{-i\omega_{k'} \left[a_{k'} e^{i(k'x-\omega_{k'} t)} - a_{k'}^\star e^{-i(k'-\omega_{k'} t)}\right]\right\} \end{aligned} \qquad (12.14)$$

Use of the orthonormality relation in Eq. (12.9) reduces this expression to[3]

$$T = \sum_k \frac{\hbar\omega_k}{4}\left[a_k a_k^\star + a_k^\star a_k - a_k a_{-k} e^{-2i\omega_k t} - a_k^\star a_{-k}^\star e^{2i\omega_k t}\right] \quad (12.15)$$

The potential energy contribution is evaluated in exactly the same manner

$$\begin{aligned} V = \frac{\sigma c^2}{2}\sum_k \sum_{k'} \int_0^L dx\, \frac{\hbar}{2\sigma L(\omega_k \omega_{k'})^{1/2}} \times \\ \left\{ik \left[a_k e^{i(kx-\omega_k t)} - a_k^\star e^{-i(kx-\omega_k t)}\right]\right\} \times \\ \left\{ik' \left[a_{k'} e^{i(k'x-\omega_{k'} t)} - a_{k'}^\star e^{-i(k'-\omega_{k'} t)}\right]\right\} \end{aligned} \qquad (12.16)$$

This reduces to

$$V = \sum_k \frac{\hbar k^2 c^2}{4\omega_k}\left[a_k a_k^\star + a_k^\star a_k + a_k a_{-k} e^{-2i\omega_k t} + a_k^\star a_{-k}^\star e^{2i\omega_k t}\right] \quad (12.17)$$

With the observation that $k^2c^2 = \omega_k^2$, the total energy becomes

$$E = T + V = \sum_k \hbar\omega_k \frac{1}{2}\left(a_k a_k^\star + a_k^\star a_k\right) \qquad ;\ \text{normal modes} \quad (12.18)$$

[3]It follows from Eq. (12.9) that $\int_0^L dx\, q_k(x,t) q_{k'}(x,t) = \delta_{k',-k} e^{-2i\omega_k t}$; one also has the complex conjugate relation. Note that while (T, V) are individually time-dependent, the total energy is a constant of the motion.

The problem has been reduced to *normal modes*, where one has the following features:

- There is an infinite, discrete set of normal-mode frequencies

$$\omega_k = |k|c = \frac{2\pi c}{L}|p| \qquad ; \; p = 0, \pm 1, \pm 2, \cdots \tag{12.19}$$

- The energy receives an additive, uncoupled, contribution from each normal mode;
- The contribution to the energy from each normal mode is just that of a *simple harmonic oscillator*!

12.1.3 *Quantization*

The quantization of the simple harmonic oscillator is carried out in detail in Prob. 4.17. It is shown there that the hamiltonian of the s.h.o. can be recast in the form

$$H = \hbar\omega\frac{1}{2}(a^\dagger a + aa^\dagger) \tag{12.20}$$

where the operators $(a, a^\dagger)$, which are just linear combinations of (p, q), obey the commutation relation

$$[a, a^\dagger] = 1 \tag{12.21}$$

The number operator is defined by

$$N \equiv a^\dagger a \tag{12.22}$$

It follows that the hamiltonian can be written in terms of the number operator as

$$H = \hbar\omega\left(N + \frac{1}{2}\right) \tag{12.23}$$

It is shown in Prob. 4.17 that the number operator and hamiltonian have the following eigenvalues and eigenstates

$$\begin{aligned} N\psi_n &= n\,\psi_n & &; \; n = 0, 1, 2, \cdots, \infty \\ H\psi_n &= E_n\psi_n = \hbar\omega\left(n + \frac{1}{2}\right)\psi_n \end{aligned} \tag{12.24}$$

This reproduces the familiar Planck spectrum of the s.h.o. (chapter 3), now properly including the zero-point energy.

It is shown in Prob. 4.18 that $(a, a^\dagger)$ are the *destruction* and *creation* operators, whose effects on the eigenstates are given by

$$\begin{aligned} a\,\psi_n &= \sqrt{n}\,\psi_{n-1} \\ a^\dagger\,\psi_n &= \sqrt{n+1}\,\psi_{n+1} \end{aligned} \tag{12.25}$$

The quantization of the string then proceeds through the identification of the classical normal-mode amplitudes $(a_k, a_k^\star)$ with the operators $(a_k, a_k^\dagger)$ for each of the infinite set of uncoupled simple harmonic oscillators. The operators referring to different modes commute. Thus the commutation relations for all of the creation and destruction operators are now

$$\begin{aligned} [a_k, a_{k'}^\dagger] &= \delta_{kk'} \\ [a_k, a_{k'}] &= [a_k^\dagger, a_{k'}^\dagger] = 0 \end{aligned} \tag{12.26}$$

The hamiltonian becomes[4]

$$\hat{H} = \sum_k \hbar\omega_k \left(\hat{N}_k + \frac{1}{2}\right) \tag{12.27}$$

The eigenstates for the entire system are then obtained from the direct product of the states for the individual oscillators

$$\psi_{n_1,n_2,\cdots,n_\infty} = \psi_{n_1}\psi_{n_2}\cdots\psi_{n_\infty} \tag{12.28}$$

Here we have simply ordered all possible values of the wavenumber $k = 2\pi p/L$. The eigenvalues for the whole system are given by

$$\begin{aligned} \hat{H}\psi_{n_1,n_2,\cdots,n_\infty} &= E_{n_1,n_2,\cdots,n_\infty}\,\psi_{n_1,n_2,\cdots,n_\infty} \\ E_{n_1,n_2,\cdots,n_\infty} &= \sum_k \hbar\omega_k \left(n_k + \frac{1}{2}\right) \end{aligned} \tag{12.29}$$

The quantity n_k is the number of quanta in the k*th* mode, and in analogy to the quantization of light, we shall refer to these quanta of the sound waves in a string as *phonons*.

[4]We henceforth use a hat over a quantity to indicate an operator in the many-particle occupation number space (except for the creation and destruction operators themselves, where this is obvious); correspondingly, unit vectors will no longer carry hats.

12.1.4 *The Quantum Field*

Once the normal-mode oscillators are quantized, the general solution to the wave equation for the string $q(x,t)$ in Eq. (12.10) becomes an *operator*

$$\hat{q}(x,t) = \sum_k \left(\frac{\hbar}{2\omega_k \sigma L}\right)^{1/2} \left[a_k e^{i(kx-\omega_k t)} + a_k^\dagger e^{-i(kx-\omega_k t)}\right]$$

; quantum field (12.30)

This is known as a *quantum field.* It has the following properties:

- It is a *local* operator since it is defined at each (x,t). As such, it serves as a basis for constructing local quantities such as the energy density in Eq. (12.6);[5]
- As an operator, it *creates and destroys phonons* since
  - a_k destroys a phonon in the k*th* mode;
  - $a_k^\dagger$ creates a phonon in the k*th* mode;
- As an operator, it satisfies certain *canonical commutation relations.*

To elaborate on the last point, we note that the kinetic momentum $\pi(x,t)dx$ in the little element of string in Fig. 2.9 is given by

$$\pi(x,t)dx = \sigma \frac{\partial q(x,t)}{\partial t} dx \quad ; \text{ momentum}$$

$$\text{or;} \qquad \pi(x,t) = \sigma \frac{\partial q(x,t)}{\partial t} \tag{12.31}$$

The quantity $\pi(x,t)$ is known as the *momentum density.* It follows from Eq. (12.30) that the momentum density field operator is

$$\hat{\pi}(x,t) = \frac{1}{i} \sum_k \left(\frac{\hbar\omega_k \sigma}{2L}\right)^{1/2} \left[a_k e^{i(kx-\omega_k t)} - a_k^\dagger e^{-i(kx-\omega_k t)}\right]$$

; momentum density (12.32)

A simple calculation using Eqs. (12.26) then leads to the following equal-time commutation relations (Prob. 12.9)

$$[\hat{q}(x,t), \hat{\pi}(x',t')]_{t=t'} = \frac{i\hbar}{L} \sum_k e^{ik(x-x')}$$

$$[\hat{q}(x,t), \hat{q}(x',t')]_{t=t'} = [\hat{\pi}(x,t), \hat{\pi}(x',t')]_{t=t'} = 0 \tag{12.33}$$

---

[5] Also the lagrangian density — see Prob. 12.11.

As we have seen many times, in the limit $L \to \infty$ the sum can be converted to an integral with the result

$$\begin{aligned} [\hat{q}(x,t), \hat{\pi}(x',t')]_{t=t'} &= i\hbar\delta(x-x') & ;\ \text{C.C.R.} \\ [\hat{q}(x,t), \hat{q}(x',t')]_{t=t'} &= [\hat{\pi}(x,t), \hat{\pi}(x',t')]_{t=t'} = 0 & (12.34) \end{aligned}$$

Here $\delta(x - x')$ is the Dirac delta function. These are the *canonical commutation relations* for the field. The first is the analogue for a continuous system of the simple particle relation $[q, p] = i\hbar$. The argument can now be carried out in reverse order, and the field expansions in Eqs. (12.30, 12.32) taken as a simple *representation* of these canonical commutation rules.

Quantum field theory provides a framework for describing the creation and destruction of particles. For example, suppose one attaches an additional very weak, massless spring to the string at $x = x_0$ so that there is an additional term in the string hamiltonian of the form

$$\begin{aligned} H' &= \int dx\, H'(x) \\ H'(x) &= \frac{1}{2}\kappa\, q^2(x,t)\delta(x-x_0) \qquad (12.35) \end{aligned}$$

When the field operator $\hat{q}(x,t)$ is inserted in this expression, the perturbation creates and destroys phonons, with a rate that can be calculated using the analysis in appendix I. (See also Prob. 12.1.)

12.2 Electromagnetic Field

The quantization of the electromagnetic field proceeds in direct analogy to the above. We work in a big cubical box of volume $\Omega = L^3$, and apply periodic boundary conditions.

12.2.1 *Normal Modes*

The total energy in the free electromagnetic field in the box is obtained through the sum of the squares of the electric and magnetic fields. In SI units where $e = e_{\text{SI}}$ and $\alpha = e^2/4\pi\varepsilon_0\hbar c$, it is given by[6]

$$E = \frac{1}{2}\int_\Omega d^3x \left(\varepsilon_0 \mathbf{E}^2 + \frac{1}{\mu_0}\mathbf{B}^2\right) \qquad ;\ \text{field energy} \qquad (12.36)$$

[6]This is derived through the construction of the energy-momentum tensor $T_{\mu\nu}$ for the electromagnetic field, which we leave for a future course in E&M.

One has the freedom of choosing a gauge for the electromagnetic potentials (see Prob. 8.9), and here we work in the *Coulomb gauge.* This gauge has the great advantage that there is a one-to-one correspondence of the resulting quanta with physical photons. For free fields, the Coulomb gauge is defined by

$$\boldsymbol{\nabla}\cdot\mathbf{A}=0 \qquad ;\ \Phi=0 \qquad ;\ \text{Coulomb gauge} \tag{12.37}$$

The electric and magnetic fields are then given by (Prob. 8.9)

$$\mathbf{E}=-\frac{\partial\mathbf{A}}{\partial t} \qquad ;\ \mathbf{B}=\boldsymbol{\nabla}\times\mathbf{A} \tag{12.38}$$

With periodic boundary conditions, the normal modes are given by

$$\begin{aligned} q_{\mathbf{k}}(\mathbf{x},t) &= \frac{1}{\sqrt{\Omega}}e^{i(\mathbf{k}\cdot\mathbf{x}-\omega_k t)} \qquad &&;\ \mathbf{k}=\frac{2\pi}{L}(n_x,n_y,n_z) \\ \omega_k &= |\mathbf{k}|c &&;\ n_i=0,\pm1,\pm2,\cdots;\ i=x,y,z \end{aligned} \tag{12.39}$$

Once again, we have an infinite, discrete set of wavenumbers, and the normal modes are orthonormal

$$\int_\Omega d^3x\, q_{\mathbf{k}}^\star(\mathbf{x},t)q_{\mathbf{k}'}(\mathbf{x},t)=\delta_{\mathbf{k}\mathbf{k}'} \tag{12.40}$$

Now introduce a set of orthogonal, transverse unit vectors $\mathbf{e}_{\mathbf{k}s}$ for each $\mathbf{k}$ as shown in Fig. 12.2. They satisfy

$$\begin{aligned} \mathbf{e}_{\mathbf{k}s}\cdot\mathbf{k} &= 0 \qquad ;\ s=1,2 \\ \mathbf{e}_{\mathbf{k}s}\cdot\mathbf{e}_{\mathbf{k}s'} &= \delta_{ss'} \end{aligned} \tag{12.41}$$

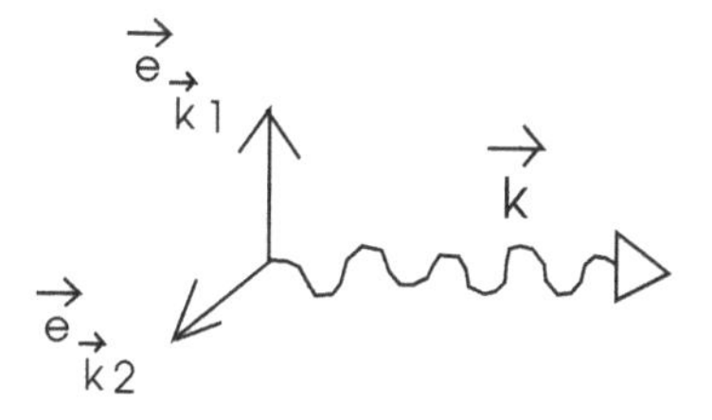

Fig. 12.2 Orthogonal, transverse unit vectors $\mathbf{e}_{\mathbf{k}s}$ for each $\mathbf{k}$.

The vector potential can now be expanded in normal modes as follows

$$\mathbf{A}(\mathbf{x},t) = \sum_{\mathbf{k}}\sum_{s=1}^{2}\left(\frac{\hbar}{2\omega_k\varepsilon_0\Omega}\right)^{1/2}\left[a_{\mathbf{k}s}\mathbf{e}_{\mathbf{k}s}e^{i(\mathbf{k}\cdot\mathbf{x}-\omega_k t)} + a^{\star}_{\mathbf{k}s}\mathbf{e}_{\mathbf{k}s}e^{-i(\mathbf{k}\cdot\mathbf{x}-\omega_k t)}\right] \tag{12.42}$$

This expansion has the following features to recommend it:

- This expression is *real*, giving rise to real $(\mathbf{E},\mathbf{B})$;
- Since only the transverse polarization vectors are used in the expansion, one has ensured that

$$\boldsymbol{\nabla}\cdot\mathbf{A} = 0 \tag{12.43}$$

- $\mathbf{A}(\mathbf{x},t)$ satisfies the wave equation, and, since the order of partial derivatives can always be interchanged, so do the fields $(\mathbf{E},\mathbf{B})$

$$\Box\,\mathbf{A}(\mathbf{x},t) = 0 \tag{12.44}$$

$$\Rightarrow \qquad \Box\,\mathbf{E}(\mathbf{x},t) = \Box\,\mathbf{B}(\mathbf{x},t) = 0 \tag{12.45}$$

- The periodic boundary conditions are obeyed;
- There is enough freedom to match the initial conditions.

The normal-mode expansion in Eq. (12.42) can now be substituted in the expression for the energy in Eq. (12.36), making use of the definition of the fields in Eqs. (12.38). Except for some straightforward manipulations of the polarization vectors, the calculation exactly parallels that in Eqs. (12.14)–(12.18), and the details will be left as an exercise (Prob. 12.2). The result is[7]

$$E = \sum_{\mathbf{k}}\sum_{s=1}^{2}\hbar\omega_k\frac{1}{2}\left(a^{\star}_{\mathbf{k}s}a_{\mathbf{k}s} + a_{\mathbf{k}s}a^{\star}_{\mathbf{k}s}\right) \qquad ;\ \text{normal modes} \tag{12.46}$$

The problem has again been reduced to normal modes. One again has an infinite, uncoupled set of simple harmonic oscillators, one for each value of the wavenumber $\mathbf{k}$ and polarization s.

---

[7]Recall $c^2 = 1/\varepsilon_0\mu_0$.

12.2.2 *Quantization*

The quantization of the electromagnetic field proceeds exactly as in Eqs. (12.20)–(12.29). The energy, which now serves as the hamiltonian for the free E-M field, takes the form

$$\hat{H} = \sum_{\mathbf{k}} \sum_{s=1}^{2} \hbar\omega_k \left(\hat{N}_{\mathbf{k}s} + \frac{1}{2} \right) \qquad ; \text{ photons} \qquad (12.47)$$

The quanta are the familiar *photons*, and all of our results on black-body radiation in chapter 3 are now reproduced. After quantization, the vector potential in Eq. (12.42) becomes a local *quantum field operator* that creates and destroys photons

$$\hat{\mathbf{A}}(\mathbf{x}, t) = \sum_{\mathbf{k}} \sum_{s=1}^{2} \left(\frac{\hbar}{2\omega_k \varepsilon_0 \Omega} \right)^{1/2} \left[a_{\mathbf{k}s} \mathbf{e}_{\mathbf{k}s} e^{i(\mathbf{k}\cdot\mathbf{x} - \omega_k t)} + a_{\mathbf{k}s}^{\dagger} \mathbf{e}_{\mathbf{k}s} e^{-i(\mathbf{k}\cdot\mathbf{x} - \omega_k t)} \right]$$

$$; \text{ quantum field} \qquad (12.48)$$

As an application, consider the interaction of a Dirac particle with the quantized electromagnetic field, where the interaction hamiltonian obtained from Eq. (9.51) is

$$\hat{H}_1 = -ec\boldsymbol{\alpha} \cdot \hat{\mathbf{A}}(\mathbf{x}, t) = -iec\gamma_4 \boldsymbol{\gamma} \cdot \hat{\mathbf{A}}(\mathbf{x}, t) \qquad ; \text{ Dirac particle} \qquad (12.49)$$

Assume this particle is bound in some potential (see, for example, Prob. 9.5). Suppose it makes a transition from a state with energy ε_i and wave function ψ_i to a state with energy ε_f and wave function ψ_f, while emitting a photon with $(\mathbf{k}, s)$ and energy $\hbar\omega_k$ through $\hat{\mathbf{A}}(\mathbf{x}, t)$. The transition rate is calculated from Eqs. (I.35) and (I.32), where the required matrix element is given by

$$\langle \phi_f | H_1 | \phi_i \rangle = -ie \left(\frac{\hbar c^2}{2\omega_k \varepsilon_0 \Omega} \right)^{1/2} \int_{\Omega} d^3x \, \bar{\psi}_f \boldsymbol{\gamma} \psi_i \cdot \mathbf{e}_{\mathbf{k}s} e^{-i\mathbf{k}\cdot\mathbf{x}} \qquad (12.50)$$

The resulting expression for the transition rate is found in Prob. 12.3; it is exact to $O(\alpha)$, where $\alpha = 1/137.0$ is the fine-structure constant.

12.2.3 *Stimulated Emission*

When applied to photons, the second of Eqs. (12.25) states that

$$a_{\mathbf{k}s}^{\dagger} \, \psi_{n_{\mathbf{k}s}} = (n_{\mathbf{k}s} + 1)^{1/2} \, \psi_{n_{\mathbf{k}s}+1} \qquad (12.51)$$

If there are photons already present in the mode $(\mathbf{k}, s)$, then when another photon is emitted into this mode, the transition amplitude is enhanced by a factor $(n_{\mathbf{k}s} + 1)^{1/2}$. This is known as *stimulated emission.* In this manner, with a collection of correctly prepared excited systems, huge numbers of photons can be emitted into a given mode, giving rise to mono-frequency, coherent, classical electromagnetic fields. This is the basis of both *lasers* ("light amplification by stimulated emission of radiation") and *masers* ("microwave amplification by stimulated emission of radiation"), which provide powerful experimental tools with a myriad of applications. Indeed, lasers are now ubiquitous in everyday life.

12.3 Dirac Field

It is evident from Eqs. (12.24) and (12.25) that in the previous discussion there can be any number of quanta of the fields in any given state. These identical particles, either phonons or photons, are *bosons.* When we turn to a Dirac field, we first have to consider the Pauli exclusion principle and ensure that only one *fermion* can occupy any given state.

12.3.1 *Anticommutation Relations*

The Pauli principle for fermions is enforced by imposing *anticommutation* relations, rather than *commutation* relations, on the creation and destruction operators. Consider one mode. Instead of the relation in Eq. (12.21), one quantizes with the conditions[8]

$$\begin{aligned} \{a, a^\dagger\} &\equiv aa^\dagger + a^\dagger a = 1 \\ \{a^\dagger, a^\dagger\} &= \{a, a\} = 0 \end{aligned} \tag{12.52}$$

Note that the second set of relations imply $a^\dagger a^\dagger = aa = 0$. To examine the further consequences of these relations, again define the number operator and its eigenstates by

$$\begin{aligned} N &\equiv a^\dagger a \\ N\psi_n &= n\psi_n \end{aligned} \tag{12.53}$$

[8]The anticommutator of two operators is defined by $\{A, B\} \equiv AB + BA$.

Then, utilizing Eqs. (12.52), one has

$$N^2 = a^\dagger a a^\dagger a = a^\dagger a(1 - aa^\dagger) = a^\dagger a = N$$
$$\text{or;} \quad N(N-1) = 0 \tag{12.54}$$

When acting on the eigenstate in Eqs. (12.53) this relation implies

$$N(N-1)\psi_n = n(n-1)\psi_n = 0$$
$$\text{or;} \quad n(n-1) = 0 \tag{12.55}$$

This implies that the eigenvalues of the number operator are $n = 0$ and $n = 1$, as required. The action of the creation and destruction operators on the corresponding eigenstates states is then given by (see Prob. 12.8)

$$a^\dagger \psi_0 = \psi_1 \qquad ; \; a\psi_1 = \psi_0 \tag{12.56}$$

12.3.2 *Dirac Field*

Once again, we assume a large cubical box with periodic boundary conditions. Then, with the aid of the free-particle Dirac spinors in Probs. 9.6 and 9.7, the Dirac field is defined by

$$\hat{\psi}(\mathbf{x}, t) = \frac{1}{\sqrt{\Omega}} \sum_{\mathbf{k}} \sum_{\lambda} \left[a_{\mathbf{k}\lambda} u_\lambda(\mathbf{k}) e^{i(\mathbf{k}\cdot\mathbf{x} - \omega_k t)} + b_{\mathbf{k}\lambda}^\dagger v_\lambda(-\mathbf{k}) e^{-i(\mathbf{k}\cdot\mathbf{x} - \omega_k t)} \right]$$
$$\hbar\omega_k = \sqrt{(\hbar \mathbf{k} c)^2 + (m_0 c^2)^2} \qquad ; \text{ Dirac field} \tag{12.57}$$

where $\lambda = (\uparrow, \downarrow)$ denotes the helicity. The field $\hat{\psi}(\mathbf{x}, t)$ is a 4-component spinor since (u, v) are; we again suppress the underlining of these column vectors. The anticommutation relations for the particle and antiparticle creation and destruction operators for all modes are defined by

$$\{a_{\mathbf{k}\lambda}, a_{\mathbf{k}'\lambda'}^\dagger\} = \{b_{\mathbf{k}\lambda}, b_{\mathbf{k}'\lambda'}^\dagger\} = \delta_{\mathbf{k}\mathbf{k}'} \delta_{\lambda\lambda'}$$
$$; \text{ all other anticommutators vanish} \tag{12.58}$$

Since expressions for products of operators referring to different modes now depend on the ordering of the operators, one now has to keep careful track of the *signs* of such expressions.

12.3.3 *Some Applications*

The total energy in the free Dirac field, which plays the role of free-field hamiltonian, is obtained by sandwiching the one-body Dirac hamiltonian of Eq. (9.21) between Dirac fields and then integrating over the box

$$\hat{H}_0 = \int_\Omega d^3x\, \hat{\psi}^\dagger \left[c\boldsymbol{\alpha}\cdot\mathbf{p} + \beta m_0 c^2\right] \hat{\psi} \tag{12.59}$$

The Dirac field in Eq. (12.57) can now be substituted in this expression, and the integration performed. The calculation is not difficult, and the result is (Prob. 12.5)

$$\hat{H}_0 = \sum_{\mathbf{k}}\sum_{\lambda} \hbar\omega_k \left(a^\dagger_{\mathbf{k}\lambda}a_{\mathbf{k}\lambda} + b^\dagger_{\mathbf{k}\lambda}b_{\mathbf{k}\lambda} - 1\right) \tag{12.60}$$

The zero-particle state now defines the *vacuum*. With respect to that state, the operator $a^\dagger_{\mathbf{k}\lambda}$ creates a one-particle state with energy increased by $\hbar\omega_k$, while the operator $b^\dagger_{\mathbf{k}\lambda}$ creates an *antiparticle* state with positive energy increase of $\hbar\omega_k$ — this is now hole theory done correctly.

To obtain the interaction of a Dirac field with the electromagnetic field, one can just extend Eq. (12.59) through the use of the full one-body Dirac hamiltonian in Eq. (9.51),

$$\begin{aligned}\hat{H} \equiv \hat{H}_0 + \hat{H}_1 &= \int_\Omega d^3x\, \hat{\psi}^\dagger \left[c\boldsymbol{\alpha}\cdot(\mathbf{p} - e\mathbf{A}) + \beta m_0 c^2 + e\Phi\right] \hat{\psi} \\ \hat{H}_1 &= \int_\Omega d^3x\, \hat{\psi}^\dagger \left[-ec\boldsymbol{\alpha}\cdot\mathbf{A} + e\Phi\right] \hat{\psi}\end{aligned} \tag{12.61}$$

Consider the scattering of a Dirac electron with $(\mathbf{k}, \lambda)$ into a state with $(\mathbf{k}', \lambda')$ through the Coulomb interaction with the nuclear Coulomb field described by the potential

$$\Phi(\mathbf{x}) = \frac{Ze_p}{4\pi\varepsilon_0}\int d^3x'\, \frac{\rho_N(\mathbf{x}')}{|\mathbf{x} - \mathbf{x}'|} \tag{12.62}$$

The operator $a_{\mathbf{k}\lambda}$ in the field $\hat{\psi}$ destroys the initial particle, while the operator $a^\dagger_{\mathbf{k}'\lambda'}$ in $\hat{\psi}^\dagger$ creates the final particle. The result is that the matrix element in Prob. I.1(a) gets replaced by

$$\langle\phi_f|H_1|\phi_i\rangle = \frac{zZe^2}{\varepsilon_0\Omega}\frac{F(\mathbf{q})}{\mathbf{q}^2}u^\dagger_{\lambda'}(\mathbf{k}')u_\lambda(\mathbf{k}) \tag{12.63}$$

The calculation in Prob. I.1 can be repeated for an electron in the ERL, and it is only necessary to replace $mc^2 \to \hbar kc$ to obtain the answer

$$\frac{d\sigma}{d\Omega_k} = z^2 Z^2 \sigma_{\rm M} |F(\mathbf{q})|^2$$

$$\sigma_{\rm M} = \left| \frac{\alpha}{2k \sin^2 \theta/2} \right|^2 |u_{\lambda'}^\dagger(\mathbf{k}') u_\lambda(\mathbf{k})|^2 \tag{12.64}$$

One must sum over the helicity of the final electron if that is not observed, since particles with both helicities get into the detector, and if the initial beam is unpolarized, one must average over the initial helicities to get the cross section. It is demonstrated in Prob. 12.6 that in the ERL

$$\frac{1}{2} \sum_\lambda \sum_{\lambda'} |u_{\lambda'}^\dagger(\mathbf{k}') u_\lambda(\mathbf{k})|^2 = \cos^2 \frac{\theta}{2} \qquad ; \text{ ERL} \tag{12.65}$$

The Rutherford cross section in Prob. I.1(e) then becomes the Mott cross section

$$\sigma_{\rm M} = \frac{\alpha^2 \cos^2 \theta/2}{4k^2 \sin^4 \theta/2} \qquad ; \text{ Mott} \tag{12.66}$$

The rate of photon emission arising from the interaction of a Dirac particle with the quantized radiation field in Eq. (12.61) is calculated in Prob. 12.3.

12.4 Many-Particle Systems

The basis states in Eqs. (12.28) are capable of describing any number of particles (quanta), and quantum field theory provides the most concise framework for formulating the quantum many-body problem. The non-relativistic many-body hamiltonian for a collection of identical particles, each with kinetic energy $T = \mathbf{p}^2/2m = -\hbar^2 \boldsymbol{\nabla}^2/2m$, and interacting through an instantaneous two-body potential of the form $V(\mathbf{x}, \mathbf{y})$, can be written as

$$\hat{H} = \int d^3x\, \hat{\psi}^\dagger(\mathbf{x}) T \hat{\psi}(\mathbf{x}) + \frac{1}{2} \int d^3x \int d^3y\, \hat{\psi}^\dagger(\mathbf{x}) \hat{\psi}^\dagger(\mathbf{y}) V(\mathbf{x}, \mathbf{y}) \hat{\psi}(\mathbf{y}) \hat{\psi}(\mathbf{x}) \tag{12.67}$$

Here the quantum field is defined by

$$\hat{\psi}(\mathbf{x}) \equiv \sum_k a_k \psi_k(\mathbf{x}) \tag{12.68}$$

where the $\psi_k(\mathbf{x})$ form a complete set of solutions to a one-body Schrödinger equation appropriate, as a starting basis, for the problem at hand; we suppress all spin indices. This framework is then applicable to either a collection of identical bosons or fermions, depending on whether commutation relations or anticommutation relations are imposed on the creation and destruction operators $(a_k^\dagger, a_k)$. The fields satisfy the canonical (anti)commutation relations[9]

$$[\hat{\psi}(\mathbf{x}), \hat{\psi}^\dagger(\mathbf{x}')]_\mp = \delta^{(3)}(\mathbf{x} - \mathbf{x}') \tag{12.69}$$

The time evolution of the many-particle system is now governed by the many-body Schrödinger equation[10]

$$i\hbar \frac{\partial \Psi}{\partial t} = \hat{H}\Psi \tag{12.70}$$

The quantum theory of many-particle systems is developed in [Fetter and Walecka (2003a)], where it is proven that these equations provide the point of departure for that work.

[9] Here $[A, B]_\mp \equiv AB \mp BA$.

[10] In the three previous examples, the quantum fields were given the free-field time dependence. Here, in the interacting case, the problem is formulated in the Schrödinger picture, where the operators are time-independent and all the time dependence is in the wave function. This is the approach to quantum mechanics employed throughout this book.

Chapter 13

Problems

2.1 Write, or obtain, a computer program that iterates the finite difference Eqs. (2.6) for a given set of initial conditions. It is always good practice to first convert the problem to dimensionless variables. Examine the following cases:

(a) A one-dimensional simple harmonic oscillator moving in the x-direction with $F = -\kappa x$ and initial conditions $x_0 = a$ and $[dx/dt]_0 = 0$. Use variables $\zeta = x/a$ and $\tau = \sqrt{\kappa/m}\, t$;

(b) A particle performing vertical motion in the upward z-direction in a gravitational field with $F = -mg$ and initial conditions $z_0 = 0$ and $[dz/dt]_0 = v_0$. Use variables $\zeta = zg/v_0^2$ and $\tau = gt/v_0$;

(c) Investigate the limit as the interval $h \to 0$ in each case.

2.2 A particle of mass m and charge e moves along the x-axis subject to an electric field $\mathcal{E}$ which points along that axis. Define $x \equiv q$ and $mdx/dt \equiv p$. Assume that at $t = 0$ the particle starts from the origin $(p, q)_0 = 0$. Calculate and plot the phase orbit $p(q)$ in phase space.

2.3 The following is a trigonometric identity

$$\cos a + \cos b = 2\cos\left(\frac{a+b}{2}\right)\cos\left(\frac{a-b}{2}\right)$$

Consider the superposition of two traveling waves of slightly different wavelengths $\lambda_1 \equiv \lambda$ and $\lambda_2 \equiv \lambda + \Delta\lambda$.

$$q(x,t) = \cos\frac{2\pi}{\lambda_1}(x - ct) + \cos\frac{2\pi}{\lambda_2}(x - ct)$$

Use the above identity to show that the resulting disturbance oscillates with a wavelength

$$\frac{1}{\lambda_{\text{osc}}} = \frac{1}{2}\left(\frac{1}{\lambda_1} + \frac{1}{\lambda_2}\right) \approx \frac{1}{\lambda}$$

Show there is an *amplitude modulation* of the disturbance with wavelength

$$\frac{1}{\lambda_{\text{amp}}} = \frac{1}{2}\left(\frac{1}{\lambda_1} - \frac{1}{\lambda_2}\right) \approx \frac{\Delta\lambda}{2\lambda^2}$$

2.4 A string of length L is stretched around a cylinder and the ends joined. It is free to oscillate without friction in a direction parallel to the axis of the cylinder (Fig. 13.1).

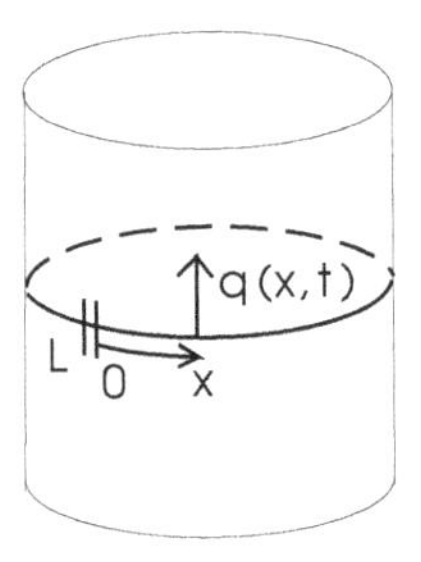

Fig. 13.1 String of length L stretched around a cylinder free to oscillate without friction in a direction parallel to the axis of the cylinder. Illustration of periodic boundary conditions.

(a) Show that the physical situation is described by *periodic boundary conditions*

$$q(x+L) = q(x) \qquad ; \text{ periodic boundary conditions}$$

(b) Show that the eigenfunctions and eigenvalues in this case can be taken as

$$\phi_n(x) = \frac{1}{\sqrt{L}} \exp(ik_n x)$$
$$k_n = \frac{2\pi}{L} n \qquad ; \; n = 0, \pm 1, \pm 2, \cdots$$

(c) Show that these eigenfunctions satisfy the orthonormality relation

$$\int_0^L \phi_n^\star(x)\phi_m(x)\, dx = \delta_{mn}$$

Here $\phi^\star$ is the complex conjugate, and δ_{mn} is the Kronecker delta.

(d) Show that if one had chosen to specify the coordinate interval by $-L/2 \leq x \leq L/2$ rather than by $0 \leq x \leq L$, these eigenfunctions continue to satisfy the orthonormality relation

$$\int_{-L/2}^{L/2} \phi_n^\star(x)\phi_m(x)\,dx = \delta_{mn}$$

2.5 A uniform string under tension τ with fixed endpoints is plucked in the middle giving rise to an initial displacement that can be approximated by (here $h \ll L$)

$$\begin{aligned} q(x,0) &= \frac{2hx}{L} && ;\ 0 \leq x \leq \frac{1}{2}L \\ &= \frac{2h(L-x)}{L} && ;\ \frac{1}{2}L \leq x \leq L \end{aligned}$$

It is released from rest. Determine the Fourier coefficients in Eqs. (2.49).

2.6 Consider other boundary conditions for the string in Fig. 2.9. Suppose the end of the string at $x = L$ is attached to a mass m_0 and also to a massless restoring spring with force constant κ. Assume that it is constrained to move without friction along a vertical wire while under a tension τ.

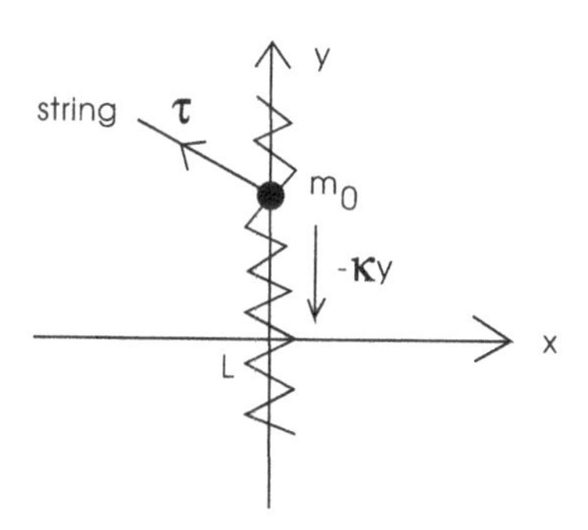

Fig. 13.2 End of string at $x = L$ attached to a mass m_0 and massless spring with force constant κ, constrained to move without friction along a vertical wire while under a tension τ.

(a) Write Newton's second law for the mass m_0. Assume small displacements, and show

$$m_0\frac{d^2y}{dt^2} = -\kappa y - \tau\frac{dy}{dx}$$

(b) Look for a normal mode, with angular frequency ω, of the whole system of mass m_0 and string. Show

$$\left[\frac{1}{y}\frac{dy}{dx}\right]_{x=L} = \frac{m_0\omega^2 - \kappa}{\tau} \qquad ; \text{ homogeneous B.C.}$$

$$\to \frac{-\kappa}{\tau} \qquad ; \; m_0 \to 0$$

The second line follows in the limit that the mass m_0 becomes vanishingly small. This is known as a *homogeneous boundary condition* on the end of the string.

(c) Suppose one goes to the limit $(m_0, \kappa) \to 0$ at finite τ. Show the boundary condition becomes

$$y'|_{x=L} = 0 \qquad ; \text{ free end}$$

Here $y' = dy/dx$. This is known as a *free end.*

2.7 Suppose one has a string with a fixed end at $x = 0$, while the end at $x = L$ is free [Prob. 2.6(c)].

(a) Show the eigenfunctions and eigenvalues are

$$q_n(x) = \left(\frac{2}{L}\right)^{1/2} \sin\frac{(2n+1)\pi x}{2L} \qquad ; \; n = 0, 1, 2, \cdots \infty$$

$$\omega_n = \frac{(2n+1)\pi c}{2L}$$

Sketch the first few eigenfunctions.

(b) Show the eigenfunctions are again orthonormal.

2.8 The hamiltonian for a classical, one-dimensional simple harmonic oscillator is given by Eq. (2.12). According to classical statistics, the differential probability in phase space at a temperature T is

$$dP(p,q) = \frac{e^{-H(p,q)/k_{\rm B}T}\, dpdq}{\int_{-\infty}^{\infty} dq \int_{-\infty}^{\infty} dp\, e^{-H(p,q)/k_{\rm B}T}}$$

(a) Show by explicit evaluation of the integrals that the mean value of the energy is

$$\langle E\rangle = \langle H\rangle = k_{\rm B}T/2 + k_{\rm B}T/2 = k_{\rm B}T$$

(b) Relate this result to the classical equipartition theorem.

2.9 (a) Use Gauss' theorem in Eq. (F.7) to derive Gauss' law from the first of Maxwell's Eqs. (2.90)

$$\oint_S d\mathbf{S}\cdot\mathbf{E} = \frac{1}{\varepsilon_0}Q \qquad ;\ \text{Gauss' law}$$

Here Q is the total charge contained in the volume V.

(b) What is the corresponding physical content of the second of Maxwell's equation?

2.10 Use Stokes' theorem in Eq. (F.10) to derive Faraday's law of induction from the third of Maxwell's Eqs. (2.90)

$$\Delta\mathcal{V} = \oint_C \mathbf{E}\cdot d\mathbf{l} = -\frac{d}{dt}\Phi_{\text{mag}} \qquad ;\ \text{Faraday's law}$$

Here $\Delta\mathcal{V}$ is the change in voltage around the closed loop C, and Φ_{mag} is the total magnetic flux through that loop.

2.11 (a) Neglect the last term in the last of Maxwell's Eqs. (2.90) (his "displacement current"), and use Stokes' theorem in Eq. (F.10) to derive Ampere's law

$$\oint_C \mathbf{B}\cdot d\mathbf{l} = \mu_0 i \qquad ;\ \text{Ampere's law}$$

Here i is the total current flowing through the area bounded by the curve C.

(b) Under what conditions can that second term be neglected?

2.12 Show that the plane wave radiation field in Eqs. (2.94) provides a solution to Maxwell's Eqs. (2.90) in free space.

2.13 Show that the cavity standing wave in Eqs. (2.96) provides a solution to Maxwell's Eqs. (2.90) in free space.

2.14 A collection of particles with masses m_j located at $\mathbf{r}_j$ interact through forces $\mathbf{F}_{kj}(\mathbf{r}_j - \mathbf{r}_k)$ and obey Newton's laws in an inertial frame. Consider the transformation to a new frame moving with constant velocity $\mathbf{V}$ relative to the first so that $\mathbf{r}'_j = \mathbf{r}_j - \mathbf{V}t$. Show that Newton's laws continue hold in the new frame, and hence conclude that it is again inertial.

3.1 N molecules of a diatomic gas A-B are in contact with a heat bath at temperature T. In addition to the motion of the molecules themselves in the gas, they are free to perform internal one-dimensional simple harmonic motion of frequency ν_{B} along the line joining them.

(a) What is the additional mean vibrational energy of the molecules according to the classical equipartition theorem?

(b) What is the additional vibrational contribution to the specific heat of the gas?

(c) What is the additional mean vibrational energy of the molecules in the Einstein model?

(d) What is the additional vibrational contribution to the specific heat in the Einstein model?

3.2 (a) Derive the Debye expression for the molar heat capacity

$$C_v = 9N_{\rm A}k_{\rm B}\left(\frac{T}{\theta_{\rm D}}\right)^3 \int_0^{\theta_{\rm D}/T} \frac{u^4\, e^u\, du}{(e^u-1)^2} \qquad ; \text{ Debye theory}$$

[*Hint*: start from Eq. (3.39).]

(b) What are the high-temperature and low-temperature limits of this result?

3.3 A Debye temperature of $\theta_{\rm D} = 1890\,^{\circ}{\rm K}$ is found to give an excellent fit to the molar heat capacity of diamond (Fig. 3.8). Diamond is pure carbon with an atomic weight of 12 (so that 12 gm = 1 mole), and the measured mass density of diamond is approximately $3.25\,{\rm gm/cm}^3$. Use Debye's theory to deduce the speed of sound c_s in diamond in m/s.

3.4 The velocities of the transverse and longitudinal waves in a solid are actually distinct. In the counting of normal modes, one should therefore replace $3/c_s^3 \to 2/c_t^3 + 1/c_l^3$. A single $\nu_{\rm max}$ is still employed.

(a) What is the corresponding modification of Eq. (3.41) and Fig. 3.8?

(b) What is now the content of the second of Eqs. (3.38)?

3.5 An energy spectrum $u(\nu, T)$, known to be that of a black body, is plotted against ν and found to have its maximum at $\nu_{\rm B}$. Determine the corresponding temperature of the distribution. [*Hint*: see Fig. 3.6.]

3.6 Classically, when light is scattered from a free electron, the electron is first accelerated according to [compare Eq. (3.43)]

$$\frac{d^2y}{dt^2} = \frac{e\mathcal{E}_0}{m}\cos{(\omega t)}$$

The accelerated electron then radiates. The result is the Thomson cross section for the scattering of light by an electron

$$\frac{d\sigma}{d\Omega} = r_0^2 \frac{1}{2}\left(1 + \cos^2\theta\right) \qquad ; \text{ Thomson cross section}$$

$$r_0 = \frac{e^2}{4\pi\varepsilon_0}\frac{1}{m_e c^2} = 2.819 \times 10^{-15}\,\text{m} \qquad ; \text{ classical electron radius}$$

Discuss, in qualitative terms, the dependence of the frequency shift of the scattered light on the incident light *intensity*, which one might expect in this picture. Here, there is a radiation pressure on the electron and it is recoiling as it re-radiates. Compare with the corresponding paradox in the classical treatment of the photoelectric effect.

3.7 Consider the Compton effect for back-scattered light. What is the fractional shift in wavelength for light of incident wavelength $10^{-10}\,$m? Of 4000Å?

3.8 The Balmer formula (1885) is an empirical relation for the frequencies of light observed in atomic transitions. For the Lyman-α series in hydrogen, it can be re-written in terms of a photon energy as

$$\varepsilon_n = -13.61\,\text{eV}\left(1 - \frac{1}{n^2}\right) \qquad ; \text{ Balmer formula}$$

$$n = 2, 3, \cdots$$

(a) Calculate the corresponding frequencies (in s^{-1}), and wavelengths (in m and Å) of the radiation from the $n = 2$ and $n = 3$ states;

(b) Explain how the Bohr model and Bohr hypotheses yield this result.

3.9 A completely isolated hydrogen atom in space can have arbitrarily large bound orbits. Assume the Bohr model with $l = n\hbar$.

(a) How large must n be to get an orbit of radius $r = 0.529\,$m?

(b) What is the binding energy of the orbit in part (a)?[1]

3.10 The muon is a particle just like the electron except that it has a mass $m_\mu = 206.8\, m_e$. What is the ratio of the radius of the first Bohr orbit of a muon relative to that of an electron for a nucleus with charge Ze_p?

3.11 An alternate method of counting the number of normal modes with fixed boundaries is as follows: The wave number vector is defined as $\mathbf{k} = (\pi/L)\mathbf{n}$ with $\mathbf{n} = (n_x, n_y, n_z)$. One can uniquely assign a *unit cube in n-space* to each triplet of integers (n_x, n_y, n_z) by using, say, the point in the most positive direction in each cube. If $L \to \infty$, then for a given

[1]Such huge atoms are indeed created and studied with modern laser spectroscopy.

wavenumber volume d^3k around $\mathbf{k}$, one is counting a very large number of normal modes, or equivalently, one is computing a very large volume in n-space. Work in this limit.

(a) Show that the total volume of the first octant in n-space out to a given $n = |\mathbf{n}|$ is[2]

$$\mathcal{N} = g\,\frac{1}{8}\left(\frac{4\pi n^3}{3}\right)$$

(b) Rewrite this expression in terms of $k = |\mathbf{k}|$, and then differentiate it with respect to k to obtain

$$\frac{d\mathcal{N}}{V} = \frac{g}{(2\pi)^3}4\pi k^2\,dk$$

Compare with Eq. (3.16).

3.12 There exists a cosmic microwave background (CMB) that fills all of space. It remains from the early hot era of the universe after the radiation decoupled and adiabatically expanded with time. The CMB exhibits an almost perfect *black-body spectrum* with an energy density

$$d\varepsilon_\gamma = \frac{8\pi\nu^2\,d\nu}{c^3}\frac{h\nu}{e^{h\nu/k_{\rm B}T}-1}$$

and a current temperature $T(t_p) = 2.73\,^{\circ}\mathrm{K}$ [Ohanian (1995)]. Show that when integrated over frequencies, this gives rise to a radiation density of

$$\varepsilon_\gamma = \frac{4\sigma}{c}T^4 \qquad ; \; \sigma = \frac{\pi^2 k_{\rm B}^4}{60\hbar^3 c^2}$$

Here σ is the Stefan-Boltzmann constant, with the value in cgs units of $\sigma = 5.670 \times 10^{-5}\,\mathrm{erg/sec\text{-}cm^2\text{-}^{\circ}K^4}$.

4.1 A particle moves in a real potential $V(x)$. Show that the continuity equation and probability flux are unchanged from those of a free particle in Eq. (4.46).

4.2 This problem demonstrates that there is enough flexibility in the Fourier transform to match an arbitrary set of initial conditions in the clas-

[2]Here we include the degeneracy factor g.

sical string problem. Consider a superposition of the solutions in Eq. (4.14)

$$\begin{aligned} q(x,t) &= \text{Re} \int_{-\infty}^{\infty} dk\, a(k) \exp\{i(kx - \omega_k t)\} \\ &= \frac{1}{2}\int_{-\infty}^{\infty} dk\, \left[a(k) \exp\{i(kx - \omega_k t)\} + a^\star(k) \exp\{-i(kx - \omega_k t)\}\right] \end{aligned}$$

Here we take the frequency $\omega_k = |k|c$ to be positive, and this real expression evidently satisfies the wave equation.[3]

(a) Show that this function and its time derivative at the initial time $t = 0$ are given by

$$\begin{aligned} q(x,0) &= \frac{1}{2}\int_{-\infty}^{\infty} dk\, \left[a(k) + a^\star(-k)\right] e^{ikx} \\ \dot{q}(x,0) \equiv \left.\frac{\partial q(x,t)}{\partial t}\right|_{t=0} &= \frac{1}{2}\int_{-\infty}^{\infty} dk\, (-i\omega_k)\left[a(k) - a^\star(-k)\right] e^{ikx} \end{aligned}$$

(b) Invert these Fourier transforms, and solve for $a(k)$ to obtain

$$a(k) = \frac{1}{2\pi}\int_{-\infty}^{\infty} dx\, \left[q(x,0) + \frac{i}{\omega_k}\dot{q}(x,0)\right] e^{-ikx}$$

This result uniquely determines the complex Fourier amplitude $a(k)$ in terms of the initial conditions of the string.

4.3 (a) Substitute the second of Eqs. (4.41) into the first to obtain the following relation[4]

$$q(x) = \int_{-\infty}^{\infty} dy \left[\frac{1}{2\pi}\int_{-\infty}^{\infty} dk\, e^{ik(x-y)}\right] q(y)$$

(b) The r.h.s. must reproduce $q(x)$ at the point x no matter what $q(y)$ does at every other point $y \neq x$ in the integral. Hence, conclude that the "kernel" in this relation must have the form

$$\frac{1}{2\pi}\int_{-\infty}^{\infty} dk\, e^{ik(x-y)} = \delta(x-y) \qquad \text{; Dirac delta function}$$

---

[3]Note that the solution now contains waves propagating in both directions.
[4]It is assumed here that $q(x)$ is a well-behaved function.

where the *Dirac delta function* is defined to have the following properties:

$$\begin{aligned} \delta(x-y) &= 0 \qquad & ; \; x \neq y \\ &\neq 0 & x = y \end{aligned}$$

$$\int_{-\infty}^{\infty} dy\, q(y)\delta(x-y) = q(x)$$

This is an expression of the *completeness* of the Fourier transform.[5]

(c) Use an analogous argument to show

$$\frac{1}{2\pi}\int_{-\infty}^{\infty} dx\, e^{i(k-k')x} = \delta(k-k')$$

(d) Use Eqs. (4.41) and the result in part (c) to prove *Parseval's theorem*

$$\int_{-\infty}^{\infty} dx\, |q(x)|^2 = \int_{-\infty}^{\infty} dk\, |a(k)|^2$$

4.4 This problem concerns *partial integration*. Start from the identity

$$\frac{d(uv)}{dx} = u\frac{dv}{dx} + \frac{du}{dx}v$$

Now integrate this expression on x between the boundaries $[a,b]$ appropriate to the problem at hand.

(a) Explain how the boundary contribution $[uv]_a^b$ *vanishes* under the following boundary conditions: localized disturbance, fixed endpoints, periodic boundary conditions;

(b) Show that under the conditions in (a), the derivative can be transferred to the other term in the integrand, with a minus sign

$$\int_a^b dx \left(u\frac{dv}{dx}\right) = -\int_a^b dx \left(\frac{du}{dx}v\right)$$

4.5 In quantum mechanics, classical quantities become *operators*. For

---

[5] Two additional useful properties of the Dirac delta function are

$$\begin{aligned} \delta(-x) &= \delta(x) \\ \delta(ax) &= \frac{1}{|a|}\delta(x) \qquad & ; \; a \text{ real} \end{aligned}$$

example,

$$\begin{aligned}
p &\rightarrow \frac{\hbar}{i}\frac{\partial}{\partial x} \\
T(p) = \frac{p^2}{2m} &\rightarrow -\frac{\hbar^2}{2m}\frac{\partial^2}{\partial x^2} \\
H(p,x) = T(p) + V(x) &\rightarrow -\frac{\hbar^2}{2m}\frac{\partial^2}{\partial x^2} + V(x) \qquad ; \text{ etc.}
\end{aligned}$$

Given a wave function $\Psi(x,t)$, the *expectation value* of an operator $O(p,x)$ is defined as

$$\langle O\rangle \equiv \frac{\int_a^b dx\, \Psi^\star(x,t)\, O(p,x)\, \Psi(x,t)}{\int_a^b dx\, \Psi^\star(x,t)\Psi(x,t)} \qquad ; \text{ expectation value}$$

An operator is said to be *hermitian* if it has the following property[6]

$$\int_a^b dx\, \Psi^\star(x,t)\, O(p,x)\, \Psi(x,t) = \int_a^b dx\, [O(p,x)\, \Psi(x,t)]^\star\, \Psi(x,t) \qquad ; \text{ hermitian}$$

The boundaries (a,b) on x are, again, appropriate to the problem at hand.

Show that under the conditions in Prob. 4.4, the momentum p, kinetic energy $T(p)$, and hamiltonian $H(p,x) = T(p) + V(x)$ with a real $V(x)$, are *hermitian operators.*

4.6 (a) A stationary state is an *eigenstate* of the hamiltonian with eigenvalue E [see Eq. (4.69)]

$$H(p,x)\,\psi(x) = E\,\psi(x)$$

Show that if $H(p,x)$ is hermitian (Prob. 4.5), then the eigenvalue E is *real*;

(b) Show that the expectation value of *any* hermitian operator is a real quantity.

4.7 Operators can be characterized by their *commutation relations.* The commutator of two operators A and B is defined by

$$[A,B] \equiv AB - BA \qquad ; \text{ commutator}$$

[6] More generally, this relation must hold for any acceptable initial and final wave functions in the integrand.

where this expression has meaning when applied to a wave function

$$[A, B]\Psi = (AB - BA)\Psi$$

One does not necessarily get the same result when these operators are applied in opposite order.

Show that the canonical commutation relation between the coordinate x and the momentum $p = (\hbar/i)\partial/\partial x$ is

$$[p, x] = \frac{\hbar}{i} \qquad ; \text{ canonical commutation relation}$$

4.8 This problem generalizes the discussion of *momentum space*. Introduce the following Fourier transform relations at a given instant in time

$$\Psi(x,t) = \frac{1}{\sqrt{2\pi}}\int_{-\infty}^{\infty} dk\, A(k,t)\, e^{ikx} \qquad ; \text{ given } t$$
$$A(k,t) = \frac{1}{\sqrt{2\pi}}\int_{-\infty}^{\infty} dx\, \Psi(x,t)\, e^{-ikx}$$

Furthermore, extend the assumption that appears above Eq. (4.55) to read

The quantity $|A(k,t)|^2\, dk$ is the probability of finding the particle with momentum between $\hbar k$ and $\hbar k + d(\hbar k)$.

(a) Show that for a free particle

$$A(k,t) = a(k)\, e^{-i\omega_k t} \qquad ; \text{ free particle}$$

Note that in the presence of a potential this relation is no longer valid.

(b) Use the completeness relations in Prob. 4.3 to prove the following:

$$\int_{-\infty}^{\infty} |A(k,t)|^2\, dk = \int_{-\infty}^{\infty} |\Psi(x,t)|^2\, dx$$
$$\int_{-\infty}^{\infty} A^\star(k,t)\, \hbar k\, A(k,t)\, dk = \int_{-\infty}^{\infty} \Psi^\star(x,t) \frac{\hbar}{i}\frac{\partial\Psi(x,t)}{\partial x}\, dx$$
$$\int_{-\infty}^{\infty} A^\star(k,t)\, i\frac{\partial A(k,t)}{\partial k}\, dk = \int_{-\infty}^{\infty} \Psi^\star(x,t)\, x\, \Psi(x,t)\, dx$$

The first relation is the generalization of Parseval's theorem.

(c) Use the results in (b) to make the following identification of the operators (p, q) in coordinate space and momentum space

$$(p,q) \to \left(\frac{\hbar}{i}\frac{\partial}{\partial x}, x\right) \qquad ; \text{ coordinate space}$$
$$\to \left(\hbar k, i\frac{\partial}{\partial k}\right) \qquad ; \text{ momentum space}$$

(e) Show that the canonical commutation relation of Prob. 4.7 holds in either case

$$[p,q] = \frac{\hbar}{i} \qquad ; \text{ canonical commutation relation}$$

Thus, coordinate space and momentum space simply provide different *representations* of these commutation relations.

4.9 Start from the definition of the expectation value in Prob. 4.5.

(a) Take the time derivative, and use the Schrödinger equation to establish *Ehrenfest's theorem* [recall Eq. (4.48)]

$$\frac{d}{dt}\langle O\rangle = \left\langle \frac{i}{\hbar}[H,O]\right\rangle \qquad ; \text{ Ehrenfest's theorem}$$

(b) Use the canonical commutation relation in Prob. 4.7 to show

$$\frac{i}{\hbar}[H,x] = \frac{p}{m} \qquad ; \quad \frac{i}{\hbar}[H,p] = -\frac{\partial}{\partial x}V(x)$$

where $H = T(p) + V(x)$.

(c) Conclude from the above results that a well-localized wave packet obeys Newton's second law, thus satisfying the *correspondence principle.*

4.10[7] (a) Prove Schwarz's inequality, which for two complex functions (f, g) states that[8]

$$\int |g(x)|^2 dx \cdot \int |f(x)|^2 dx \geq \left|\int g^\star(x) f(x)\, dx\right|^2$$

[7]Problems 4.10 and 4.11 are more difficult, but they are also very instructive.

[8]*Hints*: Start from

$$\int \left| f(x) - g(x)\frac{\int g^\star(y) f(y)\, dy}{\int |g(y)|^2 dy}\right|^2 dx \geq 0$$

Also, in (c) use

$$|a|^2 + |b|^2 \equiv \frac{1}{2}\left[|a-b|^2 + |a+b|^2\right] \geq \frac{1}{2}|a-b|^2$$

(b) Define the mean-square deviations of (x, p) for a localized wave packet as (see Prob. 4.5)

$$(\Delta x)^2 \equiv \langle (x - \langle x \rangle)^2 \rangle$$
$$(\Delta p)^2 \equiv \langle (p - \langle p \rangle)^2 \rangle$$

Use Schwarz's inequality, the hermiticity of (x, p), and the reality of $(\langle x \rangle, \langle p \rangle)$, to prove

$$(\Delta x)^2 (\Delta p)^2 \geq \frac{1}{4} \left| \langle [p, x] \rangle \right|^2$$

(c) Now use the canonical commutation relation in Prob. 4.7 to establish the rigorous form of the uncertainty principle for a localized wave packet

$$\Delta x \cdot \Delta p \geq \frac{\hbar}{2} \qquad ; \text{ uncertainty principle}$$

4.11 In contrast to the classical string problem where a small-amplitude pulse propagates without change in shape, a quantum wave packet satisfying the Schrödinger equation *spreads* with time. To illustrate this, consider a specific form of the Fourier amplitude in Eq. (4.20)[9]

$$a(k) = \exp\{-\alpha^2 (k - k_0)^2\} \qquad ; \text{ minimal packet}$$
$$\Psi(x, t) = \int_{-\infty}^{\infty} dk\, a(k) \exp\{i (kx - \omega_k t)\}$$

(a) Complete the square in the exponent in the Fourier transform, and show that[10]

$$\Psi(x, 0) = \frac{\sqrt{\pi}}{\alpha} \exp\left\{ik_0 x - \frac{x^2}{4\alpha^2}\right\}$$
$$|\Psi(x, 0)|^2 = \frac{\pi}{\alpha^2} \exp\left\{-\frac{x^2}{2\alpha^2}\right\}$$

[9]Although we leave the demonstration of the fact to the dedicated reader, at $t = 0$ this *minimal packet* actually achieves the lower bound in the uncertainty relation in Prob. 4.10(c); here α is real.

[10]Make use of the following definite integral

$$\int_{-\infty + iy}^{\infty + iy} dz\, e^{-\alpha^2 z^2} = \int_{-\infty}^{\infty} dx\, e^{-\alpha^2 x^2} = \frac{\sqrt{\pi}}{\alpha}$$

You will learn later in your career how to evaluate such an integral, as well as that for complex α required in part (b), using complex variable techniques (see [Fetter and Walecka (2003)]). For now, just accept the definite integrals, or go on to Prob. 4.12.

(b) Now repeat the calculation for finite time in $\Psi(x,t)$, and show

$$|\Psi(x,t)|^2 = \frac{\pi}{\alpha\alpha_t}\exp\left\{-\frac{(x-v_{\rm gp}t)^2}{2\alpha_t^2}\right\}$$

$$v_{\rm gp} = \frac{\hbar k_0}{m} = \frac{p_0}{m} \qquad ; \; \alpha_t^2 = \alpha^2 + \frac{\hbar^2 t^2}{4m^2\alpha^2}$$

4.12 (a) Sketch the behavior of the probability density for the packet in Prob. 4.11(b) as a function of time. Notice that while it indeed moves with the group velocity, the additional terms in the second of Eqs. (4.24) lead to a broadening of the packet as it moves along.

(b) Take α_t^2 as a measure of the width of the packet in coordinate space

$$(\Delta x)^2 \sim \alpha_t^2 \sim \alpha^2 + \frac{\hbar^2 t^2}{4m^2\alpha^2}$$

Minimize this expression with respect to α^2 and show that the best one can do to localize the particle at the time t, if it starts off as a free particle at $t = 0$, is

$$(\Delta x)^2_{\rm min} = \frac{\hbar t}{m}$$

Evaluate the quantity $\hbar/m$ for the following masses: $1\,{\rm gm}, m_p, m_e$. Discuss.

4.13 Make a good plot, to scale, of the energy levels, wave functions, and probability densities (absolute squares of the wave functions) for the first four energy levels of a particle in a one-dimensional box of length L.

4.14 Consider the following wave function for a particle in a one-dimensional box of length L

$$\Psi(x,t) = a_1\sqrt{\frac{2}{L}}\,e^{-i\omega_1 t}\sin\frac{\pi}{L} + a_2\sqrt{\frac{2}{L}}\,e^{-i\omega_2 t}\sin\frac{2\pi}{L}$$

$$a_1 = a_2 = \frac{1}{\sqrt{2}} \qquad ; \; \hbar\omega_n = \frac{\hbar^2\pi^2}{2mL^2}\,n^2$$

(a) Use the general principle of linear superposition to demonstrate that this is a solution to the time-dependent Schrödinger equation;

(b) Plot the probability density at the initial time $t = 0$;

(c) Calculate and plot the probability density at the subsequent times $(\omega_2 - \omega_1)t = m\pi/2$ for $m = 1, 2, 3, 4$.

4.15 Consider a potential which vanishes for $x < 0$ and is $V = -V_0$ for $x \geq 0$, as illustrated in Fig. 13.3.

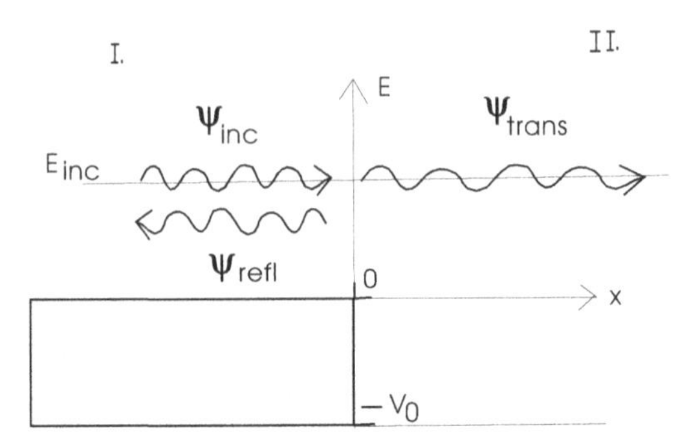

Fig. 13.3 Scattering state in a potential which vanishes for $x < 0$ and is $V = -V_0$ for $x \geq 0$.

A particle of mass m is prepared in an incident plane wave state, with momentum $\hbar k$ and energy $E_{\rm inc} \equiv E = \hbar^2 k^2/2m$, far to the left of the potential drop. In addition to the incident wave there will be both reflected and transmitted waves

$$
\begin{aligned}
\psi_{\rm inc} &= e^{ikx} && ;\ k^2 = \frac{2mE}{\hbar^2} \\
\psi_{\rm refl} &= r\, e^{-ikx} && \\
\psi_{\rm trans} &= t\, e^{i\kappa x} && ;\ \kappa^2 = \frac{2m}{\hbar^2}(E + V_0)
\end{aligned}
$$

(a) Write the boundary conditions at $x = 0$ in terms of (r, t, k, κ);
(b) Determine r and t;
(c) Compute the transmission and reflection coefficients of Eqs. (4.84);
(d) Show $\mathcal{T} + \mathcal{R} = 1$, and interpret this result;
(e) Show that as $V_0 \to \infty$ at fixed E, then $\mathcal{T} \to 0$. Hence conclude that whereas a particle will always fall into a potential well in classical mechanics, it will never fall into the well in quantum mechanics if only the well is deep enough![11]

4.16 A particle of mass m is prepared in a scattering state and sent against a potential barrier of height $V_0 > E_{\rm inc}$ as discussed in the text and illustrated in Fig. 4.18, only now the potential is of a finite thickness d (see Fig. 13.4). Now instead of just two distinct regions, there are three, and

[11] This effect is actually observed in the scattering of low-energy neutrons from the deep, attractive nuclear potential.

the scattering wave function has the form

$$
\begin{aligned}
\psi_I &= e^{ikx} + r\,e^{-ikx} &&; x < 0 &&; k = \left(2mE/\hbar^2\right)^{1/2} \\
\psi_{II} &= a\,e^{-\kappa x} + b\,e^{\kappa x} &&; 0 < x < d &&; \kappa = \left[2m(V_0 - E)/\hbar^2\right]^{1/2} \\
\psi_{III} &= t\,e^{ikx} &&; d < x
\end{aligned}
$$

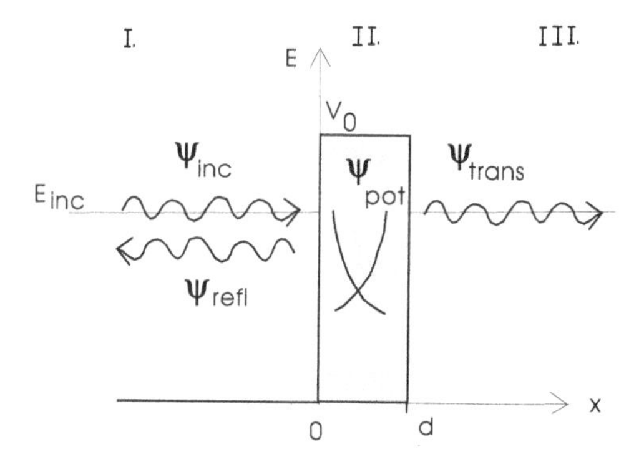

Fig. 13.4 Particle with $E_{\rm inc} < V_0$ in a scattering state with a potential of finite thickness.

(a) Write the boundary conditions at $x = 0$ and $x = d$ in terms of (r, t, a, b, k, κ). How many equations are there, and how many unknowns?[12]

(b) Solve for t. What is the transmission coefficient? Discuss.

4.17 The hamiltonian for the one-dimensional simple harmonic oscillator is

$$H = \frac{1}{2m}p^2 + \frac{1}{2}m\omega^2 q^2$$

where both (p, q) are hermitian and satisfy the canonical commutation relations in Prob. 4.8(e). Introduce the new operators[13]

$$
\begin{aligned}
a &\equiv \frac{p}{(2m\hbar\omega)^{1/2}} - iq\left(\frac{m\omega}{2\hbar}\right)^{1/2} \\
a^\dagger &\equiv \frac{p}{(2m\hbar\omega)^{1/2}} + iq\left(\frac{m\omega}{2\hbar}\right)^{1/2} \\
\mathcal{N} &\equiv a^\dagger a
\end{aligned}
$$

[12]Note that it is now necessary to keep solutions with *both* roots for κ in the region of the potential in order to satisfy the boundary conditions. There is no argument that rules out the solution $e^{+\kappa x}$ *within the potential.*

[13]$(a^\dagger, a)$ are known as the *creation and destruction operators*, while $\mathcal{N}$ is the *number operator.*

(a) Show the hamiltonian takes the form

$$H = \hbar\omega\frac{1}{2}(a^\dagger a + aa^\dagger)$$

(b) Derive the following properties of the new operators

$$\begin{aligned}
[a, a^\dagger] &= 1 \\
[\mathcal{N}, a] &= -a \\
[\mathcal{N}, a^\dagger] &= a^\dagger \\
H &= \hbar\omega\,(\mathcal{N} + 1/2)
\end{aligned}$$

(c) Label the eigenstates and eigenvalues of the hermitian operator $\mathcal{N}$ (H is hermitian) by

$$\mathcal{N}\psi_n = n\,\psi_n$$

Show from the results in part (b) that $a\,\psi_n$ is an eigenstate of $\mathcal{N}$ with eigenvalue lowered by 1, and $a^\dagger\psi_n$ is an eigenstate of $\mathcal{N}$ with eigenvalue raised by 1;

(d) Prove from the structure of H and the results in Prob. 4.5 that the eigenvalues of H must be non-negative. Hence, conclude that eventually the application of a must lead to a state where the eigenvalue n *vanishes*, otherwise further application of a would lead to an eigenstate and eigenvalue that violates this positivity condition. Hence there must be a *ground state* where

$$a\,\psi_0 = 0 \qquad ; \text{ ground state}$$

(e) Since repeated application of $a^\dagger$ on the state ψ_0 raises the eigenvalue of $\mathcal{N}$ by one, conclude that the operators $(\mathcal{N}, H)$ have the following eigenvalue spectrum

$$\begin{aligned}
\mathcal{N}\psi_n &= n\,\psi_n \qquad ; \; n = 0, 1, 2, \cdots, \infty \\
H\psi_n &= \hbar\omega\,(n + 1/2)\,\psi_n
\end{aligned}$$

(f) What is it in the above derivation that leads to the discrete eigenvalue spectrum for these operators?

4.18 One can actually construct the eigenstates of the harmonic oscillator in Prob. 4.17 as follows (with an appropriate choice of phases)

$$\psi_n = \frac{1}{\sqrt{n!}}(a^\dagger)^n\,\psi_0$$

(a) Use the hermiticity of (p, q) to show

$$\int_{-\infty}^{\infty} dq \, \left[a^\dagger \psi_m\right]^\star \psi_n = \int_{-\infty}^{\infty} dq \, \psi_m^\star a \, \psi_n$$

(b) Use the commutation relations and definition of the ground state in Prob. 4.17 to show that the wave function ψ_n is correctly normalized;

(c) Prove the following

$$\begin{aligned} a \, \psi_n &= \sqrt{n} \, \psi_{n-1} \\ a^\dagger \, \psi_n &= \sqrt{n+1} \, \psi_{n+1} \end{aligned}$$

These relations for the creation and destruction operators acting on the eigenstates of the number operator play an important role in quantum mechanics.

4.19 Make a good plot, to scale, of the energy levels, wave functions, and probability densities (absolute squares of the wave functions) for the first four energy levels of a particle in a one-dimensional simple harmonic oscillator. The analytic form of the wave functions can be found in any good book on quantum mechanics, for example [Schiff (1968)].

4.20 Start from the three-dimensional Schrödinger Eq. (4.109), assume a real potential V, and follow the arguments leading to Eqs. (4.46) to derive the three-dimensional form of the continuity Eq. (4.111).

4.21 Make good to-scale plots, by any means you find the most informative (density of points, contours, *etc.*), of the probability densities for the first three stationary states of a particle of mass m in a three-dimensional box. Assume $a < b < c$ (see Fig. 4.22), and examine the states $(n_1, n_2, n_3) = (1, 1, 1), (1, 1, 2), (1, 2, 1)$.

4.22 (a) Show that the eigenfunctions that are given in Eqs. (4.115) for a particle in the box in Fig. 4.22 satisfy the following orthonormality relation

$$\int_0^a dx \int_0^b dy \int_0^c dz \; \psi_{n_1 n_2 n_3}^\star(\mathbf{x}) \psi_{n'_1 n'_2 n'_3}(\mathbf{x}) = \delta_{n_1 n'_1} \delta_{n_2 n'_2} \delta_{n_3 n'_3}$$

(b) Consider the eigenfunctions obeying periodic boundary conditions for a free particle in a cubical box of side L centered on the origin [Eqs. (4.118)]. Show that they satisfy the orthonormality relation

$$\int_{-L/2}^{L/2} \int_{-L/2}^{L/2} \int_{-L/2}^{L/2} d^3x \; \psi_{\mathbf{k}}^\star(\mathbf{x}) \psi_{\mathbf{k}'}(\mathbf{x}) = \delta_{n_x n'_x} \delta_{n_y n'_y} \delta_{n_z n'_z}$$

Here $d^3x \equiv dxdydz$, and the triple integral goes over all three components.

4.23 (a) Show that the general solution to the Schrödinger equation for a particle of mass m in the box in Fig. 4.22 is given in terms of the normal-mode solutions in Eqs. (4.115) and (4.116) as

$$\Psi(\mathbf{x},t) = \sum_{n_1n_2n_3} A_{n_1n_2n_3}\, \psi_{n_1n_2n_3}(\mathbf{x}) \exp\left\{-\frac{i}{\hbar}E_{n_1n_2n_3}t\right\}$$

(b) Use the results in Prob. 4.22 to determine the complex amplitudes $A_{n_1n_2n_3}$ in terms of the initial conditions $\Psi(\mathbf{x},0)$.

4.24 (a) Show that the appropriate generalization to three dimensions of the completeness relation in Prob. 4.3 is

$$\frac{1}{(2\pi)^3}\int d^3k\, e^{i\mathbf{k}\cdot(\mathbf{x}-\mathbf{y})} = \delta^{(3)}(\mathbf{x}-\mathbf{y}) \qquad \text{; Dirac delta function}$$

where $\int d^3k$ represents a triple integral over all values of the components.[14] In this expression

$$\begin{aligned} \delta^{(3)}(\mathbf{x}-\mathbf{y}) &= 0 & ;\ \mathbf{x} \neq \mathbf{y} \\ &\neq 0 & \mathbf{x} = \mathbf{y} \end{aligned}$$

$$\int d^3y\, q(\mathbf{y})\delta^{(3)}(\mathbf{x}-\mathbf{y}) = q(\mathbf{x})$$

(b) Use the result in (a) to establish the three-dimensional Fourier transform relations

$$\begin{aligned} f(\mathbf{x}) &= \frac{1}{(2\pi)^{3/2}}\int d^3k\, e^{i\mathbf{k}\cdot\mathbf{x}}\, a(\mathbf{k}) \\ a(\mathbf{k}) &= \frac{1}{(2\pi)^{3/2}}\int d^3x\, e^{-i\mathbf{k}\cdot\mathbf{x}}\, f(\mathbf{x}) \end{aligned}$$

4.25 It follows from the definition of the angular momentum in Eq. (4.127) that

$$L_x = \frac{1}{i}\left(y\frac{\partial}{\partial z} - z\frac{\partial}{\partial y}\right) \qquad \text{; and cylic permutations of } (x,y,z)$$

(a) Prove the following commutation relations between the components of the angular momentum (see Prob. 4.7)

$$[L_x, L_y] = iL_z \qquad \text{; and cylic permutations of } (x,y,z)$$

[14] We now simply suppress the multiple integral signs.

(b) The square of the total angular momentum is $\mathbf{L}^2 = L_x^2 + L_y^2 + L_z^2$. Show

$$[\mathbf{L}^2, L_i] = 0 \qquad ; \; i = (x, y, z)$$

(c) Show that the components of the angular momentum are hermitian operators (see Prob. 4.5).

(d) Use the result in Prob. 4.10(b) to deduce an uncertainty relation between the components of the angular momentum.

4.26 (a) Make a change of variables from cartesian to polar coordinates in two dimensions

$$x = \rho \cos\phi \qquad ; \; y = \rho \sin\phi$$

Use the chain rule for differentiating an implicit function $f(x, y) = f[x(\rho, \phi), y(\rho, \phi)] \equiv \bar{f}(\rho, \phi)$ to show that the laplacian in Eq. (4.108) is expressed in polar coordinates as [compare Eqs. (4.130)]

$$\nabla^2 = \frac{1}{\rho}\frac{\partial}{\partial \rho}\rho\frac{\partial}{\partial \rho} + \frac{1}{\rho^2}\frac{\partial^2}{\partial \phi^2}$$

(b) Repeat part (a) for the change from cartesian to spherical coordinates in three dimensions, where (see Fig. 4.26)

$$x = r\sin\theta\cos\phi \qquad ; \; y = r\sin\theta\sin\phi \qquad ; z = r\cos\theta$$

Show[15]

$$\begin{aligned}
\nabla^2 &= \frac{1}{r^2}\frac{\partial}{\partial r}r^2\frac{\partial}{\partial r} + \frac{1}{r^2}\left[\frac{1}{\sin\theta}\frac{\partial}{\partial \theta}\sin\theta\frac{\partial}{\partial \theta} + \frac{1}{\sin^2\theta}\frac{\partial^2}{\partial \phi^2}\right] \\
&= \frac{1}{r}\frac{\partial^2}{\partial r^2}r + \frac{1}{r^2}\left[\frac{1}{\sin\theta}\frac{\partial}{\partial \theta}\sin\theta\frac{\partial}{\partial \theta} + \frac{1}{\sin^2\theta}\frac{\partial^2}{\partial \phi^2}\right]
\end{aligned}$$

(c) Use the fact that the spherical harmonics satisfy the following equation (see, for example, [Fetter and Walecka (2003)])

$$-\left[\frac{1}{\sin\theta}\frac{\partial}{\partial \theta}\sin\theta\frac{\partial}{\partial \theta} + \frac{1}{\sin^2\theta}\frac{\partial^2}{\partial \phi^2}\right] Y_{lm}(\theta, \phi) = l(l+1)\, Y_{lm}(\theta, \phi)$$

to derive the radial Eq. (4.142).

4.27 This problem illustrates some applications of the general boundary conditions (1)–(2) in quantum mechanics stated above Eq. (4.66).

[15]The algebra here is a little more intense, but worth doing once in your life; for an alternate derivation, see appendix B of [Fetter and Walecka (2003)].

(a) Consider the azimuthal solutions in Eq. (4.134). Take a linear combination $\Phi(\phi) = a_1\phi_{m_1}(\phi) + a_2\phi_{m_2}(\phi)$ of any two solutions, and demand that the corresponding *probability density* $|\Phi|^2$ be single-valued under the transformation $\phi \to \phi + 2\pi$ (see Fig. 4.25). Show that $\Delta m = m_1 - m_2 = m$, where $m \equiv$ integer. Show that if one demands that $\Phi =$ constant be an acceptable solution, then *all the m_i must be integers.*[16]

(b) The radial component of the gradient in Eqs. (4.111)–(4.112) is just $\partial/\partial r$. Let $l = 0$, and consider the radial wave function $R(r)$ in Eqs. (4.141) and (4.143) as $r \to 0$. Suppose $R(r) = a$ where a is some complex constant is an acceptable solution. Now consider a linear combination of solutions $a + b/r$. Show that the probability density arising from such a wave function is *finite* when multiplied by the volume element $4\pi r^2 dr$; however, show that the origin acts as a *point source of probability* when the radial probability flux is integrated over the area $4\pi r^2$. Hence conclude that a radial solution that goes as b/r at the origin is too singular in quantum mechanics.

4.28 (a) The analytic expressions for the radial wave functions for the first two s-states ($l = 0$) in the point Coulomb potential are [Schiff (1968)]

$$R_{10}(r) = 2\left(\frac{Z}{a_0}\right)^{3/2} e^{-Zr/a_0} \quad ; \; R_{20}(r) = \left(\frac{Z}{2a_0}\right)^{3/2}\left(2 - \frac{Zr}{a_0}\right) e^{-Zr/2a_0}$$

Here $a_0 \equiv \hbar^2/m_0(e^2/4\pi\varepsilon_0)$ is the Bohr radius. Make an accurate to-scale plot of the radial densities $|R_{\bar{n}0}(r)|^2$ for these states in units of Zr/a_0.[17]

(b) The analytic expression for the wave function for the first p-state ($l = 1$) in the point Coulomb potential is[18]

$$\psi_{21m}(r, \theta, \phi) = R_{21}(r) Y_{1m}(\theta, \phi)$$
$$R_{21}(r) = \left(\frac{Z}{2a_0}\right)^{3/2}\left(\frac{Zr}{a_0\sqrt{3}}\right) e^{-Zr/2a_0}$$

Make good to-scale plots, by any means you find the most informative (density of points, contours, *etc.*), of the probability densities for the three states with $m = 0, \pm 1$.

---

[16] This is a very important point, for when we start talking about half-integer angular momenta, we will find that the wave functions themselves are *double-valued* under the transformation $\phi \to \phi + 2\pi$.

[17] Remember the volume element in spherical coordinates is $d^3x = r^2 dr \sin\theta \, d\theta \, d\phi$.

[18] Note that whereas the newtonian orbit problem in a central potential in three dimensions can always be reduced to motion in a two-dimensional plane (see [Fetter and Walecka (2003)]), this is no longer true in quantum mechanics, where the three-dimensional problem is intrinsically different [compare Eqs. (4.132) and (4.137)]. The s-states with $l = 0$, for example, have a *uniform probability distribution* in both θ and ϕ.

4.29 Assume a collection of non-interacting fermions in their ground state in a large cubical box, of side L, with rigid walls. Use the results in Eqs. (4.115)–(4.116). Work in the limit $L \to \infty$. Write $k_i = (\pi/L)n_i$ and convert the sum over states to an integral over wavenumbers. Show one obtains the *same results* for $(N/V, k_{\rm F}, E/N)$ as in Eqs. (4.168)–(4.170) where periodic boundary conditions are employed. Discuss.

4.30 Consider the ground state of a system of N spin-zero bosons of mass m in a large cubical box, of side L, with rigid walls.

(a) Show the energy of the ground state is

$$E = \frac{\hbar^2}{2m}\frac{3\pi^2}{L^2}N$$

(b) Show the corresponding pressure is

$$P = \frac{1}{N^{2/3}}\left(\frac{\hbar^2\pi^2}{m}\right)n^{5/3}$$

where $n = N/V$ is the particle density.

(c) Hence conclude that, at fixed particle density, the pressure of this Bose gas goes to zero as the size of the box goes to infinity.

4.31 Consider a collection of N spin-zero bosons of mass m in a big cubical box of side L satisfying p.b.c. At high temperature, the chemical potential is negative (see Prob. E.3). As the temperature is lowered, the chemical potential increases, and one will eventually reach a point where the chemical potential vanishes. The chemical potential for the Bose gas must then remain zero for $T < T_0$, or the distribution function in the second of Eqs. (4.178) will become singular at finite ε_k. What happens below T_0 is that a finite fraction of the bosons begin to occupy the single state with $\mathbf{k} = 0$.

(a) Show the temperature T_0 at which the chemical potential first vanishes, and hence the temperature for the phase transition in this non-interacting Bose gas, is determined from the equation

$$\frac{N}{V} = \int \frac{d^3k}{(2\pi)^3}\frac{1}{e^{\hbar^2k^2/2mk_{\rm B}T_0} - 1}$$

(b) Go to dimensionless variables, and show this equation can be rewritten as

$$k_{\rm B}T_0 = \frac{1}{\gamma^{2/3}}\frac{\hbar^2 n^{2/3}}{2m} \qquad ; \; \gamma \equiv \frac{1}{(2\pi)^2}\int_0^\infty \frac{\sqrt{u}\,du}{e^u - 1}$$

Here $n = N/V$ is the particle density.

(c) Liquid ^{4}He has a mass density $\rho = 0.145\,\text{g/cm}$. Compute T_0. Compare with the observed superfluid transition temperature T_λ in Fig. 11.1.[19]

4.32 Consider a non-interacting, non-relativistic Fermi gas of spin-1/2 particles in its ground state at a given density $n = N/V$ (electrons in a metal, neutrons in a neutron star, *etc.*). The particles have mass m and magnetic moment μ_0. A uniform magnetic field $\mathbf{B}$ is applied. Treat the system as two separate Fermi gases, one with magnetic moment aligned with the field, and one with magnetic moment opposed. Parameterize the number of particles of each type as

$$N_\uparrow = \frac{N}{2}(1+\delta) \qquad ; N_\downarrow = \frac{N}{2}(1-\delta)$$

(a) What is the contribution to the energy of the system coming from the interaction of the spin magnetic moments with the magnetic field for a given δ?

(b) The kinetic energy of the system must increase to achieve a configuration with finite δ. Let $\varepsilon_{\mathrm{F}}^0$ be the value of the Fermi energy with $\delta = 0$. Show the increase in kinetic energy of the system is given to order δ^2 by[20]

$$\Delta E = \frac{\delta^2}{3}\varepsilon_{\mathrm{F}}^0 N$$

(c) Construct the total change in the energy as a sum of the two contributions in (a) and (b). Minimize with respect to δ to find the new ground state. Show

$$\delta = \frac{3\mu_0 B}{2\varepsilon_{\mathrm{F}}^0}$$

(d) The magnetic spin susceptibility is defined in terms of the magnetic dipole moment per unit volume $\mathbf{M}$ according to $\mathbf{M} = \chi_{\mathrm{P}}\mathbf{B}$. Hence derive the expression of the *Pauli paramagnetic spin susceptibility*

$$\chi_{\mathrm{P}} = \frac{3\mu_0^2}{2\varepsilon_{\mathrm{F}}^0} n$$

4.33 The spin wave functions for a spin-1/2 fermion in Eq. (4.160) form a two-component column matrix (a "spinor").

---

[19] Use $m_{\mathrm{He}^4} \approx 4m_p$.

[20] Use the Taylor series expansion $(1+x)^n = 1 + nx + n(n-1)x^2/2! + \cdots$, which holds for $|x| < 1$ and any n (integer or non-integer).

(a) Use the results in appendix B to show that these spinors are *orthonormal* in the sense that

$$\underline{\eta}_\lambda^\dagger \underline{\eta}_{\lambda'} = \delta_{\lambda\lambda'} \qquad ; \ (\lambda, \lambda') = (\uparrow, \downarrow)$$

(b) Consider a more general spinor of the form

$$\underline{\eta} = \begin{pmatrix} \alpha \\ \beta \end{pmatrix}$$

where (α, β) are complex numbers. Show the normalization $\underline{\eta}^\dagger \underline{\eta} = 1$ of this spinor implies $|\alpha|^2 + |\beta|^2 = 1$. In this case, one can interpret $|\alpha|^2$ as the *probability* that the particle has spin up along the z-axis and $|\beta|^2$ as the probability that it has spin down. Verify that this interpretation is consistent with the wave functions in (a).

(c) Show the most general hermitian operator in this two-dimensional spin space must be of the form $a1 + \mathbf{b} \cdot \boldsymbol{\sigma}$ where $(1, \boldsymbol{\sigma})$ are the unit and Pauli matrices (see appendix B), and $(a, \mathbf{b})$ are real.

5.1 Special relativity (see later) says that there is a magnetic field $c\mathbf{B} = (\mathbf{v}/c) \times \mathbf{E}$ associated with an electric field moving with velocity $\mathbf{v}$.

(a) Show the electric field at the position of an electron in an atom arising from the nucleus is $\mathbf{E} = (Ze_p/4\pi\varepsilon_0 r^3)\,\mathbf{r}$ where $\mathbf{r}$ points from the nucleus to the electron.

(a) Use the fact that the nucleus is moving with velocity $-\mathbf{v}$ with respect to the electron to show that the effective magnetic field at the position of the electron arising from the nuclear current is

$$\mathbf{B}_{\text{eff}} = \frac{1}{m_e c^2}\left(\frac{Ze_p}{4\pi\varepsilon_0 r^3}\right)\mathbf{r} \times (m_e \mathbf{v})$$

(b) Hence derive an expression for V_{so} appearing in Eq. (5.24)

$$V_{\text{so}}(r) = \frac{g_s}{2}\frac{Ze^2}{4\pi\varepsilon_0 \hbar c}\left(\frac{\hbar}{m_e c}\frac{1}{r}\right)^3 m_e c^2$$

(c) Evaluate the expectation value of this quantity [see the second of Eqs. (5.26)] using the radial wave function for the $2p$-state in hydrogen given in Prob. 4.28(b).[21]

---

[21] *Note*: The expression in part (b) is a factor of two larger than that obtained through a correct relativistic treatment of the electron provided by the Dirac equation (Prob. 9.3); this is due to the fact that the spinning electron is not moving with a uniform velocity $\mathbf{v}$, but is actually accelerating in its orbit. (There is an additional "Thomas precession".)

5.2 Verify Eqs. (5.54).

5.3 Write or obtain a program to integrate a second-order non-linear differential equation and reproduce the results in Fig. 5.11 and Eq. (5.70).

5.4 For an electron with $s = 1/2$, one has $j = l \pm 1/2$. Assume $g_s = 2$.
(a) Show that the Landé g-factor takes the form

$$g_{\text{Landé}} = \left(1 \pm \frac{1}{2l+1}\right) \qquad ; \, j = l \pm \frac{1}{2}$$

(b) Show that if $l = 0$, the answer is $g_{\text{Landé}} = 2$. Interpret this result.

5.5 If the atom is neutral in T-F theory, it must contain exactly Z electrons. Start from Eqs. (5.52) and (5.54) and prove

$$\int d^3r \, n(r) = Z$$

5.5 The unphysical nature of the long-range tail on the electron distribution can be avoided if one studies an *ionized* atom with z electrons removed. In this case, the shielding function will vanish at some finite radius x_R.
(a) For the ionized atom, show the T-F equation remains valid with the electrostatic potential inside the atom given by

$$\Phi(r) - \frac{ze_p}{4\pi\varepsilon_0 R} = \frac{Ze_p}{4\pi\varepsilon_0 r}\chi(x) \qquad ; \, r \leq R$$

(b) Show the radius of the atom is determined from

$$\chi(x_R) = 0 \qquad ; \, x_R = \frac{Z^{1/3}R}{ba_0}$$

(c) Show the degree of ionization is determined by (compare Prob. 5.5)

$$\int_0^R 4\pi r^2 n(r) \, dr = Z\int_0^{x_R} \chi(x)^{3/2} x^{1/2} dx = Z - z$$

5.6 Express the coefficients (c_0, c_1) in the T-F density functional $F(n)$ in Eq. (5.69) in terms of the physical constants of the atomic system.

5.7 (a) Give the anticipated electron configurations for the following elements: $_{55}$Cs, $_{70}$Yb, $_{85}$At, $_{86}$Rn;
(b) Discuss the expected chemical properties of these elements;
(c) Go to the website in Fig. 5.18, click on the elements, and check your answers to parts (a) and (b).

5.8 Identify the next alkali, halogen, and inert gas in Fig. 5.17. Compare with Fig. 5.18.

6.1 (a) The hamiltonian for two non-relativistic particles interacting through a potential that only depends on their relative coordinate is

$$H(\mathbf{r}_1, \mathbf{r}_2, \mathbf{p}_1, \mathbf{p}_2) = \frac{\mathbf{p}_1^2}{2m_1} + \frac{\mathbf{p}_2^2}{2m_2} + V(|\mathbf{r}_2 - \mathbf{r}_1|)$$

Introduce the following change of variables

$$\mathbf{R} = \frac{m_1\mathbf{r}_1 + m_2\mathbf{r}_2}{m_1 + m_2} \qquad ; \ \mathbf{r} = \mathbf{r}_2 - \mathbf{r}_1$$

$$\mathbf{P} = \mathbf{p}_1 + \mathbf{p}_2 \qquad ; \ \mathbf{p} = \frac{m_1\mathbf{p}_2 - m_2\mathbf{p}_1}{m_1 + m_2}$$

Show the hamiltonian takes the form

$$H = \frac{\mathbf{P}^2}{2M} + \frac{\mathbf{p}^2}{2\mu} + V(r) \qquad ; \ \frac{1}{\mu} \equiv \frac{1}{m_1} + \frac{1}{m_2}$$

Here $M = m_1 + m_2$ is the total mass, and μ is the *reduced mass.*

(b) Show that the commutation relations in quantum mechanics that follow from the definitions in part (a) are[22]

$$[p_i, x_j] = \frac{\hbar}{i}\delta_{ij} \qquad ; \ [P_i, X_j] = \frac{\hbar}{i}\delta_{ij}$$

$$[p_i, X_j] = [P_i, x_j] = 0$$

Hence, justify the following replacements in part (a)

$$p_i \to \frac{\hbar}{i}\frac{\partial}{\partial x_i} \qquad ; \ P_i \to \frac{\hbar}{i}\frac{\partial}{\partial X_i}$$

[*Note*: These relations can be checked with an explicit change of variables in the laplacians involved. We leave this as an exercise for the dedicated reader.]

(c) Show the solution to the stationary-state Schrödinger equation for two particles in a large box with periodic boundary conditions then *factors* into $\Psi(\mathbf{r}_1, \mathbf{r}_2) = \psi_{\mathbf{K}}(\mathbf{R})\phi(\mathbf{r})$ where the C-M wave function is that of Eq. (4.118). What is the remaining Schrödinger equation for $\phi(\mathbf{r})$?

6.2 The binding energy of the deuteron is $|E| = 2.22\,\text{MeV}$.

---

[22] See Prob. 4.7. Here the operators for the two particles are independent, and they commute with each other.

(a) Given the potential in Eq. (6.21) and the second of Eqs. (6.29), and this value of the binding energy, solve the eigenvalue Eq. (6.24) numerically to determine the new V_0;

(b) Determine and plot the corresponding wave function $u_{10}(r)$;

(c) Plot the corresponding probability density. Discuss.

6.3 Determine the correct normalization for the wave function in Fig. (6.6) for small, non-zero γ. Show that it reduces to Eq. (6.33) as $\gamma \to 0$.

6.4 Suppose the potential in Eq. (6.21) is modified to contain an infinite repulsion ("hard core") at short distances so that it is augmented with the relation $V = +\infty$ for $r \leq c < R$.[23]

(a) How is the eigenvalue Eq. (6.24) changed?

(b) Suppose $c = 0.4\,\mathrm{F}$ and $R = 2\,\mathrm{F}$, two fairly realistic values. How deep must V_0 be to produce just one bound state at zero energy?

6.5 The parity operator P in quantum mechanics is defined to reflect the spatial coordinate $P\Psi(\mathbf{r},t) \equiv \eta_P \Psi(-\mathbf{r},t)$, where, for the present purposes, $\eta_P = +1$. Suppose one has a state that at a time t_0 is an *eigenstate* of parity so that $P\Psi(\mathbf{r},t_0) \equiv \Psi(-\mathbf{r},t_0) = \pi\,\Psi(\mathbf{r},t_0)$, where π is the eigenvalue. Solution to the time-dependent Schrödinger equation involves repeated application of the hamiltonian as a function of time. Prove that if the hamiltonian *commutes* with the parity operator (*i.e.* $[P,H]=0$), then $\Psi(\mathbf{r},t)$ remains an eigenstate of parity for all time.

6.6 Prove the following theorem: Of two isobars of given B, with Z differing by one, at least one is unstable under the β-decay interactions in Eqs. (6.2)–(6.3). Assume massless neutrinos.

6.7 Suppose a radial wave function $\psi(r)$ is of the form $\psi(r) = u(r)/r$. Show that the boundary conditions that (ψ, ψ') be continuous at finite r are equivalent to (u, u') being continuous.

6.8 Use the atomic mass tables [Lawrence Berkeley Laboratory (2003)], and consider the following reactions in the center-of-mass system:

$$
\begin{aligned}
{}^{4}_{2}\mathrm{He} + {}^{12}_{6}\mathrm{C} &\to {}^{16}_{8}\mathrm{O} + \gamma \\
{}^{28}_{14}\mathrm{Si} &\to {}^{14}_{7}\mathrm{N} + {}^{14}_{7}\mathrm{N} \\
{}^{238}_{92}\mathrm{U} &\to {}^{104}_{42}\mathrm{Mo} + {}^{134}_{50}\mathrm{Sn} \\
{}^{12}_{6}\mathrm{C}^{\star}(7.6\,\mathrm{MeV}) &\to {}^{4}_{2}\mathrm{He} + {}^{4}_{2}\mathrm{He} + {}^{4}_{2}\mathrm{He}
\end{aligned}
$$

[23]The nucleon-nucleon potential exhibits such a short-range repulsion.

Here γ is a massless photon, and ${}^{12}_{6}C^*(7.6\,\text{MeV})$ represents an excited state at the indicated energy.

Determine which of these nuclear reactions is exothermic, and, if so, determine the energy release.

6.9 This problem examines the Coulomb interaction energy of two charges e_p uniformly distributed over a sphere (drop) of radius R. One makes use of Gauss' law, which states that the electric field at a radius r arising from a uniform spherical distribution of charge comes from the charge *inside* of the sphere of radius r, all of that charge acting as a point source at the origin (see chapter 5).

(a) Show that the electric field at a radius r inside the drop coming from the first particle is then given by

$$E_r = \frac{e_p}{4\pi\varepsilon_0}\frac{r}{R^3} \qquad ; r \leq R$$

(b) The field is derivable from a potential as $E_r = -\partial\Phi/\partial r$, and at the radius of the drop, the potential is known to be $e_p/4\pi\varepsilon_0 R$. Hence show that inside of the drop the potential is given by

$$\Phi = \frac{e_p}{4\pi\varepsilon_0 R}\left(\frac{3}{2} - \frac{1}{2}\frac{r^2}{R^2}\right) \qquad ; r \leq R$$

(c) Show the interaction energy of that part of the second charge in the shell $4\pi r^2\,dr$ with the first charge is given by

$$dW = \frac{e_p^2}{4\pi\varepsilon_0}\frac{3}{R^4}\left(\frac{3}{2} - \frac{1}{2}\frac{r^2}{R^2}\right) r^2\,dr$$

(d) Show that the total interaction energy of the pair of charges is

$$W = \frac{6}{5}\frac{e_p^2}{4\pi\varepsilon_0}\frac{1}{R}$$

Hence conclude that the interaction of $Z(Z-1)/2$ pairs of charges is given by the first of Eqs. (6.53).

6.10 Make use of the following relation from Eqs. (6.53)[24]

$$a_3 = \frac{3}{5}\frac{e^2}{4\pi\varepsilon_0 r_{0c}}$$

Show that the empirical value of a_3 in Table 6.2 yields the value of r_{0c} in Eq. (6.58).

---

[24] It is assumed there that the nucleus is heavy enough so that $Z(Z-1) \approx Z^2$.

6.11 (a) Use the semi-empirical mass formula (SEMF) in Eq. (6.57) and Table 6.2 to compute the energy release in the third process in Prob. 6.8

$$ {}^{238}_{92}\text{U} \rightarrow {}^{104}_{42}\text{Mo} + {}^{134}_{50}\text{Sn} $$

(b) Compare with the answer obtained in Prob. 6.8 from the observed atomic masses.

6.12 Consider the SEMF in the case of an odd nucleus ($\lambda = 0$).

(a) Minimize with respect to Z at fixed B and show that the β-stable nuclei have a $Z(B)$ of

$$ Z(B) = \frac{B}{2 + (a_3 B^{2/3})/2a_4} \qquad ; \text{ at minimum} $$

This determines a "line of stability." Write $B = N + Z$, solve self-consistently for Z at fixed N, and make a plot of $Z(N)$ with Z as the ordinate and N as the abscissa.

(b) For a self-bound system in its ground state where the pressure and entropy vanish, the neutron chemical potential at fixed Z is obtained from Eq. (D.10) as $\mu = dE/dN$. Write $B = N + Z$, and compute the neutron chemical potential $\mu_n(Z, N)$ at fixed Z from the SEMF. A nucleus will be unbound with respect to neutron emission when the neutron chemical potential first vanishes $\mu_n(Z, N) = 0$. Solve this equation for $Z(N)$, and determine the "neutron drip line". Plot this curve on the graph in part (a).

(c) Repeat part (b) for the "proton drip line".

(d) Compare the results in parts (a)–(c) with the chart of the nuclides in [National Nuclear Data Center (2007)].[25]

6.13 Consider a uniformly charged sphere of radius R and charge Z.

(a) Compute the form factor in Eq. (6.69) by explicitly carrying out the angular and radial integrals. Show

$$ F(q) = Z\left[\frac{3j_1(qR)}{qR}\right] \qquad ; \text{ uniform sphere} $$

$$ j_1(x) \equiv \frac{\sin x}{x^2} - \frac{\cos x}{x} $$

(b) Plot this result as a function of qR. Discuss.

[25]Note that the simple 5-parameter SEMF discussed here is designed to describe nuclei along the line of stability and with small deviations about it. For a more accurate description of the entire energy surface, see the discussion of effective field theory in [Walecka (2004)].

6.14 (a) Make use of Fig. 6.23 to predict the spins and parities of the following nuclei

$$^{17}_{9}\text{F} \qquad ; \, ^{61}_{29}\text{Cu} \qquad ; \, ^{105}_{46}\text{Pd} \qquad ; \, ^{121}_{49}\text{In} \qquad ; \, ^{211}_{87}\text{Fr}$$

Compare with experiment [National Nuclear Data Center (2007)].

(b) Compute the magnetic moments of these nuclei and compare with the experimental results [Lawrence Berkeley Laboratory (2005)].

6.15 Repeat Prob. 6.14 for $^{133}_{54}\text{Xe}$. Discuss.

6.16 Consider the scattering of light from a single slit where the location of the first diffraction minimum is given for small angles by Eq. (6.66).

(a) Show that this expression can be written in the more familiar form

$$\theta = \lambda/a \qquad ; \text{ first diffraction minimum}$$

(b) Hence conclude that if one wants a diffraction pattern to show up on a laboratory screen at small, but finite angles, one wants to use light whose wavelength is comparable to the size of the slit, or $\lambda = O(a)$.

6.17 Assume that the neutron and proton densities in nuclear matter are driven apart by the Coulomb interaction of the protons. Write

$$N = \frac{B}{2}(1+\delta) \qquad ; \, Z = \frac{B}{2}(1-\delta)$$

It follows, as in Prob. 4.32, that the kinetic energy of nuclear matter at fixed baryon density, to leading order in δ, is increased by an amount

$$\frac{\Delta T}{B} = \frac{\varepsilon_{\text{F0}}}{3}\delta^2$$

Here ε_{F0} is the Fermi energy for symmetric nuclear matter.

(a) Derive this result;

(b) Show this implies a *symmetry energy coefficient* in the semi-empirical mass formula of

$$a_4 = \frac{\varepsilon_{\text{F0}}}{3}$$

(c) Use $k_{\text{F0}} = 1.42\,\text{F}^{-1}$ from Table 6.3, and compare this result with the observed value of a_4 in Table 6.2;

(c) How would you improve this calculation?

6.18 Explicitly evaluate the magnetic moment in Eqs. (6.99)–(6.100), and derive the expression for the Schmidt lines in Eq. (6.101).

7.1 Work through appendix A.1 of [Walecka (2004)]. Use the Feynman rules for scalar meson exchange given there to derive the Yukawa potential in Eq. (7.29).[26]

7.2 (a) When the laboratory is taken to be a very large cubical box of volume $\Omega = L^3$, and periodic boundary conditions are applied, the final integral giving overall momentum conservation is of the form

$$\int_\Omega d^3x\, e^{i(\mathbf{K}_i - \mathbf{K}_f)\cdot\mathbf{x}} = \Omega\, \delta_{\mathbf{K}_f,\mathbf{K}_i}$$

where $\delta_{\mathbf{K}_f,\mathbf{K}_i}$ is a Kronecker delta. Show that in the limit $\Omega \to \infty$, this relation becomes (compare Prob. 4.24)

$$\int d^3x\, e^{i(\mathbf{K}_i - \mathbf{K}_f)\cdot\mathbf{x}} = (2\pi)^3\, \delta^{(3)}(\mathbf{K}_i - \mathbf{K}_f)$$

where $\delta^{(3)}(\mathbf{K}_i - \mathbf{K}_f)$ is now a Dirac delta function. Hence justify the following replacement in this limit

$$\Omega\, \delta_{\mathbf{K}_f,\mathbf{K}_i} \to (2\pi)^3\, \delta^{(3)}(\mathbf{K}_i - \mathbf{K}_f) \qquad ; \Omega \to \infty$$

(b) Use the second of Eqs. (7.45) and the results in part (a) to show that in the limits $T \to \infty$, and then $\Omega \to \infty$, the S-matrix in Eq. (7.36) takes the form

$$S_{fi} = (2\pi)^4\, \delta^{(4)}(K_f - K_i) \frac{1}{\Omega^{(n+2)/2}} \left[\frac{-iT_{fi}}{\hbar c}\right] \qquad ; T \to \infty \quad \Omega \to \infty$$

Here

$$\delta^{(4)}(K_f - K_i) \equiv \delta^{(3)}(\mathbf{K}_f - \mathbf{K}_i)\delta(K_{f0} - K_{i0})$$
$$K_{f0} - K_{i0} \equiv (W_f - W_i)/\hbar c$$

7.3 Two particles in a large box of volume Ω with parallel velocities $\mathbf{v}(1)$ and $\mathbf{v}(2)$ interact to produce a final state f. Explain why the appropriate incident flux to be used in Eq. (7.49) is

$$I_{\text{inc}} = \frac{1}{\Omega} |\mathbf{v}(1) - \mathbf{v}(2)|$$

[26] That the meson responsible for the long-range part of the nuclear force, the pion, should be a *pseudoscalar* rather than a scalar meson was, in fact, predicted by Pauli on the basis of observed properties of the deuteron [Pauli (1948)].

7.4 Reactions can proceed with decreasing rates through the strong, electromagnetic, or weak interactions, or they can be absolutely forbidden. This is governed by appropriate conservation laws (see text). Give the leading interaction through which the following reactions can take place (assume $H_{\rm UW} = 0$):

$$
\begin{aligned}
p + n &\to n + n + \pi^+ \\
\mu^+ &\to e^+ + \nu_e + \nu_\mu \\
\pi^- + p &\to n + \pi^0 + \pi^+ \\
\Xi^- &\to \Sigma^0 + e^- + \bar{\nu}_e \\
\gamma + p &\to n + \pi^+ \\
p + \bar{p} &\to p + \rho^- + n + \bar{n}
\end{aligned}
$$

7.5 Assume $(1s)$ spatial states:

(a) Give the quark configurations for all the states in the $S = -1$ and $S = -2$ sectors in Fig. 7.1;

(b) Give the quark configurations for all the mesons in the $S = \pm 1$ sectors in Fig. 7.2.

7.6 What hadrons can be made from (scu) and (scd) quarks? Give all quantum numbers.

7.7 Show that if the four-momentum transfers of interest in Eq. (7.70) are much smaller than the inverse Compton wavelengths of the massive weak vector bosons, then the amplitude for the process in Fig. 7.21 reduces to

$$\frac{-iT_{fi}}{\hbar c} = \frac{ig_W^2}{\hbar c M_W^2} j_W j_W$$

This is the same amplitude one would get with an *effective point coupling*, and a corresponding amplitude[27]

$$\frac{-iT_{fi}}{\hbar c} = \frac{-i}{\hbar c}\left[-\frac{G_{\rm F}}{\sqrt{2}} j_W j_W\right]$$

Hence make the identification of the Fermi constant in Eq. (7.71).

7.8 Consider the scattering of a massless neutrino by a heavy nucleus. Assume the point coupling of Prob. 7.7, and replace the interaction hamil-

---

[27]Compare Prob. I.2.

tonian of Prob. I.1 by the contact interaction

$$H_1 = -\frac{G_\mathrm{F}}{\sqrt{2}} \int d^3x \, j_W(\mathbf{x}_p)\delta^{(3)}(\mathbf{x}_p - \mathbf{x})J_W(\mathbf{x})$$

Assume the situation is analogous to that in Prob. I.1, and repeat the arguments given there.

(a) Show that in this case

$$\langle \phi_f | H_1 | \phi_i \rangle = -\frac{G_\mathrm{F}}{\sqrt{2}\,\Omega} F_W(\mathbf{q}) \qquad ; \ \mathbf{q} = \mathbf{k}_i - \mathbf{k}_f$$

$$F_W(\mathbf{q}) = \int d^3x \, e^{i\mathbf{q}\cdot\mathbf{x}} \, J_W(\mathbf{x}) \qquad ; \ \text{form factor}$$

(b) Assume relativistic kinematics for the scattering particle with $\varepsilon = \hbar k c$, and show

$$\int dk \, \delta(E_f - E_i) = \frac{1}{\hbar c} \qquad ; \ k \equiv k_i = k_f$$

(c) Show the incident flux in the laboratory frame is $I_{\mathrm{inc}} = c/\Omega$.

(d) Hence show the differential cross section is given by

$$\frac{d\sigma}{d\Omega_k} = \frac{1}{8\pi^2} \frac{G_\mathrm{F}^2 \, k^2}{(\hbar c)^2} |F_W(\mathbf{q})|^2 \qquad ; \ \mathbf{q} = \mathbf{k}_i - \mathbf{k}_f$$

7.9 Consider a wave function with two components ζ_i with $i = (1, 2)$, describing, for example, a spin-1/2 particle [see Eq. (4.160)].[28] Define the following transformation on this wave function by

$$\zeta_i' = \sum_{j=1}^{2} r(\boldsymbol{\omega})_{ij} \zeta_j$$

$$\underline{r}(\boldsymbol{\omega}) = \exp\left\{\frac{i}{2}\boldsymbol{\sigma}\cdot\boldsymbol{\omega}\right\} \qquad ; \ \boldsymbol{\omega} = (\omega_1, \omega_2, \omega_3)$$

Here $\boldsymbol{\sigma} = (\sigma_1, \sigma_2, \sigma_3)$ are the Pauli matrices (which we do not underline), and $\boldsymbol{\omega}$ characterizes some spatial rotation.

(a) Expand the exponential, and show the 2×2 matrix $\underline{r}(\boldsymbol{\omega})$ can be re-written as (here $\hat{\mathbf{n}}$ is a unit vector)

$$\underline{r}(\boldsymbol{\omega}) = \underline{1} \, \cos\frac{\omega}{2} + i\hat{\mathbf{n}}\cdot\boldsymbol{\sigma} \, \sin\frac{\omega}{2} \qquad ; \ \boldsymbol{\omega} \equiv \omega\hat{\mathbf{n}}$$

---

[28] Analogous arguments hold in the internal isospin space.

(b) Show that $\underline{r}(\boldsymbol{\omega})$ is both unitary and unimodular (*i.e.* it has unit determinant)

$$\underline{r}(\boldsymbol{\omega})^{\dagger}\underline{r}(\boldsymbol{\omega}) = 1 \qquad ; \ \det \underline{r}(\boldsymbol{\omega}) = 1$$

(c) Show that the product $\underline{r}(\boldsymbol{\omega}_1)\underline{r}(\boldsymbol{\omega}_2)$ is again a unitary, unimodular matrix.[29]

(d) Show that *any* unitary unimodular matrix can be written in the form given above for *some* choice of $\boldsymbol{\omega}$. Hence show that the transformations form a three-parameter, continuous *group*[30]

$$\underline{r}(\boldsymbol{\omega}_1)\underline{r}(\boldsymbol{\omega}_2) = \underline{r}(\boldsymbol{\omega}_3)$$

This is the group $SU(2)$.

(e) Consider a two-particle direct product state $\psi_{ij}(1,2) \equiv \zeta_i(1)\zeta_j(2)$ that transforms as

$$\psi'_{i'j'}(1,2) = \sum_{i=1}^{2}\sum_{j=1}^{2} r(\boldsymbol{\omega})_{i'i}\, r(\boldsymbol{\omega})_{j'j}\, \psi_{ij}(1,2)$$

Consider linear combinations with symmetrized and antisymmetrized indices

$$\psi_{ij}^{S}(1,2) \equiv \frac{1}{\sqrt{2}}\left[\psi_{ij}(1,2) + \psi_{ji}(1,2)\right]$$
$$\psi_{ij}^{A}(1,2) \equiv \frac{1}{\sqrt{2}}\left[\psi_{ij}(1,2) - \psi_{ji}(1,2)\right]$$

Show that these states are transformed among themselves by the above transformation, and hence each provides a basis for an *irreducible represent ion* of the group.[31] How many states of each type are there? What total spin do they correspond to?

8.1 (a) Make an expansion for small v/c, and verify the expression for the difference in optical pathlength in the second of Eqs. (8.7);

(b) Verify the expression for the inverse Lorentz transformation in Eq. (8.9);

(c) Verify that the Lorentz transformation in Eqs. (8.8) leaves the quadratic form in Eq. (8.10) invariant.

(d) Verify that the three-dimensional Lorentz transformation in Eqs. (8.24) leaves the quadratic form in Eq. (8.25) invariant.

---

[29] *Hint:* Recall Prob. B.3 and use the fact that for matrices $\det \underline{A}\,\underline{B} = \det \underline{A} \det \underline{B}$.

[30] Note that the infinitesimal rotations add here just as they should $\boldsymbol{\omega}_3 = \boldsymbol{\omega}_1 + \boldsymbol{\omega}_2$.

[31] For an introduction to group theory, see [Hamermesh (1989)].

8.2 In deriving the formula for the Lorentz contraction of a meter stick, only the first of Eqs. (8.14) is used. Explain the corresponding physical content of the second of Eqs. (8.14).

8.3 Start from Eqs. (8.40) and (8.42).

(a) Verify by matrix multiplication that $\underline{x}' = \underline{a}\,\underline{x}$ gives the proper expressions for the Lorentz transformation in the z-direction of the four-vector x_μ (see Fig. 8.12);

(b) Verify by matrix multiplication that $\underline{a}^T\underline{a} = \underline{a}\,\underline{a}^T = \underline{1}$.

8.4 (a) Prove that the scalar product of two four-vectors in Eq. (8.44) is invariant under Lorentz transformations;

(b) Use this result to prove that the wave operator

$$\left(\frac{\partial}{\partial x_\mu}\right)^2 = \frac{\partial^2}{\partial x^2} - \frac{1}{c^2}\frac{\partial^2}{\partial t^2} = \Box \qquad ; \text{ D'Alembertian}$$

is invariant under Lorentz transformations;

(c) The electromagnetic current in special relativity is defined by

$$ej_\mu \equiv e\left(\frac{1}{c}\mathbf{j}\,,\,i\rho\right) \qquad ; \text{ electromagnetic current}$$

where the electric charge e is now explicity exhibited. This current transforms as a four-vector under Lorentz transformations.

Establish the invariant *continuity equation* for the electromagnetic current[32]

$$\frac{\partial j_\mu}{\partial x_\mu} = \boldsymbol{\nabla}\cdot\mathbf{j} + \frac{\partial \rho}{\partial t} = 0 \qquad ; \text{ continuity equation}$$

8.5 An electron of energy 50 GeV and rest mass $m_0c^2 = 0.5$ MeV travels down an accelerator pipe of length 1 km. How long does the pipe appear to be in the rest frame of the electron (in cm)?[33]

8.6 A light signal, and a neutrino with energy $E = 2\,\text{MeV}$ and rest mass $m_0c^2 = 2\,\text{eV}$, are emitted simultaneously from a supernova which is at a distance 10^4 light-years from earth.

(a) What is the difference in arrival times at the earth?

---

[32] As an example, consider an electron fluid with electron density n and velocity field $\mathbf{v}$. The charge density is $e\rho = en$ and the current is $e\mathbf{j} = en\mathbf{v}$. The continuity equation then reduces to the familiar statement of electron number conservation $\partial n/\partial t + \boldsymbol{\nabla}\cdot(n\mathbf{v}) = 0$.

[33] Although not required here, we note that in doing numerical calculations when v/c is very close to 1, it is often useful to write $v/c = 1 - \varepsilon$ and then $\sqrt{1 - v^2/c^2} = \sqrt{(1 - v/c)(1 + v/c)} = \sqrt{2\varepsilon}\,[1 + O(\varepsilon)]$.

(b) How long does the trip take in the neutrino's rest frame?

(c) What is the distance to earth as viewed in the neutrino's rest frame?

8.7 In special relativity, a tensor of rank two, three, *etc.* is defined by how it transforms under a Lorentz transformation

$$T'_{\mu\nu\cdots} = a_{\mu\mu'}a_{\nu\nu'}\cdots T_{\mu'\nu'\cdots} \qquad ; \text{ tensor}$$

A four-vector is then a tensor of rank one.

(a) Prove that $\delta_{\mu\nu}$ transforms as a second-rank tensor (*Hint*: Define it to have its usual value in all frames, and then show that it satisfies the tensor transformation law.)

(b) The completely antisymmetric tensor in 4-D is defined by

$$\begin{aligned} \varepsilon_{\alpha\beta\gamma\delta} &= \pm 1 && ; \ (\alpha\beta\gamma\delta) \text{ an even (odd) permutation of } (1234) \\ &= 0 && ; \ \text{otherwise} \end{aligned}$$

Show that this quantity transforms as a fourth-rank tensor under Lorentz transformations with $\det \underline{a}=1$. (*Hint*: Make use of the following definition of the determinant $a_{\mu\mu'}a_{\nu\nu'}a_{\rho\rho'}a_{\sigma\sigma'}\varepsilon_{\mu'\nu'\rho'\sigma'} = \varepsilon_{\mu\nu\rho\sigma}\det \underline{a}$.)

8.8 The electromagnetic field tensor is defined by

$$\underline{F} \equiv \begin{bmatrix} 0 & cB_3 & -cB_2 & -iE_1 \\ -cB_3 & 0 & cB_1 & -iE_2 \\ cB_2 & -cB_1 & 0 & -iE_3 \\ iE_1 & iE_2 & iE_3 & 0 \end{bmatrix} \qquad ; \text{ field tensor}$$

$$F_{\mu\nu} = [\underline{F}]_{\mu\nu} \qquad ; \ (\mu,\nu) = (1,2,3,4)$$

(a) Show that Maxwell's Eqs. (2.90) in SI units can be written in the following form

$$\begin{aligned} \frac{\partial}{\partial x_\nu}F_{\mu\nu} &= \frac{e}{\varepsilon_0}j_\mu && ; \text{ Maxwell's equations} \\ \varepsilon_{\mu\nu\rho\sigma}\frac{\partial}{\partial x_\sigma}F_{\nu\rho} &= 0 && \mu = 1,2,3,4 \end{aligned}$$

Here ej_μ is the electromagnetic current in Prob. 8.4(c).[34]

(b) Show that the statement that Maxwell's equations hold in any inertial frame is equivalent to the statement that $F_{\mu\nu}$ transforms as a second-rank tensor under Lorentz transformations.

---

[34]Note that if one defines a *dual* tensor by $G_{\mu\nu} \equiv \varepsilon_{\mu\nu\rho\sigma}F_{\rho\sigma}$, then the second set of Maxwell's equations can be written as $\partial G_{\mu\nu}/\partial x_\nu = 0$.

8.9 The electromagnetic field tensor is antisymmetric in its two indices $F_{\nu\mu} = -F_{\mu\nu}$. It can be expressed in terms of the *vector potential* A_μ according to

$$\frac{1}{c}F_{\mu\nu} \equiv \frac{\partial A_\nu}{\partial x_\mu} - \frac{\partial A_\mu}{\partial x_\nu} \qquad ; \; A_\mu = \left(\mathbf{A}, \frac{i}{c}\Phi\right)$$

(a) Show that the second set of Maxwell's equations in Prob. 8.8(a) is now satisfied identically;

(b) Compare with the expression for $F_{\mu\nu}$ in Prob. 8.8, and show that the relation of A_μ to the electric and magnetic fields is given by[35]

$$\mathbf{B} = \boldsymbol{\nabla} \times \mathbf{A}$$
$$\mathbf{E} = -\boldsymbol{\nabla}\Phi - \frac{\partial \mathbf{A}}{\partial t}$$

(c) Show that the electromagnetic fields are invariant under a *gauge transformation* of the vector potential

$$A_\mu \to A_\mu + \frac{\partial \Lambda}{\partial x_\mu} \qquad ; \text{ gauge transformation}$$

where Λ is a differentiable scalar function of x_μ. [Note Prob. F.3(a)].

8.10 One can attempt to write a covariant equation of motion for a charged particle in an electromagnetic field as follows

$$\frac{dp_\mu}{d\tau} = \frac{q}{c}F_{\mu\nu}u_\nu \qquad ; \; \mu = 1, 2, 3, 4$$

Here τ is the proper time, (u_μ, p_μ) are the four-velocity and four-momentum of Eqs. (8.54) and (8.57), and $F_{\mu\nu}$ is the electromagnetic field tensor of Prob. 8.8.

(a) Given that $F_{\mu\nu}$ is a second-rank tensor, show that this is a relation between four-vectors;[36]

(b) Show that the spatial parts of this relation give the relativistic form of Newton's second law for a particle of charge q acted upon by the Lorentz

[35] In advanced quantum mechanics, Maxwell's Eqs. (2.90) are frequently written in rationalized cgs [Heaviside-Lorentz (H-L)] units where $\varepsilon_0 = 1$, $\mu_0 = 1/c^2$, $c\mathbf{B} \to \mathbf{B}$, $\mathbf{j}/c \to \mathbf{j}$, $e\rho$ is measured in esu, and $e\mathbf{j}$ in emu, where 1 emu = 1 esu/c (see appendix K). To get from SI units to H-L units in Probs. 8.4(c), 8.8, and 8.9 make the following replacements: $\varepsilon_0 \to 1$, $c\mathbf{B} \to \mathbf{B}$, $\mathbf{j}/c \to \mathbf{j}$, $c\mathbf{A} \to \mathbf{A}$.

[36] More precisely, if $u' \cdot v' = u \cdot v$, then an expression is Lorentz *invariant*. If $u'_\mu = v'_\mu$ when $u_\mu = v_\mu$, then the expression is Lorentz *covariant* — it takes the same form in any Lorentz frame.

force of Eq. (2.83)

$$\frac{d\mathbf{p}}{dt} = q\left(\mathbf{E} + \mathbf{v} \times \mathbf{B}\right) \qquad ; \text{ Newton's second law}$$

(c) Show that with $p_\mu \equiv (\mathbf{p}, iE_p/c)$, the fourth component of the above gives the correct relativistic power equation $dE_p/dt = q\mathbf{v} \cdot \mathbf{E}$. Derive this result from the relativistic relation between energy and momentum in Eq. (8.59) and the result in part (b).

8.11 The weak vector boson Z^0 has the mass $m_z c^2 = 91.19\,\text{GeV}$. It can be produced in the reaction $e^+ + e^- \rightarrow Z^0$. What is the threshold energy for production of this particle in the C-M system? What is the threshold energy for production of this particle with a positron beam incident on electrons at rest? Discuss.

8.12 Start from Eqs. (8.83) and (8.87). Use Eqs. (8.72) and (8.73) to determine the momentum $\mathbf{p}^2(p_L)$ in the C-M system. Assume elastic scattering of equal mass particles, with $m_1 = m_2 \equiv m$.

(a) Take the limit $p_L/mc \ll 1$ and derive the non-relativistic results for $\mathbf{p}^2, p'_L$, and $\cos\theta_{CM}$ as a function of $(p_L, \cos\theta_L)$ as given in Eqs. (8.88);

(b) Take the limit $p_L/mc \gg 1$ and derive the extreme-relativistic results for $\mathbf{p}^2, p'_L$, and $\cos\theta_{CM}$ as a function of $(p_L, \cos\theta_L)$ as given in Eqs. (8.89).

8.13 Assume the proper relativistic relation between energy and momentum, write the equations for conservation of energy and momentum in the laboratory system, and show that the Compton formula of Eq. (3.54) holds without approximation. Compare with the result in Prob. 8.12(b).

8.14 Consider the behavior of the four-velocity u_μ under a Lorentz transformation to a new inertial frame moving with velocity V along the z-axis relative to the first (compare Fig. 8.12). Assume the spatial velocity $\mathbf{v}$ in the four-velocity u_μ lies in the z-direction. We know the four-velocity transforms as a four-vector so that $u'_\mu = a_{\mu\nu}(V)u_\nu$. The mixing of the z- and fourth-components follows from Eq. (8.42).

(a) Show the transformed result can be obtained through matrix multiplication as

$$\begin{bmatrix} u'_z \\ u'_4 \end{bmatrix} = \begin{bmatrix} \frac{v'}{\sqrt{1-v'^2/c^2}} \\ \frac{ic}{\sqrt{1-v'^2/c^2}} \end{bmatrix} = \begin{bmatrix} \frac{1}{\sqrt{1-V^2/c^2}} & \frac{iV/c}{\sqrt{1-V^2/c^2}} \\ \frac{-iV/c}{\sqrt{1-V^2/c^2}} & \frac{1}{\sqrt{1-V^2/c^2}} \end{bmatrix} \begin{bmatrix} \frac{v}{\sqrt{1-v^2/c^2}} \\ \frac{ic}{\sqrt{1-v^2/c^2}} \end{bmatrix}$$

(b) Show

$$\begin{bmatrix} \frac{v'}{\sqrt{1-v'^2/c^2}} \\ \frac{ic}{\sqrt{1-v'^2/c^2}} \end{bmatrix} = \frac{1}{\sqrt{1-V^2/c^2}}\frac{1}{\sqrt{1-v^2/c^2}}\begin{bmatrix} v-V \\ ic(1-Vv/c^2) \end{bmatrix}$$

(c) Show that

$$v' = \frac{v-V}{1-Vv/c^2}$$

This is the law for the addition of the velocities in special relativity.

(d) Show this reduces to the galilean transformation to the new inertial frame if terms of order $(\text{velocity}/c)^2$ are neglected.

(e) Show the component relations in part (b) are individually satisfied.

8.15 Derive the second boundary condition $f'(0) = 0$ in Eqs. (8.111) from Eq. (8.104).

8.16 In direct analogy to the calculation in the text, carry out a calculation of the properties of a white dwarf star in the non-relativistic limit (NRL), which is appropriate for small mass stars.

(a) Show the equation of state in the NRL is

$$P = \zeta\rho^{5/3} \qquad ; \; \zeta = \frac{2}{5}\frac{\hbar^2}{2m_e}\left(\frac{6\pi^2}{g}\right)^{2/3}\frac{1}{(2m_p)^{5/3}}$$

(b) Show the dimensionless equation for the density distribution is[37]

$$\frac{1}{\xi^2}\frac{d}{d\xi}\left(\xi^2\frac{d}{d\xi}F\right) = -F^{3/2}$$

$$F(1) = 0 \qquad ; \; F'(0) = 0$$

Here

$$\bar{\Lambda}^3\rho \equiv F^{3/2} \qquad ; \; \bar{\Lambda} \equiv \left(\frac{8\pi G}{5\zeta}\right)R^2$$

(c) Show the mass and radius of the star are related by

$$MR^3 = 4\pi\left(\frac{5\zeta}{8\pi G}\right)^3\int_0^1 F^{3/2}\xi^2\,d\xi$$

[37]The use of the "cold" equation of state ($T = 0$) depends on the inequality $k_BT \ll \varepsilon_F$ where ε_F is the Fermi energy of the electrons. We leave the demonstration of the precise region of applicability as an exercise for the dedicated reader.

(d) Interpolate between the NRL and ERL to make a sketch of $M(R)$ for all M up to the Chandrasekhar limit for a white dwarf star. Discuss.[38]

8.17 Write, or obtain, a program to integrate the non-linear differential Eq. (8.110) subject to the boundary conditions in Eqs. (8.111), and reproduce Fig. 8.17.

8.18 (a) Use the program from Prob. 8.17 to integrate the non-linear differential equation in Prob. 8.16(b) subject to the boundary conditions stated there. Make a plot of $F(\xi)$ versus ξ. Compare with Fig. 8.17.

(b) Show that the r.h.s of the expression for MR^3 in Prob. 8.16(c) has the correct dimensions. Calculate the required integral numerically, and then calculate the numerical value of MR^3 in the NRL.

9.1 The Dirac matrices $\gamma_\mu = (\gamma_1, \gamma_2, \gamma_3, \gamma_4)$ are defined in Eqs. (9.43).
(a) Show that they are hermitian with $\gamma_\mu^\dagger = \gamma_\mu$ (See Prob. B.3);
(b) Show that they satisfy the following Clifford algebra

$$\gamma_\mu \gamma_\nu + \gamma_\nu \gamma_\mu = 2\delta_{\mu\nu}$$

(c) Derive the standard representation for $\boldsymbol{\gamma}$

$$\boldsymbol{\gamma} = \begin{pmatrix} 0 & -i\boldsymbol{\sigma} \\ i\boldsymbol{\sigma} & 0 \end{pmatrix}$$

9.2 Consider the Dirac Eq. (9.51) in the presence of a static magnetic field $\mathbf{B} = \boldsymbol{\nabla} \times \mathbf{A}$.

(a) Show that Eqs. (9.33) are modified to read

$$c\boldsymbol{\sigma} \cdot (\mathbf{p} - e\mathbf{A})\chi + m_0 c^2 \phi = E\phi$$
$$c\boldsymbol{\sigma} \cdot (\mathbf{p} - e\mathbf{A})\phi - m_0 c^2 \chi = E\chi$$

Show that these relations imply

$$\frac{c^2}{2m_0c^2 + \varepsilon} [\boldsymbol{\sigma} \cdot (\mathbf{p} - e\mathbf{A})] [\boldsymbol{\sigma} \cdot (\mathbf{p} - e\mathbf{A})] \phi = \varepsilon\phi \qquad ; \varepsilon \equiv E - m_0 c^2$$

(b) Show that the Pauli matrices satisfy

$$\sigma_i \sigma_j = \delta_{ij} + i\varepsilon_{ijk}\sigma_k$$

[38] Note that the calculation in the text in the ERL, by itself, does not determine R!

where ε_{ijk} is the completely antisymmetric tensor in three dimensions [see Prob. 8.7(b)], and use this to prove the following relation[39]

$$[\boldsymbol{\sigma}\cdot(\mathbf{p}-e\mathbf{A})]\,[\boldsymbol{\sigma}\cdot(\mathbf{p}-e\mathbf{A})] = \mathbf{p}^2 - e(\mathbf{A}\cdot\mathbf{p}+\mathbf{p}\cdot\mathbf{A}) + e^2\mathbf{A}^2 \\ -e\hbar\,\boldsymbol{\sigma}\cdot(\boldsymbol{\nabla}\times\mathbf{A})$$

(c) Hence conclude that when $|\varepsilon|/2m_0c^2 \ll 1$, the Dirac equation reduces to the following Schrödinger equation for the upper components ϕ

$$H\phi = \varepsilon\phi$$
$$H = \frac{\mathbf{p}^2}{2m_0} - \frac{e}{2m_0}(\mathbf{A}\cdot\mathbf{p}+\mathbf{p}\cdot\mathbf{A}) + \frac{e^2\mathbf{A}^2}{2m_0} - \frac{e\hbar}{2m_0}\boldsymbol{\sigma}\cdot\mathbf{B}$$

(d) Use this result to obtain Eqs. (9.52).

9.3 Consider the Dirac Eq. (9.51) for a particle in a static, central, electric field given by $\mathbf{E} = -\boldsymbol{\nabla}\Phi(r)$, and repeat Prob. 9.2.

(a) Show that in this case the Dirac equation can be reduced to

$$\left[c\boldsymbol{\sigma}\cdot\mathbf{p}\,\frac{1}{2m_0c^2+\varepsilon-e\Phi(r)}\,c\boldsymbol{\sigma}\cdot\mathbf{p}+e\Phi(r)\right]\phi = \varepsilon\phi \qquad ;\ \varepsilon\equiv E-m_0c^2$$

(b) Show that[40]

$$\boldsymbol{\sigma}\cdot\mathbf{p}\,\Phi(r)\,\boldsymbol{\sigma}\cdot\mathbf{p} = \Phi\mathbf{p}^2 + (\mathbf{p}\Phi)\cdot\mathbf{p} + \frac{\hbar}{r}\left(\frac{d\Phi}{dr}\right)\boldsymbol{\sigma}\cdot(\mathbf{r}\times\mathbf{p})$$

(c) Hence conclude that if $|\varepsilon|/2m_0c^2 \ll 1$, and $e\Phi\left\{1+O[(p/m_0c)^2]\right\} \approx e\Phi$, then the Dirac equation reduces to the following Schrödinger equation for the upper components ϕ

$$H\phi = \varepsilon\phi$$
$$H = \frac{\mathbf{p}^2}{2m_0} + e\Phi(r) + \frac{e\hbar^2}{(2m_0c)^2}\frac{1}{r}\left(\frac{d\Phi}{dr}\right)\boldsymbol{\sigma}\cdot\mathbf{l}$$

Here the spin-dependent contribution has been retained, and the angular momentum identified as $\mathbf{r}\times\mathbf{p} = \hbar\,\mathbf{l}$.

(d) Use this result to obtain Eq. (9.53).

---

[39] We here use the shorthand convention that repeated Latin indices are summed from 1 to 3; then $[p_i, A_j] = (\hbar/i)\nabla_i A_j$ and $\varepsilon_{ijk}\sigma_k\boldsymbol{\nabla}_i A_j = \boldsymbol{\sigma}\cdot(\boldsymbol{\nabla}\times\mathbf{A})$.

[40] Note that $\boldsymbol{\nabla}\Phi(r) = (\mathbf{r}/r)(d\Phi/dr)$.

9.4 (a) Consider the covariant form of the Dirac Eq. (9.48). Make the following *gauge-invariant replacement* [see part (b)]

$$\frac{\partial}{\partial x_\mu} \to \frac{\partial}{\partial x_\mu} - \frac{ie}{\hbar}A_\mu \qquad ; \; A_\mu = (\mathbf{A}, \frac{i}{c}\Phi)$$

Equation (9.48) then becomes

$$\left[\gamma_\mu\left(\frac{\partial}{\partial x_\mu} - \frac{ie}{\hbar}A_\mu\right) + M\right]\Psi = 0 \qquad ; \text{ Dirac eqn in E-M field}$$

Verify that this is equivalent to the Dirac equation written as

$$\left[c\boldsymbol{\alpha}\cdot(\mathbf{p} - e\mathbf{A}) + \beta m_0 c^2 + e\Phi\right]\Psi = i\hbar\frac{\partial\Psi}{\partial t}$$

This is Eq. (9.51).

(b) The gauge transformation of Prob. 8.9 leaves the electromagnetic fields $(\mathbf{E}, \mathbf{B})$ unchanged, and hence it should leave physical results unchanged. It is also true that since physics is bilinear in the wave function, physical results should be independent of the overall phase of the wave function. Show that the covariant form of the Dirac equation in part (a) is unchanged under the following replacements

$$A_\mu \to A_\mu + \frac{\partial\Lambda(x)}{\partial x_\mu} \qquad ; \text{ gauge invariance}$$

$$\Psi \to \exp\left\{\frac{ie}{\hbar}\Lambda(x)\right\}\Psi$$

This is the statement of the *gauge invariance* of the theory.

9.5 A picture that has proven useful for understanding much of nuclear structure is that of Dirac nucleons moving in effective, strong, Lorentz vector and scalar fields (V_μ, S). Here the Dirac equation reads

$$\left[\gamma_\mu\left(\frac{\partial}{\partial x_\mu} - \frac{ig_v}{\hbar c}V_\mu\right) + \left(M_N - \frac{g_s}{\hbar c}S\right)\right]\Psi = 0 \qquad ; \text{ Dirac nucleons}$$

Here (g_v, g_s) are coupling constants, analogous to the electric charge e. In the static case, in isotropic or spherically symmetric systems, the vector field has only a fourth component

$$V_\mu = i\delta_{\mu 4}V_0$$

(a) Show that in such systems, the above is equivalent to a Dirac equation of the form

$$
\begin{aligned}
(H_0 + H_1)\Psi &= i\hbar\frac{\partial \Psi}{\partial t} \\
H_0 &= c\boldsymbol{\alpha}\cdot\mathbf{p} + \beta m_N c^2 \\
H_1 &= g_v V_0 - \beta g_s S
\end{aligned}
$$

(b) Show that the binding energy in the Schrödinger equation in Prob. 9.3(c) now arises from the *difference* of the strong potentials $g_v V_0 - g_s S$.

(c) Show that with central potentials $[g_v V_0(r), g_s S(r)]$, the spin-orbit interaction in Prob. 9.3(c) is given by

$$
V_{SO} = \frac{\hbar^2}{(2m_N c)^2}\frac{1}{r}\left[\frac{d}{dr}(g_v V_0 + g_s S)\right]\boldsymbol{\sigma}\cdot\mathbf{l}
$$

It is the *sum* of the strong potentials $g_v V_0 + g_s S$ that contributes to the spin-orbit interaction. In this way, one is able to understand the strong spin-orbit force of the nuclear shell model.[41]

9.6 (a) The positive-energy spinors $u(\mathbf{k})$ in Eq. (9.29) satisfy the Dirac equation

$$
\begin{aligned}
(\hbar c\,\boldsymbol{\alpha}\cdot\mathbf{k} + \beta m_0 c^2)u(\mathbf{k}) &= E_k u(\mathbf{k}) \\
E_k &= \sqrt{(\hbar \mathbf{k} c)^2 + (m_0 c^2)^2}
\end{aligned}
$$

Use the arguments in Eqs. (9.31)–(9.38) to solve this equation. Show the solutions, normalized to $u_\lambda^\dagger u_{\lambda'} = \delta_{\lambda\lambda'}$, can be written as

$$
u_\lambda(\mathbf{k}) = \left(\frac{E_k + m_0 c^2}{2E_k}\right)^{1/2}\begin{pmatrix} \eta_\lambda \\ [\hbar c\,\boldsymbol{\sigma}\cdot\mathbf{k}/(E_k + m_0 c^2)]\,\eta_\lambda \end{pmatrix}
$$

Here the η_λ, with $\lambda = (\uparrow, \downarrow)$, are the two-component spinors in Eqs. (4.160).

(b) Consider the negative-energy solutions $v(-\mathbf{k})$ with energy eigenvalue $E = -E_k$ and momentum eigenvalue $\mathbf{p} = -\hbar\mathbf{k}$, again normalized to $v_\lambda^\dagger v_{\lambda'} = \delta_{\lambda\lambda'}$. Show they can be written

$$
v_\lambda(-\mathbf{k}) = \left(\frac{E_k + m_0 c^2}{2E_k}\right)^{1/2}\begin{pmatrix} [\hbar c\,\boldsymbol{\sigma}\cdot\mathbf{k}/(E_k + m_0 c^2)]\,\eta_\lambda \\ \eta_\lambda \end{pmatrix}
$$

---

[41] Relativistic Hartree calculations in this picture were first carried out by [Horowitz and Serot (1981)]; this approach is discussed further in [Walecka (2004)].

9.7 (a) Suppose the η_λ is Prob. 9.6 are two-component eigenstates of helicity so that $\boldsymbol{\sigma} \cdot \hat{\mathbf{k}}\,\eta_\lambda = \lambda\eta_\lambda$.[42] Show that the Dirac spinors in Prob. 9.6 are then corresponding eigenstates of the Dirac helicity $\boldsymbol{\Sigma} \cdot \hat{\mathbf{k}}$.

(b) Show that the Dirac spinors in Prob. 9.6 satisfy the following orthonormality relations

$$u_{\lambda'}^\dagger(\mathbf{k})u_\lambda(\mathbf{k}) = v_{\lambda'}^\dagger(-\mathbf{k})v_\lambda^\dagger(-\mathbf{k}) = \delta_{\lambda'\lambda}$$
$$u_{\lambda'}^\dagger(\mathbf{k})v_\lambda(\mathbf{k}) = v_{\lambda'}^\dagger(\mathbf{k})u_\lambda(\mathbf{k}) = 0$$

9.8 The matrix γ_5 is defined by $\gamma_5 \equiv \gamma_1\gamma_2\gamma_3\gamma_4$. Show the following:
(a) $\gamma_5^\dagger = \gamma_5$;
(b) $\gamma_5^2 = 1$;
(c) $\gamma_5\gamma_\mu + \gamma_\mu\gamma_5 = 0$;
(d) Verify its standard representation in Eq. (9.68).

9.9 (a) The wave function ϕ is defined in terms of the Dirac wave function ψ by $\phi \equiv (1/2)(1+\gamma_5)\psi$. Use the results in Prob. 9.8, and prove the following relation satisfied by the weak-interaction vertex [recall Eq. (9.44)]

$$\bar{\psi}\gamma_\mu(1+\gamma_5)\psi = 2\bar{\phi}\gamma_\mu\phi$$

(b) There are 16 linearly independent 4×4 matrices, and they can be taken as $1, \gamma_5, \gamma_\mu, \gamma_\mu\gamma_5$, and $\sigma_{\mu\nu} \equiv (1/2i)[\gamma_\mu, \gamma_\nu]$. Show that when written in terms of the ϕ, the weak vertex is unique. Show[43]

$$\bar{\phi}\gamma_\mu\gamma_5\phi = \bar{\phi}\gamma_\mu\phi$$
$$\bar{\phi}\phi = \bar{\phi}\gamma_5\phi = \bar{\phi}\sigma_{\mu\nu}\phi = 0$$

9.10 (a) Make use of the Dirac spinor in Prob. 9.6(a) and the standard representation for the gamma matrices. Explicitly evaluate $\bar{u}(\mathbf{k}')u(\mathbf{k})$ and $\bar{u}(\mathbf{k}')\gamma_5 u(\mathbf{k})$;

(b) Note that under a spatial reflection: $(\mathbf{r}, \mathbf{p}) \to (-\mathbf{r}, -\mathbf{p})$,[44] and $(\mathbf{l}, \boldsymbol{\sigma}) \to (\mathbf{l}, \boldsymbol{\sigma})$. Hence show that $\bar{u}u$ is a scalar under the parity transformation, and $\bar{u}\gamma_5 u$ is a *pseudoscalar*;

(c) Repeat for $\bar{u}\gamma_\mu u$ and $\bar{u}\gamma_\mu\gamma_5 u$. Show the first represents an ordinary polar vector, and the second an *axial vector.*

9.11 (a) Start from the stationary-state Dirac Eq. (9.21). Multiply on the left by $\psi^\dagger\beta$, multiply the adjoint of that equation on the right by $\beta\psi$,

[42] The easiest way to ensure this is to let $\hat{\mathbf{k}}$ define the z-axis.
[43] This is basis of the V-A theory [Feynman and Gell-Mann (1958)].
[44] Or $\mathbf{k} \to -\mathbf{k}$.

and add. Show

$$m_0c^2\,\psi^\dagger\psi = E\,\bar{\psi}\psi$$

(b) Since $u^\dagger u$ represents a probability *density* (probability/volume), it is not invariant under a Lorentz transformation. The quantity $\bar{u}u$ *is* Lorentz invariant. Why is this true? Show that the Dirac spinors $U(\mathbf{k})$ with invariant norm are given by

$$U(\mathbf{k}) = \left(\frac{E_k}{m_0c^2}\right)^{1/2} u(\mathbf{k}) \qquad ; \; \bar{U}U = 1$$

Here $E_k = \hbar c\sqrt{\mathbf{k}^2 + M^2}$.

9.12 For a free particle, the combination $k_\mu^2 = \mathbf{k}^2 - k_0^2 = -M^2$ is unchanged under a Lorentz transformation. Show that this represents a two-sheeted hyperboloid in four-dimensional k-space (the "mass hyperboloid"). Proper Lorentz transformations preserve the sign of k_0, and are thus confined to one sheet of the mass hyperboloid. Now define $k_\mu^2 \equiv k^2$, and

(a) Demonstrate the following relation for the Dirac delta function

$$\delta(k^2 + M^2) = \frac{1}{2\sqrt{\mathbf{k}^2 + M^2}}\left[\delta\left(k_0 - \sqrt{\mathbf{k}^2 + M^2}\right) + \delta\left(k_0 + \sqrt{\mathbf{k}^2 + M^2}\right)\right]$$

(b) Show that $d^4k \equiv d^3k\,dk_0$ is a Lorentz invariant.[45]

(c) Show the Lorentz-invariant phase space volume is

$$\frac{d^3k}{2E_k} = \frac{1}{\hbar c}\delta(k^2 + M^2)\theta(k_0)\,d^4k \qquad ; \text{ invariant phase space}$$

where the θ-function is $\theta(k_0) = 1$ if $k_0 > 0$, and $\theta(k_0) = 0$ if $k_0 < 0$.

9.13 Consider a scattering process in a Lorentz frame where the incident momentum $\hbar\mathbf{k}_1$ and the target momentum $\hbar\mathbf{k}_2$ are parallel to each other (lab frame, C-M frame, *etc.*). Show the incident flux in such a frame can be written in terms of the four-vectors (k_1, k_2) as [recall Prob. 7.3]

$$I_{\text{inc}} = \frac{c}{\Omega}\frac{(\hbar c)^2}{E_1E_2}\left[(k_1\cdot k_2)^2 - M_1^2M_2^2\right]^{1/2}$$

9.14 Consider inelastic electron scattering from a proton where the initial and final electron four-momenta are $(k_1, k_2) = (k_1, k_1 - q)$ (compare

[45]Hint: The transformation between volume elements is $d^4k' = \frac{\partial(k_1'\cdots k_0')}{\partial(k_1\cdots k_0)}d^4k$, where $\frac{\partial(k_1'\cdots k_0')}{\partial(k_1\cdots k_0)}$ is the jacobian determinant. If this is unfamiliar, just accept part (b).

Fig. 7.10). The response surfaces $W_{1,2}$ can be written as functions of the two Lorentz scalars q^2 and $\nu = -q \cdot p/mc$ where $m = m_p$.

(a) Show that in the laboratory frame, in the ERL,

$$q^2 = 4\tilde{k}_1\tilde{k}_2 \sin^2\frac{\theta}{2} \qquad ; \ \nu = \tilde{k}_1 - \tilde{k}_2 \qquad ; \text{ lab frame}$$

Here $\tilde{k} \equiv |\mathbf{k}|$.

(b) The functions $W_{1,2}(q^2, \nu)$ can be calculated under any kinematic conditions. Consider the "$\mathbf{p} \rightarrow \infty$ frame" where the electron and proton approach each other with $v \rightarrow c$. In this frame the proton is flattened by Lorentz contraction, and any internal motion is slowed down by time dilation. Suppose a quark in the proton now carries a fraction x of the proton's four-momentum p, and that the scattering takes place incoherently from the free, point-like quark. Show from the kinematics that in the *deep-inelastic* regime, where $(q^2, \nu) \rightarrow \infty$ and all masses are negligible, the quantity x is given by

$$x = \frac{q^2}{2M\nu} \qquad ; \text{ Bjorken-x}$$

where $M = mc/\hbar$. In this regime, the response surfaces become functions of the single variable x. This is the basis of the quark-parton model [Bjorken and Paschos (1969)].[46]

10.1 This problem concerns the Doppler shift in special relativity. A light signal is sent out from a source at $\bar{O}$ moving with velocity V in the z-direction. In its rest frame the source undergoes dN oscillations in a proper time $d\tau$. All observers can agree on this number dN. The proper frequency in the rest frame of the source is $\bar{\nu} = dN/d\tau$ or $dN = \bar{\nu}\, d\tau$. Now Lorentz transform to the laboratory frame where the source is moving with velocity V.

(a) Show the corresponding time interval dt in the lab is given by the relation

$$d\tau = \left(1 - \frac{V^2}{c^2}\right)^{1/2} dt$$

(b) During the time dt, the light front has traveled a distance $dl = cdt$ in the laboratory frame and hence arrives at an origin O a distance $dl = cdt$ away as shown in Fig. 13.5(a).

[46]For an introduction to electron scattering, see [Walecka (2001)].

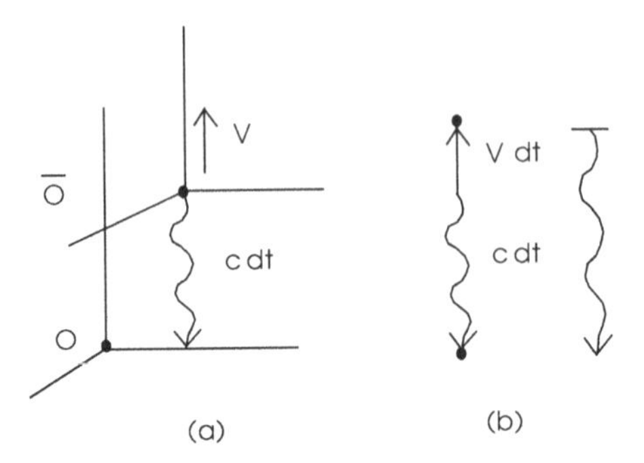

Fig. 13.5 Configuration for calculation of the Doppler shift in special relativity.

In the time dt, the source has moved to a new position a distance $dl' = cdt + Vdt$ away from O. The wavelength of the radiation received at O is thus increased as indicated in Fig. 13.5(b). Since λ is the actual distance the signal has traveled divided by the number of oscillations, show that

$$\lambda = \frac{(c+V)dt}{dN} = \bar{\lambda}\,\frac{(1+V/c)}{(1-V^2/c^2)^{1/2}}$$

Here $\bar{\lambda} = c/\bar{\nu}$ is the proper wavelength of the radiation for a source at rest.

(c) Show that in the limit $V/c \to 0$, this reduces the familiar freshman physics result. Note that $V = \pm|V|$ can have either sign in these arguments.

10.2 (a) Consider a curve $q(x)$. Show that a little element of length along the curve is given by $ds = [1 + (q')^2]^{1/2}dx$ where $q' \equiv dq(x)/dx$. Hence show that the total length of the curve S is given by

$$S = \int_{x_1}^{x_2} \left[1 + (q')^2\right]^{1/2} dx$$

(b) Use the result in Prob. 10.3 to prove that the shortest distance between two points is a straight line.

10.3 You are given an integral of the following form

$$S = \int_{x_1}^{x_2} F(q, q'; x)dx$$

where $q' \equiv dq(x)/dx$ and $F(q, q'; x)$ is a specified function of the indicated variables (the last one indicates any additional explicit x dependence). This is known as a *functional* of $q(x)$. It assigns to every function $q(x)$ a number S. Suppose you want to find that curve $q(x)$ that *minimizes* S.

(a) Let $q(x)$ be the solution to the problem. Consider small *variations* about that curve of the form $q(x) \rightarrow q(x)+\lambda\eta(x)$ where λ is an infinitesimal and $\eta(x)$ is arbitrary except for the condition of fixed endpoints $\eta(x_1) = \eta(x_2) = 0$ (see Fig. 13.6).

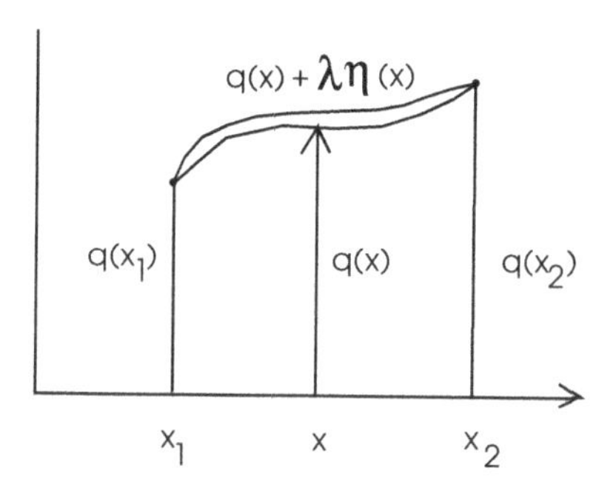

Fig. 13.6 Curve $q(x)$ that minimizes the functional S, and small variation about that curve $q(x) + \lambda\eta(x)$ where λ is an infinitesimal and $\eta(x)$ is arbitrary except for the fixed endpoints $\eta(x_1) = \eta(x_2) = 0$.

It follows that $q'(x) \rightarrow q'(x) + \lambda\eta'(x)$. Now make a Taylor series expansion in λ, and show that S takes the form

$$S(\lambda) = S_0 + \lambda \int_{x_1}^{x_2} \left[\eta(x)\left(\frac{\partial F}{\partial q}\right) + \frac{d\eta(x)}{dx}\left(\frac{\partial F}{\partial q'}\right)\right] dx + O(\lambda^2)$$

Here the partial derivatives indicate that all the other variables in $F(q, q'; x)$ are held fixed in carrying out the differentiation;

(b) If $q(x)$ is the solution to the problem, then any other curve must increase the value of S; hence $S(\lambda)$ must have a minimum at $\lambda = 0$. Show this condition becomes

$$\int_{x_1}^{x_2} \left[\eta(x)\left(\frac{\partial F}{\partial q}\right) + \frac{d\eta(x)}{dx}\left(\frac{\partial F}{\partial q'}\right)\right] dx = 0$$

Note that this condition only ensures that $S(\lambda)$ has an extremum at $\lambda = 0$ — minimum, maximum, or point of inflection; which it is can usually be argued from the physics of the problem;

(c) Carry out a partial integration on the second term using the condition of fixed endpoints, and show that the condition in part (b) becomes

$$\int_{x_1}^{x_2} \left[\left(\frac{\partial F}{\partial q}\right) - \frac{d}{dx}\left(\frac{\partial F}{\partial q'}\right)\right] \eta(x)\, dx = 0$$

(d) Since $\eta(x)$ is arbitrary, conclude that the curve $q(x)$ that minimizes the functional S is determined by the following second-order differential equation

$$\frac{d}{dx}\left(\frac{\partial F}{\partial q'}\right) - \left(\frac{\partial F}{\partial q}\right) = 0 \qquad \text{; Euler-Lagrange equation}$$

Since the solution is required to pass through the two endpoints, it is determined. This differential equation is known as the Euler-Lagrange equation for this variational problem.[47] Note that the d/dx here is a *total* derivative.

10.4 You are given a lagrangian $L(q, \dot{q};\, t) = T - V$ for a particle with generalized coordinate $q(t)$ and kinetic and potential energies (T, V) respectively. Here $\dot{q} \equiv dq(t)/dt$. The lagrangian is a given function of the indicated variables (the last denotes any additional explicit t dependence). The *action* is defined by

$$S \equiv \int_{t_1}^{t_2} L(q, \dot{q};\, t)\, dt \qquad \text{; action}$$

Hamilton's principle states that the path the particle takes between the points $q(t_1)$ and $q(t_2)$ is the one that minimizes the action (more generally, makes it *stationary*).

(a) Use the results in Prob. 10.3 to derive *Lagrange's equation* for the subsequent motion

$$\frac{d}{dt}\left(\frac{\partial L}{\partial \dot{q}}\right) - \frac{\partial L}{\partial q} = 0 \qquad \text{; Lagrange's equation}$$

(b) Show that if one has a problem with n dimensions, there will be one Lagrange equation for each linearly independent coordinate q^i with $i = 1, 2, \cdots, n$.

10.5 You are given the lagrangian in cartesian coordinates for a non-relativistic particle moving in a potential $V(\mathbf{x})$

$$L = \frac{1}{2}m(\dot{x}^2 + \dot{y}^2 + \dot{z}^2) - V(\mathbf{x})$$

Use Lagrange's equations in Prob. 10.4 to derive the equations of motion. Show they are exactly those given by Newton's second law.

10.6 The lifetime of a free muon is $\tau = 2.197 \times 10^{-6}$ sec. How close to the Schwarzschild radius, $r/R_s = 1 + \epsilon$, must a muon be to have its

[47] See the section on "calculus of variations" in [Fetter and Walecka (2003)].

laboratory lifetime extended to 1 sec?

10.7 Prove that in the Robertson-Walker metric with $k = 0$, the horizon recedes with a constant velocity $(v/c)_{\rm H} = 2/3$.

11.1 Make use of Fig. 11.7, which gives the low-temperature excitation spectrum $\varepsilon_k(k)$ in He-II, to answer the following:
(a) What is the velocity of sound $c_{\rm sound}$ in m/s?
(b) What is the critical velocity $v_{\rm critical}$ in m/s?

11.2 It is shown in [Fetter and Walecka (2003a)] that with the interaction in Eq. (11.15), and with the same set of approximations as used in the text, the excitation spectrum in the weakly interacting Bose gas is that of a sound wave with a speed of sound given by $c_{\rm sound} = (n_0 g/m)^{1/2}$. Use the results in Prob. 11.1(a), and Eqs. (11.30) and (11.26), to estimate the vortex core size in He-II. Compare with the observed value given in the text.

11.3 Write or obtain a program to integrate the dimensionless, nonlinear Gross-Pitaevskii equation for the vortex in Eqs. (11.27) and (11.28), and reproduce Fig. 11.6.

11.4 Consider a slab of bulk superconductor with a surface in the $x = 0$ plane and occupying the half-space $x \geq 0$. Assume there is a static magnetic field which for $x \ll 0$ is $\mathbf{B} = B_0\hat{\mathbf{e}}_z$. In this case, London augmented Maxwell's equations with an additional relation inside the superconductor which allows one to trace the magnetic field through the surface[48]

$$\boldsymbol{\nabla} \times \mathbf{j} = -\frac{1}{\mu_0 \lambda^2}\mathbf{B} \qquad ; \; \lambda^2 \equiv \frac{mc^2\varepsilon_0}{n_s e^2}$$

Hence, when combined with Maxwell's equations, one has for static fields

$$\begin{aligned} \boldsymbol{\nabla} \times \mathbf{B} &= \mu_0\,\mathbf{j} && ; \text{ Maxwell} \\ \boldsymbol{\nabla} \cdot \mathbf{B} &= 0 && \\ \boldsymbol{\nabla} \times \mathbf{j} &= -\frac{1}{\mu_0 \lambda^2}\mathbf{B} && ; \text{ London} \end{aligned}$$

Here $\mathbf{j} = 0$ for $x < 0$, and $\mathbf{E} = 0$ everywhere.
(a) Look for a solution to this set of equations of the form $\mathbf{B} = B(x)\hat{\mathbf{e}}_z$.

[48]See [Fetter and Walecka (2003a)]; here n_s is the density of superconducting electrons. In the London gauge, this equation is written as $\mathbf{j} = -(1/\mu_0\lambda^2)\mathbf{A}$ where $\mathbf{B} = \boldsymbol{\nabla} \times \mathbf{A}$.

Take the curl of the first equation and show that (*Hint*: see Prob. F.4)

$$\frac{d^2B(x)}{dx^2} = 0 \qquad ; \; x < 0$$

$$\frac{d^2B(x)}{dx^2} = \frac{1}{\lambda^2}B(x) \qquad ; \; x \geq 0$$

(b) Show the acceptable solution in (a) is

$$B(x) = B_0 \qquad ; \; x < 0$$

$$= B_0 e^{-x/\lambda} \qquad x \geq 0$$

Here λ is the *penetration depth*, with a typical value of $\lambda_{\text{exp}} = 510$ Å in superconducting tin.

(c) Show that for $x \geq 0$

$$\mathbf{j} = j(x)\hat{\mathbf{e}}_y \qquad ; \; j(x) = \frac{B_0}{\mu_0 \lambda} e^{-x/\lambda}$$

(d) Show that Maxwell's equations given above imply

$$\boldsymbol{\nabla} \cdot \mathbf{B} = \boldsymbol{\nabla} \cdot \mathbf{j} = 0$$

and that the obtained solutions satisfy these relations.

11.5 Carry out a partial integration on the expression for the gap Δ in Eq. (11.62) to isolate the singular part of the integral in Eq. (11.63). Make use of the fact that $\ln(t^2 - k_{\text{F}}^2)$ is integrable. Assume that $|\tilde{u}(t)|^2$ is well-behaved and that it vanishes sufficiently fast as $t \to \infty$.

11.6 Consider the ordinary solution to the eigenvalue Eq. (11.59) given in Eq. (11.61).[49] Show that it can be written as the expectation value of the potential in Eqs. (11.45) taken with the wave function in Eq. (11.46)

$$\Delta E(k) = \frac{\hbar^2}{2\mu}(\kappa^2 - k^2) = \int d^3x \, \phi_{\mathbf{k}}^\star(\mathbf{x}) V(x) \phi_{\mathbf{k}}(\mathbf{x}) \Big/ \int d^3x \, \phi_{\mathbf{k}}^\star(\mathbf{x}) \phi_{\mathbf{k}}(\mathbf{x})$$

where the potential is generalized to be non-local and separable as in Eqs. (11.53)–(11.55)

11.7 The Schrödinger equation for a particle in a magnetic field follows from the hamiltonian in Eqs. (11.68) and the gauge-invariant replacement in Eq. (11.69)

$$i\hbar\frac{\partial \Psi}{\partial t} = \frac{1}{2m}(\mathbf{p} - e\mathbf{A})^2\Psi$$

[49] We now use an equality in Eq. (11.61); note that here λ can have either sign.

Here the canonical momentum is given by $\mathbf{p} = (\hbar/i)\boldsymbol{\nabla}$ and the vector potential $\mathbf{A}(\mathbf{x})$ is real. The probability density and current then become

$$\begin{aligned}\rho &= |\Psi|^2 \\ \mathbf{s} &= \frac{1}{2m}\left\{\Psi^\star(\mathbf{p}-e\mathbf{A})\Psi + [(\mathbf{p}-e\mathbf{A})\Psi]^\star\Psi\right\}\end{aligned}$$

(a) Show that the continuity equation given in Eqs. (4.111) is satisfied.

(b) Make use of Eqs. (9.50), and generalize this result to the case of an arbitrary electromagnetic field.

12.1 Retain the next term in the expansion in Eq. (12.1), and treat the resulting change in V as a perturbation. Discuss the processes it describes when the quantum field expression in Eq. (12.30) is substituted into it.

12.2 (a) Show

$$\begin{aligned}\int_\Omega d^3x\, q_{\mathbf{k}}(\mathbf{x},t)q_{\mathbf{k}'}(\mathbf{x},t) &= \delta_{\mathbf{k}',-\mathbf{k}}\, e^{-2i\omega_k t} \\ [\mathbf{k}\times\mathbf{e}_{\mathbf{k}s}]\cdot[\mathbf{k}\times\mathbf{e}_{\mathbf{k}s'}] &= \mathbf{k}^2\, \mathbf{e}_{\mathbf{k}s}\cdot\mathbf{e}_{\mathbf{k}s'}\end{aligned}$$

where $q_{\mathbf{k}}(\mathbf{x},t)$ is the normal mode in Eq. (12.39);

(b) Now substitute the expression for the vector potential in Eq. (12.42) into the electromagnetic field energy in Eqs. (12.36) and (12.38) to obtain the normal-mode expansion in Eq. (12.46).

12.3 Consider a Dirac particle bound in some potential making a transition from a state with wave function ψ_i and energy ε_i to a state with ψ_f and ε_f, while emitting a photon of polarization $\mathbf{e}_{\mathbf{k}s}$ and energy $\hbar\omega_k$. The transition rate can be calculated from Eqs. (I.35) and (I.32) through the use of the matrix element developed in Eq. (12.50). Show this radiative transition rate is

$$d\omega_{fi} = \frac{\alpha\omega_k}{2\pi}\left|\int_\Omega d^3x\, e^{-i\mathbf{k}\cdot\mathbf{x}}\mathbf{e}_{\mathbf{k}s}\cdot\bar{\psi}_f\boldsymbol{\gamma}\psi_i\right|^2 d\Omega_k \qquad ;\ \text{photoemission}$$

where $\varepsilon_i = \varepsilon_f + \hbar\omega$, and α is the fine-structure constant.

12.4 (a) The continuity equation in non-relativistic quantum mechanics is given in Eq. (4.111). Use the argument in Eq. (9.42) to show the corresponding electromagnetic current four-vector ej_μ is obtained from

$$\begin{aligned}\frac{e}{c}\mathbf{j} &= \frac{e\hbar}{2im_0c}\left[\Psi^\star\boldsymbol{\nabla}\Psi - (\boldsymbol{\nabla}\Psi)^\star\,\Psi\right] \qquad ;\ \text{E-M current} \\ e\rho &= e\Psi^\star\Psi\end{aligned}$$

(b) Show the radiative transition rate in Prob. 12.3 now becomes

$$d\omega_{fi} = \frac{\alpha\omega_k}{2\pi} \left| \frac{\hbar}{2m_0c} \int_\Omega d^3x \, e^{-i\mathbf{k}\cdot\mathbf{x}} \mathbf{e_{k}}_s \cdot \left[\psi_f^\star \boldsymbol{\nabla} \psi_i - (\boldsymbol{\nabla}\psi_f)^\star \psi_i \right] \right|^2 d\Omega_k$$

(c) Use the result in Prob. 9.2(c) to show that this is what is obtained with a non-relativistic reduction of the Dirac equation.

12.5 Substitute the expansion of the Dirac field in Eq. (12.57) into Eq. (12.59) to obtain the normal-mode expression for the hamiltonian of the free Dirac field in Eq. (12.60). Make use of the Dirac equation satisfied by the free Dirac spinors and the orthonormality relations in Prob. 9.7.

12.6 (a) Show that in the ERL the Dirac spinor in Prob. 9.6(a) becomes

$$u_\lambda(\mathbf{k}) = \frac{1}{\sqrt{2}} \begin{pmatrix} \eta_\lambda \\ \boldsymbol{\sigma}\cdot\mathbf{e_k}\,\eta_\lambda \end{pmatrix} \qquad ; \; \mathbf{e_k} \equiv \frac{\mathbf{k}}{k}$$

(b) Use this result to derive Eq. (12.65).

12.7 (a) Prove the completeness relation for the Dirac spinors in Prob. 9.6[50]

$$\sum_\lambda \left[\underline{u}_\lambda(\mathbf{k})\, \underline{u}_\lambda(\mathbf{k})^\dagger + \underline{v}_\lambda(\mathbf{k})\, \underline{v}_\lambda(\mathbf{k})^\dagger \right]_{\alpha\beta} = \delta_{\alpha\beta}$$

(b) Use this result to derive the canonical equal-time anticommutation relation for the Dirac field

$$\left\{ \psi_\alpha(\mathbf{x},t), \psi_\beta^\dagger(\mathbf{x}',t') \right\}_{t=t'} = \delta_{\alpha\beta}\, \delta^{(3)}(\mathbf{x}-\mathbf{x}')$$

12.8 (a) Prove from Eqs. (12.52)–(12.53) that $a^\dagger\psi_0 \propto \psi_1$ and $a^\dagger\psi_1 = 0$;
(b) Prove that $a\psi_1 \propto \psi_0$ and $a\psi_0 = 0$.
The phases of the states can then be chosen so that Eqs. (12.56) are satisfied.

12.9 Derive Eqs. (12.33).

12.10 (a) Insert the field expansion in Eq. (12.68) into the hamiltonian in Eq. (12.67), and reduce that expression to a sum over modes;
(b) Discuss the processes described by the resulting hamiltonian.

[50] See Prob. B.4.

12.11 The lagrangian and action for the string are obtained as follows

$$L = T - V = \frac{\sigma}{2}\int_0^l dx \left\{ \left[\frac{\partial q(x,t)}{\partial t}\right]^2 - c^2 \left[\frac{\partial q(x,t)}{\partial x}\right]^2 \right\} \qquad ; \text{ lagrangian}$$

$$S = \int_{t_1}^{t_2} L\, dt = \int_{t_1}^{t_2} dt \int_0^l dx\, \mathcal{L}\left(\frac{\partial q}{\partial t}, \frac{\partial q}{\partial x}, q;\, t\right) \qquad ; \text{ action}$$

The quantity $\mathcal{L}$ is known as the *lagrangian density*, and to avoid confusion, we now use l for the length of the string.

(a) Use the analysis of Hamilton's principle in Prob. 10.4, together with fixed endpoints in time and periodic boundary conditions in space, to derive the continuum form of Lagrange's equation[51]

$$\frac{\partial}{\partial t}\left[\frac{\partial \mathcal{L}}{\partial(\partial q/\partial t)}\right] + \frac{\partial}{\partial x}\left[\frac{\partial \mathcal{L}}{\partial(\partial q/\partial x)}\right] - \frac{\partial \mathcal{L}}{\partial q} = 0 \qquad ; \text{ Lagrange's equation}$$

Here the partial derivatives of $\mathcal{L}$ keep all the other variables in $\mathcal{L}$ fixed, while the final partial derivatives with respect to (x, t) keep the other variable in this pair fixed;

(b) Show this reproduces the wave equation for $q(x, t)$;

(c) With the introduction of the "two-vector" $x_\mu \equiv (x, ict)$, and the convention that repeated Greek indices are summed from 1 to 2, show the lagrangian density and Lagrange's equation can be rewritten in invariant form as[52]

$$\mathcal{L} = -\frac{\tau}{2}\left(\frac{\partial q}{\partial x_\mu}\right)^2$$

$$\frac{\partial}{\partial x_\mu}\left[\frac{\partial \mathcal{L}}{\partial(\partial q/\partial x_\mu)}\right] - \frac{\partial \mathcal{L}}{\partial q} = 0$$

B.1 Verify the statement on block multiplication of matrices made in this appendix at the start of the discussion of Dirac matrices.

B.2 Verify the relations in Eqs. (B.18) by direct multiplication of the Dirac matrices in Eqs. (B.17).

B.3 Prove that the adjoint of a product of matrices is the product of the adjoints in the reverse order

$$[\underline{A}\,\underline{B}]^\dagger = \underline{B}^\dagger \underline{A}^\dagger$$

[51] Continuum mechanics is developed in [Fetter and Walecka (2003)].

[52] Recall that here c is the sound velocity with $c^2 = \tau/\sigma$.

B.4 If $\underline{a}$ and $\underline{b}$ are n-component column vectors, then $\underline{a}\,\underline{b}^\dagger$ is an $n \times n$ matrix (just use the extended rule for matrix multiplication). Prove the completeness relation for the two-component spinors in Eqs. (4.160)

$$\sum_\lambda \underline{\eta}_\lambda \underline{\eta}_\lambda^\dagger = \underline{1} \qquad ; \text{ completeness}$$

B.5 Evaluate a few of the structure constants f^{abc} in Eq. (B.20). How many are there in total?

B.6 (a) Show that it follows from Eq. (9.17) that in matrix notation $\underline{\Psi}^\dagger$ is the row vector $\underline{\Psi}^\dagger = (\psi_1^\star, \psi_2^\star, \cdots, \psi_n^\star)$ and Eq. (9.9) is then just $\rho = \underline{\Psi}^\dagger\underline{\Psi}$;

(b) Show that the new adjoint in Eq. (9.44) is $\overline{\underline{\Psi}} = (\psi_1^\star, \psi_2^\star, -\psi_3^\star, -\psi_4^\star)$.

C.1 Verify Eqs. (C.12).

C.2 Consider the Fourier cosine series in Eqs. (C.6):

(a) Show that it defines a function in the interval $-L \leq x \leq L$ that is even about the origin with $f(-x) = f(x)$;

(b) Show that the *slope* of this function vanishes on the boundaries;

(c) Show that this series defines an extension of this function to all x that is *periodic in x* with period $2L$.

(d) What are the corresponding features of the Fourier sine series in Eqs. (2.45) and (2.48)?

C.3 Compute the Fourier transform $\tilde{F}(k)$ in Eqs. (C.20) for the following functions:

(a) $F(x) = e^{-\gamma x}$ where γ is a constant;

(b) $F(x) = \sin \lambda x$ where λ is a constant. [*Hint*: Use Eq. (A.12) to write $\sin\phi = (e^{i\phi} - e^{-i\phi})/2i$.]

C.3 (a) Use Eq. (A.12) and the assumption that $f(x)$ is real and symmetric with $f(-x) = f(x)$ to rewrite the Fourier cosine series in Eqs. (C.6) in the following form

$$f(x) = \frac{1}{\sqrt{l}} \sum_{n=-\infty}^{\infty} a_e(k_n) e^{ik_n x} \qquad ; \; k_n = \frac{2\pi n}{l}$$

$$a_e(k_n) = \frac{1}{\sqrt{l}} \int_{-l/2}^{l/2} dx\, e^{-ik_n x} f(x) \qquad n = 0, \pm 1, \pm 2, \cdots, \pm\infty$$

where $l \equiv 2L$. This forms a *complex Fourier series*;

(b) Show that in this case $a_e(k_n)^\star = a_e(-k_n) = a_e(k_n)$.

C.4 Extend the results in Prob. C.3 to a real *antisymmetric* function $g(-x) = g(x)$.

(a) Show the modification of the result in Prob. C.3(a) is to replace $f(x)$ by $g(x)$, and $a_e(k_n)$ by $-ia_o(k_n)$, with $a_o(0) = 0$;

(b) Show that in this case $a_o(k_n)^\star = -a_o(-k_n) = a_o(k_n)$.

C.5 (a) Use the results in the previous two problems to extend the complex Fourier series in Prob. C.3(a) to an *arbitrary* real function $f(x)$ with Fourier coefficients $a(k_n)$;

(b) Show $a(k_n)^\star = a(-k_n)$ in this case;

(c) Show the extension of $f(x)$ to all x is then periodic with period l;[53]

(d) Take the limit $l \to \infty$, and re-derive the Fourier integral representation in Eqs. (C.20).

C.6 Use the results in Prob. 4.3 to show that the Fourier integral representation in Eqs. (C.20) can be extended to hold for an *arbitrary, well-behaved, complex function* $F(x)$.

E.1 The discussion in appendix E centers on independent *localized* particles. If the particles are *non-localized*, then a configuration that simply corresponds to an interchange of particle *labels* does not give rise to a new, distinct complexion. The number of such interchanges is $N!$, and thus one should replace $\Omega \to \Omega/N!$ in these arguments. Show the only effect of this change is to replace the Helmholtz free energy in Eq. (E.31) by

$$A(N, V, T) = -k_{\rm B}T \ln \frac{(\text{p.f.})^N}{N!} \qquad ; \text{ non-localized particles}$$

E.2 The classical partition function for a particle in a box of volume V in three dimensions is given by the first of Eqs. (E.33) as

$$(\text{p.f.}) = \frac{1}{h^3} \int \cdots \int e^{-(p_1^2+p_2^2+p_3^2)/2mk_{\rm B}T}\, dp_1 \cdots dp_3\, dq_1 \cdots dq_3$$

Show

$$(\text{p.f.}) = V \left(\frac{2\pi m k_{\rm B} T}{h^2} \right)^{3/2} \qquad ; \text{ classical limit}$$

E.3 Use the results in Probs. E.1 and E.2, and Eqs. (D.13), to show that the classical limit of the chemical potential for an ideal gas of density

[53] As in the physical string problem, we are content here to assume that $f(x)$ is everywhere continuous with a continuous first derivative; the Fourier series is capable of representing a broader class of functions.

$n = N/V$ is given by

$$\frac{\mu}{k_{\rm B}T} = -\ln\left\{\frac{1}{n}\left(\frac{2\pi m k_{\rm B}T}{h^2}\right)^{3/2}\right\} \qquad ; \text{ classical limit}$$

E.4 The quantum partition function for a particle in a box is given by

$$(\text{p.f.}) = \sum_{n_1=1}^{\infty}\sum_{n_2=1}^{\infty}\sum_{n_3=1}^{\infty} e^{-E_{n_1n_2n_3}/k_{\rm B}T}$$

where $E_{n_1n_2n_3}$ is the eigenvalue in Eq. (4.116). Introduce the quantities $u_i \equiv \hbar\pi n_i/\sqrt{2mk_{\rm B}T}$ with $i = (1,2,3)$. In the high-temperature limit where $T \to \infty$, the sum over n_i can be replaced by an integral over u_i, using the technique employed many times in this text.

(a) Show that the high-temperature limit of this quantum partition function is the classical result in Prob. E.2.

(b) Verify the identification of the constant h, here put in by hand in the classical partition function, with *Planck's constant.*.

E.5 (a) Consider the periodic orbit in phase space for the simple harmonic oscillator of energy E defined by Eqs. (2.12) and (2.15). The area of an ellipse is given by $A = \pi ab$ where (a, b) are the semi-major and semi-minor axes respectively. Show the area of the orbit in phase space is

$$A = \frac{2\pi E}{\omega} \qquad ; \kappa \equiv m\omega^2$$

(b) When quantized, the energy spectrum of the oscillator is that in Eq. (4.102). Show that the area in phase space between two adjacent levels is given by

$$A_{n+1} - A_n = h$$

where h is Planck's constant. This is a special case of the early Bohr-Sommerfeld quantization condition for periodic motion

$$\oint p\,dq = nh \qquad ; \text{ Bohr-Sommerfeld}$$

While not an exact result, it does hold quite generally for large quantum numbers, or in the classical limit.

(c) At low temperatures, the partition function in Eqs. (E.31) receives a contribution from only a few states, while at high T, many states contribute. In this case, one can group the contributions by using a mean value

$e^{-\varepsilon(p,q)/k_{\rm B}T}$ times the number of states $dn = dA/h$ in the small area of phase space $dA = dpdq$ from part (b). Thus, justify the first of Eqs. (E.32), and derive the first of Eqs. (E.33) for the partition function with $s = 1$ in the high T, or classical limit.

E.6 Use the equipartition theorem to show that the internal energy of an ideal gas of N structureless constituents at a temperature T is given by $E(T) = 3Nk_{\rm B}T/2$. Note that this only depends on the temperature.

F.1 (a) Complete the demonstration of Gauss' theorem in two dimensions in Eq. (F.6) by extending the analysis to include the other three quadrants;

(b) Complete the demonstration of Stokes' theorem in two dimensions in Eq. (F.9) by extending the analysis to include the other three quadrants.

F.2 Extend the proof of Gauss' theorem from two to three dimensions and derive Eq. (F.7). It is sufficient here to confine the analysis to the first octant.

F.3 Assume $\phi(x, y, z)$ is some differentiable scalar function of position, and use the definition of the gradient $\boldsymbol{\nabla}$ in Eq. (4.112).

(a) Show

$$\boldsymbol{\nabla} \times \boldsymbol{\nabla}\phi = 0$$

(b) Let $d\mathbf{l} = \hat{\mathbf{e}}_x\, dx + \hat{\mathbf{e}}_y\, dy + \hat{\mathbf{e}}_z\, dz$ be an infinitesimal displacement. Show the total differential of ϕ is given by

$$d\phi = \boldsymbol{\nabla}\phi \cdot d\mathbf{l}$$

(c) Show that in cylindrical coordinates (z, r, θ) the gradient $\boldsymbol{\nabla}$ takes the form

$$\boldsymbol{\nabla} = \hat{\mathbf{e}}_z \frac{\partial}{\partial z} + \hat{\mathbf{e}}_r \frac{\partial}{\partial r} + \hat{\mathbf{e}}_\theta \frac{1}{r}\frac{\partial}{\partial \theta}$$

F.4 Let $\mathbf{v}(\mathbf{x})$ be an arbitrary vector field. Work in cartesian coordinates, and prove the following vector identity

$$\boldsymbol{\nabla} \times (\boldsymbol{\nabla} \times \mathbf{v}) = \boldsymbol{\nabla}(\boldsymbol{\nabla} \cdot \mathbf{v}) - \nabla^2 \mathbf{v}$$

G.1 Start from Eqs. (12.39)–(12.42) and derive Eqs. (3.24) and (3.23) for the free radiation field in a big box with periodic boundary conditions.

H.1 The symmetrizing operator $\sum_{(\mathcal{P})} \mathcal{P}$ produces a wave function that is symmetric under the interchange of any pair of identical bosons. Assume three spin-zero bosons and an appropriate set of single-particle wave function $\phi_\alpha(\mathbf{x})$.

(a) Write the wave function $\Phi_{\alpha_1\alpha_2\alpha_3}(\mathbf{x}_1, \mathbf{x}_2, \mathbf{x}_3)$ when the α_i are distinct;

(b) Repeat part (a) for $\alpha_1 = \alpha_2$;

(c) Determine the normalization of the wave function in both cases;

(d) Write the expectation value of the kinetic energy in both cases.

H.2 Derive Eq. (H.11).

H.3 The Hartree-Fock (H-F) approximation determines the "best" single-particle wave functions to use in a Slater determinant to describe an arbitrary system of many identical fermions. The resulting H-F eqns can be derived from a variational calculation that minimizes the energy functional in Eq. (H.31) and (H.34). We forgo that derivation here, and simply state the H-F equations. In the notation of appendix H they read

$$\begin{aligned}
&t\underline{\phi}_\alpha(\mathbf{x}) + \sum_\beta \left[\int \underline{\phi}^\dagger_\beta(\mathbf{y})V(|\mathbf{x}-\mathbf{y}|)\underline{\phi}_\beta(\mathbf{y})d^3y\right]\underline{\phi}_\alpha(\mathbf{x}) \\
&\quad - \sum_\beta \left[\int \underline{\phi}^\dagger_\beta(\mathbf{y})V(|\mathbf{x}-\mathbf{y}|)\underline{\phi}_\alpha(\mathbf{y})d^3y\right]\underline{\phi}_\beta(\mathbf{x}) = \varepsilon_\alpha\underline{\phi}_\alpha(\mathbf{x}) \qquad ; \text{ H-F eqns}
\end{aligned}$$

Here the indices (α, β) run over the occupied states, and ε_α is an eigenvalue.[54] As they stand, these form a set of N coupled, non-linear, integro-differential equations with appropriate boundary conditions. The first potential term represents the *direct* interaction, and the second the *exchange* interaction.

Show that if the exchange term is neglected, the Hartree-Fock equations reduce to the Hartree equations of chapter 5.

H.4 (a) Multiply the H-F eqns of Prob. H.3 by $\underline{\phi}^\dagger_\gamma(\mathbf{x})$, and then carry out $\int d^3x$, to derive the following result (in the notation of appendix H)

$$\langle\gamma|t|\alpha\rangle + \sum_\beta [\langle\gamma\beta|V|\alpha\beta\rangle - \langle\gamma\beta|V|\beta\alpha\rangle] = \varepsilon_\alpha\langle\gamma|\alpha\rangle$$

(b) Assume that t and V are hermitian (Prob. 4.5). Derive the following relations on the one- and two-body matrix elements of Eqs. (H.30) and

[54] An additional one-body potential U can always be included by replacing $t \to t + U$.

(H.33)

$$\begin{aligned}\langle\gamma|t|\alpha\rangle^\star &= \langle\alpha|t|\gamma\rangle \\ \langle\alpha\beta|V|\gamma\delta\rangle^\star &= \langle\gamma\delta|V|\alpha\beta\rangle \\ \langle\alpha\beta|V|\gamma\delta\rangle &= \langle\beta\alpha|V|\delta\gamma\rangle\end{aligned}$$

(c) Set $\gamma = \alpha$ in part (a), take the complex conjugate, assume normalized single-particle wave functions with $\langle\alpha|\alpha\rangle = 1$, and prove that the eigenvalues in the H-F eqns are *real*

$$\varepsilon_\alpha^\star = \varepsilon_\alpha \qquad ; \text{ eigenvalues real}$$

(d) Now interchange $\gamma \leftrightarrow \alpha$ in part (a), take the complex conjugate, and subtract the resulting expression. Show the solutions to the H-F eqns corresponding to different eigenvalues are *orthogonal.* Since the solutions are to be normalized (and one can always orthogonalize degenerate solutions), show that

$$\langle\alpha|\beta\rangle = \delta_{\alpha\beta} \qquad ; \text{ solutions orthonormal}$$

H.5 Set $\alpha = \gamma$ in Prob. H.4(a), and use the result in Eq. (H.34) to show that the H-F single-particle energy and total energy of the system can be written as

$$\begin{aligned}\varepsilon_\alpha &= t_\alpha + v_\alpha \qquad ; \text{ H-F energy} \\ E &= \sum_\alpha t_\alpha + \frac{1}{2}\sum_\alpha v_\alpha\end{aligned}$$

where

$$\begin{aligned}t_\alpha &\equiv \langle\alpha|t|\alpha\rangle \\ v_\alpha &\equiv \sum_\beta \left[\langle\alpha\beta|V|\alpha\beta\rangle - \langle\alpha\beta|V|\beta\alpha\rangle\right]\end{aligned}$$

H.6 Assume a uniform system of identical spin-1/2 fermions in a big box of volume Ω with p.b.c., and a spin-independent two-particle potential of finite range. Show that the single-particle wave functions of Eqs. (4.159) provide the *solutions to the H-F eqns* in the form[55]

$$[t + V_D - V_E(\mathbf{k})]\,\underline{\phi}_{\mathbf{k}\lambda}(\mathbf{x}) = \varepsilon_{\mathbf{k}}\,\underline{\phi}_{\mathbf{k}\lambda}(\mathbf{x})$$

[55]This result follows from the *translational invariance* of this system.

In this expression the direct and exchange potentials are given by

$$V_D = \frac{1}{\Omega}\sum_{\mathbf{k}'\lambda'}^{k_F} \int V(z)\, d^3z = n\int V(z)\, d^3z$$

$$V_E(\mathbf{k}) = \frac{1}{\Omega}\sum_{\mathbf{k}'\lambda'}^{k_F} \delta_{\lambda\lambda'} \int e^{i(\mathbf{k}-\mathbf{k}')\cdot\mathbf{z}}\, V(z)\, d^3z$$

where $n = N/V$ is the particle density. The eigenvalues are then

$$\varepsilon_{\mathbf{k}} = \frac{\hbar^2\mathbf{k}^2}{2m} + \frac{1}{\Omega}\sum_{\mathbf{k}'\lambda'}^{k_F} \int V(z)\, d^3z \left[1 - \delta_{\lambda\lambda'}\, e^{i(\mathbf{k}-\mathbf{k}')\cdot\mathbf{z}}\right]$$

H.7 Start from the result for $\varepsilon_{\mathbf{k}}$ in Prob. H.6:

(a) Show that for "short-range" potentials, the exchange contribution to the H-F energy will be comparable to the direct contribution. What does "short-range" mean?

(b) Give an argument that for "long-range" potentials, the exchange term can be expected to be much smaller than the direct term. Use this argument to justify the Hartree approximation used for atoms in chap. 5. [*Hint*: Consider the role of $\exp\{i(\mathbf{k}-\mathbf{k}')\cdot\mathbf{z}\}$ in the integrand.]

I.1 This problem presents one application of Eqs. (I.35) and (I.36). Consider the scattering of a spin-zero particle of mass m, charge ze_p, and coordinate $\mathbf{x}_p$ through the Coulomb interaction with the nuclear charge distribution $Z\rho_N(\mathbf{x})$. Here H_1 is given by

$$H_1 = \frac{zZe^2}{4\pi\varepsilon_0}\int d^3x\, \frac{\rho_N(\mathbf{x})}{|\mathbf{x}_p - \mathbf{x}|}$$

The initial and final wave functions for the particle are

$$\phi_i(\mathbf{x}_p) = \frac{1}{\sqrt{\Omega}}e^{i\mathbf{k}_i\cdot\mathbf{x}_p} \qquad ; \ \phi_f(\mathbf{x}_p) = \frac{1}{\sqrt{\Omega}}e^{i\mathbf{k}_f\cdot\mathbf{x}_p}$$

(a) Show the matrix element of the perturbation is[56]

$$\langle\phi_f|H_1|\phi_i\rangle = \frac{zZe^2}{\varepsilon_0\Omega}\frac{F(\mathbf{q})}{\mathbf{q}^2} \qquad ; \ \mathbf{q} = \mathbf{k}_i - \mathbf{k}_f$$

$$F(\mathbf{q}) = \int d^3x\, e^{i\mathbf{q}\cdot\mathbf{x}}\, \rho_N(\mathbf{x}) \qquad ; \text{ form factor}$$

[56] Use $\int d^3x\, e^{i\mathbf{q}\cdot\mathbf{x}}/x = \mathrm{Lim}_{\mu\to 0}\int d^3x\, e^{i\mathbf{q}\cdot\mathbf{x}}\, e^{-\mu x}/x = \mathrm{Lim}_{\mu\to 0}\, 4\pi/(\mathbf{q}^2+\mu^2) = 4\pi/\mathbf{q}^2$.

(b) Assume non-relativistic kinematics for the scattering particle, and show

$$\int dk_f \, \delta(E_f - E_i) = \frac{m}{\hbar^2 k} \qquad ; \; k \equiv k_i = k_f$$

(c) Show the incident flux in the laboratory frame is $I_{\rm inc} = \hbar k / m\Omega$;
(d) Hence show the differential cross section is given by

$$\frac{d\sigma}{d\Omega_k} = Z^2 z^2 \frac{4\alpha^2}{\mathbf{q}^4} \left(\frac{mc}{\hbar}\right)^2 |F(\mathbf{q})|^2 \qquad ; \; \mathbf{q} = \mathbf{k}_i - \mathbf{k}_f$$

Here α is the fine structure constant, $d\Omega_k$ is the differential solid angle for the scattered particle, and $\mathbf{q}$ is the momentum transfer. Note that the quantization volume Ω cancels from this result.

(e) Show the result in part (d) can be rewritten as

$$\frac{d\sigma}{d\Omega_k} = z^2 Z^2 \sigma_{\rm Ruth} |F(\mathbf{q})|^2$$

$$\sigma_{\rm Ruth} = \left| \frac{\hbar c \, \alpha}{4E \sin^2 \theta/2} \right|^2 \qquad ; \; \text{Rutherford}$$

Here $E = \hbar^2 k^2 / 2m$ is the incident energy, and $\sigma_{\rm Ruth}$ is the Rutherford cross section (see [Fetter and Walecka (2003)]).

I.2 Compare Eqs. (I.32) and (I.35) with Eq. (7.46), and conclude that to within an overall phase, the T-matrix for a two-body scattering process is given to first order in the interaction H_1 by[57]

$$\frac{\Omega \, \delta_{\mathbf{K}_f, \mathbf{K}_i}}{\Omega^2} T_{fi}^{(1)} = \langle \phi_f | H_1 | \phi_i \rangle$$

K.1 (a) Show that $e_{\rm cgs} = 4.803 \times 10^{-10}$ esu
(b) How many esu are there in 1 C?
(c) What are the dimensions of e and $e_{\rm cgs}$?

K.2 Use the Lorentz force equation for an electron in a magnetic field in both SI and cgs units to show that $1\,\text{T} = 10^4$ Gauss.

---

[57]The proper treatment of the volume factors Ω in this case proceeds as in chapter 7.

Appendix A

Complex Variables—A Primer

This appendix briefly reviews the main features of complex analysis, essential to, among other things, the development of quantum mechanics.

Real Variables. We start from the following power series expansions with real variables (See, for example, [Arfken and Weber (1995)])[58]

$$
\begin{aligned}
e^x &= 1 + x + \frac{x^2}{2!} + \frac{x^3}{3!} + \frac{x^4}{4!} + \cdots \\
\cos x &= 1 - \frac{x^2}{2!} + \frac{x^4}{4!} + \cdots \\
\sin x &= x - \frac{x^3}{3!} + \frac{x^5}{5!} + \cdots
\end{aligned}
\tag{A.1}
$$

These expansions converge for all finite x. The first can be taken to define the exponential function. It follows directly from algebraic manipulations of the series, and from the definition of the derivative, that

$$
\begin{aligned}
e^x e^y &= e^{x+y} \\
\frac{d}{dx} e^x &= e^x
\end{aligned}
\tag{A.2}
$$

Complex Variables. A complex variable is a number pair (x, y) defined by the relation

$$
z = x + iy \qquad ; \; i^2 = -1 \tag{A.3}
$$

Here i is the imaginary root of $i^2 = -1$.

The quantity z has the properties of a two-dimensional vector running from the origin to the point (x, y) as illustrated in Fig. A.1. The geometric

[58]A thorough presentation of mathematical methods in theoretical physics is contained in [Morse and Feshbach (1953)].

length $(x^2 + y^2)^{1/2}$ is readily expressed as $|z| \equiv \sqrt{|z|^2}$, where

$$|z|^2 \equiv z^\star z = (x - iy)(x + iy) = x^2 + y^2 \tag{A.4}$$

Here $|z|$ is called the *modulus* of z, and $z^\star \equiv x - iy$ is its complex conjugate.

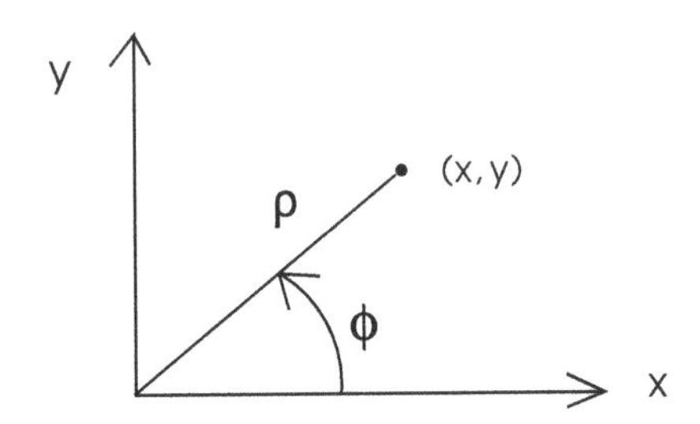

Fig. A.1 A representation of the complex number $z = x + iy$. We refer to this as the *complex plane.*

Two complex numbers can be added by adding the real and imaginary parts just like a vector

$$z_1 + z_2 = x_1 + x_2 + i(y_1 + y_2) \tag{A.5}$$

The multiplication of two complex numbers is more complicated but follows from the rules of algebra as

$$z_1 z_2 = (x_1 + iy_1)(x_2 + iy_2) = x_1 x_2 - y_1 y_2 + i(x_1 y_2 + x_2 y_1) \tag{A.6}$$

Division can be defined by using multiplication to rationalize the fraction

$$\frac{1}{z} = \frac{1}{x + iy} = \frac{1}{x + iy}\,\frac{x - iy}{x - iy} = \frac{x - iy}{|z|^2} \tag{A.7}$$

All the arsenal of algebra can now be employed.

As with the ordinary notion of functions, a function of a complex variable $f(z)$ is a mapping that assigns to the complex number z another complex number $f(z)$ with real and imaginary parts given by

$$f(z) = u(x, y) + iv(x, y) \tag{A.8}$$

The derivative of $f(z)$ is also defined in the usual fashion

$$\frac{d}{dz} f(x) \equiv \text{Lim}_{\,\delta \to 0} \left[\frac{f(z + \delta) - f(z)}{\delta} \right] \tag{A.9}$$

Analytic functions have a derivative that is independent of just how this limit is taken in the complex z plane. *Integration* follows from the notion of multiplication and the addition of small elements.[59]

The exponential function of a complex variable can now be defined exactly as in the first of Eqs. (A.1)

$$e^z \equiv 1 + z + \frac{z^2}{2!} + \frac{z^3}{3!} + \frac{z^4}{4!} + \frac{z^5}{5!} + \cdots \tag{A.10}$$

This series also converges for all finite z. It again follows from algebraic manipulation of this series, and from the definition of the derivative, that

$$\begin{aligned} e^{z_1}e^{z_2} &= e^{z_1+z_2} \\ \frac{d}{dz}e^z &= e^z \end{aligned} \tag{A.11}$$

A central relation can now be obtained by considering the function

$$\begin{aligned} e^{i\phi} &= 1 + (i\phi) + \frac{(i\phi)^2}{2!} + \frac{(i\phi)^3}{3!} + \frac{(i\phi)^4}{4!} + \frac{(i\phi)^5}{5!} + \cdots \\ &= 1 - \frac{\phi^2}{2!} + \frac{\phi^4}{4!} + \cdots + i\left(\phi - \frac{\phi^3}{3!} + \frac{\phi^5}{5!} + \cdots\right) \\ &= \cos\phi + i\sin\phi \end{aligned} \tag{A.12}$$

Here the real relations in Eqs. (A.1) have been used to obtain the final equality. This relation provides a particularly useful representation of the complex number z, since it is evident from Fig. (A.1) that

$$\begin{aligned} x &= \rho\cos\phi \qquad\qquad ; \; y = \rho\sin\phi \\ z &= \rho(\cos\phi + i\sin\phi) = \rho e^{i\phi} \\ |z|^2 &= \rho^2 \end{aligned} \tag{A.13}$$

Now multiplication and division are simple, for

$$\begin{aligned} z_1 z_2 &= \rho_1\rho_2 e^{i(\phi_1+\phi_2)} \\ \frac{z_1}{z_2} &= \frac{\rho_1}{\rho_2}e^{i(\phi_1-\phi_2)} \end{aligned} \tag{A.14}$$

They are the usual operations with the moduli, and in one case the phases are added, and in the other, subtracted.

Examples. We give a few examples:

---

[59]There is an extensive introduction to the theory of functions of a complex variable, one of the loveliest parts of mathematics, in appendix A of [Fetter and Walecka (2003)].

(1) It follows from Eq. (A.12), but more directly from simply considering where the phase ϕ takes you in the complex plane, that

$$
\begin{aligned}
e^{i\pi} &= e^{-i\pi} = -1 \\
e^{2\pi i} &= 1 \\
e^{n\pi i} &= \left(e^{i\pi}\right)^n = (-1)^n \qquad ; \ n \text{ integer} \\
e^{i\pi/2} &= i \qquad \text{etc.}
\end{aligned} \tag{A.15}
$$

(2) Consider the integral

$$
\begin{aligned}
I_{mn} &= \frac{1}{L}\int_{-L/2}^{L/2} e^{i(k_m-k_n)x}\,dx \\
k_p &= \frac{2\pi p}{L} \qquad ; \ p = 0, \pm 1, \pm 2, \cdots, \pm\infty
\end{aligned} \tag{A.16}
$$

If $m = n$, then the exponent vanishes and the integral is 1. If the integers $m \neq n$, then the ordinary rules of integration give

$$
\begin{aligned}
I_{mn} &= \frac{1}{iL(k_m-k_n)}\left[e^{i(k_m-k_n)x}\right]_{-L/2}^{L/2} \\
&= \frac{1}{iL(k_m-k_n)}\left[e^{i(m-n)\pi} - e^{-i(m-n)\pi}\right] \\
&= \frac{1}{iL(k_m-k_n)}\left[(-1)^{(m-n)} - (-1)^{(m-n)}\right] \\
&= 0 \qquad ; \ m \neq n
\end{aligned} \tag{A.17}
$$

Hence the integral I_{mn} in Eq. (A.16) given by

$$
I_{mn} = \delta_{mn} \tag{A.18}
$$

where δ_{mn} is the usual Kronecker delta defined by

$$
\begin{aligned}
\delta_{mn} &= 1 \qquad ; \ \text{if } m = n \\
&= 0 \qquad ; \ \text{if } m \neq n
\end{aligned} \tag{A.19}
$$

Appendix B

Matrices

Start from the following set of n linear, inhomogeneous algebraic equations

$$\begin{aligned} a_{11}x_1 + a_{12}x_2 + \cdots + a_{1n}x_n &= y_1 \\ a_{21}x_1 + a_{22}x_2 + \cdots + a_{2n}x_n &= y_2 \\ \vdots \qquad\qquad\qquad\qquad \vdots &= \vdots \\ a_{n1}x_1 + a_{n2}x_2 + \cdots + a_{nn}x_n &= y_n \end{aligned} \tag{B.1}$$

Here the a_{ij} are numerical coefficients, the y_i are given, and one is to solve for the unknowns x_i. These equations can be recast in *matrix form.* One defines the $n \times n$ matrix $\underline{a}$ with $[\underline{a}]_{ij} \equiv a_{ij}$, where the first index indicates the *row* and the second index indicates the *column.* Thus

$$\underline{a} = \begin{bmatrix} a_{11} & a_{12} & \cdots & a_{1n} \\ a_{21} & a_{22} & \cdots & a_{2n} \\ \vdots & \vdots & \vdots & \vdots \\ a_{n1} & a_{n2} & \dots & a_{nn} \end{bmatrix} \tag{B.2}$$

Simultaneously, one introduces a column vector $\underline{x}$ with $[\underline{x}]_i \equiv x_i$, so that

$$\underline{x} \equiv \begin{bmatrix} x_1 \\ x_2 \\ \vdots \\ x_n \end{bmatrix} \tag{B.3}$$

and similarly for $\underline{y}$

The ith element of the *matrix product* $\underline{a}\,\underline{x}$ is defined by summing the elements of the ith row of $\underline{a}$ with each element of the column of $\underline{x}$

$$[\underline{a}\,\underline{x}]_i \equiv \sum_{j=1}^{n} a_{ij}x_j \tag{B.4}$$

Thus the linear Eqs. (B.1) can be recast in matrix form as

$$\begin{aligned} [\underline{a}\,\underline{x}]_i &= [\underline{y}]_i \qquad ; \; i = 1, 2, \cdots, n \\ \text{or ;} \qquad \underline{a}\,\underline{x} &= \underline{y} \end{aligned} \tag{B.5}$$

Matrix Multiplication. Motivated by the above, one defines the multiplication of two matrices as follows: The product of two $n \times n$ matrices is again an $n \times n$ matrix where the i-jth element is obtained by taking the sum of the products of each element of the ith row of the first matrix times each element of the jth column of the second. Thus

$$[\underline{a}\,\underline{b}]_{ij} \equiv \sum_{k=1}^{n} a_{ik}b_{kj} \tag{B.6}$$

Pictorially, this is easy to remember—just run across the appropriate row of the first matrix and appropriate column of the second, taking products and adding the result as you go.[60]

The Inverse Matrix. The problem of solving the original set of algebraic Eqs. (B.1) is now reduced to finding the *inverse* matrix $\underline{a}^{-1}$ of the matrix $\underline{a}$ defined by

$$\underline{a}^{-1}\,\underline{a} = \underline{a}\,\underline{a}^{-1} = \underline{1} \tag{B.7}$$

Here $\underline{1}$ is the unit matrix with 1's down the diagonal and 0's everywhere else.[61] The solution to the original equations is evidently

$$\underline{x} = \underline{a}^{-1}\,\underline{y} \tag{B.8}$$

The procedure for finding the inverse summarizes the usual procedure of manipulation of the original linear equations by multiplication and addition to find a solution—it can be found in any good textbook on linear algebra (see, for example, [Hildebrand (1992)]).

Unitary and Hermitian Matrices. The complex conjugate of the transpose of a matrix (obtained by interchanging rows and columns) is called

[60] This rule is readily extended to matrices of other shapes.

[61] The unit matrix is frequently suppressed in writing matrix equations.

the *adjoint matrix* and is indicated with a dagger. Thus

$$U_{ij}^{\dagger} \equiv U_{ji}^{\star} \qquad ; \text{ adjoint matrix} \tag{B.9}$$

A matrix which is equal to its adjoint is said to be *hermitian*. A matrix whose adjoint is equal to its inverse is said to be *unitary*. Thus

$$\begin{aligned} \underline{U}^{\dagger} &= \underline{U} && ; \text{ hermitian} \\ \underline{U}^{\dagger} &= \underline{U}^{-1} && ; \text{ unitary} \end{aligned} \tag{B.10}$$

Eigenvalues. Let $\underline{H}$ be a hermitian matrix. The eigenvalues λ and eigenvectors $\underline{u}$ (simply a column matrix) of $\underline{H}$ are defined by the relation

$$\underline{H}\,\underline{u} = \lambda\,\underline{u} \tag{B.11}$$

The eigenvalues λ are real numbers if $\underline{H}$ is hermitian (compare Prob. 4.6).

Pauli Matrices. A set of hermitian 2×2 matrices, which play a central role in quantum mechanics, are the Pauli matrices defined by

$$\underline{\sigma}_x \equiv \begin{pmatrix} 0 & 1 \\ 1 & 0 \end{pmatrix} \qquad \underline{\sigma}_y \equiv \begin{pmatrix} 0 & -i \\ i & 0 \end{pmatrix} \qquad \underline{\sigma}_z \equiv \begin{pmatrix} 1 & 0 \\ 0 & -1 \end{pmatrix} \tag{B.12}$$

Here the subscripts simply label the matrices. Readers can easily convince themselves, through the rules for matrix multiplication given above, that

$$\begin{aligned} \underline{\sigma}_i^{\dagger} &= \underline{\sigma}_i && ; \; i = x, y, z \\ \underline{\sigma}_i\,\underline{\sigma}_j + \underline{\sigma}_i\,\underline{\sigma}_j &= 2\delta_{ij} && \\ \underline{\sigma}_x\,\underline{\sigma}_y - \underline{\sigma}_y\,\underline{\sigma}_x &= 2i\underline{\sigma}_z && ; \text{ and cyclic permutations of } (x, y, z) \end{aligned} \tag{B.13}$$

Here δ_{ij} is the Kronecker delta, and the unit matrix is again suppressed. With a redefinition of the *spin* operator as $\underline{S}_i \equiv \underline{\sigma}_i/2$, one has

$$\begin{aligned} \underline{S}_i &\equiv \frac{1}{2}\underline{\sigma}_i && ; \; i = x, y, z \\ \underline{S}_x\underline{S}_y - \underline{S}_y\underline{S}_x &= i\underline{S}_z && ; \text{ and cyclic permutations of } (x, y, z) \end{aligned} \tag{B.14}$$

These are just the commutation relations of *angular momentum* (see Prob. 4.25). The space here is two-dimensional and this represents spin-1/2.

Dirac Matrices. First note that it is possible to subdivide matrices into blocks and then simply multiply the blocks using the rules of matrix multiplication. This gives the correct result for the multiplication of the entire matrix.

Thus the standard representation of the 4×4 Dirac matrices is given in terms of 2×2 blocks as[62]

$$\underline{\alpha}_x \equiv \begin{pmatrix} 0 & \sigma_x \\ \sigma_x & 0 \end{pmatrix} \quad \underline{\alpha}_y \equiv \begin{pmatrix} 0 & \sigma_y \\ \sigma_y & 0 \end{pmatrix} \quad \underline{\alpha}_z \equiv \begin{pmatrix} 0 & \sigma_z \\ \sigma_z & 0 \end{pmatrix} \quad \underline{\beta} \equiv \begin{pmatrix} 1 & 0 \\ 0 & -1 \end{pmatrix} \tag{B.15}$$

It follows, for example, that

$$\underline{\alpha}_x \, \underline{\alpha}_y = \begin{pmatrix} \sigma_x \sigma_y & 0 \\ 0 & \sigma_x \sigma_y \end{pmatrix} = \begin{pmatrix} i\sigma_z & 0 \\ 0 & i\sigma_z \end{pmatrix} \tag{B.16}$$

Explicitly, in 4×4 form, the Dirac matrices are given by

$$\underline{\alpha}_x = \begin{bmatrix} 0 & 0 & 0 & 1 \\ 0 & 0 & 1 & 0 \\ 0 & 1 & 0 & 0 \\ 1 & 0 & 0 & 0 \end{bmatrix} \qquad \underline{\alpha}_y = \begin{bmatrix} 0 & 0 & 0 & -i \\ 0 & 0 & i & 0 \\ 0 & -i & 0 & 0 \\ i & 0 & 0 & 0 \end{bmatrix}$$

$$\underline{\alpha}_z = \begin{bmatrix} 0 & 0 & 1 & 0 \\ 0 & 0 & 0 & -1 \\ 1 & 0 & 0 & 0 \\ 0 & -1 & 0 & 0 \end{bmatrix} \qquad \underline{\beta} = \begin{bmatrix} 1 & 0 & 0 & 0 \\ 0 & 1 & 0 & 0 \\ 0 & 0 & -1 & 0 \\ 0 & 0 & 0 & -1 \end{bmatrix} \tag{B.17}$$

These matrices satisfy the following relations

$$\begin{aligned} \underline{\beta}^\dagger &= \underline{\beta} \\ \underline{\alpha}_i^\dagger &= \underline{\alpha}_i \qquad\qquad ; \; i = x, y, z \\ \underline{\beta}^2 &= 1 \\ \underline{\alpha}_i \, \underline{\beta} + \underline{\beta} \, \underline{\alpha}_i &= 0 \\ \underline{\alpha}_i \, \underline{\alpha}_j + \underline{\alpha}_j \, \underline{\alpha}_i &= 2\delta_{ij} \end{aligned} \tag{B.18}$$

The reader should again verify these relations and verify that the same results are obtained from the 2×2 form, which is always much easier to manipulate.

Gell-Mann Matrices. There are eight three-by-three, traceless, hermitian, Gell-Mann matrices $\underline{\lambda}^a$—the analogues of the Pauli matrices.[63] The

[62] We now suppress the underlining of the 2×2 matrices.

[63] The *trace* of a matrix is the sum of its diagonal elements.

matrices $\underline{\lambda}^a$ for $a = 1, \cdots, 8$ are given in order by

$$\begin{pmatrix} & 1 & \\ 1 & & \\ & & \end{pmatrix} \begin{pmatrix} & -i & \\ i & & \\ & & \end{pmatrix} \begin{pmatrix} 1 & & \\ & -1 & \\ & & \end{pmatrix} \begin{pmatrix} & & 1 \\ & & \\ 1 & & \end{pmatrix} \begin{pmatrix} & & -i \\ & & \\ i & & \end{pmatrix}$$

$$\begin{pmatrix} & & \\ & & 1 \\ & 1 & \end{pmatrix} \begin{pmatrix} & & \\ & & -i \\ & i & \end{pmatrix} \begin{pmatrix} 1/\sqrt{3} & & \\ & 1/\sqrt{3} & \\ & & -2/\sqrt{3} \end{pmatrix} \tag{B.19}$$

These matrices satisfy the Lie algebra of $SU(3)$[64]

$$[\frac{1}{2}\underline{\lambda}^a, \frac{1}{2}\underline{\lambda}^b] = if^{abc}\frac{1}{2}\underline{\lambda}^c \tag{B.20}$$

Here the f^{abc} are the structure constants of the group; they are antisymmetric in the indices (abc), and the repeated Latin index is here summed from 1 to 8. Problem B.5 provides some facility with these matrices.

[64]Note $[\underline{A}, \underline{B}] \equiv \underline{A}\underline{B} - \underline{B}\underline{A}$.

Appendix C

Fourier Series and Fourier Integrals

We first demonstrate the orthonormality of the eigenfunctions in Eq. (2.45)

$$\int_0^L dx\, \phi_m(x)\phi_n(x) = \frac{2}{L}\int_0^L dx\, \sin\left(\frac{m\pi x}{L}\right)\sin\left(\frac{n\pi x}{L}\right) = \delta_{mn} \qquad (C.1)$$

Use

$$\sin a\, \sin b = \frac{1}{2}\left[\cos(a-b) - \cos(a+b)\right] \qquad (C.2)$$

Hence the integral in Eq. (C.1) takes the form

$$\begin{aligned}
&\frac{2}{L}\int_0^L dx\, \sin\left(\frac{m\pi x}{L}\right)\sin\left(\frac{n\pi x}{L}\right) \\
&\quad = \frac{1}{L}\int_0^L dx\left[\cos\frac{(m-n)\pi x}{L} - \cos\frac{(m+n)\pi x}{L}\right] \\
&\quad = \frac{1}{L}\left[\frac{\sin(m-n)\pi x/L}{(m-n)\pi/L} - \frac{\sin(m+n)\pi x/L}{(m+n)\pi/L}\right]_0^L \\
&\quad = 0 \qquad\qquad ;\ \text{if } m \neq n
\end{aligned} \qquad (C.3)$$

If $m = n$ and both are non-zero, then the second term in the second line of the above still vanishes; however, the first term now gives 1. This establishes Eq. (C.1).

It is evident from this analysis that the use of the identity

$$\cos a\, \cos b = \frac{1}{2}\left[\cos(a-b) + \cos(a+b)\right] \qquad (C.4)$$

also leads to the orthonormality relation

$$\frac{2}{L}\int_0^L dx\, \cos\left(\frac{m\pi x}{L}\right)\cos\left(\frac{n\pi x}{L}\right) = \delta_{mn} \qquad ;\ (m,n) = 1, 2, \cdots, \infty \quad (C.5)$$

The integral vanishes if either $m = 0$ or $n = 0$ and the other index is non-zero; if both indices are zero so that $m = n = 0$, the integral is not 1, but 2.

Consider, then, the following Fourier series, and its inversion through the use of this orthonormality relation,

$$
\begin{aligned}
f(x) &= \sum_{n=1}^{\infty} a_n \left(\frac{2}{L}\right)^{1/2} \cos\left(\frac{n\pi x}{L}\right) + a_0 \left(\frac{1}{L}\right)^{1/2} \\
a_n &= \left(\frac{2}{L}\right)^{1/2} \int_0^L dx\, f(x) \cos\left(\frac{n\pi x}{L}\right) \qquad ; n = 1, 2, \cdots, \infty \\
a_0 &= \left(\frac{1}{L}\right)^{1/2} \int_0^L dx\, f(x)
\end{aligned}
\tag{C.6}
$$

From the first of Eqs. (C.6), $f(x)$ as defined here is an even function of x about the origin

$$
f(-x) = f(x) \qquad ; \text{ even} \tag{C.7}
$$

Both $a_n \equiv a(k_n)$ and $\cos(k_n x)$ are even functions of $k_n \equiv n\pi/L$, and hence, so is their product.

Now take the limit $L \to \infty$, and assume the function $f(x)$ is integrable so that

$$
\int_0^{\infty} dx\, f(x) = I \qquad ; \text{ finite} \tag{C.8}
$$

The last term in the first of Eqs. (C.6), which then goes as I/L, is negligible in this limit. Equation (2.55) can now be used to convert the sum over k_n in the remaining term to an integral over k

$$
\begin{aligned}
f(x) &= \frac{L}{2\pi} \int_{-\infty}^{\infty} dk\, a(k) \left(\frac{2}{L}\right)^{1/2} \cos kx \qquad ; L \to \infty \\
a(k) &= \left(\frac{2}{L}\right)^{1/2} \int_0^{\infty} dx\, f(x) \cos kx \\
&= \left(\frac{2}{L}\right)^{1/2} \frac{1}{2} \int_{-\infty}^{\infty} dx f(x) \cos kx
\end{aligned}
\tag{C.9}
$$

With the redefinition

$$
\left(\frac{L}{\pi}\right)^{1/2} a(k) \equiv \tilde{f}(k) \tag{C.10}
$$

Eqs. (C.9) can be rewritten in the limit $L \to \infty$ as

$$\begin{aligned} f(x) &= \frac{1}{\sqrt{2\pi}} \int_{-\infty}^{\infty} dk \tilde{f}(k) \cos kx \\ \tilde{f}(k) &= \frac{1}{\sqrt{2\pi}} \int_{-\infty}^{\infty} dx f(x) \cos kx \qquad ; \ f(x) \text{ even} \end{aligned} \tag{C.11}$$

We say that $[f(x), \tilde{f}(k)]$ form a *cosine Fourier transform* pair. Knowledge of one implies the other. The first relation can, in fact, be considered an *integral equation* for $\tilde{f}(k)$, whose solution is given by the second (and *vice versa*). That the functions must be reproduced for all values of the arguments is what allows one to invert these relations.

Consider an *odd* function $g(-x) = -g(x)$. Similar arguments then yield the sine Fourier transform pair

$$\begin{aligned} g(x) &= \frac{1}{\sqrt{2\pi}} \int_{-\infty}^{\infty} dk\, \tilde{g}(k) \sin kx \\ \tilde{g}(k) &= \frac{1}{\sqrt{2\pi}} \int_{-\infty}^{\infty} dx\, g(x) \sin kx \qquad ; \ g(x) \text{ odd} \end{aligned} \tag{C.12}$$

Suppose $F(x)$ is a real function. Introduce the complex quantity

$$\begin{aligned} \tilde{F}(k) &= \frac{1}{\sqrt{2\pi}} \int_{-\infty}^{\infty} e^{-ikx} F(x)\, dx \\ &= \mathrm{Re}\, \tilde{F}(k) + i \mathrm{Im}\, \tilde{F}(k) \end{aligned} \tag{C.13}$$

In the second line, $\tilde{F}(k)$ is separated into its real and imaginary parts. The use of Eq. (A.12) allows this expression to be written as

$$\tilde{F}(k) = \frac{1}{\sqrt{2\pi}} \int_{-\infty}^{\infty} (\cos kx - i \sin kx) F(x)\, dx \tag{C.14}$$

Hence one can identify

$$\begin{aligned} \mathrm{Re}\, \tilde{F}(k) &= \frac{1}{\sqrt{2\pi}} \int_{-\infty}^{\infty} F(x) \cos kx\, dx \\ -\mathrm{Im}\, \tilde{F}(k) &= \frac{1}{\sqrt{2\pi}} \int_{-\infty}^{\infty} F(x) \sin kx\, dx \end{aligned} \tag{C.15}$$

By inspection, $\mathrm{Re}\, \tilde{F}(k)$ is even about $k = 0$ and $-\mathrm{Im}\, \tilde{F}(k)$ is odd.

The function $F(x)$ is uniquely split into parts that are even and odd about $x = 0$ by writing

$$F(x) \equiv \frac{1}{2}[F(x) + F(-x)] + \frac{1}{2}[F(x) - F(-x)]$$
$$= F_{\rm E}(x) + F_{\rm O}(x) \tag{C.16}$$

Since any *totally odd* term in the integrand integrates to zero, Eqs. (C.15) can be rewritten as

$$\begin{aligned}\text{Re}\,\tilde{F}(k) &= \frac{1}{\sqrt{2\pi}}\int_{-\infty}^{\infty} F_{\rm E}(x)\cos kx\,dx \\ -\text{Im}\,\tilde{F}(k) &= \frac{1}{\sqrt{2\pi}}\int_{-\infty}^{\infty} F_{\rm O}(x)\sin kx\,dx\end{aligned} \tag{C.17}$$

Consider next the quantity

$$\begin{aligned}&\frac{1}{\sqrt{2\pi}}\int_{-\infty}^{\infty} e^{ikx}\tilde{F}(k)\,dk \\ &\qquad = \frac{1}{\sqrt{2\pi}}\int_{-\infty}^{\infty}[\text{Re}\,\tilde{F}(k) + i\text{Im}\,\tilde{F}(k)](\cos kx + i\sin kx)dk \\ &\qquad = \frac{1}{\sqrt{2\pi}}\int_{-\infty}^{\infty}[\text{Re}\,\tilde{F}(k)\,\cos kx - \text{Im}\,\tilde{F}(k)\,\sin kx]dk\end{aligned} \tag{C.18}$$

Here those terms with odd integrands have again been discarded in the last line. Now make use of the previous results in Eqs. (C.11) and (C.12) for the inverse Fourier transforms of the real even and odd functions $[\text{Re}\,\tilde{F}(k), -\text{Im}\,\tilde{F}(k)]$ in Eqs. (C.17). This gives

$$\frac{1}{\sqrt{2\pi}}\int_{-\infty}^{\infty} e^{ikx}\tilde{F}(k)\,dk = F_{\rm E}(x) + F_{\rm O}(x) = F(x) \tag{C.19}$$

In *summary*, we have derived the *complex form* of the Fourier transform relation for a real function $F(x)$

$$\begin{aligned}F(x) &= \frac{1}{\sqrt{2\pi}}\int_{-\infty}^{\infty} e^{ikx}\tilde{F}(k)\,dk \\ \tilde{F}(k) &= \frac{1}{\sqrt{2\pi}}\int_{-\infty}^{\infty} e^{-ikx}F(x)\,dx\end{aligned} \tag{C.20}$$

The symmetry of these relations makes it evident that they also hold for a real function $\tilde{F}(k)$.

The Fourier integral representation in Eqs. (C.20) is extended to an arbitrary, well-behaved, complex function $F(x)$ in Prob. C.6.

Appendix D

Some Thermodynamics

We review some of the basic principles of thermodynamics, and we first concentrate on closed systems with a fixed number N of constituents in a volume V.

First Law. The first law of thermodynamics expresses the principle of the conservation of energy. It states that

$$dE = đQ - đW \tag{D.1}$$

Here $đQ$ represents an infinitesimal element of *heat* supplied to the system and $đW$ an infinitesimal element of *work* done by the system (Fig. D.1).

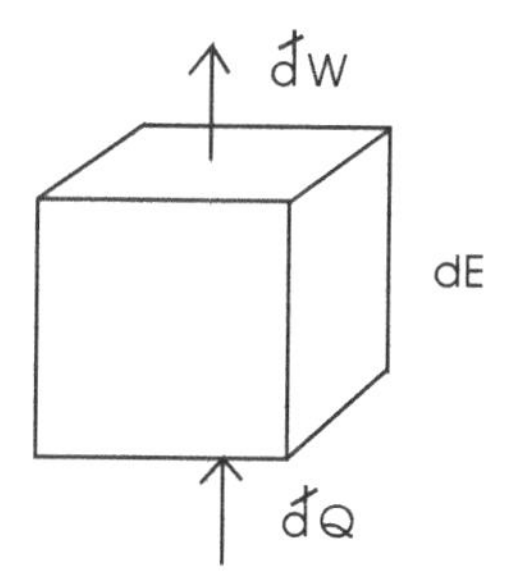

Fig. D.1 Quantities appearing in the first law of thermodynamics.

In general, both of these quantities depend on just how the process is carried out (this is the reason for the đ notation). In contrast, dE is the differential of a *state function*, the *internal energy*, which depends only on the thermodynamic variables characterizing the system, for example $E(N, V, T)$ where T is the temperature.

For a perfect gas at a pressure P, the element of work done in the surroundings under a small increase in volume is given by (Fig. D.2)

$$đW = PdV \qquad ; \text{ perfect gas} \tag{D.2}$$

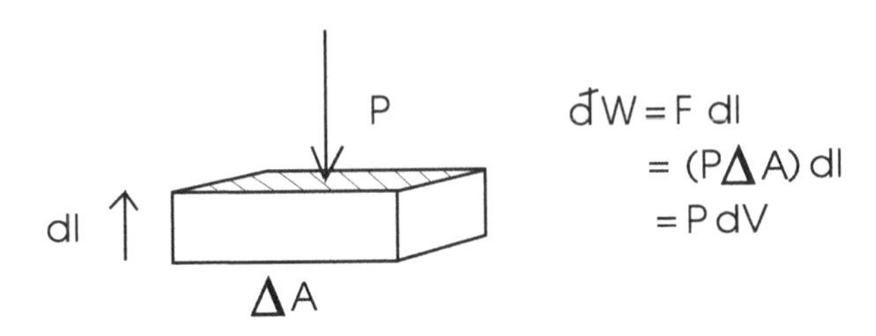

Fig. D.2 Work done by a perfect gas at a pressure P under an increase in volume dV.

With only pressure-volume work, the first law of thermodynamics states that the increase in internal energy is given by the difference

$$dE = đQ - PdV \tag{D.3}$$

Second Law. For a particular type of process, one that is *quasi-static and reversible*, the ratio of heat flow $đQ_{\rm R}$ to the absolute temperature T defines the differential of a second state function, the *entropy*[65]

$$\frac{đQ_{\rm R}}{T} = dS \qquad ; \text{ entropy} \quad \text{second law} \tag{D.4}$$

This state function depends on the thermodynamic variables of the system, say $S(N, V, E)$, with a total differential given by (Fig. D.3)

$$dS(N,V,E) = \left(\frac{\partial S}{\partial N}\right)_{\rm V,E} dN + \left(\frac{\partial S}{\partial V}\right)_{\rm N,E} dV + \left(\frac{\partial S}{\partial E}\right)_{\rm N,V} dE \tag{D.5}$$

For quasi-static, reversible processes, a combination of the first law in Eq. (D.3) with the second law in Eq. (D.4) then gives in the case of pressure-volume work

$$dE = TdS - PdV \tag{D.6}$$

[65] We do not go through the details of the very lovely, and impressive, classical argument that this quantity forms a state function. It can be found in any good book on thermodynamics, for example [Zemansky (1968)], and it deserves to be studied.

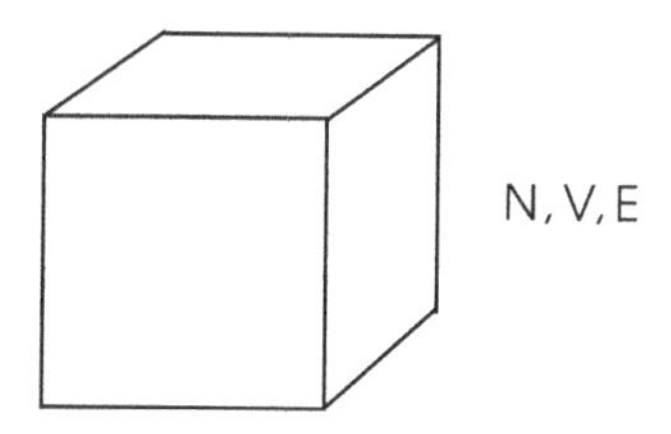

Fig. D.3 The entropy as a state function $S(N, V, E)$.

Third Law. The third law states that (for most systems) the entropy vanishes at absolute zero

$$S \to 0 \qquad ; \text{ as } T \to 0 \qquad \text{third law} \tag{D.7}$$

The *Helmholtz free energy* is another state function defined by

$$A \equiv E - TS \qquad ; \text{ Helmholtz free energy} \tag{D.8}$$

For a closed system doing pressure-volume work, its total differential follows from this definition and Eq. (D.6)

$$dA = -SdT - PdV \tag{D.9}$$

For an *open system* where one is free to increase or decrease the number of particles, say through a semi-permeable membrane, one augments the first and second laws in Eq. (D.6) with a term μdN where μ is the *chemical potential*

$$dE = TdS - PdV + \mu dN \qquad ; \text{ open system} \tag{D.10}$$

Equation (D.9) is then modified to read

$$dA = -SdT - PdV + \mu dN \tag{D.11}$$

The Helmholtz free energy is a state function, and if expressed in terms (N, V, T), it has a total differential given by

$$dA(N, V, T) = \left(\frac{\partial A}{\partial N}\right)_{V,T} dN + \left(\frac{\partial A}{\partial V}\right)_{N,T} dV + \left(\frac{\partial A}{\partial T}\right)_{N,V} dT \tag{D.12}$$

If $A(N, V, T)$ is known by some means (see appendix E), then one can compare with Eq. (D.11) to determine (S, P, μ) by differentiation

$$S = -\left(\frac{\partial A}{\partial T}\right)_{N,V} \quad ; \; P = -\left(\frac{\partial A}{\partial V}\right)_{N,T} \quad ; \; \mu = \left(\frac{\partial A}{\partial N}\right)_{V,T} \tag{D.13}$$

Appendix E

Some Statistical Mechanics

The goal of statistical mechanics is to relate the equilibrium thermodynamic properties of a system to the properties of its constituents. Furthermore, we seek to describe the *distribution* of the constituents at equilibrium.[66] The key is that in applications of interest, we are describing the behavior of a vast number ($\approx 10^{23}$) of these constituents.

Let us start with a simple problem. Consider a set of N independent, localized particles occupying a discrete set of energy levels $(\varepsilon_1, \varepsilon_2, \cdots, \varepsilon_\infty)$ as illustrated in Fig. E.1.[67]

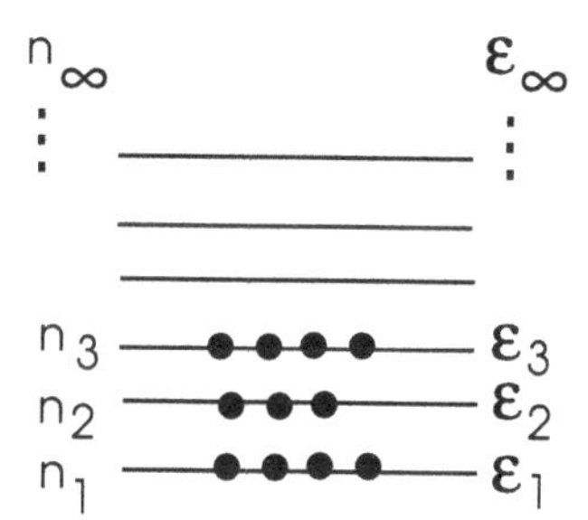

Fig. E.1 N independent, localized particles occupying a discrete set of energy levels $(\varepsilon_1, \varepsilon_2, \cdots, \varepsilon_\infty)$. There are n_1 in the first state, n_2 in the second state, *etc.*

We make the two basic statistical hypotheses:

I. *Assume all microscopic configurations ("complexes") consistent with (N, V, E) are a priori equally probable.*

---

[66] For an introduction to statistical mechanics, see [Walecka (2000)]; see also [Lifshitz and Landau (1984); Huang (1987)].

[67] For example, atoms moving independently at the sites of a crystal lattice.

II. *Assume the entropy is given by*

$$S = k_{\rm B} \ln \Omega$$
$$\text{where ;} \qquad \Omega \equiv \text{total number of complexions} \tag{E.1}$$

We comment on this second assumption:

- It is due to Boltzmann, and $k_{\rm B}$ is Boltzmann's constant;
- It is a remarkable relation that relates a (rather mysterious) thermodynamic function to an enumeration of the total number of distinct configurations of the system;
- It is consistent with everything we know;
- From the first and second laws of thermodynamics, one then has for a quasi-static reversible process with pressure-volume work

$$dE = TdS - PdV \tag{E.2}$$

Let us evaluate Ω for a system of N independent, localized particles, each of which can occupy one of a discrete set of energy levels $(\varepsilon_1, \varepsilon_2, \cdots, \varepsilon_\infty)$. Suppose that n_1 of them occupy ε_1, n_2 occupy ε_2, *etc.*, as illustrated in Fig. E.1. Since the total number of particles and total energy (N, E) are specified, one has

$$\sum_{i=1}^{\infty} n_i = N \qquad ; \sum_{i=1}^{\infty} n_i \varepsilon_i = E \qquad ; \text{ fixed} \tag{E.3}$$

There are N ways to choose the first particle to put into the first level. Then $N - 1$ ways to choose the second particle, and so on. Thus there are $N!$ ways to distribute the particles in Fig. E.1. Now rearrangement of the labels within any given level does not give a new, distinguishable configuration, since all that matters is that a given set of particles occupy that level. There are $n_i!$ such rearrangements for the ith level. Thus one has to correct the original overcounting by dividing by this quantity. This must be done for all the levels. One then simply has to sum over all possible distributions of $(n_1, n_2, n_3, \cdots, n_\infty)$ to get the total number of complexions Ω. Thus

$$\Omega = \sum_{n_1} \sum_{n_2} \cdots \sum_{n_\infty} \frac{N!}{n_1! n_2! \cdots n_\infty!} \qquad ; \text{ total number of complexions}$$
$$\sum_{i=1}^{\infty} n_i = N \qquad ; \sum_{i=1}^{\infty} n_i \varepsilon_i = E \qquad ; \text{ constraints} \tag{E.4}$$

One sums over all values of n_i consistent with the constraints. Note that there is no problem if a level is empty, or if there is only one particle in a level, since $0! = 1! = 1$. One is now faced with the mathematical problem of doing the sum in the first line of Eqs. (E.4) subject to the two constraints in the second line.

In order to gain insight, consider the simpler problem of a system with only two such energy levels (Fig. E.2).

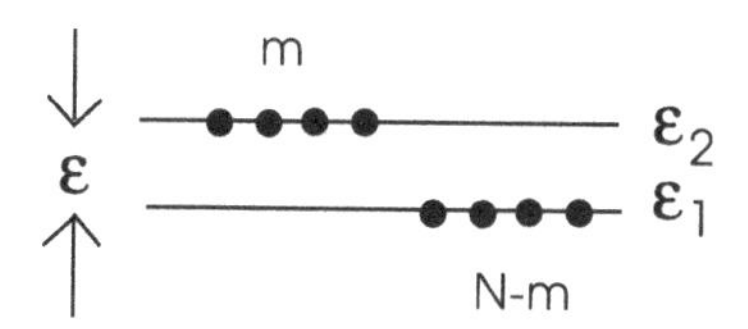

Fig. E.2 N independent, localized particles in a system with only two discrete energy levels $(\varepsilon_1, \varepsilon_2)$. In this configuration, there are m particles in the second level and $N-m$ in the first.

Here m particles have been placed in the upper level and $N-m$ in the lower level, so the first constraint in Eqs. (E.4) is automatically satisfied. Now assume that $N\varepsilon \ll E$, where ε is the level spacing in Fig. E.2. This allows the second constraint to be ignored. In this case, the sum in Eq. (E.4) reduces to the unconstrained expression

$$\Omega = \sum_{m=0}^{N} \frac{N!}{m!(N-m)!} \tag{E.5}$$

But this sum can be done exactly! Simply use the binomial theorem

$$(x_1 + x_2)^N = \sum_{m=0}^{N} \frac{N!}{m!(N-m)!} x_1^m x_2^{N-m} \tag{E.6}$$

and set $x_1 = x_2 = 1$ to obtain

$$\begin{aligned} \Omega &= 2^N \\ S &= k_{\mathrm{B}} N \ln 2 \end{aligned} \tag{E.7}$$

This result is easily understood since each particle can be in either one of two states.

Now suppose we try to evaluate Ω another way by *keeping just the largest term in the sum* in Eq. (E.5). To find that term, we maximize

$$t(m) = \frac{N!}{m!(N-m)!} \tag{E.8}$$

This is done by employing Stirling's formula, which approximates $\ln N!$ for large N

$$\ln N! = N \ln N - N + O(\ln N) \qquad ; \text{Stirling's formula} \tag{E.9}$$

Then, instead of maximizing $t(m)$, we can just as well maximize $\ln t(m)$

$$\begin{aligned} \ln t(m) &= \ln N! - \ln m! - \ln (N-m)! \\ &\cong N \ln N - N - m \ln m + m - (N-m) \ln (N-m) + (N-m) \\ &= N \ln N - m \ln m - (N-m) \ln (N-m) \end{aligned} \tag{E.10}$$

To maximize this expression, set $\partial \ln t(m)/\partial m = 0$

$$\begin{aligned} 0 &= -\ln m - 1 + \ln (N-m) + 1 \\ \Rightarrow \qquad m^\star &= \frac{N}{2} \qquad\qquad ; \text{for maximum} \end{aligned} \tag{E.11}$$

Here $m^\star$ represents the *most probable distribution.* The largest term then gives

$$\begin{aligned} \ln t(m^\star) &= \ln \frac{N!}{(N/2)!\,(N/2)!} \\ &\cong N \ln N - N - 2\left(\frac{N}{2} \ln \frac{N}{2} - \frac{N}{2}\right) \\ &= N \ln 2 \end{aligned} \tag{E.12}$$

Hence

$$S \cong k_{\rm B} \ln t(m^\star) = k_{\rm B} N \ln 2 \tag{E.13}$$

This gives exactly the same answer!

The largest term in the sum for Ω gives exactly the same answer as the sum itself. How can this be? The answer is that N is *very large*, and what we are interested in for the entropy is $\ln \Omega$. Since the sum contains N positive terms, it is clearly bounded by

$$t(m^\star) \leq \Omega \leq N\, t(m^\star) \tag{E.14}$$

Hence

$$\begin{aligned} \ln t(m^\star) \ &\leq \ \ln \Omega \ \leq \ \ln t(m^\star) + \ln N \\ N \ln 2 \ &\leq \ \ln \Omega \ \leq \ N \ln 2 + \ln N \end{aligned} \tag{E.15}$$

Here $N \gg \ln N$, and the last term is completely negligible.[68]

Let us now return to the problem of evaluating the sum in Eqs. (E.4) subject to the stated constraints. Based upon the above argument, we will be content to approximate $\ln \Omega$ by keeping just the largest term in the sum, corresponding to the *most probable* distribution. This implies

$$\begin{aligned} \ln \Omega &\cong \ln t(n^\star) \\ t(n) &\equiv \frac{N!}{n_1! n_2! \cdots n_\infty!} \end{aligned} \tag{E.16}$$

We seek to maximize

$$\ln t(n) = N \ln N - N - \sum_{i=1}^{\infty} (n_i \ln n_i - n_i) \tag{E.17}$$

subject to the constraints

$$\sum_{i=1}^{\infty} n_i = N \qquad ; \sum_{i=1}^{\infty} n_i \varepsilon_i = E \tag{E.18}$$

This is again a well-defined mathematical problem. Let each of the n_i change by a small amount δn_i. Then the maximum is obtained from the relations

$$\begin{aligned} \delta t(n) = -\delta \sum_{i=1}^{\infty} (n_i \ln n_i - n_i) = -\sum_{i=1}^{\infty} (\ln n_i)\, \delta n_i &= 0 \qquad ; \text{ for maximum} \\ \sum_{i=1}^{\infty} \delta n_i &= 0 \\ \sum_{i=1}^{\infty} \varepsilon_i \, \delta n_i &= 0 \end{aligned} \tag{E.19}$$

Multiply the second equation by a constant α, and the third by a constant

[68]For example, if $N = 10^{23}$, then $\ln N = 53$. The reader is urged to spend some time pondering these numbers. Note that the additional term in the last expression in Eqs. (E.15) is of the same order as those neglected in the use of Stirling's formula.

β, and add to the first. The result is

$$\sum_{i=1}^{\infty}(\alpha + \beta\varepsilon_i - \ln n_i)\,\delta n_i = 0 \tag{E.20}$$

The constraints have here been incorporated with *Lagrange multipliers.* Choose (α, β) so that the coefficients of the first two variations $(\delta n_1, \delta n_2)$ vanish. The remaining variations $(\delta n_3, \delta n_4, \cdots, \delta n_\infty)$ are now *linearly independent*, and hence their coefficients must also vanish. Thus *all* the coefficients of the δn_i must be zero

$$\alpha + \beta\varepsilon_i - \ln n_i^\star = 0 \qquad ; \; i = 1, 2, \cdots, \infty \tag{E.21}$$

An exponentiation then gives the solution for the maximization of $\ln t(n)$

$$n_i^\star = e^{\alpha} e^{\beta\varepsilon_i} \tag{E.22}$$

This gives the *most probable distribution*, or the *equilibrium distribution*, of the particles. The constants (α, β) are now chosen to satisfy the constraint equations. First, α is determined from

$$\begin{aligned} \sum_{i=1}^{\infty} n_i^\star &= N \\ \Rightarrow \qquad e^{\alpha} &= \frac{N}{\sum_i e^{\beta\varepsilon_i}} \end{aligned} \tag{E.23}$$

Then β follows from

$$\begin{aligned} \sum_{i=1}^{\infty} \varepsilon_i n_i^\star &= E \\ \Rightarrow \qquad \frac{E}{N} &= \frac{\sum_i \varepsilon_i\, e^{\beta\varepsilon_i}}{\sum_i e^{\beta\varepsilon_i}} \end{aligned} \tag{E.24}$$

This equation determines $\beta(N, V, E)$, where the V dependence arises implicitly from the dependence of the single-particle energy levels ε_i on the volume.[69]

In summary

$$\begin{aligned} \frac{E}{N} &= \frac{\sum_i \varepsilon_i\, e^{\beta\varepsilon_i}}{\sum_i e^{\beta\varepsilon_i}} && ; \text{ total energy E} \\ \frac{n_i^\star}{N} &= \frac{e^{\beta\varepsilon_i}}{\sum_i e^{\beta\varepsilon_i}} && ; \text{ equilibrium distribution} \end{aligned} \tag{E.25}$$

[69]See the chapter on quantum mechanics.

In order to explicitly determine β from the thermodynamic properties of the system, we go back to

$$\begin{aligned}
S &= k_{\rm B} \ln \Omega \cong k_{\rm B} \ln t(n^\star) \\
&= k_{\rm B} \left[N \ln N - N - \sum_{i=1}^{\infty} (n_i^\star \ln n_i^\star - n_i^\star) \right] \\
&= k_{\rm B} \left[N \ln N - \sum_{i=1}^{\infty} (\alpha + \beta\varepsilon_i)\, n_i^\star \right] \\
&= k_{\rm B} \left[N \ln N - \alpha N - \beta E \right] \\
&= k_{\rm B} \left\{ N \ln N - N \left[\ln N - \ln \left(\sum_{i=1}^{\infty} e^{\beta\varepsilon_i} \right) \right] - \beta E \right\}
\end{aligned} \tag{E.26}$$

Here Eq. (E.21) has been used in the third line, and Eq. (E.23) in the last. Thus

$$S(N, V, E) = k_{\rm B} \left[N \ln \left(\sum_{i=1}^{\infty} e^{\beta\varepsilon_i} \right) - \beta E \right] \tag{E.27}$$

Now make use of the thermodynamic relation in Eq. (E.2) which holds for a closed system with fixed N undergoing pressure-volume work

$$\begin{aligned}
dE &= TdS - PdV \\
\Rightarrow \quad \left(\frac{\partial S}{\partial E} \right)_{N,V} &= \frac{1}{T}
\end{aligned} \tag{E.28}$$

This equation introduces the absolute temperature T. Constant volume implies constant ε_i, and since β is an implicit function of E, one has

$$\begin{aligned}
\frac{1}{k_{\rm B}T} &= \left[\frac{N}{(\sum_i e^{\beta\varepsilon_i})} \left(\sum_{i=1}^{\infty} \varepsilon_i e^{\beta\varepsilon_i} \right) \frac{\partial \beta}{\partial E} - E \frac{\partial \beta}{\partial E} - \beta \right] = -\beta \\
\Rightarrow \quad \beta &= -\frac{1}{k_{\rm B}T}
\end{aligned} \tag{E.29}$$

Hence the Lagrange multiplier β is simply the negative of the (dimensionless) inverse temperature $1/k_{\rm B}T$.

In *summary*, it follows from our basic statistical assumptions that for

N independent, localized systems in a volume V and at a temperature T

$$\begin{aligned}
\frac{n_i^\star}{N} &= \frac{e^{-\varepsilon_i/k_BT}}{\sum_i e^{-\varepsilon_i/k_BT}} && ;\ \text{Boltzmann distribution} \\
\frac{E}{N} &= \frac{\sum_i \varepsilon_i\, e^{-\varepsilon_i/k_BT}}{\sum_i e^{-\varepsilon_i/k_BT}} && ;\ \text{total energy E} \qquad (E.30)
\end{aligned}$$

Equation (E.27) can be rearranged to yield the following expression for the *Helmholtz free energy* $A \equiv E - TS$

$$\begin{aligned}
A(N,V,T) &= -k_BT\,\ln(\text{p.f.})^N && ;\ \text{Helmholtz free energy} \\
(\text{p.f.}) &\equiv \sum_{i=1}^{\infty} e^{-\varepsilon_i/k_BT} && ;\ \text{partition function} \qquad (E.31)
\end{aligned}$$

This last relation defines the *partition function.* The thermodynamic quantities (μ, P, S) can then be determined from $A(N,V,T)$ by differentiation, as indicated in Eqs. (D.13).

Classical Partition Function. Start with a free particle in phase space (p,q) as discussed in chapter 2. Classical statistical mechanics assumes

Equal a priori weighting of each point in phase space.

One correspondingly replaces the above sums by integrals

$$\begin{aligned}
h\sum_{i=1}^{\infty} &\leftrightarrow \int\!\!\int dpdq && ;\ \text{one-dimension} \\
h^s\sum_{i=1}^{\infty} &\leftrightarrow \int\cdots\int dp_1\cdots dp_s\,dq_1\cdots dq_s && ;\ \text{s-dimensions} \qquad (E.32)
\end{aligned}$$

Here h is just a constant, which has to be put in by hand in classical physics, to make the dimensions of the cell size $dpdq$ come out correctly. This constant cancels in the ratios $n_i^\star/N$ and E/N; it only enters into the calculation of the absolute entropy. We will see that this result comes out of the classical limit of quantum mechanics, where h can be identified with Planck's constant.[70]

In the classical limit, the one-particle partition function and Boltzmann

[70]See Probs. E.4 and E.5.

distribution function for a system at a temperature T thus take the forms

$$
\begin{aligned}
(\text{p.f.}) &= \frac{1}{h^s}\int\cdots\int e^{-\varepsilon(p,q)/k_{\text{B}}T}\,dp_1\cdots dp_s\,dq_1\cdots dq_s \\
&\qquad\qquad ;\ \text{classical partition function} \\
\frac{n_i^\star}{N} &= \frac{1}{h^s}\frac{1}{(\text{p.f.})}e^{-\varepsilon(p,q)/k_{\text{B}}T}\,dp_1\cdots dp_s\,dq_1\cdots dq_s \\
&\qquad\qquad ;\ \text{Boltzmann distribution function} \qquad \text{(E.33)}
\end{aligned}
$$

Appendix F

Some Vector Calculus

In this appendix we prove two theorems from vector calculus that allow one to convert Maxwell's equations from differential to integral form, which is the way they are usually presented in introductory E&M courses. For simplicity, we prove the theorems in two dimensions for a convex region of area A surrounded by a curve C (see Fig. F.1), and leave the generalizations to the dedicated reader.

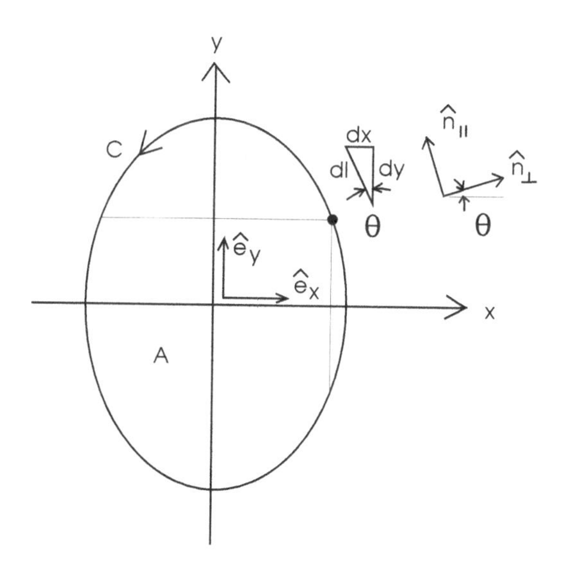

Fig. F.1 Configuration at the point $[x, y_{\max}(x)]$ or $[x_{\max}(y), y]$ in the first quadrant for a convex two-dimensional region of area A bounded by the curve C. Here $(\hat{\mathbf{n}}_\perp, \hat{\mathbf{n}}_\parallel)$ are unit vectors normal and tangent to the curve at that point.

Surface and Line Elements. Consider the first quadrant. At a given x, the curve is described by $y_{\rm max}(x)$. Similarly, at a given y, the curve is described by $x_{\rm max}(y)$. Let dl be an infinitesimal length tangent to the curve at that point and (dx, dy) be the corresponding magnitudes of the cartesian displacements

$$dx = dl\,\sin\theta \qquad ; \; dy = dl\,\cos\theta \tag{F.1}$$

Define unit vectors normal and tangent to the curve at that point by $(\hat{\mathbf{n}}_\perp, \hat{\mathbf{n}}_\parallel)$ respectively, as indicated in Fig. F.1. These are obtained from the cartesian unit vectors $(\hat{\mathbf{e}}_x, \hat{\mathbf{e}}_y)$ by a rotation through the angle θ. We define the *surface element* and *line element* by

$$\begin{aligned} d\mathbf{S} &\equiv \hat{\mathbf{n}}_\perp\, dl = dl\,(\hat{\mathbf{e}}_x \cos\theta + \hat{\mathbf{e}}_y \sin\theta) \qquad && ; \text{ surface element} \\ d\mathbf{l} &\equiv \hat{\mathbf{n}}_\parallel\, dl = dl\,(-\hat{\mathbf{e}}_x \sin\theta + \hat{\mathbf{e}}_y \cos\theta) \qquad && ; \text{ line element} \end{aligned} \tag{F.2}$$

Gauss' Theorem. Let $\mathbf{v}(x,y)$ be a vector field

$$\mathbf{v}(x,y) = v_x(x,y)\,\hat{\mathbf{e}}_x + v_y(x,y)\,\hat{\mathbf{e}}_y \qquad ; \text{ vector field} \tag{F.3}$$

Gauss' theorem converts the integral of the divergence of $\mathbf{v}(x,y)$ over the area to a surface integral over the bounding surface, in this case the curve C

$$\int_A d^2x\, \nabla\cdot\mathbf{v} = \oint_C \mathbf{v}\cdot d\mathbf{S} \qquad ; \text{ Gauss' theorem} \tag{F.4}$$

Here $d^2x \equiv dxdy$. We prove this and the subsequent theorem in the first quadrant, leaving the extension to include the other three quadrants as problems for the dedicated reader. Write out the l.h.s. of this relation in detail

$$\int_A d^2x\, \nabla\cdot\mathbf{v} = \int_A dxdy \left(\frac{\partial v_x}{\partial x} + \frac{\partial v_y}{\partial y}\right) \tag{F.5}$$

The integral over x at fixed y can now be immediately carried out in the first term, and over y at fixed x in the second. The result is

$$\begin{aligned} \int_A d^2x\, \nabla\cdot\mathbf{v} &= \int dy\, v_x[x_{\rm max}(y), y] + \int dx\, v_y[x, y_{\rm max}(x)] + \cdots \\ &= \int_C dl\,\{\cos\theta\; v_x[x_{\rm max}(y), y] + \sin\theta\; v_y[x, y_{\rm max}(x)]\} + \cdots \\ &= \oint_C dl\,\hat{\mathbf{n}}_\perp\cdot\mathbf{v} = \oint_C d\mathbf{S}\cdot\mathbf{v} \end{aligned} \tag{F.6}$$

Here the relations between dl and (dx, dy) in Eqs. (F.1) have been introduced in the second line, and that defining the element of surface area $d\mathbf{S}$ in Eqs. (F.2) in the last line. The result is Gauss' theorem.

An extension to three dimensions, again left as a problem, gives

$$\int_V d^3x \, \nabla \cdot \mathbf{v} = \oint_S d\mathbf{S} \cdot \mathbf{v} \tag{F.7}$$

where $d\mathbf{S} = \hat{\mathbf{n}} \, dS$ with $\hat{\mathbf{n}}$ an outward pointing normal unit vector on the bounding surface, and dS an element of surface area.

Stokes' Theorem. Consider the corresponding integral over the area A of the curl of the vector field $\mathbf{v}(x, y)$ dotted into $d\mathbf{A} = \hat{\mathbf{e}}_z \, d^2x$ where $\hat{\mathbf{e}}_z$ is the third member of a right-handed set of cartesian unit vectors in Fig. F.1 (it comes out of the page). Write out this expression in detail

$$\int_A d\mathbf{A} \cdot \nabla \times \mathbf{v} = \int_A d^2x \left(\frac{\partial v_y}{\partial x} - \frac{\partial v_x}{\partial y} \right) \qquad ; \; d\mathbf{A} = \hat{\mathbf{e}}_z \, d^2x \tag{F.8}$$

Again, the integral on x at fixed y can be done in the first term, and on y at fixed x in the second. The result is

$$\begin{aligned}
\int_A d\mathbf{A} \cdot \nabla \times \mathbf{v} &= \int dy \, v_y[x_{\max}(y), y] - \int dx \, v_x[x, y_{\max}(x)] + \cdots \\
&= \int_C dl \, \{\cos\theta \; v_y[x_{\max}(y), y] - \sin\theta \; v_x[x, y_{\max}(x)]\} + \cdots \\
&= \oint_C dl \, \hat{\mathbf{n}}_{\parallel} \cdot \mathbf{v}
\end{aligned} \tag{F.9}$$

Here the relations between dl and (dx, dy) in Eqs. (F.1) have been introduced in the second line, and that defining $\hat{\mathbf{n}}_{\parallel}$ in Eqs. (F.2) in the last line. If the expression defining the line element $d\mathbf{l}$ in Eqs. (F.2) is now introduced, one has

$$\int_A d\mathbf{A} \cdot \nabla \times \mathbf{v} = \oint_C d\mathbf{l} \cdot \mathbf{v} \qquad ; \text{Stokes' theorem} \tag{F.10}$$

The result is Stokes' theorem, which relates the integral of the z-component of the curl of $\mathbf{v}$ over the area A to a line integral of $\mathbf{v}$ around the bounding curve C.

With the observation that contributions in opposite directions along a bounding curve cancel, Stokes' theorem can be extended to regions of arbitrary shape by covering them with many small regions where the above result holds.

Appendix G

Black-Body Flux

In this appendix we calculate the energy flux out of a black body in terms of the electromagnetic energy density $u(\nu, T)$ in the cavity. Consider a small volume in the cavity of size $d^3r = r^2\, dr \sin\theta\, d\theta\, d\phi$ located at (r, θ, ϕ). The energy in this little region will be radiated equally in all directions with velocity c. How much goes through a small area A in the wall?

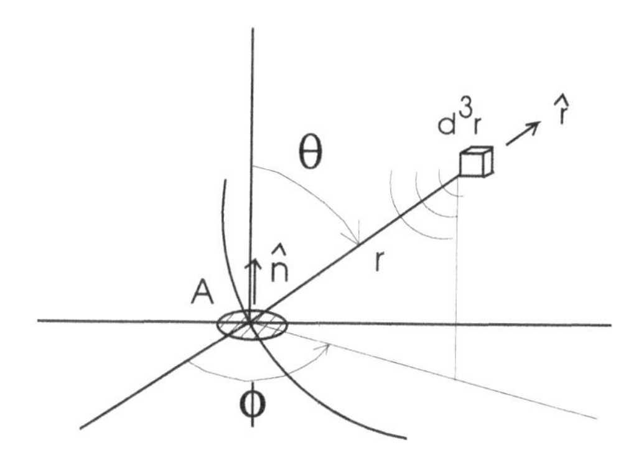

Fig. G.1 Energy flux out of a black body in terms of the electromagnetic energy density $u(\nu, T)$ in the cavity.

The small area A must first be projected onto a sphere of radius r (see Fig. G.1). This projection is

$$A\,\hat{\mathbf{n}} \cdot \hat{\mathbf{r}} = A\cos\theta \qquad ;\ \text{projected area} \tag{G.1}$$

The fraction of the radiation passing through A is then

$$\text{fraction through A} = \frac{A\cos\theta}{4\pi r^2} \tag{G.2}$$

Now sum this over all small regions d^3r in the cavity whose light reaches the surface. The integrated energy transmission through A in a time interval Δt is then

$$\begin{aligned} \text{energy through A} &= \int_0^{c\Delta t} r^2\,dr \int_0^{\pi/2} \sin\theta\,d\theta \int_0^{2\pi} d\phi \frac{A\cos\theta}{4\pi r^2} u(\nu,T) \\ &= \frac{2\pi A}{4\pi} c\,u(\nu,T)\Delta t \int_0^1 x\,dx \end{aligned} \tag{G.3}$$

where $x \equiv \sin\theta$. Hence

$$\frac{1}{A\,\Delta t} \times \text{energy through A} = \frac{1}{4} c\,u(\nu,T) \tag{G.4}$$

This is the energy flux through the surface.

Problem G.1 provides a complete derivation of the number of normal modes per unit volume of the radiation field in a large cubical box with periodic boundary conditions, which is identical to Eq. (3.24).

Appendix H

Wave Functions for Identical Particles

The Pauli exclusion principle follows from a more general principle of quantum mechanics:

> *The wave function for a system of identical bosons (fermions) must be symmetric (antisymmetric) under the interchange of any two particles.*

In this appendix, we investigate a few consequences of this principle. Introduce the following generic notation $\phi_\alpha(1)$ for the single-particle wave functions appropriate to the problem at hand. Here (1) denotes a complete set of coordinates for the first particle, and α a complete set of single-particle quantum numbers. For example, with spin-zero bosons in a central potential, one has

$$\phi_\alpha(1) = \psi_{nlm_l}(\mathbf{x}_1) \qquad ; \text{ spin-0 bosons in central potential} \quad \text{(H.1)}$$

For spin-1/2 fermions in a big box with periodic boundary conditions, one has

$$\underline{\phi}_\alpha(1) = \frac{1}{\sqrt{\Omega}}\psi_{\mathbf{k}}(\mathbf{x}_1)\,\underline{\eta}_\lambda(1) \qquad ; \text{ spin-1/2 fermions with p.b.c.} \qquad \text{(H.2)}$$

and so on. We assume that the single-particle wave functions are orthonormal so that generically

$$\int d^3x\,\phi^\dagger_{\alpha_1}(\mathbf{x})\phi_{\alpha_2}(\mathbf{x}) = \delta_{\alpha_1\alpha_2} \qquad \text{(H.3)}$$

where the dagger now implies an additional spin matrix product in the case of spin-1/2 fermions (see Prob. 4.33).

H.1 Bosons

Consider the case of two spin-zero bosons. The above principle implies that the normalized wave function for this system must take the form

$$\begin{aligned}\Phi_{\alpha_1\alpha_2}(1,2) &= \frac{1}{\sqrt{2}}\left[\phi_{\alpha_1}(1)\phi_{\alpha_2}(2) + \phi_{\alpha_1}(2)\phi_{\alpha_2}(1)\right] \quad ; \ \alpha_2 \neq \alpha_1 \\ &= \phi_{\alpha_1}(1)\phi_{\alpha_1}(2) \quad ; \ \alpha_2 = \alpha_1 \end{aligned} \tag{H.4}$$

Let the hamiltonian of the two-particle system have the form

$$H(1,2) = T(1) + T(2) + V(1,2) \tag{H.5}$$

Since the particles are identical

$$V(2,1) \equiv V(1,2) \tag{H.6}$$

The *expectation value* of the hamiltonian (see Prob. 4.5) taken with the wave function in Eqs. (H.4) is then given by

$$\begin{aligned}\langle H\rangle &= \sum_{\alpha=\alpha_1}^{\alpha_2} \langle\alpha|T|\alpha\rangle + \langle\alpha_1\alpha_2|V|\alpha_1\alpha_2\rangle + \langle\alpha_1\alpha_2|V|\alpha_2\alpha_1\rangle \ ; \ \alpha_2 \neq \alpha_1 \\ &= 2\langle\alpha_1|T|\alpha_1\rangle + \langle\alpha_1\alpha_1|V|\alpha_1\alpha_1\rangle \quad ; \ \alpha_2 = \alpha_1 \end{aligned} \tag{H.7}$$

Here the two-particle matrix element is defined by

$$\langle\alpha_1\alpha_2|V|\alpha_3\alpha_4\rangle \equiv \int d^3x_1 \int d^3x_2 \ \phi_{\alpha_1}^\dagger(\mathbf{x}_1)\phi_{\alpha_2}^\dagger(\mathbf{x}_2)V(1,2)\phi_{\alpha_3}(\mathbf{x}_1)\phi_{\alpha_4}(\mathbf{x}_2) \tag{H.8}$$

In the first of Eqs. (H.7), the first matrix element of the potential is known as the *direct* interaction, and the second is the *exchange* interaction. If one did not take into account the required symmetry of the wave function, only the direct interaction would be present.

As a second example, consider a collection of N spin-zero bosons all condensed into the $\mathbf{k} = 0$ state in a big box of volume $L^3 \equiv \Omega$ with p.b.c., where the single-particle wave functions are those in Eq. (4.118). In this case, the many-particle ground-state wave function is

$$\Phi_{\alpha_1\alpha_2\cdots\alpha_N}(1,2,\cdots,N) = \phi_0(1)\phi_0(2)\cdots\phi_0(N) \tag{H.9}$$

Assume the many-body hamiltonian has the form

$$H = \sum_{i=1}^{N} T(i) + \frac{1}{2} \sum_{i \neq j = 1}^{N} V(|\mathbf{x}_i - \mathbf{x}_j|) \tag{H.10}$$

where $V(|\mathbf{x}_i - \mathbf{x}_j|)$ is of finite range. The expectation value of this hamiltonian is then given by (Prob. H.2)

$$\begin{aligned} \langle H \rangle &= \frac{1}{2} \frac{N(N-1)}{\Omega} \tilde{V}(0) \\ \tilde{V}(0) &\equiv \int d^3x \, V(x) \end{aligned} \tag{H.11}$$

H.2 Fermions

Consider two identical spin-1/2 fermions (*e.g.* electrons). The analog of Eq. (H.4) in this case is[71]

$$\Phi_{\alpha_1 \alpha_2}(1,2) = \frac{1}{\sqrt{2}} [\phi_{\alpha_1}(1)\phi_{\alpha_2}(2) - \phi_{\alpha_1}(2)\phi_{\alpha_2}(1)] \quad ; \; \alpha_2 \neq \alpha_1 \tag{H.12}$$

The antisymmetrized wave function *vanishes identically* if $\alpha_1 = \alpha_2$, and hence *the Pauli exclusion principle is an immediate consequence of the required antisymmetry of the fermion wave function.*

The analog of Eq. (H.7) is

$$\langle H \rangle = \sum_{\alpha=\alpha_1}^{\alpha_2} \langle \alpha | T | \alpha \rangle + \langle \alpha_1 \alpha_2 | V | \alpha_1 \alpha_2 \rangle - \langle \alpha_1 \alpha_2 | V | \alpha_2 \alpha_1 \rangle \tag{H.13}$$

Several comments are relevant:

- The exchange interaction is always present for two identical fermions;
- The exchange interaction comes in with a minus sign;
- The sum of the direct and exchange interactions vanishes identically if $\alpha_1 = \alpha_2$.

There is another way to write Eq. (H.12), and that is as a 2×2 determinant

$$\Phi_{\alpha_1 \alpha_2}(1,2) = \frac{1}{\sqrt{2}} \begin{vmatrix} \phi_{\alpha_1}(1) & \phi_{\alpha_1}(2) \\ \phi_{\alpha_2}(1) & \phi_{\alpha_2}(2) \end{vmatrix} \tag{H.14}$$

[71]For convenience, we now suppress the underlining of the two-component spinors, restoring it at the end when we explicitly write the one- and two-particle matrix elements.

This determinant has the following properties:

- It changes sign under the interchange of the two columns or the two rows;
- It vanishes if the two columns or two rows are identical.

This result has an immediate generalization to the many-fermion system.

$$\Phi_{\alpha_1\alpha_2\cdots\alpha_N}(1,2,\cdots,N) = \frac{1}{\sqrt{N!}}\begin{vmatrix} \phi_{\alpha_1}(1) & \phi_{\alpha_1}(2) & \cdots & \phi_{\alpha_1}(N) \\ \phi_{\alpha_2}(1) & \phi_{\alpha_2}(2) & \cdots & \phi_{\alpha_2}(N) \\ \vdots & \vdots & \vdots & \vdots \\ \phi_{\alpha_N}(1) & \phi_{\alpha_N}(2) & \cdots & \phi_{\alpha_N}(N) \end{vmatrix} \tag{H.15}$$

This determinant again has the following properties:

- It changes sign under the interchange of any two columns or two rows;
- It vanishes if any two columns or two rows are identical.

This is known as a *Slater determinant*.[72] It is clearly advantageous to learn how to calculate with this quantity, and we develop some techniques for doing so.[73]

One can write the determinant in Eq. (H.15) in the following manner

$$\Phi_{\alpha_1\alpha_2\cdots\alpha_N}(1,2,\cdots,N) = \left(\frac{1}{\sqrt{N_P}}\sum_{(\mathcal{P})}(-1)^P\mathcal{P}\right)\prod_{i=1}^{N}\phi_{\alpha_i}(i) \tag{H.16}$$

Here:

(1) The *normal sequence* is given by

$$\prod_{i=1}^{N}\phi_{\alpha_i}(i) = \phi_{\alpha_1}(1)\phi_{\alpha_2}(2)\cdots\phi_{\alpha_N}(N) \quad ; \text{normal sequence} \tag{H.17}$$

For fermions, all single-particle quantum numbers must be distinct;

(2) $\mathcal{P}$ is an operator that produces one of the permutations of the particle labels $(1,2,\cdots,N)$;

(3) P is the "signature" of the permutation $(-1)^P = (-1)^{\#\text{ of interchanges}}$;

(4) $N_P = N!$ is the number of permutations of $(1,2,\cdots,N)$.

---

[72]After the distinguished many-particle physicist J. C. Slater. If the reader is not yet familiar with determinants, he or she can take Eq. (H.16) as the definition.

[73]The true wave function for the many-body system can always be written as a linear combination of these basis wave functions. Note that the many-boson wave function can be obtained by applying the (un-normalized) *symmetrizing* operator $\sum_{(\mathcal{P})}\mathcal{P}$ to the normal sequence; however, we do not pursue that here (see, however, Prob. H.1).

As an example, consider three particles with labels (123). The operator in Eq. (H.16) can be written in terms of a set of particle interchange operators as

$$\sum_{(\mathcal{P})}(-1)^P\mathcal{P} = 1 - \mathcal{P}_{12} - \mathcal{P}_{13} - \mathcal{P}_{23} + \mathcal{P}_{13}\mathcal{P}_{12} + \mathcal{P}_{13}\mathcal{P}_{23} \qquad \text{(H.18)}$$

If we keep track of the permutations by writing $\begin{pmatrix} 1\,2\,3 \\ i\,j\,k \end{pmatrix}$ where (ijk) is the permuted set of labels, then when acting on the label set (123), this operator produces the following expression

$$\left(\sum_{(\mathcal{P})}(-1)^P\mathcal{P}\right)(123) = \begin{pmatrix} 1\,2\,3 \\ 1\,2\,3 \end{pmatrix} - \begin{pmatrix} 1\,2\,3 \\ 2\,1\,3 \end{pmatrix} - \begin{pmatrix} 1\,2\,3 \\ 3\,2\,1 \end{pmatrix} - \begin{pmatrix} 1\,2\,3 \\ 1\,3\,2 \end{pmatrix} + \begin{pmatrix} 1\,2\,3 \\ 2\,3\,1 \end{pmatrix} + \begin{pmatrix} 1\,2\,3 \\ 3\,1\,2 \end{pmatrix}$$

$$\text{or;}\quad \left(\sum_{(\mathcal{P})}(-1)^P\mathcal{P}\right)(123) = (123) - (213) - (321) - (132) + (231) + (312) \qquad \text{(H.19)}$$

It is readily verified that this expression is now explicitly antisymmetric under the interchange of any pair of labels.

Let us demonstrate that the antisymmetry of the expression in Eq. (H.16) under the interchange of any pair of labels is a general result. Consider

$$\mathcal{P}_{ij}\sum_{(\mathcal{P})}(-1)^P\mathcal{P} = \sum_{(\mathcal{P})}(-1)^P\left(\mathcal{P}_{ij}\mathcal{P}\right) \qquad \text{(H.20)}$$

Now observe the following:

(1) It is a general property of permutations that *as $\mathcal{P}$ goes over the set of all permutations, so does $\mathcal{P}'' \equiv \mathcal{P}_{ij}\mathcal{P}$*;[74]

(2) Since $\mathcal{P}''$ has one more interchange, $(-1)^P = (-1)(-1)^{P+1} = (-1)(-1)^{P''}$;

Thus

$$\mathcal{P}_{ij}\sum_{(\mathcal{P})}(-1)^P\mathcal{P} = (-1)\sum_{(\mathcal{P}'')}(-1)^{P''}\mathcal{P}'' = (-1)\sum_{(\mathcal{P})}(-1)^P\mathcal{P} \qquad \text{(H.21)}$$

[74]The permutations form a group — the *symmetric* group.

This establishes the result.

We now compute the expectation value of a totally symmetric operator O that commutes with $\mathcal{P}$, for example $\sum_i t_i$ or $(1/2)\sum_{i\neq j} V_{ij}$.[75] Then

$$\langle O \rangle = \frac{1}{N_P}\int (d\tau) \sum_{\mathcal{P}}\sum_{\mathcal{P}'} (-1)^{P+P'} \left(\mathcal{P}\Pi^\dagger\right) O \left(\mathcal{P}'\Pi\right) \tag{H.22}$$

Here Π is the normal sequence in Eq. (H.17), and $(d\tau)$ is the many-particle volume element. A spin matrix product for pairs of spinors with the same particle label is implied. This expression can be analyzed through the following series of steps and observations:

(1) We first show that all the terms in $\sum_{\mathcal{P}}$ are *identical*. Look at one term with a given $\mathcal{P}$, call it (A)

$$(A) = \sum_{\mathcal{P}'} (-1)^{P+P'} \int (d\tau) \left(\mathcal{P}\Pi^\dagger\right) O \left(\mathcal{P}'\Pi\right) \tag{H.23}$$

(2) All the indices in the integrand in this expression are *dummy indices* — they are all integrated (or summed) over;

(3) Perform the *interchange of dummy indices* corresponding to $\mathcal{P}^{-1}$ (the inverse of $\mathcal{P}$). Then

$$\begin{aligned}(A) &= \sum_{\mathcal{P}'} (-1)^{P+P'} \int (d\tau)\Pi^\dagger \left(\mathcal{P}^{-1} O \mathcal{P}'\Pi\right) \\ &= \sum_{\mathcal{P}'} (-1)^{P+P'} \int (d\tau)\Pi^\dagger O \left(\mathcal{P}^{-1}\mathcal{P}'\Pi\right)\end{aligned} \tag{H.24}$$

Here the last line follows from the total symmetry of O;

(4) We have now returned to the original ordering in $\Pi^\dagger$ without changing the integral;

(5) Now $\left(\mathcal{P}^{-1}\mathcal{P}'\right) \equiv \mathcal{P}''$ again goes over all the permutations as $\mathcal{P}'$ does, and $(-1)^{P+P'} = (-1)^{P''}$ since there are as many interchanges in $\mathcal{P}^{-1}$ as in $\mathcal{P}$;

(6) Thus

$$(A) = \sum_{\mathcal{P}''} (-1)^{P''} \int (d\tau)\Pi^\dagger O \left(\mathcal{P}''\Pi\right) \tag{H.25}$$

Hence all the terms in $\sum_{\mathcal{P}}$ can indeed be reduced to this same expression;

---

[75] These are, after all, identical particles.

(7) Finally, $(1/N_P)\sum_{\mathcal{P}} = 1$ by definition.

We thus arrive at the lovely, simple result that the expectation value of a totally symmetric operator taken with the Slater determinant in Eq. (H.15) can be reduced to the expression

$$\langle O \rangle = \int \Pi^\dagger O \left(\sqrt{N_P}\Phi\right) d\tau \tag{H.26}$$

Here Π is the *normal sequence* in Eq. (H.17), O is the operator, and Φ is the wave function in Eq. (H.16). We have now got rid of one of the determinants in the expectation value, and the orthonormality of the single-particle wave functions makes it a very simple task to evaluate this final expression.[76]

H.3 Some Applications

Consider some applications of this result:

(1) Let $O = 1$. This is the simplest totally symmetric operator, and this calculation provides a check on the normalization of the many-particle wave function in Eq. (H.16). The only term that contributes in Eqs. (H.26) and (H.16) is Π, since any other term has at least two particles interchanged, and one then gets zero by the orthogonality of the single-particle wave functions in the normal sequence. Thus

$$\langle 1 \rangle = 1 \qquad ; \text{ normalization} \tag{H.27}$$

(2) Let

$$O = T = \sum_{i=1}^{N} t_i \qquad ; \text{ kinetic energy} \tag{H.28}$$

Or, more generally, let O be *any* one-body operator. In this case

$$\langle T \rangle = \int (d\tau)\phi^\dagger_{\alpha_1}(1)\cdots\phi^\dagger_{\alpha_N}(N)\left(\sum_{i=1}^{N} t_i\right) \times \\ \sum_{(\mathcal{P})}(-1)^P\mathcal{P}\left[\phi_{\alpha_1}(1)\cdots\phi_{\alpha_N}(N)\right] \tag{H.29}$$

[76]This derivation is taken from a course on Nuclear Theory given by Prof. V. F. Weisskopf at M.I.T. in 1956.

Now t_i operates only on the *ith* coordinate. The only term that then contributes in the sum on $\mathcal{P}$ is Π, since any other term has *two particles* interchanged, and hence *one* of the single-particle matrix elements is zero by orthogonality. Thus

$$\langle T\rangle = \sum_{i=1}^{N}\int \underline{\phi}^{\dagger}_{\alpha_i}(\boldsymbol{\xi}_i)t_i\,\underline{\phi}_{\alpha_i}(\boldsymbol{\xi}_i)\,d^3\xi_i = \sum_{i=1}^{N}\langle\alpha_i|t|\alpha_i\rangle \qquad \text{(H.30)}$$

Therefore the expectation value of the one-body kinetic energy operator takes the form

$$\langle T\rangle = \sum_{i=1}^{N}\langle\alpha_i|t|\alpha_i\rangle = \sum_{\alpha}\langle\alpha|t|\alpha\rangle \quad ;\ \text{single-particle matrix elements}$$

$$\text{sum over occupied states} \qquad \text{(H.31)}$$

The sum is over the single-particle matrix elements, and the last expression is a short-hand for the sum over the occupied states in the normal sequence.

(3) Let

$$O = V = \frac{1}{2}\sum_{i\neq j=1}^{N} V_{ij} \qquad ;\ \text{potential energy} \qquad \text{(H.32)}$$

Or, more generally, let O be any *two-body* operator. We now have *two* terms contributing in the sum over $\mathcal{P}$, the direct term coming from Π and the exchange term coming from the permutation that just interchanges i and j. This term will have one additional minus sign. Thus, with a potential $V_{ij} = V(|\mathbf{x}_i - \mathbf{x}_j|)$,

$$\langle V\rangle = \frac{1}{2}\sum_{i\neq j=1}^{N}\int\int d^3\xi_i d^3\xi_j\,\underline{\phi}^{\dagger}_{\alpha_i}(\boldsymbol{\xi}_i)\underline{\phi}^{\dagger}_{\alpha_j}(\boldsymbol{\xi}_j)V(|\boldsymbol{\xi}_i - \boldsymbol{\xi}_j|)\times$$
$$\left[\underline{\phi}_{\alpha_i}(\boldsymbol{\xi}_i)\underline{\phi}_{\alpha_j}(\boldsymbol{\xi}_j) - \underline{\phi}_{\alpha_j}(\boldsymbol{\xi}_i)\underline{\phi}_{\alpha_i}(\boldsymbol{\xi}_j)\right] \qquad \text{(H.33)}$$

Hence, with the same notation as above,

$$\langle V\rangle = \frac{1}{2}\sum_{i\neq j=1}^{N}\left[\langle\alpha_i\alpha_j|V|\alpha_i\alpha_j\rangle - \langle\alpha_i\alpha_j|V|\alpha_j\alpha_i\rangle\right]$$
$$\langle V\rangle = \frac{1}{2}\sum_{\alpha}\sum_{\beta}\left[\langle\alpha\beta|V|\alpha\beta\rangle - \langle\alpha\beta|V|\beta\alpha\rangle\right]$$

$$;\ \text{sum over occupied states} \qquad \text{(H.34)}$$

The sum is now over two-particle matrix elements, and the double sum goes over the states occupied in the normal sequence. Note that we can include the term where $\alpha = \beta$ in the final expression, since the summand vanishes identically for this term.

(4) We note that if the initial and final many-particle wave functions *differ*, then the same proof as above gives

$$\langle \Phi_f | O | \Phi_i \rangle = \int \Pi_f^\dagger O \left(\sqrt{N_P} \Phi_i \right) d\tau \tag{H.35}$$

These are all extremely useful results.

The Hartree-Fock approximation determines the "best" single-particle wave functions to use in a Slater determinant to describe an arbitrary many-fermion system. The resulting H-F equations are examined in Probs. H.3-H.7.

An alternative, very powerful, approach to the physics of many interacting identical particles based on second quantization and quantum field theory is developed in [Fetter and Walecka (2003a)].

Appendix I

Transition Rate

We start with a set of eigenstates $\phi_n(x)$ of some hamiltonian H_0, and the corresponding solutions to the Schrödinger equation $\Phi_n(x,t)$,

$$
\begin{aligned}
H_0\phi_n(x) &= E_n\phi_n(x) \\
\Phi_n(x,t) &= \phi_n(x)e^{-iE_nt/\hbar}
\end{aligned}
\tag{I.1}
$$

Now suppose that an additional term is added to the hamiltonian so that

$$
H = H_0 + H_1 \tag{I.2}
$$

The full Schrödinger equation reads

$$
i\hbar\frac{\partial}{\partial t}\Psi(x,t) = H\Psi(x,t) \tag{I.3}
$$

If the system is prepared in an eigenstate of H_0 at an initial time t_0, then as time progresses, H_1 will cause transitions to other eigenstates of H_0. The goal of this appendix is to find the *transition rate* to lowest order in H_1.

The full solution to the problem can be expanded in terms of the eigenstates of H_0 according to[77]

$$
\Psi(x,t) = \sum_n c_n(t)\phi_n(x)e^{-iE_nt/\hbar} \qquad ; \text{ full solution} \tag{I.4}
$$

It is assumed that this state is normalized so that

$$
\langle\Psi|\Psi\rangle \equiv \int dx\,\Psi^\star(x,t)\Psi(x,t) = 1 \tag{I.5}
$$

[77]This is a consequence of the *completeness* of the eigenfunctions of H_0.

By the general principles of quantum mechanics, the probability of finding the system in the state ϕ_n at the time t is then given by

$$|c_n(t)e^{-iE_nt/\hbar}|^2 = |c_n(t)|^2 \qquad ; \text{ probability in } \phi_n \tag{I.6}$$

The orthonormality of the eigenfunctions of H_0 allows this expression to be written as the absolute square of

$$\langle\phi_n|\Psi\rangle \equiv \int dx\, \phi_n^\star(x)\Psi(x,t) = c_n(t)e^{-iE_nt/\hbar} \tag{I.7}$$

Substitution of Eq. (I.4) into Eq. (I.3) leads to

$$\sum_n \left(i\hbar\frac{dc_n}{dt} + E_nc_n\right)\phi_n e^{-iE_nt/\hbar} = \sum_n c_n\left(H_0 + H_1\right)\phi_n e^{-iE_nt/\hbar}$$

$$\text{or;} \qquad \sum_n \left(i\hbar\frac{dc_n}{dt}\right)\phi_n e^{-iE_nt/\hbar} = \sum_n c_nH_1\phi_n e^{-iE_nt/\hbar} \tag{I.8}$$

Another application of the orthonormality of the eigenfunctions of H_0 then gives

$$i\hbar\frac{dc_n(t)}{dt} = \sum_m \langle\phi_n|H_1|\phi_m\rangle c_m(t)e^{i(E_n-E_m)t/\hbar}$$

$$\langle\phi_n|H_1|\phi_m\rangle \equiv \int dx\, \phi_n^\star(x)H_1\phi_m(x) \tag{I.9}$$

This presents an exact, infinite set of coupled first-order differential equations in the time for the coefficients $c_n(t)$, in terms of the matrix elements of the perturbation H_1.

Suppose $H_1 = 0$. The above equations then reduce to

$$\frac{d}{dt}c_n(t) = 0$$

$$\Rightarrow \qquad c_n(t) = c_n(t_0) \equiv c_ne^{iE_nt_0/\hbar} \qquad ; \text{ constant} \tag{I.10}$$

where the last relation defines c_n. Then

$$\Psi(x,t) = \sum_n c_n(t_0)\phi_n(x)e^{-iE_nt/\hbar} = \sum_n c_n\phi_n(x)e^{-iE_n(t-t_0)/\hbar}$$

$$\Psi(x,t_0) = \sum_n c_n\phi_n(x) \qquad ; H_1 = 0 \tag{I.11}$$

This is just the general solution to the unperturbed Schrödinger equation

$$i\hbar\frac{\partial\Psi}{\partial t} = H_0\Psi \tag{I.12}$$

that matches the initial condition $\Psi(x, t_0)$.

Now suppose that $H_1 \neq 0$, and that the system is prepared in some arbitrary state $\Psi_i(x, t_0)$,

$$\Psi_i(x, t_0) = \sum_n c_n^{(i)}(t_0)\phi_n(x)e^{-iE_n t_0/\hbar} \qquad ; \text{ prepared state at } t_0 \qquad \text{(I.13)}$$

Then

(1) If $H_1 = 0$, Eqs. (I.10) give the time development of the state as

$$\Psi_i(x, t) = \sum_n c_n^{(i)}(t_0)\phi_n(x)e^{-iE_n t/\hbar} \qquad ; \; H_1 = 0 \qquad \text{(I.14)}$$

(2) If $H_1 \neq 0$, Eq. (I.4) gives the solution to the full Schrödinger equation as

$$\Psi_i(x, t) = \sum_n c_n^{(i)}(t)\phi_n(x)e^{-iE_n t/\hbar} \qquad ; \; H_1 \neq 0 \qquad \text{(I.15)}$$

The coefficients $c_n^{(i)}(t)$ develop a slow time dependence due to the effects of H_1. To find these coefficients, one can integrate the first of Eqs. (I.9) on the time from t_0 to t, starting from the initial values $c_n^{(i)}(t_0)$[78]

$$c_n^{(i)}(t) - c_n^{(i)}(t_0) = -\frac{i}{\hbar}\sum_m \int_{t_0}^{t} dt\, \langle\phi_n|H_1|\phi_m\rangle c_m^{(i)}(t)e^{i(E_n - E_m)t/\hbar} \qquad \text{(I.16)}$$

This is an exact, infinite set of coupled linear *integral* equations for the coefficients $c_n^{(i)}(t)$, which develop from the initial set $c_n^{(i)}(t_0)$.

Now assume that H_1 is small and the transition is weak. Equations (I.16) can be *iterated* to first order in H_1 by observing that[79]

$$\begin{aligned} c_m^{(i)}(t) &= c_m^{(i)}(t_0) + O(H_1) \\ c_m^{(i)}(t) &\approx c_m^{(i)}(t_0) \qquad ; \text{ to } \mathrm{O(H_1)} \text{ on the r.h.s.} \end{aligned} \qquad \text{(I.17)}$$

Suppose the system is prepared in a definite state ϕ_i at the time t_0, then

$$c_m^{(i)}(t_0) = c_i(t_0)\delta_{mi} \qquad ; \text{ prepared state} \qquad \text{(I.18)}$$

Since this state is normalized

$$|c_i(t_0)|^2 = 1 \qquad ; \text{ normalization} \qquad \text{(I.19)}$$

[78] One performs $\int_{t_0}^{t} dt$ on the first of Eqs. (I.9).
[79] This is known as Dirac perturbation theory.

Then, by Eq. (I.16), if the state n is different from the initial state i,

$$c_n^{(i)}(t) \approx -\frac{i}{\hbar}c_i(t_0)\int_{t_0}^{t} dt\,\langle\phi_n|H_1|\phi_i\rangle e^{i(E_n-E_i)t/\hbar} \qquad ; n \neq i \quad \text{(I.20)}$$

If we are interested in transitions to a definite final state $\phi_n = \phi_f \neq \phi_i$ (Fig. I.1), then

$$c_f^{(i)}(t) \approx -\frac{i}{\hbar}c_i(t_0)\int_{t_0}^{t} dt\,\langle\phi_f|H_1|\phi_i\rangle e^{i(E_f-E_i)t/\hbar} \qquad ; f \neq i \quad \text{(I.21)}$$

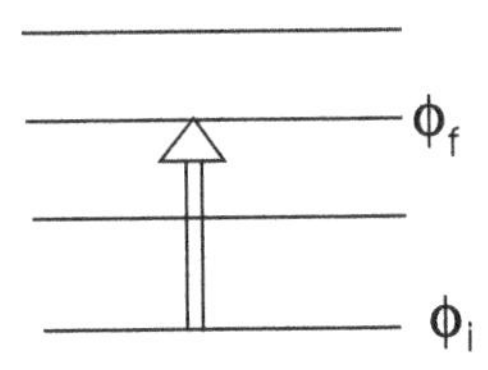

Fig. I.1 Transition from a state ϕ_i to a state ϕ_f induced by H_1.

The *probability* that the system will be in the state ϕ_f at the time t if it started in the state ϕ_i at the time t_0 now follows from Eq. (I.6) as[80]

$$P_{fi}(t,t_0) = |c_f^{(i)}(t)|^2 = \left|-\frac{i}{\hbar}\int_{t_0}^{t} dt\,\langle\phi_f|H_1|\phi_i\rangle e^{i(E_f-E_i)t/\hbar}\right|^2 \qquad \text{(I.22)}$$

This is *first-order time-dependent perturbation theory*, and this expression is exact to $O(H_1)$. The problem has been reduced to taking the finite Fourier transform of the matrix element of the perturbation with respect to the angular frequency $(E_f - E_i)/\hbar$.

The perturbation H_1 may, or may not, have an explicit time dependence. This depends on the nature of the problem at hand. Consider the case where the perturbation H_1 has no explicit time dependence. In this case, the time integral in Eq. (I.22) can be immediately performed. Define $(E_f - E_i)/\hbar \equiv \omega$ and $t - t_0 \equiv T$. Then use

$$\begin{aligned}\int_{t_0}^{t} dt\,e^{i\omega t} &= \frac{1}{i\omega}\left[e^{i\omega t} - e^{i\omega t_0}\right] = \frac{e^{i\omega t_0}}{i\omega}\left[e^{i\omega T} - 1\right] \\ &= e^{i\omega t_0}e^{i\omega T/2}\left[\frac{2\sin(\omega T/2)}{\omega}\right]\end{aligned} \qquad \text{(I.23)}$$

[80]Note Eq. (I.19).

It follows from Eq. (I.22) that

$$P_{fi}(T) = \frac{1}{\hbar^2}|\langle\phi_f|H_1|\phi_i\rangle|^2 \left[\frac{\sin^2(\omega T/2)}{(\omega/2)^2}\right] \qquad ; \omega = \frac{1}{\hbar}(E_f - E_i)$$
$$T = t - t_0 \qquad \text{(I.24)}$$

Now one expects that the probability of the transition will be proportional to the time interval over which the perturbation acts. Thus to get a sensible physical quantity for large T, we define the *transition rate* by

$$\text{transition rate} \equiv (\text{transition probability}) \;/\; (\text{time interval}) \qquad \text{(I.25)}$$

Hence, from Eq. (I.24),

$$R_{fi} \equiv \frac{P_{fi}(T)}{T}$$
$$= \frac{1}{\hbar^2}|\langle\phi_f|H_1|\phi_i\rangle|^2 \left[T\,\frac{\sin^2(\omega T/2)}{(\omega T/2)^2}\right] \qquad ; \text{transition rate} \quad \text{(I.26)}$$

Consider the function $f_T(\omega)$ defined by

$$f_T(\omega) \equiv \left[T\,\frac{\sin^2(\omega T/2)}{(\omega T/2)^2}\right] \qquad \text{(I.27)}$$

It is sketched as a function of ω in Fig. I.2.

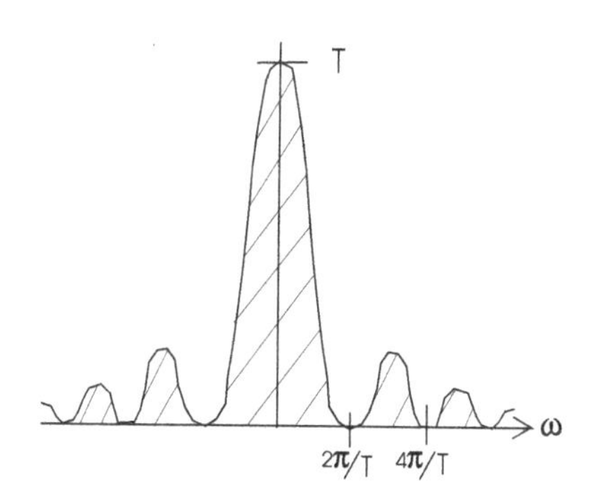

Fig. I.2 Sketch of the function $f_T(\omega)$ in Eq. (I.27).

The function $f_T(\omega)$ has the following properties:

(1) For large T, at the origin ($\omega = 0$), it grows as T

$$f_T(0) \to T \qquad ; T \to \infty \qquad \text{(I.28)}$$

(2) For large T, away from the origin ($\omega \neq 0$), it goes to zero

$$f_T(\omega) \to 0 \qquad ; T \to \infty \quad \omega \neq 0 \tag{I.29}$$

(3) When integrated over all ω, a simple change of variables $x \equiv \omega T/2$ gives

$$\int_{-\infty}^{\infty} f_T(\omega)\, d\omega = 2\int_{-\infty}^{\infty} \frac{\sin^2 x}{x^2}\, dx = 2\pi \qquad ; \text{ all } T \tag{I.30}$$

The last result is obtained from any good table of integrals.

We thus observe that in the limit of large T, $f_T(\omega)$ has all the properties of the *Dirac delta function* (see Prob. 4.3)

$$\text{Lim}_{T\to\infty} f_T(\omega) = 2\pi\delta(\omega) = 2\pi\hbar\,\delta(E_f - E_i) \tag{I.31}$$

For large T, the transition rate in Eq. (I.26) therefore becomes[81]

$$R_{fi} = \frac{P_{fi}(T)}{T} = \frac{2\pi}{\hbar}|\langle\phi_f|H_1|\phi_i\rangle|^2\delta(E_f - E_i) \quad ; \text{ transition rate} \tag{I.32}$$

This expression is indeed independent of T. It is exact to $O(H_1)$. If the perturbation has a harmonic time dependence so that $H_1(t) = H_1 e^{\pm i\omega_0 t}$, then the only change is to replace $E_f - E_i \to E_f - E_i \pm \hbar\omega_0$ in Eqs. (I.22) and (I.32). Fermi called Eq. (I.32) the "Golden Rule"; it is one of the most useful results in quantum mechanics.

As one class of examples, consider processes where there is a particle in the continuum in the final state (*e.g.* scattering, decay, *etc.*). Take the laboratory to be a huge cubical box of volume $\Omega = L^3$ (Fig. 7.7), and apply periodic boundary conditions so that the single-particle wave functions are those of Eq. (4.118) for spin-zero bosons and Eq. (4.159) for spin-1/2 fermions. Since the particle states in the big box are spaced very close together (Fig. 4.35), one is *forced* to make a measurement of transitions into a *group* of states. The transition probability will vary smoothly about the central values, and one thus employs

[81]The limit $T \to \infty$ is, in fact, a rather subtle one. For example, T cannot be too large or the initial state will be depleted, and the conditions in Eqs. (I.17)–(I.19) will no longer hold. Thus one must keep $P_{fi}(T) \ll 1$, or $|T\langle\phi_f|H_1|\phi_i\rangle| \ll \hbar$. For any finite T, however, this relation is always satisfied as $H_1 \to 0$.

$$
\begin{aligned}
R_{fi} &= \text{transition rate into one state} \\
dn_f &= \text{number of states observed in detectors}
\end{aligned} \tag{I.33}
$$

The relevant transition rate is therefore given by

$$
d\omega_{fi} = R_{fi} dn_f \qquad ; \text{ observed transition rate} \tag{I.34}
$$

With the aid of the differential form of Eq. (4.166), it follows that

$$
d\omega_{fi} = R_{fi} \frac{\Omega d^3k}{(2\pi)^3} \qquad ; \text{ observed transition rate} \tag{I.35}
$$

If one is computing a scattering cross section, then if I_{inc} is the incident flux (see text), it follows from Eq. (7.49) that

$$
I_{\text{inc}}\, d\sigma_{fi} = d\omega_{fi} \qquad ; \text{ cross section} \tag{I.36}
$$

The energy-conserving delta function in Eq. (I.32) is now handled with $\int dk\, \delta(E_f - E_i)$, and the artificial quantization volume Ω must cancel at the end in any physical result (see, for example, Prob. I.2). For one specific application, see Prob. I1.

Appendix J

Neutrino Mixing

In this appendix we give a brief discussion of two-neutrino mixing. It arises because the neutrinos entering into the weak interactions may not be the energy eigenstates of the total hamiltonian. We here consider only two flavors of Dirac neutrinos, for example (ν_e, ν_μ), and mixing in vacuum.[82] Note that in introducing neutrino mixing, we by necessity violate individual lepton number conservation.

The problem is just one in two-state quantum mechanics. Here we work in the subspace of given $(\mathbf{p}, \lambda)$, where the wave functions are those of Eqs. (4.159). Recall $\mathbf{p} = \hbar\mathbf{k}$, and the index λ is now the *helicity*, the component of the spin along the direction of motion. These quantum numbers will be suppressed in the subsequent arguments. The wave functions now carry an additional third index for the *type* of neutrino (ν_e, ν_μ), and this index will be made explicit. Just as with isospin, the wave functions for the different types of neutrinos are orthogonal

$$\langle\psi_{\nu_j}|\psi_{\nu_k}\rangle \equiv \int_\Omega d^3x\ \psi^\dagger_{\nu_j}(\mathbf{x})\psi_{\nu_k}(\mathbf{x}) = \delta_{jk} \qquad ;(j,k) = (e,\mu) \qquad \text{(J.1)}$$

The two neutrino states are produced through the weak interactions, and their corresponding leptons numbers (l_e, l_μ) are exactly conserved by all the interactions in the standard model. Suppose these are not eigenstates of the full H.[83] We then look for the two states that do satisfy

$$H\psi_s = E^{(s)}\psi_s \qquad ;\ s = 1,2 \qquad \text{(J.2)}$$

[82]The full three-flavor sector with $(\nu_e, \nu_\mu, \nu_\tau)$ is analyzed in the same fashion, but the analysis is more complicated.

[83]Which is now assumed to contain an additional ultra-weak contribution $H_{\rm UW}$.

These will be some linear combinations of ψ_{ν_e} and ψ_{ν_μ}.

$$\psi_s = \sum_{j=e,\mu} a_j^{(s)} \psi_{\nu_j} \tag{J.3}$$

Substitute this expression into Eq. (J.2), and then compute the matrix element with ψ_{ν_k}

$$\sum_{j=e,\mu} \left[H_{kj} - E^{(s)}\delta_{kj}\right] a_j^{(s)} = 0 \qquad ; \; k = e, \mu$$
$$H_{kj} \equiv \langle \psi_{\nu_k} | H | \psi_{\nu_j} \rangle \tag{J.4}$$

This is now a familiar two-dimensional matrix eigenvalue problem. One can always change the overall phase of the neutrino states without changing any of our previous physical results. Choose the relative phase so that the matrix H_{kj} is real. The coefficients $a_j^{(s)}$ can then also be assumed real. The modal matrix is constructed by placing the normalized coefficients down the columns (see [Fetter and Walecka (2003)])

$$\underline{\mathcal{M}} = \begin{pmatrix} a_1^{(1)} & a_1^{(2)} \\ a_2^{(1)} & a_2^{(2)} \end{pmatrix} \tag{J.5}$$

It is readily verified that this is a real, orthogonal matrix that diagonalizes the hamiltonian

$$\underline{\mathcal{M}}^T \underline{\mathcal{M}} = \underline{1}$$
$$\underline{\mathcal{M}}^T \underline{H} \underline{\mathcal{M}} = \underline{H}_D \; = \; \begin{pmatrix} E^{(1)} & 0 \\ 0 & E^{(2)} \end{pmatrix} \tag{J.6}$$

Any real, orthogonal, 2×2 matrix can be parameterized as

$$\underline{\mathcal{M}} = \begin{pmatrix} \cos\theta & -\sin\theta \\ \sin\theta & \cos\theta \end{pmatrix} \tag{J.7}$$

The general solution to the time-dependent Schrödinger equation within the neutrino subspace can now be written as

$$\Psi(t) = \sum_{s=1,2} c_s \, \psi_s \, e^{-iE^{(s)}t/\hbar}$$
$$i\hbar \frac{\partial}{\partial t} \Psi(t) = H\Psi(t) \tag{J.8}$$

The coefficients in this expansion are determined by the initial conditions

$$\Psi(0) = \sum_{s=1,2} c_s \psi_s$$
$$c_s = \langle \psi_s | \Psi(0) \rangle \qquad \text{(J.9)}$$

The probability to find the system in the state ψ_{ν_j} at the time t is given by quantum mechanics as

$$P_j(t) = |\langle \psi_{\nu_j} | \Psi(t) \rangle|^2 \qquad \text{(J.10)}$$

Since

$$\langle \psi_{\nu_j} | \psi_s \rangle = a_j^{(s)} \qquad \text{(J.11)}$$

the above results can be combined to write

$$P_j(t) = \left| \sum_{s=1,2} c_s a_j^{(s)} e^{-iE^{(s)}t/\hbar} \right|^2 \qquad \text{(J.12)}$$

As an example, suppose one starts with a pure ν_e beam created by the weak interactions at time $t = 0$. The new energy eigenstates are given by

$$\psi_1 = \cos\theta\, \psi_{\nu_e} + \sin\theta\, \psi_{\nu_{\mu^-}}$$
$$\psi_2 = -\sin\theta\, \psi_{\nu_e} + \cos\theta\, \psi_{\nu_\mu} \qquad \text{(J.13)}$$

These relations are readily inverted to give

$$\psi_{\nu_e} = \cos\theta\, \psi_1 - \sin\theta\, \psi_2$$
$$\psi_{\nu_\mu} = \sin\theta\, \psi_1 + \cos\theta\, \psi_2 \qquad \text{(J.14)}$$

The initial state of pure ν_e therefore has coefficients

$$c_1 = \cos\theta \qquad\qquad c_2 = -\sin\theta \qquad \text{(J.15)}$$

Let us compute the probability to find the system in the state ψ_{ν_μ} after some time T. In this case

$$a_\mu^{(1)} = \sin\theta \qquad\qquad a_\mu^{(2)} = \cos\theta \qquad \text{(J.16)}$$

Thus

$$P_{\nu_\mu \leftarrow \nu_e}(T) = \sin^2\theta \cos^2\theta \left| e^{-iE^{(1)}T/\hbar} - e^{-iE^{(2)}T/\hbar} \right|^2$$
$$= \sin^2(2\theta)\, \sin^2\left(\frac{T\Delta E}{2\hbar}\right) \qquad ; \; \Delta E \equiv E^{(1)} - E^{(2)} \qquad \text{(J.17)}$$

By conservation of probability (or direct calculation), the probability that the system remains in the state ψ_{ν_e} is

$$P_{\nu_e \leftarrow \nu_e}(T) = 1 - \sin^2(2\theta)\,\sin^2\left(\frac{T\Delta E}{2\hbar}\right) \tag{J.18}$$

This is evidently smaller than 1 if the last term is non-zero.

Now suppose that the neutrinos are highly relativistic and travel a distance L after they are produced with a velocity $\approx c$. In this case $cT = L$ and

$$\Delta E = (p^2c^2 + m_1^2c^4)^{1/2} - (p^2c^2 + m_2^2c^4)^{1/2} \approx \frac{c^4\Delta m^2}{2pc}$$
$$\Delta m^2 \equiv m_1^2 - m_2^2 \tag{J.19}$$

Here $p \equiv |\mathbf{p}|$.

In summary, in this simple two-state mixing calculation with an initially produced ψ_{ν_e} state and relativistic neutrinos that travel in vacuum a distance L from the source, one finds

$$P_{\nu_\mu \leftarrow \nu_e}(L) = \sin^2(2\theta)\,\sin^2\left(\frac{Lc^2\Delta m^2}{4\hbar p}\right)$$
$$P_{\nu_e \leftarrow \nu_e}(L) = 1 - \sin^2(2\theta)\,\sin^2\left(\frac{Lc^2\Delta m^2}{4\hbar p}\right) \tag{J.20}$$

These relations provide the basis for major experimental efforts to measure $(\Delta m^2, \theta)$.

Appendix K

Units

K.1 Standard International (SI)

In this book we use SI units where mass, distance, and time are measured in kilograms (kg), meters (m), and seconds (s). Force is then measured in Newtons (N), where 1N = 1kg-m/s^2. Charge is measured in Coulombs (C), voltage in volts (V), and current in amperes (A), where 1A = 1C/s.[84]

Coulomb's law states that the force between two static point charges e separated by a displacement $\mathbf{r} = \mathbf{r}_2 - \mathbf{r}_1$ is

$$\mathbf{F} = \frac{e^2}{4\pi\varepsilon_0}\frac{\mathbf{r}}{r^3} \tag{K.1}$$

The laws governing the fields $(\mathbf{E}, \mathbf{B})$ are summarized by Maxwell's equations. In free space, with sources $(\rho, \mathbf{j})$, in these units they are

$$\begin{aligned}
\nabla \cdot \mathbf{E} &= \frac{\rho}{\varepsilon_0} \\
\nabla \cdot \mathbf{B} &= 0 \\
\nabla \times \mathbf{E} &= -\frac{\partial \mathbf{B}}{\partial t} \\
\nabla \times \mathbf{B} &= \mu_0 \mathbf{j} + \frac{1}{c^2}\frac{\partial \mathbf{E}}{\partial t}
\end{aligned} \tag{K.2}$$

If n is the number density for the constituents of charge e, here the electrons, then the charge density and current are given by[85]

$$\rho(\mathbf{x}, t) = en(\mathbf{x}, t) \qquad ; \ \mathbf{j}(\mathbf{x}, t) = en(\mathbf{x}, t)\mathbf{v}(\mathbf{x}, t) \tag{K.3}$$

[84] These are now the units of choice in most introductory physics courses.

[85] Strictly speaking, $\mathbf{j}$ is the current *density*; however, we shall always refer to $\mathbf{j}$ as the *current*.

where $\mathbf{v}$ is the velocity field of the charges. The conservation law for the number density reads

$$\frac{\partial n}{\partial t} + \boldsymbol{\nabla} \cdot (n\mathbf{v}) = 0 \tag{K.4}$$

The velocity of light is given by $c = 1/\sqrt{\varepsilon_0 \mu_0}$. The Lorentz force acting on a particle with charge e moving in $(\mathbf{E}, \mathbf{B})$ is given in these units by

$$\mathbf{F} = e\,(\mathbf{E} + \mathbf{v} \times \mathbf{B}) \tag{K.5}$$

The fine-structure constant is dimensionless, and it is given in SI units by

$$\alpha = \left(\frac{e^2}{4\pi\varepsilon_0}\right)\frac{1}{\hbar c} = \frac{1}{137.0} \tag{K.6}$$

K.2 Heaviside-Lorentz (rationalized cgs)

The unit of charge in SI units, the Coulomb, is defined to be the charge carried by a certain number of elementary constituents (*i.e.* electrons) $Q = N_0|e|$. The dimensionless number $1/4\pi\varepsilon_0$ then simply provides the constant of proportionality between the measured (Coulomb force)$\times$(distance)^2 on the unit (charge)^2 and the unit value of (force)$\times$(distance)^2 in mks units. One could use a different amount of charge N_0 in the unit, and this would simply change the number ε_0. *A particularly convenient unit of charge is that which gives $\varepsilon_0 = 1$ in cgs units.* The charge on the electron in these units is $e_{\rm HL}$. The dimensional quantity $\mu_0 = 1/c^2\varepsilon_0 \to 1/c^2$ relates the magnetic field $\mathbf{B}$ to $\mathbf{j}$, and $\mathbf{B}$ is in turn related to a force through the Lorentz force equation. With the redefinition $c\mathbf{B} \to \mathbf{B}$, Coulomb's Law and the Lorentz force equation become

$$\mathbf{F} = \frac{e_{\rm HL}^2}{4\pi}\frac{\mathbf{r}}{r^3} \qquad ; \ \mathbf{F} = e_{\rm HL}\left(\mathbf{E} + \frac{\mathbf{v}}{c} \times \mathbf{B}\right) \tag{K.7}$$

and Maxwell's equations now read

$$\begin{aligned}
\nabla \cdot \mathbf{E} &= \rho \\
\nabla \cdot \mathbf{B} &= 0 \\
\nabla \times \mathbf{E} &= -\frac{1}{c}\frac{\partial \mathbf{B}}{\partial t} \\
\nabla \times \mathbf{B} &= \mathbf{j} + \frac{1}{c}\frac{\partial \mathbf{E}}{\partial t}
\end{aligned} \tag{K.8}$$

The charge density and current are given by

$$\rho(\mathbf{x},t) = e_{\rm HL}n(\mathbf{x},t) \qquad ; \; \mathbf{j}(\mathbf{x},t) = \frac{e_{\rm HL}}{c}n(\mathbf{x},t)\mathbf{v}(\mathbf{x},t) \qquad \text{(K.9)}$$

Note that the *dimensions* of both $(\mathbf{B},\mathbf{j})$ have now changed.

The fine-structure constant is left invariant in H-L units since $e^2/\varepsilon_0 \equiv e_{\rm HL}^2$, and since it is dimensionless, the fine-structure constant does not depend on the choice of units. Hence $\alpha = e_{\rm HL}^2/4\pi\hbar c = 1/137.0$. These are known as Heaviside-Lorentz, or rationalized cgs, units. They are used in most books that involve advanced quantum mechanics.[86]

K.3 cgs

A further simplification occurs if one redefines the unit of charge to eliminate the $1/4\pi$ in Coulomb's law. With the new electron charge $e_{\rm cgs}$, Maxwell's equations become

$$\begin{aligned}
\nabla\cdot\mathbf{E} &= 4\pi\rho \\
\nabla\cdot\mathbf{B} &= 0 \\
\nabla\times\mathbf{E} &= -\frac{1}{c}\frac{\partial\mathbf{B}}{\partial t} \\
\nabla\times\mathbf{B} &= 4\pi\mathbf{j} + \frac{1}{c}\frac{\partial\mathbf{E}}{\partial t}
\end{aligned} \qquad \text{(K.10)}$$

The charge density and current are now given by

$$\rho(\mathbf{x},t) = e_{\rm cgs}n(\mathbf{x},t) \qquad ; \; \mathbf{j}(\mathbf{x},t) = \frac{e_{\rm cgs}}{c}n(\mathbf{x},t)\mathbf{v}(\mathbf{x},t) \qquad \text{(K.11)}$$

The Lorentz force is

$$\mathbf{F} = e_{\rm cgs}\left(\mathbf{E} + \frac{\mathbf{v}}{c}\times\mathbf{B}\right) \qquad \text{(K.12)}$$

and the fine-structure constant is simply $\alpha = e_{\rm cgs}^2/\hbar c = 1/137.0$. One says the charge density is now measured in esu, and the current in emu where 1 emu = 1 esu/c. These are known as cgs units. They are used in most books based on non-relativistic quantum mechanics.[87]

Conversion. A way to remember the conversion of units is the following:

---

[86] For example, [Walecka (2004)]. One often further sets $\hbar = c = 1$, which implies that all momenta and energies become inverse lengths.

[87] For example, [Fetter and Walecka (2003a)], where the $1/c$ in the current is left in Maxwell's equations.

(1) To go from SI to H-L units, replace

$$\begin{aligned}
&\varepsilon_0 \rightarrow 1 \quad &; \mu_0 \rightarrow 1/c^2 \quad &; e \rightarrow e_{\rm HL} \\
&\mathbf{E} \rightarrow \mathbf{E}_{\rm HL} \quad &; \mathbf{B} \rightarrow \mathbf{B}_{\rm HL}/c &
\end{aligned} \tag{K.13}$$

(2) To go from H-L to cgs units, replace

$$e_{\rm HL} \rightarrow \sqrt{4\pi} e_{\rm cgs} \quad ; \sqrt{4\pi}\mathbf{E}_{\rm HL} \rightarrow \mathbf{E}_{\rm cgs} \quad ; \sqrt{4\pi}\mathbf{B}_{\rm HL} \rightarrow \mathbf{B}_{\rm cgs} \tag{K.14}$$

Appendix L

Fundamental Constants

$k_\mathrm{B} = 1.381 \times 10^{-23}\,\mathrm{J/^oK}$	Boltzmann's constant
$N_\mathrm{A} = 6.022 \times 10^{23}/\mathrm{mole}$	Avagadro's number
$R = N_\mathrm{A}\, k_\mathrm{B} = 1.987\,\mathrm{cal/mole\text{-}^oK}$	gas constant
$\lvert e\rvert = 1.602 \times 10^{-19}\,\mathrm{C}$	electron charge
$1/4\pi\varepsilon_0 = 8.988 \times 10^{9}\,\mathrm{Nm^2/C^2}$	Coulomb's law
$\mu_0/4\pi = 1.000 \times 10^{-7}\,\mathrm{Ns^2/C^2} = 1.000 \times 10^{-7}\,\mathrm{Tm/A}$	Ampere's law
$c = 1/\sqrt{\varepsilon_0\mu_0} = 2.998 \times 10^{8}\,\mathrm{m/s}$	velocity of light
$h = 6.626 \times 10^{-34}\,\mathrm{Js}$	Planck's constant
$\hbar = h/2\pi = 1.055 \times 10^{-34}\,\mathrm{Js}$	Planck's constant/2π
$\hbar/m_e c = 3.862 \times 10^{-13}\,\mathrm{m}$	electron Compton wavelength
$\hbar/m_p c = 2.103 \times 10^{-16}\,\mathrm{m}$	proton Compton wavelength
$a_0 = \hbar^2/m_e(e^2/4\pi\varepsilon_0) = 0.5292 \times 10^{-10}\,\mathrm{m} = 0.5292\,\mathrm{\AA}$	Bohr radius
$\alpha = e^2/(4\pi\varepsilon_0)\hbar c = 1/137.0$	fine-structure constant
$m_e c^2 = 0.5110\,\mathrm{MeV}$	electron rest mass
$m_p c^2 = 938.3\,\mathrm{MeV}$	proton rest mass
$\mathcal{R} = \alpha^2 m_e c^2/2 = 13.61\,\mathrm{eV}$	Rydberg
$\lvert e\rvert\hbar/2m_e = 9.274 \times 10^{-24}\,\mathrm{J/T}$	Bohr magneton
$\lvert e\rvert\hbar/2m_p = 5.051 \times 10^{-27}\,\mathrm{J/T}$	nucleon magneton

$h/2|e| = 2.068 \times 10^{-15}\,\text{Js/C} = 2.068 \times 10^{-15}\,\text{Tm}^2$ flux quantum

$\sigma = \pi^2 k_B^4/60\hbar^3c^2 = 5.670 \times 10^{-8}\,\text{J/s-m}^2\text{-}^\circ\text{K}^4$ Stefan-Boltzmann constant

$\text{G} = 6.673 \times 10^{-11}\,\text{m}^3/\text{kg-s}^2$ gravitational constant

$1\,\text{cal} = 4.184\,\text{J}$ heat equivalent

L.1 Conversion Factors

$$
\begin{aligned}
1\,\text{eV} &= 1.602 \times 10^{-19}\,\text{J} \\
1\,\text{MeV}/c^2 &= 1.783 \times 10^{-30}\,\text{kg} \\
1\,\text{J} &= 10^7\,\text{erg} \\
1\,\text{N} &= 10^5\,\text{dyne} \\
1\,\text{Å} &= 10^{-10}\,\text{m} \\
1\,\text{F} &\equiv 1\,\text{fm} = 10^{-15}\,\text{m} \\
1\,\text{C} &= 2.998 \times 10^9\,\text{esu} \\
1\,\text{T} &= 10^4\,\text{Gauss} \\
1\,\text{J/T} &= 10^3\,\text{erg/Gauss} \\
1\,\text{eV}/hc &= 8066\,\text{cm}^{-1} \\
\hbar c &= 197.3\,\text{MeV-F} \\
\hbar^2/2m_p &= 20.74\,\text{MeV-F}^2 \\
Gm_p^2/\hbar c &= 5.905 \times 10^{-39}
\end{aligned}
$$

Appendix M

Some Significant Names for Theoretical Physics

Isaac Newton (1642-1727)
Jean Le Rond d'Alembert (1717-1783)
Joseph Louis Lagrange (1736-1813)
Jean Baptiste Joseph Fourier (1768-1830)
Karl Friedrich Gauss (1777-1855)
Augustin-Louis Cauchy (1789-1857)
William Rowan Hamilton (1805-1865)
George Gabriel Stokes (1819-1903)
Hermann Ludwig Ferdinand Helmholtz (1821-1894)
George Friedrich Bernhard Riemann (1826-1866)
James Clerk Maxwell (1831-1879)
John William Strutt (Lord Rayleigh) (1842-1919)
Ludwig Eduard Boltzmann (1844-1906)
Hendrik Antoon Lorentz (1853-1927)
Joseph John Thomson (1856-1940)
Max Karl Ernst Ludwig Planck (1858-1947)
Ernest Rutherford (1871-1937)
Albert Einstein (1879-1955)
Max Born (1882-1970)
Peter Joseph William Debye (1884-1966)
Niels Henrik David Bohr (1885-1962)
Erwin Rudolf Josef Alexander Schrödinger (1885-1962)
Arthur Holly Compton (1892-1962)
Louis Victor Pierre Raymond duc de Broglie (1892-1987)
Satyendra Nath Bose (1894-1974)
Douglas Rayner Hartree (1897-1958)
Wolfgang Ernst Pauli (1900-1958)

Enrico Fermi (1901-1954)
Werner Karl Heisenberg (1901-1976)
Paul Adrien Maurice Dirac (1902-1984)
Eugene Paul Wigner (1902-1995)
Felix Bloch (1905-1983)
Maria Goeppert-Mayer (1906-1972)
Hans Albrecht Bethe (1906-2005)
Hideki Yukawa (1907-1981)
John Bardeen (1908-1991)
Lev Davidovich Landau (1908-1968)
Victor Frederick Weisskopf (1908-2002)
Subrahmanyan Chandrasekhar (1910-1995)
Julian Seymour Schwinger (1918-1994)
Richard Phillips Feynman (1919-1988)
Aage Niels Bohr (1922-)
Chen Ning Yang (1922-)
Walter Kohn (1923-)
Tsung-Dao Lee (1926-)
Benjamin Roy Mottleson (1926-)
Murray Gell-Mann (1929-)
Leon Nathan Cooper (1930-)
John Robert Schrieffer (1931-)
Sheldon Lee Glashow (1932-)
Steven Weinberg (1933-)
Jeffrey Goldstone (1933-)
James Daniel Bjorken (1934-)
Kenneth Geddes Wilson (1936-)
David Jonathan Gross (1941-)
Hugh David Politzer (1949-)
Frank Wilczek (1951-)

Bibliography

Arfken, G. B., and Weber, H. C., (1995). *Mathematics for Physicists, 4th ed.*, Academic Press, New York, NY

Bardeen, J., Cooper, L. N., and Schrieffer, J. R., (1957). *Phys. Rev.*, **106**, 162; *Phys. Rev.*, **108**, 1175

Beiser, A., (2002). *Concepts of Modern Physics, 6th ed.*, McGraw-Hill, New York, NY

Bernstein, J., Fishbane, P. M., and Gasiorowicz, S. G., (2000). *Modern Physics*, Prentice-Hall, Englewood Cliffs, NJ

Bethe, H. A., and Goldstone, J., (1957). *Proc. Roy. Soc. (London)* **A238**, 551

Bhaduri, R. K., (1988). *Models of the Nucleon*, Addison-Wesley, Reading, MA

Bjorken, J. D., and Drell, S. D., (1964). *Relativistic Quantum Mechanics*, McGraw-Hill, New York, NY

Bjorken, J. D., and Drell, S. D., (1965). *Relativistic Quantum Fields*, McGraw-Hill, New York, NY

Bjorken, J. D., and Paschos, E. A., (1969). *Phys. Rev.* **185**, 1975

Blatt, J. M., and Weisskopf, V. F., (1952). *Theoretical Nuclear Physics*, John Wiley and Sons, New York, NY

Bohr, A., and Mottelson, B. R., (1969). *Nuclear Structure Vol. I, Single-Particle Motion*, W. A. Benjamin, Reading, MA

Bohr, A., and Mottelson, B. R., (1969). *Nuclear Structure Vol. II, Nuclear Deformations*, W. A. Benjamin, Reading, MA

Born, M., (1926). *Zeit. f. Physik* **37**, 863; (1927) *Nature*, **119**, 354

Born, M., (1989). *Atomic Physics, 8th ed.*, Dover Publications, Mineola, NY

Carlson, T. A., Nestor, C. W., Wasserman, N., and McDowell, T. D., (1970). *Atomic Data* **2**, 63

Cheng, T.-P., and Li, L.-F., (1984). *Gauge Theory of Elementary Particle Physics*, Clarendon Press, Oxford, UK

CNST, (2007). *Flux lattice*, http://www.cnst.nist.gov/epg/Projects/STM/super cond.html

Colorado, (2007). *The Bose-Einstein Condensate*, http://www.colorado.edu/physics/2000/bec

Commons, E. D., and Buchsbaum, P. H., (1983). *Weak Interactions of Leptons and Quarks*, Cambridge University Press, Cambridge, UK

Cooper, L. N., (1956). *Phys. Rev.*, **104**, 1189

de Shalit, A., and Talmi, I., (1963). *Nuclear Shell Theory*, Academic Press, New York, NY

de Shalit, A., and Feshbach, H., (1974). *Theoretical Nuclear Physics Vol. I, Nuclear Structure*, John Wiley and Sons, New York, NY

Dirac, P. A. M., (1926). *Proc. Roy. Soc.*, **117**, 610

Edmonds, A. R., (1974). *Angular Momentum in Quantum Mechanics*, 3rd printing, Princeton University Press, Princeton, NJ

Einstein, A. (1905). "Zur Electrodynamik bewegter Körper," *Annalen der Physik* **17**, 891

Einstein, A. (1916). "Die Grundlage der Allgemeinen Relativitäts Theorie," *Annalen der Physik* **49**, 50

Eisberg, R., and Resnick, R., (1985). *Quantum Physics of Atoms, Molecules, Solids, Nuclei, and Particles, 2nd ed.*, John Wiley and Sons, New York, NY

Feshbach, H., (1991). *Theoretical Nuclear Physics Vol. II, Nuclear Reactions*, John Wiley and Sons, Inc., New York, NY

Fetter, A. L., and Walecka, J. D., (2003). *Theoretical Mechanics of Particles and Continua*, McGraw-Hill, New York, NY (1980); reissued by Dover Publications, Mineola, NY

Fetter, A. L., and Walecka, J. D., (2003a). *Quantum Theory of Many-Particle Systems*, McGraw-Hill, New York, NY (1971); reissued by Dover Publications, Mineola, NY

Feynman, R. P., Metropolis, N., and Teller, E., (1949). *Phys. Rev.* **75**, 1561

Feynman, R. P., (1949). *Phys. Rev.* **76**, 749; *Phys. Rev.* **76**, 769

Feynman, R. P., and Gell-Mann, M., (1958). *Phys. Rev.* **109**, 193

Feynman, R. P., Leighton, R. B., and Sands, M., (1965). *The Feynman Lectures on Physics*, Addison-Wesley, Reading, MA

Friedman, J. I., and Kendall, H. W., (1972). *Ann. Rev. Nucl. Sci.* **22**, 203

Gasser, J., and Leutwyler, H., (1982). *Phys. Rep.* **87**, 77

Gell-Mann, M., and Ne'eman, Y., (1963). *The Eightfold Way*, W. A. Benjamin, Reading, MA

Glashow, S. L., Iliopoulos, J., and Maiani, L., (1970). *Phys. Rev.* **D2**, 1285

Green, A. E. S., (1954). *Phys. Rev.* **95**, 1006

Griffiths, D. J., (1998). *Introduction to Electrodynamics, 3rd ed.*, Benjamin Cummings, San Francisco, CA

Griffiths, D. J., (2004). *Introduction to Quantum Mechanics, 2nd ed.*, Benjamin

Cummings, San Francisco, CA

Gross, D. J., and Wilczek, F., (1973). *Phys. Rev. Lett.* **30**, 1343

Halzen, F., and Martin, A. D., (1984). *Quarks and Leptons*, John Wiley and Sons, New York, NY

Hamermesh, M., (1989). *Group Theory and Its Applications to Physical Problems*, Dover Publications, Mineola, NY

Hartree, D. R., (1928). *Proc. Cambridge. Phil. Soc.* **24**, 89, 111

Heisenberg, W., (1927). *Zeit. f. Physik* **43**, 172

Hildebrand, F. B., (1992). *Methods of Applied Mathematics*, Dover Publications, Mineola, NY

Hofstadter, R., (1956). *Rev. Mod. Phys.* **28**, 214

Horowitz, C. J., and Serot, B. D., (1981). *Nucl. Phys.* **A368**, 503

Huang, K., (1987). *Statistical Mechanics, 2nd ed.*, John Wiley and Sons, New York, NY

Itzykson, C., and Zuber, J.-B., (1980). *Quantum Field Theory*, McGraw-Hill, New York, NY

Jackson, J. D., (1998). *Classical Electrodynamics, 3rd ed.*, John Wiley and Sons, New York, NY

Kohn, W., (1999). *Rev. Mod. Phys.* **71**, 1253

Krane, K., (1996). *Modern Physics, 2nd ed.*, John Wiley and Sons, New York, NY

Kuhn, H. G., (1969). *Atomic Spectra, 2nd ed.*, Academic Press, New York, NY

Landau, L. D., (1941). *J. Phys. (USSR)*, **5**, 71; (1947) *J. Phys. (USSR)*, **11**, 91

Landau, L. D., and Lifshitz, E. M., (1981). *Quantum Mechanics, 3rd ed.*, Butterworth-Heinemann, New York, NY

Lawrence Berkeley Laboratory, (2003). *Atomic Masses*, http://ie.lbl.gov/toi2003/MassSearch.asp

Lawrence Berkeley Laboratory, (2005). *Nuclear Moments*, http://ie.lbl.gov/toipdf/mometbl.pdf

Lee, T. D., and Yang, C. N., (1956). *Phys. Rev.* **104**, 254

LHC, (2008). *Large Hadron Collider*, http://www.cern.ch/lhc/

Lifshitz, E. M., and Landau, L. D., (1984). *Statistical Physics, 3rd ed.*, Butterworth-Heinemann, New York, NY

Llewellyn, R. A., and Tipler, P. A., (2002). *Modern Physics, 4th ed.*, W. H. Freeman and Co., New York, NY

Martin, B. R., (2006). *Nuclear and Particle Physics: An Introduction*, John Wiley and Sons, New York, NY

Mayer, M. G., and Jensen, J. H. D., (1955). *Elementary Theory of Nuclear Shell Structure*, John Wiley and Sons, New York, NY

Merzbacher, E., (1997). *Quantum Mechanics*, John Wiley and Sons, New York, NY

Morse, P. M., and Feshbach, H., (1953). *Methods of Theoretical Physics*, McGraw-Hill, New York, NY

National Nuclear Data Center, (2007). *Nuclear Data*, http://www.nndc.bnl.gov/

NIST, (2007). *Atomic Data*, http://www.physics.nist.gov/PhysRefData/

Ohanian, H. C., (1995). *Modern Physics, 2nd ed.*, Prentice-Hall, Upper Saddle River, NJ

Particle Data Group, (2006). *Particle Data Tables 2006*, http://pdg.lbl.gov/2007/tables/contents_tables.html

Pauli, W., (1948). *Meson Theory of Nuclear Forces, 2nd ed.*, Interscience, New York, NY

Perkins, D. H., (2000). *Introduction to High Energy Physics, 4th ed.*, Cambridge University Press, Cambridge, UK

Planck, M., (1901). *Annalen der Physik* **4**, 533

Politzer, H. D., (1973). *Phys. Rev. Lett.* **30**, 1346

Preston, M. A., and Bhaduri, R. K., (1982). *Structure of the Nucleus*, 2nd printing, Addison-Wesley, Reading, MA

Resnick, R., (1968). *Introduction to Special Relativity*, John Wiley and Sons, New York, NY

Resnick, R., and Halliday, D., (1992). *Basic Concepts in Relativity and Early Quantum Theory*, McMillan, New York, NY

Salam, A., and Ward, J. C., (1964). *Phys. Lett.* **13**, 168

Schiff, L. I., (1968). *Quantum Mechanics, 3rd ed.*, McGraw-Hill, New York, NY

Schrödinger, E., (1926). *Annalen der Physik*, **79**, 489; **81**, 109

Schwarzschild, K., (1916). "Über das Gravitationsfeld eines Massenpunktes nach der Einsteinschen Theorie," *Sitzungsberichte der Königlich Preussischen Akademie der Wissenschaften* **1**, 189

Schwinger, J., (1958). *Selected Papers on Quantum Electrodynamics*, ed. J. Schwinger, Dover Publications, Mineola, NY

Serway, R. A., Moses, C. J., and Moyer, C. A., (1997). *Modern Physics, 2nd ed.*, Saunders College Publishing, Philadelphia, PA

Sobel'man, I. I., (1972). *Introduction to the Theory of Atomic Spectra*, Pergamon Press, New York, NY

Superconductors, (2007). *Superconductors*, http://www.superconductors.org/

Taylor, J. R., Zafiratos, C. D., and Dobson, M., (2003). *Modern Physics for Scientists and Engineers, 2nd ed.*, Prentice-Hall, Englewood Cliffs, NJ

TJNAF, (2007). *Thomas Jefferson National Accelerator Facility*, http://www.jlab.org/

Walecka, J. D., (2000). *Fundamentals of Statistical Mechanics: Manuscript and Notes of Felix Bloch, Prepared by J. D. Walecka*, Stanford U. Press (1989), Stanford, CA; reissued by World Scientific Publishing Company, Singapore

Walecka, J. D., (2001). *Electron Scattering for Nuclear and Nucleon Structure*,

Cambridge University Press, Cambridge, UK

Walecka, J. D., (2004). *Theoretical Nuclear and Subnuclear Physics, 2nd ed.*, World Scientific Publishing Company, Singapore

Walecka, J. D., (2007). *Introduction to General Relativity*, World Scientific Publishing Company, Singapore

Weinberg, S., (1967). *Phys. Rev. Lett.* **18**, 188; (1972) *Phys. Rev.* **D5**, 1412

Wentzel, G., (1949). *Quantum Theory of Fields*, Interscience, New York, NY

Weyl, H., (1950). *The Theory of Groups and Quantum Mechanics*, Dover Publications, Mineola, NY

Wigner, E. P., (1937). *Phys. Rev.* **51**, 106

Wikipedia, (2007). *Periodic Table*, http://en.wikipedia.org/wiki/Periodic_table

Wilson, K., (1974). *Phys. Rev.* **D10**, 2445

Wong, Samuel S. M., (1999). *Introductory Nuclear Physics, 2nd ed.*, Wiley-Interscience, New York, NY

Yukawa, H., (1935). *Proc. Phys.-Math. Soc. Jpn. Ser.3* **17**, 48

Zemansky, M. W., (1968). *Heat and Thermodynamics: an Intermediate Textbook, 5th ed.*, McGraw-Hill, New York, NY

Index

About the Author

Professor John Dirk Walecka is Governor's Distinguished CEBAF Professor of Physics, Professor Emeritus at Stanford University and Professor of Physics at the College of William and Mary. He was the Scientific Director of the Continuous Electron Beam Accelerator Facility (CEBAF) in its initial stage (from 1986 to 1992). He was awarded the Bonner Prize for Nuclear Physics by the American Physical Society and was a Distinguished Schiff Lecturer and Primakoff Lecturer. For his many contributions to research, administration, and teaching, he was awarded the Virginia Lifetime Achievement in Science.